AF693674

PRACTICAL ELECTROMAGNETICS

BICENTENNIAL 1807 WILEY 2007 BICENTENNIAL

The Wiley Bicentennial–Knowledge for Generations

Each generation has its unique needs and aspirations. When Charles Wiley first opened his small printing shop in lower Manhattan in 1807, it was a generation of boundless potential searching for an identity. And we were there, helping to define a new American literary tradition. Over half a century later, in the midst of the Second Industrial Revolution, it was a generation focused on building the future. Once again, we were there, supplying the critical scientific, technical, and engineering knowledge that helped frame the world. Throughout the 20th Century, and into the new millennium, nations began to reach out beyond their own borders and a new international community was born. Wiley was there, expanding its operations around the world to enable a global exchange of ideas, opinions, and know-how.

For 200 years, Wiley has been an integral part of each generation's journey, enabling the flow of information and understanding necessary to meet their needs and fulfill their aspirations. Today, bold new technologies are changing the way we live and learn. Wiley will be there, providing you the must-have knowledge you need to imagine new worlds, new possibilities, and new opportunities.

Generations come and go, but you can always count on Wiley to provide you the knowledge you need, when and where you need it!

William J. Pesce
President and Chief Executive Officer

Peter Booth Wiley
Chairman of the Board

PRACTICAL ELECTROMAGNETICS

FROM BIOMEDICAL SCIENCES TO WIRELESS COMMUNICATION

DEVENDRA K. MISRA
University of Wisconsin — Milwaukee
Milwaukee, Wisconsin

BICENTENNIAL 1807 WILEY 2007 BICENTENNIAL

WILEY-INTERSCIENCE
A JOHN WILEY & SONS, INC., PUBLICATION

Copyright © 2007 by John Wiley & Sons, Inc. All rights reserved.

Published by John Wiley & Sons, Inc., Hoboken, New Jersey
Published simultaneously in Canada

No part of this publication may be reproduced, stored in a retrieval system, or transmitted in any form or by any means, electronic, mechanical, photocopying, recording, scanning, or otherwise, except as permitted under Section 107 or 108 of the 1976 United States Copyright Act, without either the prior written permission of the Publisher, or authorization through payment of the appropriate per-copy fee to the Copyright Clearance Center, Inc., 222 Rosewood Drive, Danvers, MA 01923, 978-750-8400, fax 978-750-4470, or on the web at www.copyright.com. Requests to the Publisher for permission should be addressed to the Permissions Department, John Wiley & Sons, Inc., 111 River Street, Hoboken, NJ 07030, 201-748-6011, fax 201-748-6008, or online at http://www.wiley.com/go/permission.

Limit of Liability/Disclaimer of Warranty: While the publisher and author have used their best efforts in preparing this book, they make no representations or warranties with respect to the accuracy or completeness of the contents of this book and specifically disclaim any implied warranties of merchantability or fitness for a particular purpose. No warranty may be created or extended by sales representatives or written sales materials. The advice and strategies contained herein may not be suitable for your situation. You should consult with a professional where appropriate. Neither the publisher nor author shall be liable for any loss of profit or any other commercial damages, including but not limited to special, incidental, consequential, or other damages.

For general information on our other products and services or for technical support, please contact our Customer Care Department within the United States at 877-762-2974, outside the United States at 317-572-3993 or fax 317-572-4002.

Wiley also publishes its books in a variety of electronic formats. Some content that appears in print may not be available in electronic formats. For more information about Wiley products, visit our web site at www.wiley.com.

Library of Congress Cataloging-in-Publication Data:

Misra, Devendra, 1949-
Practical electromagnetics : from biomedical sciences to wireless communication / by Devendra K. Misra
p. cm.
Includes bibliographical references and index.
ISBN-13 0-471-74865-6
ISBN-10 0-471-74865-X (acid-free paper)
1. Electromagnetism- -Textbooks. I. Title.
QC760.2.M57 2006
537- -dc22 2006008776

Printed in the United States of America

10 9 8 7 6 5 4 3 2 1

CONTENTS

PREFACE

Electromagnetic fields is one of the classical disciplines that has been around for some time. However, still today new scientific and technological applications in this area keep on growing. Therefore electromagnetic fields is a core subject not only for electrical engineering and physics majors, but also for students coming from many other disciplines. There are a number of excellent textbooks that were developed to meet the requirements of traditional curricula. These are mostly suited for introductory one- and two-semester sequences in electrical engineering or physics programs. Since many of the new areas developed lately warrant integration into the electrical engineering curriculum, it is often difficult to accommodate a traditional two-semester sequence in electromagnetic fields along with the suitable prerequisite courses. Further there is almost no room left for an upper-level course that includes some contemporary applications of electromagnetic fields. This book is attempting to address these issues.

The subject matter of this book is arranged such that it is suitable for students with the basic calculus and basic physics that are required at the college level. After a brief review of the fundamental units and prefixes, electrical sources and fundamental quantities are introduced in Chapter 1. This is followed by the characteristics of sinusoidal waves. A working introduction of complex numbers is also included in this chapter. Chapter 2 begins with the definitions of vectors and the manipulation techniques of vectors in rectangular, cylindrical, and spherical coordinates. The concepts of scalar and vector fields are introduced. Vector calculus is briefly reviewed along with the gradient of scalar fields, and the divergence and curl of vector fields. The relevant theorems are covered as well in this chapter.

Maxwell's equations in their large-scale form are introduced in Chapter 3 as experimental facts expressed mathematically. A number of examples are used to illustrate their importance. After introducing the equation of continuity, Maxwell's equations in point form are found via the divergence and Stokes theorems. Sections on the constitutive relations and the boundary conditions follow. The chapter ends with introductions of the Lorentz force equation and the Poynting theorem. Chapter 4 presents the propagation characteristics of uniform plane waves along with some simple applications. The formulation of Helmholtz's equation is summarized and used in the construction of uniform plane waves propagating in various media of infinite extent. The concepts of reflection and transmission of these waves at an interface with another medium are also discussed in this chapter.

Chapter 5 begins with the fundamentals of signal propagation in frequency domain. The Smith chart technique is presented for analyzing transmission line circuits. This chapter concludes with an introduction to time-domain analysis of signals traveling on a transmission line. The concepts of magnetic charge and current are the subject of Chapter 6. After a brief discussion of the modified Maxwell equations, vector and scalar potentials are introduced. The techniques covered in this chapter are used to construct solutions of electromagnetic problems in rectangular, cylindrical, and spherical coordinate systems.

Chapter 7 begins with an analysis of electromagnetic fields produced by an infinitesimal current element in an unbounded medium. After a discussion of antenna parameters, the radiation behavior of a linear antenna is presented. Following this, the concepts of antenna arrays are presented along with broadside and end-fire arrays. This chapter includes a section on the Friis transmission formula and the radar range equation.

Electrostatic and magnetostatic fields are considered in Chapters 8 and 9 as a special case of time-varying electrical sources. A number of electrostatic problems are analyzed via Poisson's and Laplace's equations. Energy storage in electrical capacitors and inductors is introduced along with a few application examples. The fundamentals of magnetic materials and the magnetic circuits are also discussed in Chapter 9.

Further applications of the solution techniques of Chapter 6 are found in Chapter 10 where the signal propagation is studied in cylindrical waveguides of rectangular and circular cross sections. This chapter includes sections on rectangular and circular cylindrical cavity resonators. Chapter 11, the last chapter of this book briefly explains two widely used numerical techniques to solve electromagnetic problems. The concepts of finite difference are illustrated by a few simple examples. A section on the method of moments is included in this chapter. The problem formulation required for the method of moments is the topic of the last section of this chapter, and it covers the scattering of a planar electromagnetic wave from a conducting cylinder as well.

There are eight appendixes providing relevant mathematical formulas, Bessel functions, and Legendre functions. An appendix is also devoted to the concept of the delta function along with its application in finding electromagnetic fields in an unbounded medium.

Some of the highlights of the book are as follows:

- The fundamentals of complex numbers and vector analysis are presented at the beginning along with a number of solved examples. Readers already familiar with these topics can easily skip the first two chapters.
- Maxwell's equations are introduced early in Chapter 3 as mathematical representations of experiments. After a brief presentation of constitutive relations and boundary conditions, the Lorentz force equation is introduced. A number of examples are used to illustrate applications of the various formulas.
- Uniform plane waves are presented in Chapter 4. The chapter includes a number of simple application examples.
- Transmission line circuits are presented in Chapter 5. This discussion can be easily moved to the beginning of a semester's work if an instructor prefers that.
- The rest of the material can be used for upper level technical elective or a beginning graduate level courses. Also the last section of Chapter 4 can be easily replaced with Chapters 8 and 9 in a first course.
- There are some 170 fully solved examples that are closely connected to the material presented in each chapter. The book includes over 225 practice problems as well.

ACKNOWLEDGMENTS

So many scientists, mathematicians, and engineers have enriched my studies of electromagnetic fields over the years that it is not possible to list everyone here in this short space. I have benefited also from many researchers and professionals working in this area. I gratefully acknowledge the contributions of each and every one of them for their generosity and kindness. Also I would like to thank the anonymous reviewers who provided several valuable suggestions. I deeply appreciate the support I received from my wife, Ila, son, Shashank, and daughter-in-law, Julie, during the course of this project. This project got started because of the enthusiastic support I received from Val Moliere and her staff at Wiley. I feel fortunate for continued current support from Paul Petralia and his assistant Whitney A. Lesch.

DEVENDRA K. MISRA

University of Wisconsin — Milwaukee

1

INTRODUCTION

The area of electromagnetic fields plays a central role in numerous scientific and technical applications. It includes television and audio broadcasting, optical communication, and wireless communications. We not only can communicate from virtually anywhere on the planet but can get detailed pictures of distant planets at the speed of light. Civilian and military personnel in aviation, law enforcement, remote sensing, weather forecast, to name a few, currently use a variety of radars. Physicians can get a detailed image and better diagnostics of the patient through devices and instruments developed via the principles of electromagnetic fields. Since electrical signals generate and propagate inside the human body to manage various functions, these electrical fields (e.g., the electrocardiogram) are monitored in diagnostics as well.

The concept of electromagnetic fields is believed to have originated in ancient Greece. The Greeks observed electric and magnetic phenomenon perhaps as early as 800 BC. They noticed that a piece of amber attracts pieces of feather after rubbing. In Greek, amber is called "elecktron" and the word "electric" comes from that. Further they found that certain naturally found stones attracted pieces of iron. These are now called magnetite, perhaps after the shepherd named Magnes who got stuck due to the iron nails in his shoes. Magnetite was found in a northern central district of Greece named Magnesia, the word "magnetic" comes from that. It was found through systematic experiments that there are two different kinds of electrical charges. Benjamin Franklin named them positive and negative. Perhaps most significant developments in electromagnetic fields

Practical Electromagnetics: From Biomedical Sciences to Wireless Communication.
By Devendra K. Misra
Copyright © 2007 John Wiley & Sons, Inc.

TABLE 1.1 Fundamental SI Units

Quantity	Unit	Symbol
Length	meter	m
Mass	kilogram	kg
Time	second	s
Current	ampere	A
Temperature	Kelvin	K
Luminous intensity	candela	cd

TABLE 1.2 Multiplying Prefixes for Units

Multiplying Prefix	Prefix	Symbol	Multiplying Prefix	Prefix	Symbol
10^{18}	exa	E	10^{-2}	centi	c
10^{15}	peta	P	10^{-3}	milli	m
10^{12}	tera	T	10^{-6}	micro	μ
10^{9}	giga	G	10^{-9}	nano	n
10^{6}	mega	M	10^{-12}	pico	p
10^{3}	kilo	k	10^{-15}	femto	f
			10^{-18}	atto	a

happened during nineteenth century when it was established that electric and magnetic fields are related phenomena. Principles of electrical generators and motors were established, and the light being an electromagnetic wave was recognized. In 1873 Maxwell elegantly summarized various observations and experimental facts through a set of equations that formed the foundation of technological advances of today.

The SI system of units is followed in this book, as it is standard practice in modern science and engineering. Fundamental SI units are given in Table 1.1 along with their symbols. Other units that we will need to express electromagnetic field quantities can be expressed in terms of these, and therefore those units are regarded as secondary. Further sometimes we come across so small or so large a quantity that usual unit becomes inconvenient. This can be avoided via the prefixes that are listed in Table 1.2.

1.1 ELECTRICAL SOURCES AND FUNDAMENTAL QUANTITIES

Now we know that there are positive and negative electrical charges as protons and electrons in an atom. Electrical charge cannot be created or destroyed, but it can be transferred. In the case of biological systems, such as the human body, positive and negative ions of sodium, potassium, and chloride play role in transferring information from one point to the other. For instance, the action potentials propagate through the nerve fibers, carrying information to or from the brain. Similarly the pacemaker cells found at SA node of the human heart generates an electrical signal that propagates through cells to control the blood flow.

Electrical charge is measured in coulombs (denoted as C), after the great French physicist Charles Coulomb. For example, the charge on an electron is $-1.602 \cdot 10^{-19}$ C. Thus 1 C of charge will require over $6 \cdot 10^{18}$ electrons. The electrical current is found when the electrical charge moves. A unit of electric current is an ampere (denoted as A), which is equivalent to coulomb per second. A volume charge density ρ is defined as the charge per unit volume. Note that the volume has to be small enough to represent point-to-point variations accurately; at the same time it should be large enough to contain a fairly large number of discrete charges. Mathematically

$$\rho = \lim_{\delta V \to 0} \frac{\delta q}{\delta V} \ \mathrm{C/m^3}, \tag{1.1.1}$$

where δq is the charge in coulomb that exists in a small volume δV.

In certain physical situations the charge may exist only on an area element δs or even on a length element $\delta \ell$. Surface or length charge densities are more appropriate in such cases. These densities are defined as

$$\rho_s = \lim_{\delta s \to 0} \frac{\delta q}{\delta s} \ \mathrm{C/m^2} \tag{1.1.2}$$

and

$$\rho_\ell = \lim_{\delta \ell \to 0} \frac{\delta q}{\delta \ell} \ \mathrm{C/m}. \tag{1.1.3}$$

The electrical current I is defined as follows:

$$I = \frac{dQ}{dt} \mathrm{A}. \tag{1.1.4}$$

Since current must flow through a finite area, it is not defined at a point. An associated point function used in electromagnetic fields is the current density $\vec{J}$ that is defined as the current per unit area:

$$\vec{J} = \lim_{\delta A \to 0} \frac{\delta I}{\delta A} \hat{n} = \rho \vec{v} \ \mathrm{A/m^2}, \tag{1.1.5}$$

where $\hat{n}$ is the unit vector normal to the elemental area δA (therefore in the direction of current flow) and $\vec{v}$ is velocity of the charge flow.

The surface current is defined as

$$\vec{J}_s = \lim_{\delta\ell \to 0} \frac{\delta I}{\delta\ell} \text{ A/m.} \tag{1.1.6}$$

Electric Field Intensity

It is found that like charges repel and unlike charges attract one another. Electric field intensity $\vec{E}$ (also known as the electric field strength) at a point in space is defined as the electric force acting on a positive test charge placed at that point divided by the magnitude of that test charge. It is assumed that the test charge is small enough such that it does not disturb the original charge distribution. Hence

$$\vec{E} = \lim_{q \to 0} \frac{\vec{F}}{q}. \tag{1.1.7}$$

According to this definition, a unit of the electric field intensity is Newton per coulomb. A more commonly used unit is volt per meter, which is defined via the electrical potential. The work done by an external source to move a charge of 1 C from one point to another is defined as the potential difference:

$$W_{ab} = \Delta\Phi = -\int_a^b \vec{E} \cdot d\vec{\ell} \text{ J/C.} \tag{1.1.8}$$

The joule per coulomb unit for the potential difference $\Delta\Phi$ is defined as volt (V), and therefore electric field intensity is expressed in volt per meter.

Electric Flux Density

Michael Faraday was the first to develop the concept of electric flux by way of an ice pail experiment. His experiments showed that if two concentric spheres, isolated from each other, are used and a positive charge is placed on the inner sphere, then the positive charge induces a negative charge of equal amount. He concluded that there was some electric flux proportional to the charge. Today electric flux density $\vec{D}$ is defined as the electric flux passing through per unit area. Sometimes it is also called the electric displacement. It is expressed in coulombs per square meter (C/m^2). This provides a clue to understanding the interaction of electric fields with material media.

Magnetic Flux Density

As mentioned earlier, Greeks knew about magnetism from around 700 BC. Pierre de Maricourt used a spherical natural magnet in 1269 to map out the direction that a needle took when placed at various points on a spherical surface. He found that the directional lines originated from one point and terminated at an entirely opposite

TABLE 1.3 Fundamental Field Quantities

Field Vectors		Symbol	SI Unit
Electric field	Electric field intensity or electric field strength	$\vec{E}$	V/m
	Electric flux density or electric displacement	$\vec{D}$	C/m^2
Magnetic field	Magnetic flux density	$\vec{B}$	T or Wb/m^2
	Magnetic field intensity or magnetic field strength	$\vec{H}$	A/m

point. He named these points the poles of the magnet. It was found later on that these poles existed for all magnets and so were termed north and south poles. Further it was found that the two poles could not be separated the same way as negative and positive electrical charges. Electrical lines of force emanate from positive charges and terminate at negative charges. Similarly magnetic lines of force originate from the north pole and sink at the south pole.

Experiments showed that the magnetic field cannot change the velocity of a moving electrical charge, but it can change the direction of its motion. The force per unit charge was found to be proportional to the velocity as well as a magnetic quantity that is called the magnetic flux density $\vec{B}$. It is also known as the magnetic induction and expressed in weber per square meter (Wb/m^2) or more commonly in tesla (T).

Magnetic Field Intensity

A current-carrying conductor has an encircling magnetic field. If a conducting wire loop of enclosed area A has current I, then its vector magnetic moment is defined as IA with its direction normal to the loop. Since there are electrons orbiting and spinning in atoms, they produce equivalent magnetic moments. The magnetization vector $\vec{M}$ describes the magnetic state of a substance. The magnitude of this magnetization vector is defined as the magnetic moment per unit volume of the substance. When there are magnetic fields due to the current flow as well as those produced by the substance, it becomes convenient to introduce another vector field quantity $\vec{H}$ that represents the magnetic field intensity or the magnetic field strength. It is expressed in amperes per meter (A/m). Fundamental electromagnetic field quantities are listed in Table 1.3 along with their symbols and units.

1.2 STATIC AND DYNAMIC FIELDS

Electrical charges that are constant with time produce electrical fields that do not vary with time, and therefore these fields are called electrostatic fields. Similarly a constant current produces a magnetostatic field. The static fields are independent of each other. On the other hand, time-varying sources—both currents and charges—produce electric as well as magnetic fields that change with time, and therefore these are called dynamic fields. Dynamic electric fields are related to dynamic magnetic fields. In other words, time-varying electric fields are not independent of magnetic fields generated by the source. In general, the time variation of these fields can have any form. However, if they change with time following a sine function, then we call them the time-harmonic fields. Since even nonsinusoidal signals can be expressed as a sum of many (theoretically infinite) sinusoidal signals, we will focus mainly on time-harmonic fields. It is also easier to analyze time-harmonic fields than other dynamic fields.

Propagating Sinusoidal Waves

When a string is disturbed at its one end, the signal travels along its length. This may be called a one-dimensional propagation of the wave. Similarly a wave propagating over a drumhead is two-dimensional. In case of electromagnetic fields we deal with three-dimensional waves. The direction of propagation is normal to a surface over which electrical characteristics are the same. If this surface represents a plane, then it is called a uniform plane wave. If the surface is cylindrical, then it is a cylindrical wave. Similarly the spherical waves have uniform characteristics over the spherical surfaces.

A wave traveling through a medium that does not absorb its energy is expressed as

$$v(z, t) = A\cos(\omega t - \beta z + \theta), \tag{1.2.1}$$

where A is known as amplitude of the wave; ω is the angular frequency in rad/s and t is the time variable in s; β is called the propagation constant in rad/m and z is the space variable in m, and θ is an arbitrary constant phase angle in rad. Figure 1.1(a) shows typical characteristics of v as a function of time t when keeping

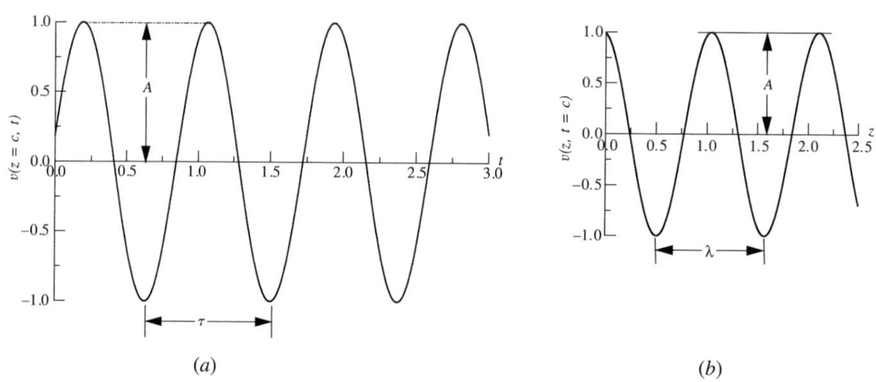

Figure 1.1 (a) $v(z, t)$ as a function of t with z constant and (b) as a function of z with t constant.

the distance z constant (i.e., at a fixed location). The time over which a wave goes through a 2π rad (i.e., 360°) phase change is known as the wave's time period T. The inverse of the time period in seconds is known as the wave's frequency f, which is expressed in hertz (Hz). Angular frequency ω is related with this as $2\pi f$, and it is expressed in radian per second (rad/s). Figure 1.1(b) shows the characteristics of v as a function of distance z while time t is kept fixed. Thus it represents a snapshot of v. The amplitude of the wave is the same as that in Figure 1.1(a), so the general behavior looks the same. However, the amplitude is a function of distance now. The distance over which the wave goes through a 2π rad change is called the wavelength λ of the wave. The wavelength is expressed in meters. The propagation constant β is equal to $2\pi/\lambda$. The ratio of the angular velocity to its propagation constant gives the phase velocity v_p of the wave:

$$v_p = \frac{\omega}{\beta} = \frac{2\pi f}{2\pi/\lambda} = f\lambda. \tag{1.2.2}$$

The velocity of any electromagnetic wave in free space is the same as that of the speed of light, which is approximately equal to $3 \cdot 10^8$ m/s. The electromagnetic spectrum is given in Table 1.4. Note that the visible light, infrared, X rays, gamma rays, and radio waves all belong to the family of electromagnetic waves. Table 1.5 lists commercial broadcast TV and radio channels that are also part of this family.

Example 1.1

Wavelength of a laser signal in air is found to be 632.8 nm. Find its frequency in Hz and in rad/s, and the time period.

From (1.2.2) we have

$$f = \frac{3 \cdot 10^8}{632.8 \cdot 10^{-9}} = 474.08 \cdot 10^{12}\ \text{Hz} = 474.08\ \text{THz},$$

$$\omega = 2\pi f = 2.9787 \cdot 10^{15}\ \text{rad/s} = 2.9787\ \text{Prad/s},$$

and

$$T = \frac{1}{f} = \frac{1}{474.08 \cdot 10^{12}} = 2.11 \cdot 10^{-15}\ \text{s} = 2.11\ \text{fs}.$$

Example 1.2

The dipole antenna shown in Figure 1.2 is found to be very effective in sensing electromagnetic radiation when its length is close to one-half of the signal wavelengths. Find a suitable length that works for the FM band.

TABLE 1.4 Electromagnetic Spectrum

Band Designation	Frequency Range	Wavelength Range (in Free Space)	Selected Applications
Extremely low frequency (ELF)	3–30 Hz	10–100 Mm	Detection of buried conductors
Super low frequency (SLF)	30–300 Hz	1–10 Mm	Electric power, submarine communication
Ultra low frequency (ULF)	0.3–3 kHz	0.1–1 Mm	Audio, telephone
Very low frequency (VLF)	3–30 kHz	10 km–100 km	Sonar, navigation
Low frequency (LF)	30–300 kHz	1 km–10 km	Radio beacon, navigation
Radio frequency			
MF	300–3000 kHz	100 m–1 km	AM radio
High frequency (HF)	3–30 MHz	10 m–100 m	Shortwave radio
Very high frequency (VHF)	30–300 MHz	1 m–10 m	TV and FM radio, air-traffic control
Ultra high frequency (UHF)	300–3000 MHz	10 cm–1 m	TV, radar, navigation, ISM
Microwave			
Super high frequency (SHF)	3–30 GHz	1 cm–10 cm	Satellite communication, radar, remote sensing
Extremely high frequency (EHF)	30–300 GHz	0.1 cm–1 cm	
Optic waves			
Far infrared (Far IR)	0.3–60 THz	5–1000 μm	Optical communication, photobiology, bioimaging, bionanophotonics
Mid-infrared (Mid IR)	60–120 THz	2.5–5 μm	
Near infrared (Near IR)	120–429 THz	0.7–2.5 μm	
Visible	~429–750 THz	400–700 nm	
Ultraviolet (UV)	750–3000 THz	100–400 nm	
X rays	3 PHz–3 EHz	0.1–100 nm	Medical diagnosis, imaging, NDE
Gamma rays	3–300 EHz	0.1–1 pm	Food irradiation, cancer therapy

TABLE 1.5 Frequency Bands Used in Commercial Broadcasting

	Channels	Frequency Range	Wavelength Range
AM	107	535 kHz–1605 kHz	186.92 m–560.75 m
TV	2–4	54 MHz–72 MHz	4.17 m–5.56 m
	5–6	76 MHz–88 MHz	3.41 m–3.95 m
FM	100	88 MHz–108 MHz	2.78 m–3.41 m
TV	7–13	174 MHz–216 MHz	1.39 m–1.72 m
	14–83	470 MHz–890 MHz	33.7 cm–63.83 cm

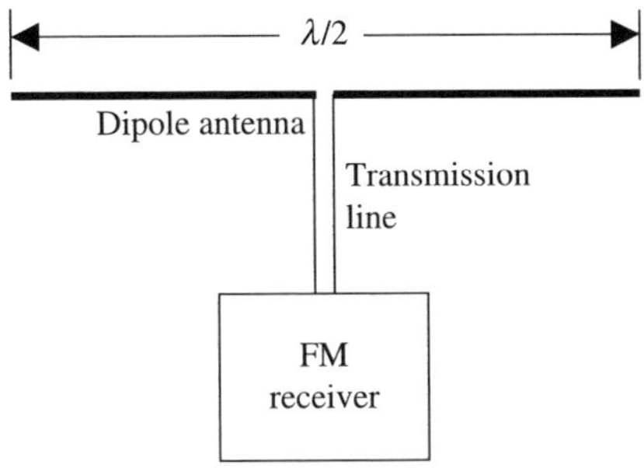

Figure 1.2 A dipole antenna with a receiver.

From Table 1.5, the FM band includes 88 MHz to 108 MHz. The signal wavelengths at these two extremes are found to be

$$\lambda_{88} = \frac{3 \cdot 10^8}{88 \cdot 10^6} = 3.41 \text{ m}$$

and

$$\lambda_{108} = \frac{3 \cdot 10^8}{108 \cdot 10^6} = 2.78 \text{ m}.$$

Therefore, if we select 1.7 m long antenna, then its length is exactly one-half wavelength at 88 MHz, and it will be longer at other frequencies in the band. On the other hand, a 1.39 m long antenna will be exactly one-half wavelength at 108 MHz but shorter than one-half at lower frequencies in the band. A 1.5 m long antenna should be a good compromise for the band.

Example 1.3

A satellite is being used for the wireless communication at 2.5 GHz. If its altitude is 10,370 km, find minimum possible time lag in signal reception between the two ground stations.

Since all electromagnetic signals propagate with speed of light, the time lag Δt by the signal to reach at the receiver is

$$\Delta t = \frac{2 \cdot 10370 \cdot 10^3}{3 \cdot 10^8} = 6.9133 \cdot 10^{-2}\,\text{s} = 69.13\,\text{ms}$$

1.3 WORKING WITH COMPLEX NUMBERS AND FUNCTIONS

Complex numbers are used to simplify mathematical analysis in many areas involving sinusoidal signals. This includes the electrical circuits and electromagnetic fields. Therefore this section reviews the complex numbers and functions briefly. Consider a point P on the complex plane as shown in Figure 1.3. Its coordinates are (a, b). This point may be identified in polar coordinates as $c\angle\phi$. Using complex algebra, we can express this as follows:

$$c = a + jb, \tag{1.3.1}$$

where

$$j = \sqrt{-1} \longrightarrow j^2 = -1 \tag{1.3.2}$$

From Figure 1.3 we find that

$$|c| = \sqrt{a^2 + b^2} \tag{1.3.3}$$

and

$$\phi = \tan^{-1}\left(\frac{b}{a}\right). \tag{1.3.4}$$

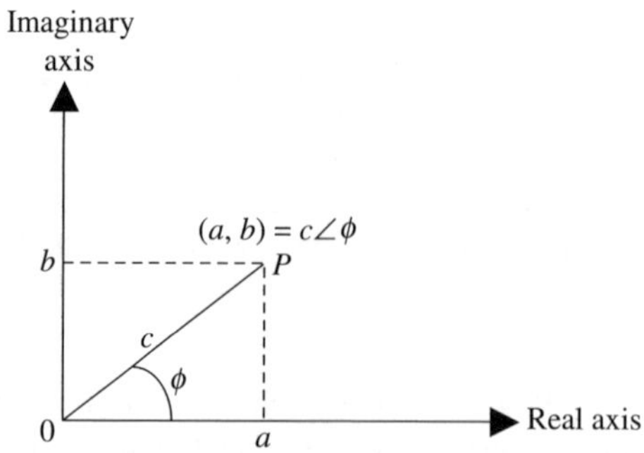

Figure 1.3 Graphical representation of the complex number.

Similarly

$$a = |c| \cos \phi \tag{1.3.5}$$

and

$$b = |c| \sin \phi. \tag{1.3.6}$$

Therefore

$$c = |c|\angle\phi = a + jb = |c| \cos \phi + j|c| \sin \phi = |c|(\cos \phi + j \sin \phi) = |c|e^{j\phi} \tag{1.3.7}$$

We can switch back and forth as needed in the analysis. Equation (1.3.7) includes De Moivre's formula, which is written as follows:

$$e^{\pm jnx} = \cos nx \pm j \sin nx, \tag{1.3.8}$$

where the + sign on the left goes with the + on the right. Similarly two minus signs go together. Also

$$(e^{jx})^n = (\cos x + j \sin x)^n \tag{1.3.9}$$

From (1.3.8) we find that

$$\sin nx = \frac{e^{jnx} - e^{-jnx}}{2j} \tag{1.3.10}$$

and

$$\cos nx = \frac{e^{inx} + e^{-jnx}}{2}. \tag{1.3.11}$$

Addition and Subtraction of Complex Numbers

Consider two complex numbers c and h that can be expressed as follows:

$$c = a + jb = |c|\angle\phi = |c|e^{j\phi_1} \tag{1.3.12}$$

and

$$h = f + jg = |h|\angle\phi_2 = |h|e^{j\phi_2}. \tag{1.3.13}$$

Addition of these two complex numbers is done by adding the real and imaginary parts of c and h as follows:

$$c + h = (a + jb) + (f + jg) = (a + f) + j(b + g). \tag{1.3.14}$$

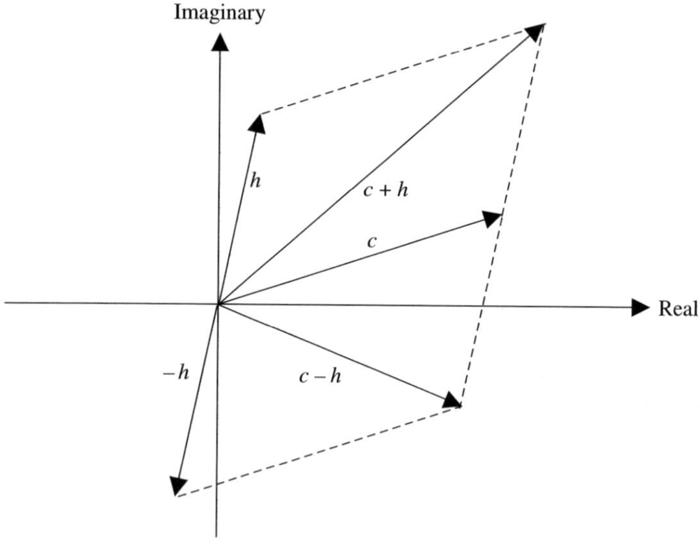

Figure 1.4 Graphical representation of addition and subtraction of two complex numbers.

The addition process is depicted graphically in Figure 1.4. If c and h form the two sides of a parallelogram, then its diagonal represents the sum of two complex numbers. Subtraction is just an extension of this addition. It is defined as follows:

$$c - h = (a + jb) - (f + jg) = (a - f) + j(b - g). \tag{1.3.15}$$

Complex Conjugate of a Complex Number

Complex conjugate of c is denoted by c^*. It is defined as follows:

$$c^* = a - jb. \tag{1.3.16}$$

Multiplication and Division of Complex Numbers

The product of two complex numbers is found after multiplying the numbers term by term as follows:

$$c \cdot h = (a + jb) \cdot (f + jg) = af + jbf + jag + j^2bg.$$

Using (1.3.2), we find that

$$c \cdot h = (af - bg) + j(bf + ag). \tag{1.3.17}$$

Using polar representations for complex numbers, we find the equivalent result to be

$$c \cdot h = |c|e^{i\phi_1} \cdot |h|e^{j\phi_2} = |c| \cdot |h|e^{j(\phi_1+\phi_2)}. \tag{1.3.18}$$

Therefore, if a complex number is multiplied by its conjugate, then we have

$$c \cdot c^* = (a + jb) \cdot (a - jb) = a^2 + b^2 = |c|^2. \tag{1.3.19}$$

Division of a complex number with another is found after multiplying the numerator and the denominator by the complex conjugate of the denominator as follows:

$$\frac{c}{h} = \frac{a + jb}{f + jg}\frac{f - jg}{f - jg} = \frac{(af + bg) + j(bf - ag)}{f^2 + g^2}. \tag{1.3.20}$$

Using polar representations, we find the equivalent result to be

$$\frac{c}{h} = \frac{|c|e^{j\phi_1}}{|h|e^{j\phi_2}} = \frac{|c|}{|h|}e^{j(\phi_1 - \phi_2)}. \tag{1.3.21}$$

Example 1.4

For $c = 1 + j2$ and $h = -2 - j1$, find (a) $c + h$, (b) $c - h$, (c) $c \cdot h$, and (d) c/h.
We can find polar representations for the both as follows:

$$c = 1 + j2 = 2.2361e^{j1.1071}$$

and

$$h = -2 - j1 = 2.2361e^{-j2.6779}.$$

(a) $c + h = 1 + j2 + (-2 - j1) = -1 + j1.$
(b) $c - h = 1 + j2 - (-2 - j1) = 3 + j3.$
(c) $c \cdot h = (1 + j2)(-2 - j1) = -2 - j4 - j1 - j^2 2 = -j5.$
Alternatively,

$$c \cdot h = 2.2361e^{j1.1071} \cdot 2.2361e^{-j2.6779} = 5e^{-j1.5708} = -j5.$$

(d) $$\frac{c}{h} = \frac{1 + j2}{-2 - j1} \cdot \frac{-2 - j1}{-2 - j1} = \frac{-2 - j4 + j1 - 2}{(-2)^2 + (-1)^2} = \frac{-4 - j3}{4 + 1} = -\frac{4}{5} - j\frac{3}{5}$$

$$= -0.8 - j0.6.$$

Alternatively,

$$\frac{c}{h} = \frac{2.2361e^{j1.1071}}{2.2361e^{-j2.6779}} = 1e^{j3.7850} = -0.8 - j0.6.$$

Square Root of a Complex Number

The square root of a complex number is multivalued because

$$c = |c|e^{j\phi} = |c|e^{j(\phi+2n\pi)}, \qquad n = 0, 1, 2, \ldots . \tag{1.3.22}$$

Therefore

$$\sqrt{c} = [|c|e^{j(\phi+2n\pi)}]^{1/2} = \sqrt{|c|}e^{j[(\phi+2n\pi)/2]}, \qquad n = 0, 1, 2, \ldots . \tag{1.3.23}$$

Example 1.5

Find the square root(s) of a complex number $c = 1 + j2$.

$$c = 1 + j2 = 2.2361e^{j1.1071} = 2.2361\, e^{j(2n\pi+1.1071)}, \qquad n = 0, 1, 2, \ldots .$$

Therefore

$$\sqrt{c} = [2.2361e^{j(2n\pi+1.1071)}]^{1/2} = \sqrt{2.2361}e^{j(2n\pi+1.1071)/2}$$
$$= 1.4954\, e^{j(n\pi+0.5536)}.$$

For $n = 0$,

$$\sqrt{c} = 1.4954e^{j0.5536} = 1.2720 + j0.7862.$$

For $n = 1$,

$$\sqrt{c} = 1.4954e^{j(\pi+0.5536)} = -1.2720 - j0.7862.$$

These two roots repeat for higher order n.

Complex Representation of Time-Harmonic Signals

A time-harmonic real physical quantity $v(t)$ is expressed mathematically as follows:

$$v(t) = V_o \cos(\omega t + \theta), \tag{1.3.24}$$

where V_o is called the amplitude, $\omega = 2\pi f$ is the angular frequency in radian per second, f is frequency in Hz, and θ is the phase angle of $v(t)$.

We can also write (1.3.24) using complex representation as follows:

$$v(t) = \mathrm{Re}[V_o \cos(\omega t + \theta) + jV_o \sin(\omega t + \theta)].$$

Note that Re stands for "real part of".

Using (1.3.7), we find that

$$v(t) = \mathrm{Re}\lfloor V_o e^{j(\omega t+\theta)} \rfloor = \mathrm{Re}\lfloor V_o e^{j\theta} e^{j\omega t} \rfloor = \mathrm{Re}\lfloor V e^{i\omega t} \rfloor, \tag{1.3.25}$$

where $V = V_o e^{j\theta}$ is called the phasor representation of $v(t)$. In general, the phasor is a complex number, whereas the actual quantity is a purely real function of time.

In practice, Re and $e^{j\omega t}$ are suppressed for simplicity. However, whenever we need an expression for $v(t)$ from a given V, the complex quantity V is multiplied by $e^{j\omega t}$ and then the real part of that is taken as the desired $v(t)$.

Addition and Subtraction of Phasors

If $u(t) = U_o \cos(\omega t + \theta_1)$ and $v(t) = V_o \cos(\omega t + \theta_2)$, then

$$u(t) + v(t) = \mathrm{Re}\{U e^{j\omega t}\} + \mathrm{Re}\{V e^{j\omega t}\} = \mathrm{Re}\{(U + V) e^{j\omega t}\}, \tag{1.3.26}$$

where $U = U_o e^{j\theta_1}$ and $V = V_o e^{j\theta_2}$ are the phasors of $u(t)$ and $v(t)$, respectively. Therefore the phasor equivalent for $u(t) + v(t)$ is $U + V$. Similarly it can be verified that the phasor equivalent for $u(t) - v(t)$ is $U - V$.

Derivative and the Phasor

For $v(t) = V_o \cos(\omega t + \theta)$ we find that

$$\frac{\partial}{\partial t} v(t) = \frac{\partial}{\partial t}[V_o \cos(\omega t + \theta)] = -V_o \omega \sin(\omega t + \theta).$$

If we use the phasor representation, then we find that

$$\begin{aligned}\frac{\partial}{\partial t} v(t) &= \frac{\partial}{\partial t}[V_o \cos(\omega t + \theta)] = \mathrm{Re}\frac{\partial}{\partial t}[V_o e^{j\theta} e^{j\omega t}] = \mathrm{Re}[j\omega V e^{i\omega t}] \\ &= \mathrm{Re}[j\omega V_o e^{j(\omega t+\theta)}] = \mathrm{Re}[j\omega V_o\{\cos(\omega t + \theta) + j \sin(\omega t + \theta)\}] \\ &= -\omega V_o \cdot \sin(\omega t + \theta).\end{aligned}$$

Therefore the phasor equivalent for $\partial v(t)/\partial t$ is $j\omega V$.

Integration and the Phasor

For $v(t) = V_o \cos(\omega t + \theta)$ we find that

$$\int v(t) dt = \int V_o \cos(\omega t + \theta) dt = \frac{V_o}{\omega} \sin(\omega t + \theta).$$

If we use the phasor representation, then we find that

$$\int v(t)\,dt = \int V_o \cos(\omega t + \theta)\,dt = \mathrm{Re}\int V_o e^{j(\omega t+\theta)} = \mathrm{Re}\left[\frac{V_o e^{j\theta} e^{j\omega t}}{j\omega}\right] = \mathrm{Re}\left[\frac{V}{j\omega} e^{j\omega t}\right]$$

$$= \mathrm{Re}\left[\frac{V_o \cos(\omega t + \theta) + jV_o \sin(\omega t + \theta)}{j\omega}\right] = \frac{V_o}{\omega}\sin(\omega t + \theta)$$

Therefore the phasor for $\int v(t)\,dt$ is $V/j\omega$.

Example 1.6

Current $i(t)$ flowing through a capacitor C produces the voltage $v(t) = V_o \cos(\omega t + \theta)$ across it, as shown in Figure 1.5. Establish the relation between current and voltage phasors.

If $q(t)$ is the electrical charge on one side of the capacitor, then its relation with the capacitance C and the voltage $v(t)$ is given as follows:

$$C = \frac{q(t)}{v(t)} \longrightarrow q(t) = Cv(t) \longrightarrow \frac{dq(t)}{dt} = i(t) = C\frac{dv(t)}{dt}. \tag{1.3.27}$$

For $v(t) = V_o \cos(\omega t + \theta)$ we have

$$i(t) = C\frac{d}{dt}[V_o \cos(\omega t + \theta)] = -\omega C V_o \sin(\omega t + \theta) = \omega C V_o \cos\left(\frac{\pi}{2} + \omega t + \theta\right).$$

If we use phasors, then we get

$$i(t) = \mathrm{Re}\left[\omega C V_o e^{j[\theta+(\pi/2)]} e^{j\omega t}\right] = \mathrm{Re}[j\omega C V_o e^{j\theta} e^{j\omega t}] = \mathrm{Re}[j\omega C V e^{j\omega t}].$$

Therefore

$$I = j\omega C V,$$

where I and V are current and voltage phasors, respectively. Note that the same result is found from (1.3.27) after switching to phasors and replacing the time derivative by $j\omega$. Reactance X_c of a capacitor is defined as

$$X_c = \frac{V}{I} = \frac{1}{j\omega C}.$$

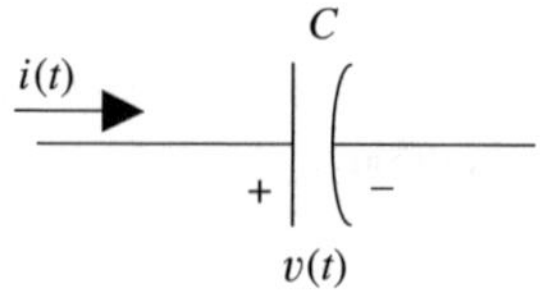

Figure 1.5 Current and voltage on a capacitor.

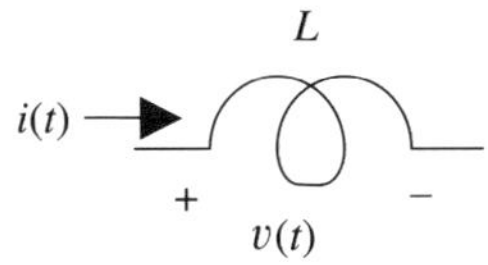

Figure 1.6 Voltage and current in an inductor.

Example 1.7

Current $i(t)$ flowing through an inductor L produces the voltage $v(t) = V_o \cos(\omega t + \theta)$ across it, as shown in Figure 1.6. Establish the relation between current and voltage phasors.

Current $i(t)$ and voltage $v(t)$ in an inductor are related as follows:

$$i(t) = \frac{1}{L}\int^t v(\tau)d\tau = \frac{1}{L}\int^t V_o\ \cos(\omega\tau + \theta)d\tau = \frac{V_o}{\omega L}\sin(\omega t + \theta)$$
$$= \frac{V_o}{\omega L}\cos\left(\omega t + \theta - \frac{\pi}{2}\right).$$

If we use phasors, then we have

$$i(t) = \frac{1}{L}\int^t v(\tau)d\tau = \frac{1}{L}\int^t V_o\cos(\omega\tau + \theta)d\tau = \mathrm{Re}\left[\frac{V}{j\omega L}e^{j\omega t}\right]. \tag{1.3.28}$$

Therefore

$$I = \frac{V}{j\omega L},$$

where I and V are current and voltage phasors, respectively. Note that the same result is found from (1.3.28) after switching to phasors and replacing the time integral by $1/j\omega$. Reactance X_L of an inductor is defined as

$$X_L = \frac{V}{I} = j\omega L.$$

Time Averages

The time average of a periodic signal is found by averaging the signal over a full period. Therefore for $u(t) = U_o \cos(\omega t + \theta_1)$, the time average is found to be

$$\langle u(t)\rangle = \frac{1}{T}\int_0^T u(t)dt = \frac{U_o}{T}\int_0^T \cos(\omega t + \theta_1)\ dt = \frac{U_o}{\omega T}\sin(\omega t + \theta_1)_0^T = 0.$$

As expected, the time average of a sinusoidal signal is zero.

On the other hand, the time average of the product of two sinusoidal signals is nonzero. For example, if we have another sinusoidal signal $v(t)$ such that $v(t) = V_o \cos(\omega t + \theta_2)$, then

$$\begin{aligned}
\langle u(t)v(t)\rangle &= \frac{U_o V_o}{T}\int_0^T \cos(\omega t + \theta_1)\cos(\omega t + \theta_2)\, dt \\
&= \frac{U_o V_o}{T}\int_0^T \frac{\cos(2\omega t + \theta_1 + \theta_2) + \cos(\theta_1 + \theta_2)}{2}\, dt \\
&= \frac{U_o V_o}{2}\cos(\theta_1 + \theta_2).
\end{aligned}$$

PROBLEMS

1.1. The wavelength of a laser signal in air is found to be 589.3 nm. Find its frequency in Hz and in rad/s, and the time period.

1.2. The dipole antenna shown in Figure 1.2 is found to be very effective in sensing electromagnetic radiation when its length is close to one-half of the signal wavelengths. Find a suitable length that works for the VHF band.

1.3. A satellite is being used for the wireless communication at 5.9 GHz. If its altitude is 36,000 km, find minimum possible time lag in signal reception between the two ground stations.

1.4. For $c = 3 + j4$ and $h = -4 + j3$, find (a) $c + h$, (b) $c - h$, (c) $c \cdot h$, and (d) c/h.

1.5. For $c = 3 - j4$ and $h = -4 - j3$, find (a) $c + h$, (b) $c - h$, (c) $c \cdot h$, and (d) c/h.

1.6. Find the square root(s) of a complex number $c = 3 - j4$.

1.7. Find the square root(s) of a complex number $c = -3 - j4$.

1.8. Voltage $v(t) = 2\cos(120\pi t + 0.15)$ V is measured across a 2 μF capacitor. Find the phasor current flowing through it.

1.9. Voltage $v(t) = 2\cos(2000\pi t)$ V is measured across a 2 mH inductor. Find its reactance and the phasor current.

2

VECTORS AND FIELDS

As mentioned in the preceding chapter, certain electromagnetic quantities require two associated numbers (its magnitude and the direction) to uniquely specify it at a given point while only one (the magnitude) suffices for others. These quantities can be categorized as vector and scalar quantities, respectively. The representation of quantities as vector and scalar quantities not only simplifies the notation, it also helps in the analysis of an electromagnetic system. A scalar quantity f is completely specified by a single number along with its dimensional unit, whereas two numbers are associated with a vector quantity; one specifies its magnitude and the other its direction. For example, electrical charge and potential are scalar, while electric field intensity and magnetic field intensity are vector quantities. Other examples of scalar quantities are the volume or mass of an object or the temperature at a given point because these require only one number along with its appropriate unit. The velocity and the force are vector quantities because these require one number specifying the magnitude and the other its direction. Similarly a scalar function $f(x, y, z)$ is completely specified by a single number at the point (x, y, z), whereas a vector function $\vec{A}(x, y, z)$ requires two numbers, one for its magnitude and the other for its direction at a given point (x, y, z). The analysis and manipulation techniques of vectors and scalars are presented in this chapter.

Practical Electromagnetics: From Biomedical Sciences to Wireless Communication.
By Devendra K. Misra
Copyright © 2007 John Wiley & Sons, Inc.

2.1 WORKING WITH VECTORS

Unit Vectors and Vector Components

Unit vectors are helpful in analyzing systems involving vectors. A unit vector always has unit length (or the magnitude) in a given direction. Since the length of an arbitrary vector $\vec{A}$ can be expressed as $|\vec{A}| = A$, a unit vector $\hat{a}$ along this vector will have $|\hat{a}| = 1$, while its direction will be the same as that of the vector $\vec{A}$. Mathematically

$$\hat{a} = \frac{\vec{A}}{|\vec{A}|}. \tag{2.1.1}$$

Figure 2.1 shows a vector $\vec{A}$ in the rectangular (Cartesian) coordinate system. In the figure $\hat{x}$, $\hat{y}$, and $\hat{z}$ are the unit basis vectors along the x, y, and z axes, respectively. The vector can be resolved first into two components, A_y along the y axis and A_p that is on the x-z plane. A_p can be resolved subsequently into A_x and A_y along the x and y axes, respectively. Thus the components of a vector in a given coordinate system are the set of its projections along the three unit basis vectors:

$$\vec{A} = A_x\hat{x} + A_y\hat{y} + A_z\hat{z}. \tag{2.1.2}$$

The magnitude (or length) of this vector is clearly

$$|\vec{A}| = A = \sqrt{A_x^2 + A_y^2 + A_z^2}. \tag{2.1.3}$$

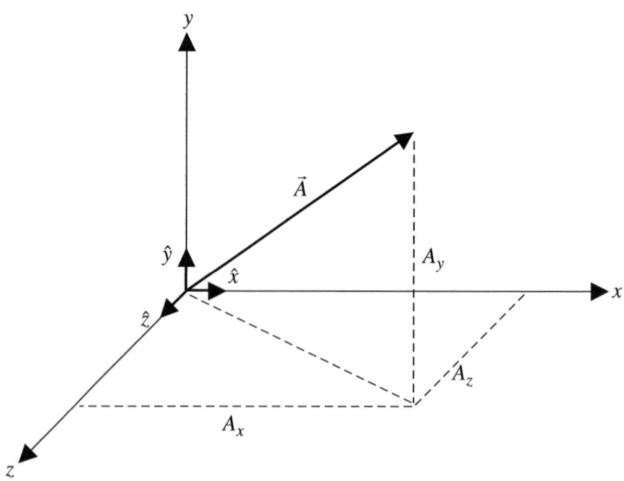

Figure 2.1 Projections of a vector along the three axes of the rectangular coordinates.

Example 2.1

For $\vec{A} = 4\hat{x} - 6\hat{y} + 12\hat{z}$, find its magnitude $|\vec{A}|$ and a unit vector in the direction of the vector.

Find the magnitude as

$$|\vec{A}| = \sqrt{4^2 + (-6)^2 + 12^2} = 14.$$

The unit vector along $\vec{A}$ is then

$$\hat{a} = \frac{4\hat{x} - 6\hat{y} + 12\hat{z}}{14} = \frac{2}{7}\hat{x} - \frac{3}{7}\hat{y} + \frac{6}{7}\hat{z}.$$

Position Vector

As shown in Figure 2.2, a point $P(x, y, z)$ in the rectangular coordinate system can be expressed uniquely via a vector $\vec{r}$ as follows:

$$\vec{r} = x\hat{x} + y\hat{y} + z\hat{z}. \tag{2.1.4}$$

This vector $\vec{r}$ is called the position vector of point P. Hence

$$P(x, y, z) = P(\vec{r}). \tag{2.1.5}$$

Scalar Fields

A scalar quantity that varies with the coordinates (x, y, and z in the rectangular system) can be expressed in terms of a variable position vector $\vec{r}$ as $f(\vec{r})$. The

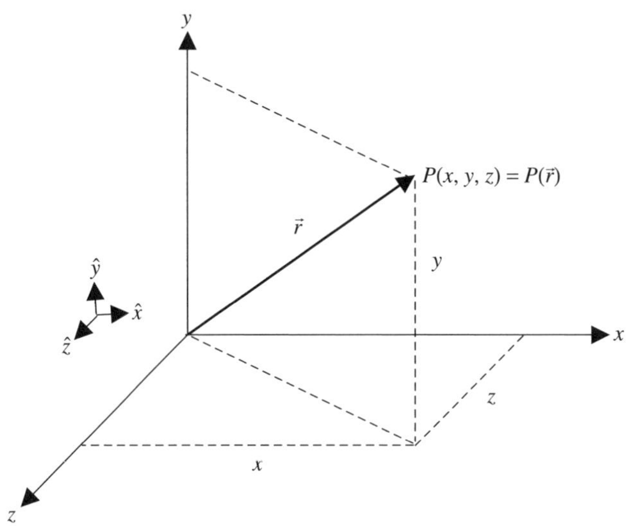

Figure 2.2 Position vector of a point $P(x, y, z)$ in the rectangular coordinate system.

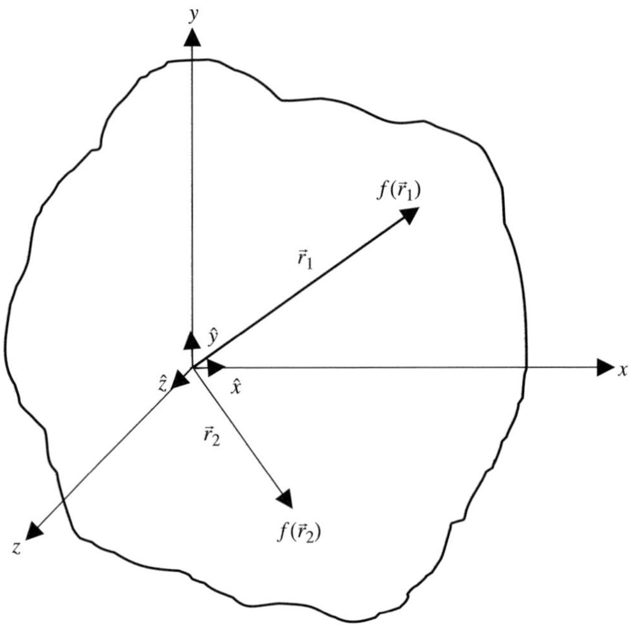

Figure 2.3 A scalar field $f(\vec{r})$ in the rectangular coordinate system.

scalar field is $f(\vec{r})$ with continuous first partial derivatives at almost every point in its domain of definition V. As illustrated in Figure 2.3, a scalar field assigns a number $f(\vec{r})$ to each point $\vec{r}$ in its domain V. Points where $f(\vec{r})$ or its derivatives are discontinuous are the source points of the field. Examples of scalar fields include temperature, pressure, electric charge, and electrical potential.

Vector Fields

A vector quantity that varies with the coordinates (x, y, and z in the rectangular system) can be expressed in terms of a variable position vector $\vec{r}$ as $\vec{A}(\vec{r})$. The vector field is $\vec{A}(\vec{r})$ with continuous first partial derivatives at almost every point in its domain of definition V. Points where $\vec{A}(\vec{r})$ or its derivatives are discontinuous are the source points of the vector field. As illustrated in Figure 2.4, a vector field can be expressed in terms of three scalar component fields as follows:

$$\vec{A}(\vec{r}) = \hat{x}A_x(\vec{r}) + \hat{y}A_y(\vec{r}) + \hat{z}A_z(\vec{r}). \tag{2.1.6}$$

Thus a vector field assigns a vector $\vec{A}(\vec{r})$ to each point $\vec{r}$ in its domain V. Examples of vector fields include current density, electric field intensity, magnetic field intensity, electric flux density, magnetic flux density, force, and the velocity.

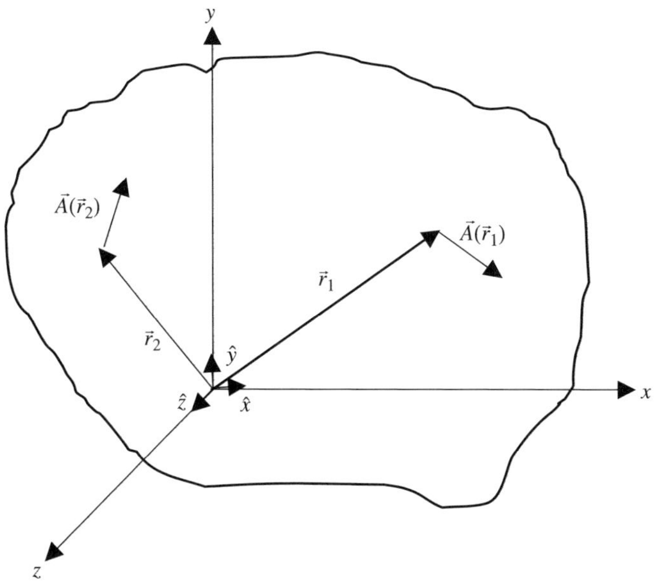

Figure 2.4 A vector field $\vec{A}(\vec{r})$ in the rectangular coordinate system.

Addition and Subtraction of Vectors

The sum of two vectors $\vec{A}$ and $\vec{B}$ is expressed as $\vec{C} = \vec{A} + \vec{B}$. It is defined as the componentwise sum of the two vectors. If

$$\vec{A} = \sum_{\alpha} \hat{u}_{\alpha} A_{\alpha} \tag{2.1.7}$$

and

$$\vec{B} = \sum_{\alpha} \hat{u}_{\alpha} B_{\alpha}, \tag{2.1.8}$$

then

$$\vec{A} + \vec{B} = \sum_{\alpha} \hat{u}_{\alpha}(A_{\alpha} + B_{\alpha}) = \vec{C}. \tag{2.1.9}$$

Consider vectors $\vec{A}$ and $\vec{B}$ in rectangular coordinate system, as illustrated in Figure 2.5. These vectors can be expressed as follows:

$$\vec{A} = A_x\hat{x} + A_y\hat{y} + A_z\hat{z} \tag{2.1.10}$$

and

$$\vec{B} = B_x\hat{x} + B_y\hat{y} + B_z\hat{z}. \tag{2.1.11}$$

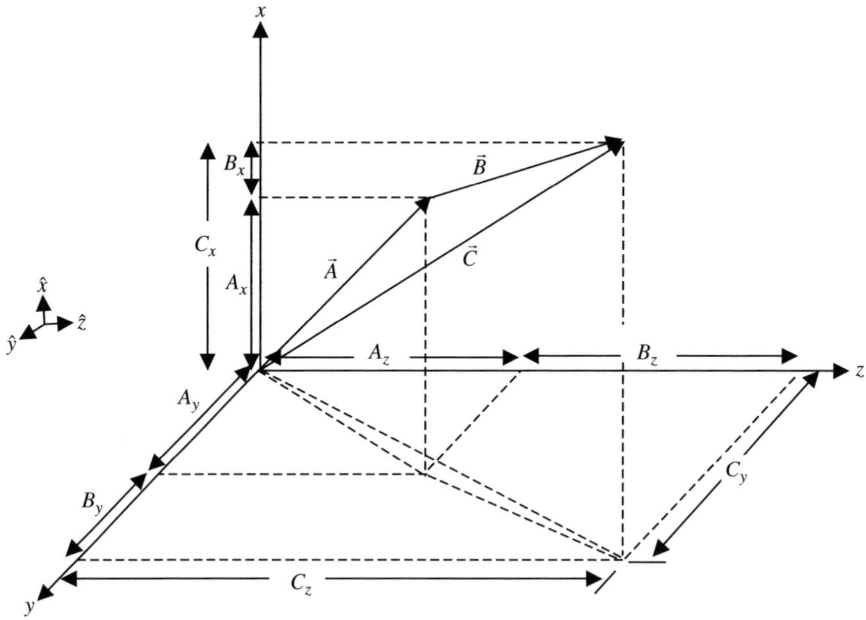

Figure 2.5 Graphical addition of vectors $\vec{A}$ and $\vec{B}$ in the rectangular coordinate system.

Therefore

$$\vec{C} = \vec{A} + \vec{B} = (A_x + B_x)\hat{x} + (A_y + B_y)\hat{y} + (A_z + B_z)\hat{z}$$
$$= C_x\hat{x} + C_y\hat{y} + C_z\hat{z}, \quad (2.1.12)$$

where

$$C_x = A_x + B_x, \quad (2.1.13)$$
$$C_y = A_y + B_y, \quad (2.1.14)$$

and

$$C_z = A_z + B_z. \quad (2.1.15)$$

Figure 2.5 illustrates this mathematical process graphically. Note that a single vector equation is equivalent to a set of three scalar equations. This feature of vector equations helps to keep the electromagnetic field analysis compact and simple.

The following relations can be verified easily using the addition procedure of the vectors.

$$\vec{A} + \vec{B} = \vec{B} + \vec{A}, \quad (2.1.16)$$
$$\vec{A} + (\vec{B} + \vec{C}) = (\vec{A} + \vec{B}) + \vec{C}. \quad (2.1.17)$$

Thus we find that the vector addition follows the commutative and the associative laws. Further the negative of a vector has the same length as the original vector, but its direction is reversed. This property can be used to find the difference of two vectors as follows:

$$\vec{C} = \vec{A} - \vec{B} = \vec{A} + (-\vec{B}). \quad (2.1.18)$$

In other words, if

$$\vec{A} = A_x\hat{x} + A_y\hat{y} + A_z\hat{z} \quad (2.1.19)$$

and

$$\vec{B} = B_x\hat{x} + B_y\hat{y} + B_z\hat{z}, \quad (2.1.20)$$

then

$$\vec{C} = \vec{A} - \vec{B} = (A_x - B_x)\hat{x} + (A_y - B_y)\hat{y} + (A_z - B_z)\hat{z}. \quad (2.1.21)$$

Therefore the subtraction of vectors may be interpreted as a generalization of the addition, with the direction of the vector to be subtracted reversed.

Example 2.2

A vector $\vec{A}$ is defined in the rectangular coordinate system as being directed from $(0, -1, 3)$ to $(5, 1, -2)$. Find (a) a vector expression for $\vec{A}$, (b) the magnitude of $\vec{A}$, and (c) a unit vector pointing in the direction of $\vec{A}$.

(a) As shown in Figure 2.6, position vectors for the two points are found to be

$$\vec{r}_1 = -\hat{y} + 3\hat{z}$$

and

$$\vec{r}_2 = 5\hat{x} + \hat{y} - 2\hat{z}.$$

Therefore

$$\vec{A} = \vec{r}_2 - \vec{r}_1 = 5\hat{x} + 2\hat{y} - 5\hat{z}.$$

(b) From (a),

$$|\vec{A}| = \sqrt{5^2 + 2^2 + (-5)^2} = \sqrt{54} \approx 7.35.$$

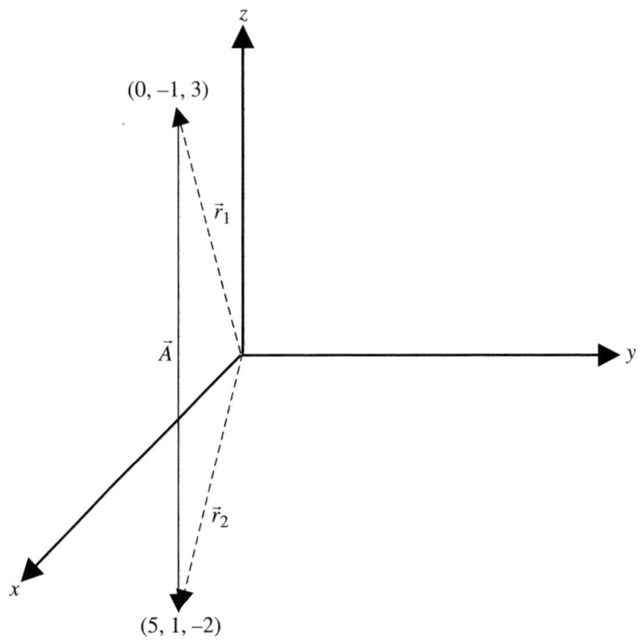

Figure 2.6 Points and related vectors for Example 2.2.

(c) Unit vector along

$$\vec{A} = \hat{a} = \frac{\vec{A}}{|\vec{A}|} = \frac{5\hat{x} + 2\hat{y} - 5\hat{z}}{\sqrt{54}} \approx 0.68\hat{x} + 0.27\hat{y} - 0.68\hat{z}.$$

Scalar and Vector Products

Two different kinds of multiplications are defined for the vectors. Scalar (or dot) multiplication of two vectors produces a scalar quantity, whereas a vector (or cross) product results in another vector that is perpendicular to the plan that contains the two vectors. The scalar or dot product of two vectors is a scalar quantity defined as follows:

$$\vec{A} \cdot \vec{B} = |\vec{A}||\vec{B}| \cos(\theta), \tag{2.1.22}$$

where θ is the smaller angle between the two vectors, as illustrated in Figure 2.7.

The scalar product is commutative as well as distributive. This can be proved easily via the definition of the dot products. These properties can be written as follows:

$$\vec{A} \cdot \vec{B} = \vec{B} \cdot \vec{A} \tag{2.1.23}$$

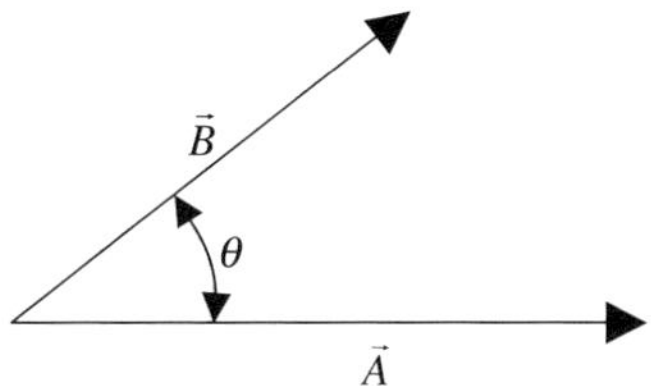

Figure 2.7 Two vectors $\vec{A}$ and $\vec{B}$ used for scalar multiplication.

and

$$(\vec{A} + \vec{B}) \cdot \vec{C} = \vec{A} \cdot \vec{C} + \vec{B} \cdot \vec{C}. \tag{2.1.24}$$

As an example, consider the unit basis vectors $(\hat{x}, \hat{y}, \hat{z})$ in rectangular coordinate system (x, y, z). Note that the unit basis vectors are mutually orthogonal (i.e., $\hat{x} \perp \hat{y} \perp \hat{z}$). From (2.1.22) and the Figure 2.7, we find that

$$\cos(\theta = 0) = 1 \longrightarrow \hat{x} \cdot \hat{x} = \hat{y} \cdot \hat{y} = \hat{z} \cdot \hat{z} = 1 \tag{2.1.25}$$

and

$$\cos\left(\theta = \frac{\pi}{2}\right) = 0 \longrightarrow \hat{x} \cdot \hat{y} = \hat{y} \cdot \hat{z} = \hat{z} \cdot \hat{x} = 0. \tag{2.1.26}$$

Further

$$\begin{aligned}
\vec{A} \cdot \vec{B} &= (\hat{x}A_x + \hat{y}A_y + \hat{z}A_z) \cdot (\hat{x}B_x + \hat{y}B_y + \hat{z}B_z) \\
&= (\hat{x} \cdot \hat{x})\, A_x B_x + (\hat{y} \cdot \hat{y})\, A_y B_y + (\hat{z} \cdot \hat{z})\, A_z B_z \\
&\quad + (\hat{x} \cdot \hat{y})(A_x B_y + A_y B_x) + (\hat{y} \cdot \hat{z})(A_y B_z + A_z B_y) \\
&\quad + (\hat{x} \cdot \hat{z})(A_x B_z + A_z B_x).
\end{aligned}$$

Therefore

$$\vec{A} \cdot \vec{B} = A_x B_x + A_y B_y + A_z B_z. \tag{2.1.27}$$

As a special case, consider the scalar product of a vector with itself. It may be found from (2.1.27) as follows:

$$\vec{A} \cdot \vec{A} = |\vec{A}|^2 \cos(0) = |\vec{A}|^2 = A^2 = A_x^2 + A_y^2 + A_z^2. \tag{2.1.28}$$

As shown in Figure 2.8, the scalar product of two vectors can be interpreted as the product of the length of one vector with the component of other vector that is parallel to it. Further, if $\vec{A} \neq 0$ and $\vec{B} \neq 0$, then $\vec{A} \cdot \vec{B} = 0$ implies that the two vectors are mutually orthogonal (i.e., $\vec{A} \perp \vec{B}$).

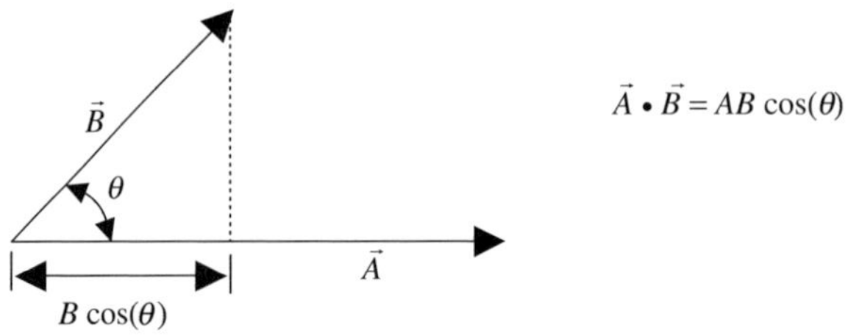

Figure 2.8 Physical interpretation of the scalar multiplication.

As mentioned earlier, the vector or cross product of two vectors $\vec{A}$ and $\vec{B}$ is a vector quantity written as $\vec{A} \times \vec{B}$. As illustrated in Figure 2.9, the direction of this new vector is perpendicular to the plane containing the two vectors and its sense is such that if the right-handed screw at $\vec{A}$ is rotated from $\vec{A}$ to $\vec{B}$ following the smaller in-between angle (θ), then this will advance in the direction of $\vec{A} \times \vec{B}$. The length of this vector is found to be

$$|\vec{A} \times \vec{B}| = |\vec{A}||\vec{B}| \sin(\theta). \tag{2.1.29}$$

As illustrated in Figure 2.10, geometrically $|\vec{A} \times \vec{B}|$ represents the area of a parallelogram formed by the two vectors. Note that if $\vec{A} \neq 0$ and $\vec{B} \neq 0$ then $\vec{A} \times \vec{B} = 0$ implies that the two vectors are parallel or collinear.

It can be proved easily via its definition that the cross product is anticommutative but follows the distributive law. Mathematically

$$\vec{A} \times \vec{B} = -\vec{B} \times \vec{A} \tag{2.1.30}$$

and

$$(\vec{A} + \vec{B}) \times \vec{C} = \vec{A} \times \vec{C} + \vec{B} \times \vec{C}. \tag{2.1.31}$$

Consider the unit basis vectors $(\hat{x}, \hat{y}, \hat{z})$ in rectangular coordinate system (x, y, z). Note that the unit basis vectors are mutually orthogonal (i.e., $\hat{x} \perp \hat{y} \perp \hat{z}$). From (2.1.29) we find that

$$\sin(\theta = 0) = 0 \rightarrow \hat{x} \times \hat{x} = \hat{y} \times \hat{y} = \hat{z} \times \hat{z} = 0 \tag{2.1.32}$$

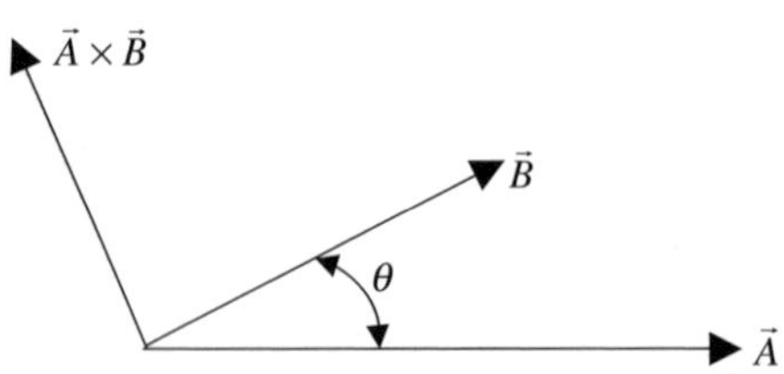

Figure 2.9 Two vectors $\vec{A}$ and $\vec{B}$ used for vector multiplication.

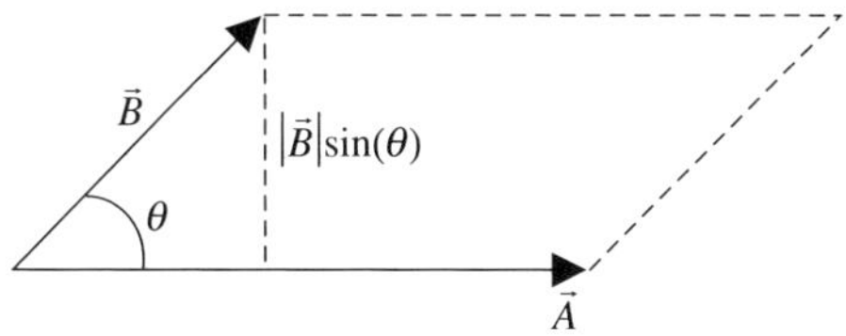

Figure 2.10 Physical interpretation of the vector multiplication.

and

$$\sin\left(\theta = \frac{\pi}{2}\right) = 1 \rightarrow \hat{x} \times \hat{y} = \hat{z}, \qquad \hat{y} \times \hat{z} = \hat{x}, \qquad \hat{z} \times \hat{x} = \hat{y}. \tag{2.1.33}$$

Therefore

$$\begin{aligned}\vec{A} \times \vec{B} &= (\hat{x}A_x + \hat{y}A_y + \hat{z}A_z) \times (\hat{x}B_x + \hat{y}B_y + \hat{z}B_z) \\ &= (\hat{x} \times \hat{x})A_xB_x + (\hat{y} \times \hat{y})A_yB_y + (\hat{z} \times \hat{z})A_zB_z + (\hat{x} \times \hat{y})(A_xB_y - A_yB_x) \\ &\quad + (\hat{y} \times \hat{z})(A_yB_z - A_zB_y) + (\hat{z} \times \hat{x})(A_zB_x - A_xB_z).\end{aligned}$$

This equation can be simplified, via (2.1.32) and (2.1.33), as follows:

$$\vec{A} \times \vec{B} = \hat{x}(A_yB_z - A_zB_y) + \hat{y}(A_zB_x - A_xB_z) + \hat{z}(A_xB_y - A_yB_x). \tag{2.1.34}$$

Note that this result can be obtained by expanding the following determinant about its first row:

$$\vec{A} \times \vec{B} = \begin{vmatrix} \hat{x} & \hat{y} & \hat{z} \\ A_x & A_y & A_z \\ B_x & B_y & B_z \end{vmatrix}. \tag{2.1.35}$$

Example 2.3

Find the position vectors for points $P_1(2, 5, 7)$ and $P_2(4, -3, 6)$. If a parallelogram is formed with these two position vectors as its two adjacent sides, then prove that the diagonals of this parallelogram bisect each other.

Points P_1 and P_2 are located using the rectangular coordinates as shown in Figure 2.11. This includes the corresponding position vectors $\vec{r}_1$ and $\vec{r}_2$, and the associated parallelogram. Vectors $\vec{D}_1$ and $\vec{D}_2$ are drawn along the diagonal of the parallelogram. Thus the two position vectors can be found as follows:

$$\vec{r}_1 = 2\hat{x} + 5\hat{y} + 7\hat{z}$$

and

$$\vec{r}_2 = 4\hat{x} - 3\hat{y} + 6\hat{z}.$$

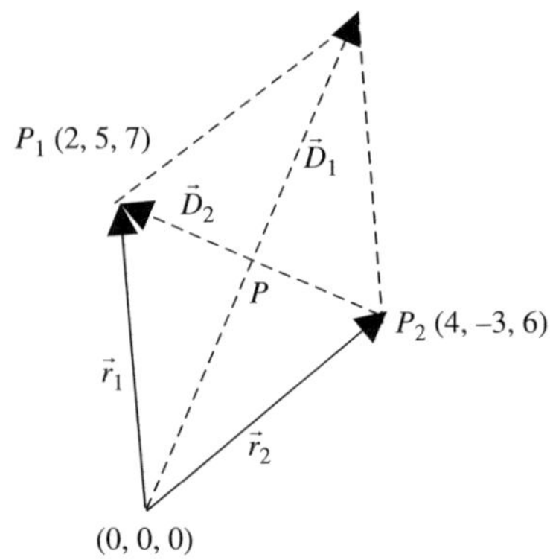

Figure 2.11 Points and related vectors for Example 2.3.

Therefore

$$\vec{D}_1 = \vec{r}_1 + \vec{r}_2 = 6\hat{x} + 2\hat{y} + 13\hat{z}$$

and

$$\vec{D}_2 = \vec{r}_1 - \vec{r}_2 = -2\hat{x} + 8\hat{y} + \hat{z}.$$

The position vector for the point P at which the two diagonals intersect can be expressed as $\alpha\vec{D}_1$ or $\vec{r}_2 + \beta\vec{D}_2$, where α and β are unknown constants. These can be evaluated as follows:

For $\alpha\vec{D}_1 = \vec{r}_2 + \beta\vec{D}_2$,

$$\alpha(6\hat{x} + 2\hat{y} + 13\hat{z}) = 4\hat{x} - 3\hat{y} + 6\hat{z} + \beta(-2\hat{x} + 8\hat{y} + \hat{z})$$
$$= (4 - 2\beta)\hat{x} + (-3 + 8\beta)\hat{y} + (6 + \beta)\hat{z}.$$

Equating coefficients of each component on the two sides, we get

$$6\alpha = 4 - 2\beta,$$
$$2\alpha = -3 + 8\beta,$$

and

$$13\alpha = 6 + \beta.$$

Coefficients α and β of any two of this set of equations can be easily evaluated to be 0.5. This proves that the two diagonals bisect each other.

Example 2.4

Find a vector $\vec{B}$ that is perpendicular to $\vec{A} = -8\hat{x} + 9\hat{y} - \hat{z}$, has no z component, and has a magnitude of unity.

Since the vector has no z component, it can be assumed to be

$$\vec{B} = x\hat{x} + y\hat{y}.$$

Further, it is perpendicular to $\vec{A}$, therefore

$$\vec{A} \cdot \vec{B} = 0 \longrightarrow (-8\hat{x} + 9\hat{y} - \hat{z}) \cdot (x\hat{x} + y\hat{y}) = -8x + 9y = 0 \longrightarrow x = \frac{9}{8}y.$$

We need one more independent relation between x and y to evaluate the unknown vector. This can be found using the condition that this vector has a magnitude of unity (hence it is a unit vector):

$$|\vec{B}| = 1 \longrightarrow \sqrt{x^2 + y^2} = 1 \longrightarrow y^2 = 1 - x^2 = 1 - \left(\frac{9}{8}y\right)^2 \longrightarrow \left(1 + \frac{81}{64}\right)y^2 = 1.$$

Solving for y and x, we find that

$$y = \pm\frac{8}{\sqrt{145}}$$

and

$$x = \pm\frac{9}{\sqrt{145}}.$$

Therefore the desired vector is

$$\vec{B} = \pm\frac{1}{\sqrt{145}}(9\hat{x} + 8\hat{y}).$$

Example 2.5

If $\vec{A} = \hat{x} + 2\hat{y} - 3\hat{z}$ and $\vec{B} = 2\hat{x} - \hat{y} + \hat{z}$ then find (a) the component of $\vec{B}$ in direction of $\vec{A}$, (b) the smallest angle between the two vectors, and (c) a unit vector perpendicular to the plane that contains two vectors.

(a) Unit vector along $\vec{A}$,

$$\vec{A} = \hat{a} = \frac{\vec{A}}{|\vec{A}|} = \frac{\hat{x} + 2\hat{y} - 3\hat{z}}{\sqrt{1 + 2^2 + (-3)^2}} = \frac{\hat{x} + 2\hat{y} - 3\hat{z}}{\sqrt{14}} = 0.27\hat{x} + 0.53\hat{y} - 0.8\hat{z}.$$

Then the component of $\vec{B}$ along $\vec{A}$, B_A, is found to be

$$B_A = |\vec{B} \cdot \hat{a}| = |(2\hat{x} - \hat{y} + \hat{z}) \cdot (0.27\hat{x} + 0.53\hat{y} - 0.8\hat{z})| = 0.8.$$

(b) If angle between the two vectors is θ, then

$$\cos(\theta) = \frac{\vec{A}\bullet\vec{B}}{|\vec{A}||\vec{B}|} = \frac{\hat{a}\bullet\vec{B}}{|B|} = \frac{-3}{\sqrt{14}\sqrt{6}} = -0.33 \rightarrow \theta = \cos^{-1}(-0.33) = 109^\circ.$$

(c) Let $\vec{C}$ be perpendicular to the plane that contains $\vec{A}$ and $\vec{B}$, then

$$\vec{C} = \vec{A}\times\vec{B} = \begin{vmatrix} \hat{x} & \hat{y} & \hat{z} \\ 1 & 2 & -3 \\ 2 & -1 & 1 \end{vmatrix} = -\hat{x} - 7\hat{y} - 5\hat{z}.$$

Therefore the unit vector parallel to $\vec{C}$ is

$$\hat{c} = \pm\frac{\vec{C}}{|C|} = \pm\frac{-\hat{x} - \hat{7} - 5\hat{z}}{\sqrt{1+7^2+5^2}} = \pm(0.1155\hat{x} + 0.8083\hat{y} - 0.5774\hat{z})$$

Example 2.6

A triangle ABC is found by connecting the points A $(-1, 0, 2)$, B $(0, 1, 0)$, and C $(1, -1, 0)$. Find the point D that makes the plane a parallelogram $ABDC$.

Various points and associated vectors are illustrated in Figure 2.12. The position vectors for points A, B, and C are found to be

$$\vec{r}_A = -\hat{x} + 2\hat{z},$$
$$\vec{r}_B = \hat{y},$$

and

$$\vec{r}_C = \hat{x} - \hat{y}.$$

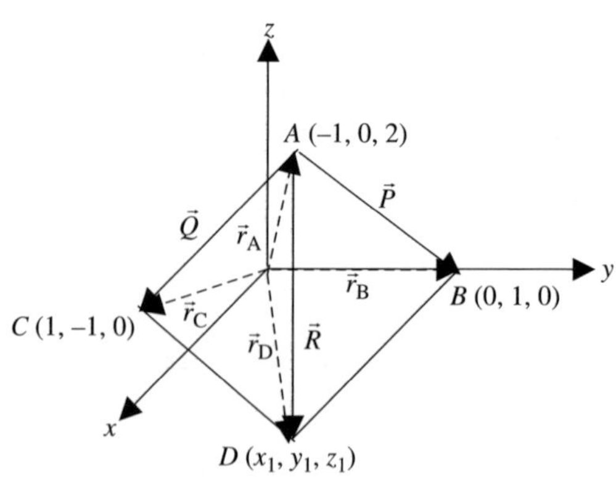

Figure 2.12 Points and related vectors for Example 2.6.

Therefore

$$\vec{P} = \vec{r}_B - \vec{r}_A = \hat{x} + \hat{y} - 2\hat{z}$$

and

$$\vec{Q} = \vec{r}_C - \vec{r}_A = 2\hat{x} - \hat{y} - 2\hat{z}.$$

Since $\vec{P}$ and $\vec{Q}$ are two sides of the parallelogram $ABDC$, vector $\vec{R}$ along the diagonal must be

$$\vec{R} = \vec{P} + \vec{Q} = 3\hat{x} - 4\hat{z}.$$

Now the position vector $\vec{r}_D$ for point D can be determined as follows:

$$\vec{r}_D = \vec{r}_A + \vec{R} = 2\hat{x} - 2\hat{z}$$

Therefore the coordinates of point D are (2, 0, −2).

Scalar and Vector Triple Products

The scalar triple product of three vectors $\vec{A}$, $\vec{B}$, and $\vec{C}$ may be shown to be

$$\vec{A} \cdot (\vec{B} \times \vec{C}) = (\vec{A} \times \vec{B}) \cdot \vec{C} = \vec{C} \cdot (\vec{A} \times \vec{B}) = (\vec{C} \times \vec{A}) \cdot \vec{B}. \tag{2.1.36}$$

The vector triple product of these three vectors $\vec{A}$, $\vec{B}$, and $\vec{C}$ is given as

$$\vec{A} \times (\vec{B} \times \vec{C}) = \vec{B}(\vec{A} \cdot \vec{C}) - \vec{C}(\vec{A} \cdot \vec{B}). \tag{2.1.37}$$

As obvious from the right-hand side, this product sometimes referred as the "*BAC CAB*" rule.

Example 2.7

Verify (2.1.36) for $\vec{A} = 4\hat{x} + 3\hat{y} + 2\hat{z}$, $\vec{B} = 2\hat{x} + 5\hat{z}$, and $\vec{C} = -7\hat{y}$.

From (2.1.35),

$$\vec{B} \times \vec{C} = \begin{vmatrix} \hat{x} & \hat{y} & \hat{z} \\ 2 & 0 & 5 \\ 0 & -7 & 0 \end{vmatrix} = 35\hat{x} - 14\hat{z}.$$

Now using (2.1.27), we get

$$\vec{A} \cdot (\vec{B} \times \vec{C}) = (4\hat{x} + 3\hat{y} + 2\hat{z}) \cdot (35\hat{x} - 14\hat{z}) = 140 - 28 = 112.$$

Similarly

$$\vec{A} \times \vec{B} = \begin{vmatrix} \hat{x} & \hat{y} & \hat{z} \\ 4 & 3 & 2 \\ 2 & 0 & 5 \end{vmatrix} = 15\hat{x} - 16\hat{y} - 6\hat{z}$$

and

$$\vec{C} \cdot (\vec{A} \times \vec{B}) = -7\hat{y} \cdot (15\hat{x} - 16\hat{y} - 6\hat{z}) = 112.$$

Repeating one more time for the last term of (2.1.36), we get

$$\vec{C} \times \vec{A} = \begin{vmatrix} \hat{x} & \hat{y} & \hat{z} \\ 0 & -7 & 0 \\ 4 & 3 & 2 \end{vmatrix} = -14\hat{x} + 28\hat{z}$$

and

$$\vec{B} \cdot (\vec{C} \times \vec{A}) = (\vec{C} \times \vec{A}) \cdot \vec{B} = (-14\hat{x} + 28\hat{z}) \cdot (2\hat{x} + 5\hat{z}) = -28 + 140 = 112.$$

Thus we find that all the sides of (2.1.36) produce 112 for the given vectors. This verifies the relation.

Example 2.8

Verify (2.1.37) for $\vec{A} = 4\hat{x} + 3\hat{y} + 2\hat{z}$, $\vec{B} = 2\hat{x} + 5\hat{z}$, and $\vec{C} = -7\hat{y}$.

Since the given vectors are same as in Example 2.7, we already know that

$$\vec{B} \times \vec{C} = 35\hat{x} - 14\hat{z}.$$

Using (2.1.35) along with this result, we find that

$$\vec{A} \times (\vec{B} \times \vec{C}) = \begin{vmatrix} \hat{x} & \hat{y} & \hat{z} \\ 4 & 3 & 2 \\ 35 & 0 & -14 \end{vmatrix} = -42\hat{x} + 126\hat{y} - 105\hat{z}.$$

Since $\vec{A} \cdot \vec{C} = (4\hat{x} + 3\hat{y} + 2\hat{z}) \cdot (-7\hat{y}) = -21$,

$$\vec{B}(\vec{A} \cdot \vec{C}) = (2\hat{x} + 5\hat{z})(-21) = -42\hat{x} - 105\hat{z}.$$

Similarly

$$\vec{A} \cdot \vec{B} = (4\hat{x} + 3\hat{y} + 2\hat{z}) \cdot (2\hat{x} + 5\hat{z}) = 8 + 10 = 18$$

and

$$\vec{C}(\vec{A} \cdot \vec{B}) = -7\hat{y}(18) = -126\hat{y}.$$

Therefore

$$\vec{B}(\vec{A} \cdot \vec{C}) - \vec{C}(\vec{A} \cdot \vec{B}) = -42\hat{x} - 105\hat{z} + 126\hat{y} = \vec{A} \times (\vec{B} \times \vec{C}).$$

2.2 COORDINATE SYSTEMS

Systems of orthogonal coordinates are useful in the analysis and manipulation of vectors. We have already used orthogonal coordinates in the preceding section for rectangular coordinates. This section presents more detail on rectangular coordinates along with two other commonly used coordinate systems, namely cylindrical (ρ, ϕ, z) and spherical (r, θ, ϕ) coordinates.

Rectangular Coordinates

We showed extensively how to use these coordinates in the preceding section. Mathematical expressions are relatively simple to manipulate in the rectangular coordinates system, especially because the unit basis vectors do not change directions from point to point. Figure 2.13 shows incremental length $d\vec{\ell}$ projected along the three axes. This incremental length can be expressed mathematically as follows:

$$d\vec{\ell} = \hat{x}dx + \hat{y}dy + \hat{z}dz. \tag{2.2.1}$$

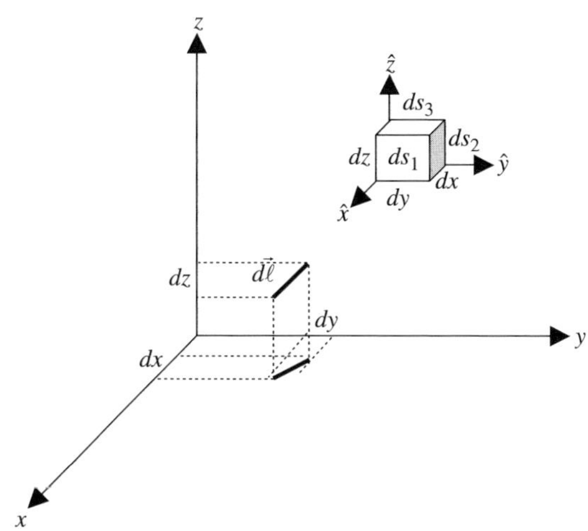

Figure 2.13 Incremental length and an infinitesimal volume in the rectangular coordinates system.

The figure also gives a detailed representation of the small rectangular volume's six bounding surfaces $\pm ds_1$, $\pm ds_2$, and $\pm ds_3$. The surface areas of the incremental volume can be expressed mathematically as follows:

$$d\vec{s}_1 = \pm\hat{x}dydz, \tag{2.2.2}$$

$$d\vec{s}_2 = \pm\hat{y}dzdx, \tag{2.2.3}$$

$$d\vec{s}_3 = \pm\hat{z}dxdy, \tag{2.2.4}$$

and

$$dv = dxdydz. \tag{2.2.5}$$

Cylindrical Coordinates

If a structure under consideration has a cylindrical geometry, then employing the rectangular coordinates system can make the analysis too complex and even impossible to solve. In such cases a cylindrical coordinates system is used. As illustrated in Figure 2.14, the unit basis vectors in cylindrical coordinates are $\hat{\rho}$, $\hat{\phi}$, and $\hat{z}$. The angle ϕ is specified with respect to the $+x$ axis. The expressions for an incremental unit of length along the surfaces bounding a cylinder in the shape of a pillbox, and

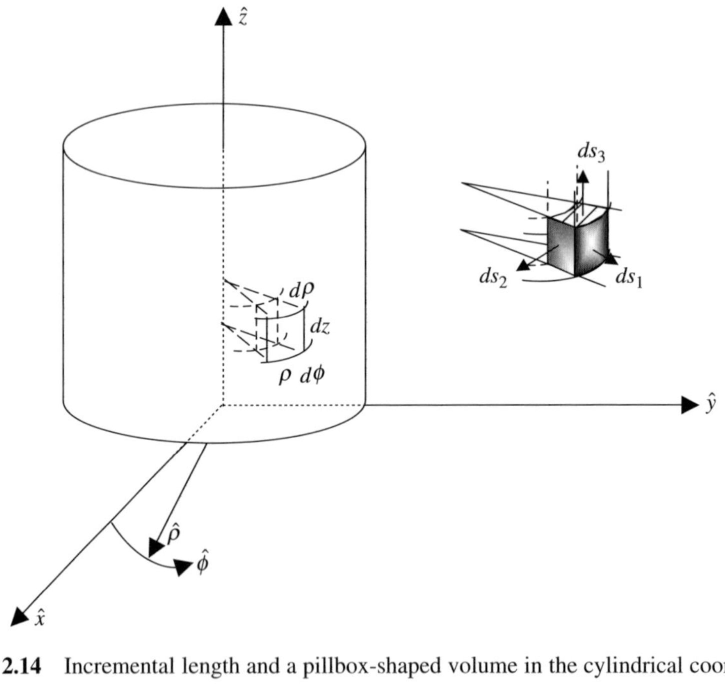

Figure 2.14 Incremental length and a pillbox-shaped volume in the cylindrical coordinates system.

the volume of that pillbox, can be found easily as follows:

$$d\vec{\ell} = \hat{\rho}d\rho + \hat{\phi}\rho d\phi + \hat{z}dz, \tag{2.2.6}$$

$$d\vec{s}_1 = \pm\hat{\rho}\rho d\phi dz, \tag{2.2.7}$$

$$d\vec{s}_2 = \pm\hat{\phi}d\rho dz, \tag{2.2.8}$$

$$d\vec{s}_3 = \pm\hat{z}\rho d\rho d\phi, \tag{2.2.9}$$

and

$$dv = \rho d\rho d\phi dz. \tag{2.2.10}$$

Note that in this coordinates system the two unit vectors $\hat{\rho}$ and $\hat{\phi}$ change directions from point to point, whereas $\hat{z}$ stays in same direction. The mathematical manipulation rules described in the preceding section are still applicable after appropriate modifications. The scalar multiplications of the unit basis vectors expressed by (2.1.25) and (2.1.26) take the following form in cylindrical coordinates:

$$\cos(\theta = 0) = 1 \rightarrow \hat{\rho}\cdot\hat{\rho} = \hat{\phi}\cdot\hat{\phi} = \hat{z}\cdot\hat{z} = 1 \tag{2.2.11}$$

and

$$\cos\left(\theta = \frac{\pi}{2}\right) = 0 \rightarrow \hat{\rho}\cdot\hat{\phi} = \hat{\phi}\cdot\hat{z} = \hat{z}\cdot\hat{\rho} = 0. \tag{2.2.12}$$

Similarly the cross products of unit vectors expressed by (2.1.32) and (2.1.33) become

$$\sin(\theta = 0) = 0 \rightarrow \hat{\rho}\times\hat{\rho} = \hat{\phi}\times\hat{\phi} = \hat{z}\times\hat{z} = 0 \tag{2.2.13}$$

and

$$\sin\left(\theta = \frac{\pi}{2}\right) = 1 \rightarrow \hat{\rho}\times\hat{\phi} = \hat{z}, \quad \hat{\phi}\times\hat{z} = \hat{\rho}, \quad \hat{z}\times\hat{\rho} = \hat{\phi}. \tag{2.2.14}$$

Spherical Coordinates

Figure 2.15 shows another orthogonal coordinates system that facilitates the study of systems with spherical geometry. As illustrated, the unit basis vectors in this case are $\hat{r}$, $\hat{\theta}$, and $\hat{\phi}$. Note that the angle θ is specified with respect to the +z axis, whereas ϕ is with respect to the x axis (the same as in the cylindrical coordinates system). Further all three unit vectors change directions from point to point. The rules described in the preceding section for mathematical manipulations are then applicable after appropriate modifications. The scalar multiplications of the unit basis vectors expressed by (2.1.25) and (2.1.26) take the following form in spherical coordinates:

$$\cos(\theta = 0) = 1 \rightarrow \hat{r}\cdot\hat{r} = \hat{\theta}\cdot\hat{\theta} = \hat{\phi}\cdot\hat{\phi} = 1 \tag{2.2.15}$$

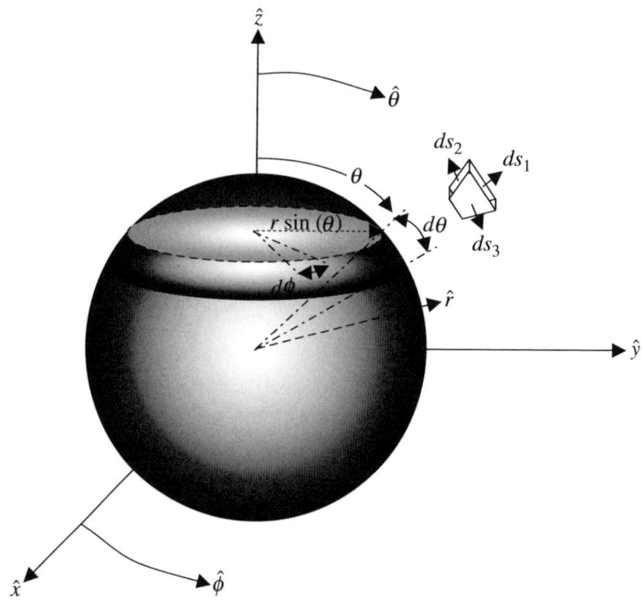

Figure 2.15 Incremental length and a small unit of volume in the spherical coordinates system.

and

$$\cos\left(\theta = \frac{\pi}{2}\right) = 0 \rightarrow \hat{r} \cdot \hat{\theta} = \hat{\theta} \cdot \hat{\phi} = \hat{\phi} \cdot \hat{r} = 0. \tag{2.2.16}$$

Similarly the cross products of the unit vectors expressed by (2.1.32) and (2.1.33) become

$$\sin(\theta = 0) = 0 \rightarrow \hat{r} \times \hat{r} = \hat{\theta} \times \hat{\theta} = \hat{\phi} \times \hat{\phi} = 0 \tag{2.2.17}$$

and

$$\sin\left(\theta = \frac{\pi}{2}\right) = 1 \rightarrow \hat{r} \times \hat{\theta} = \hat{\phi}, \qquad \hat{\theta} \times \hat{\phi} = \hat{r}, \quad \hat{\phi} \times \hat{r} = \hat{\theta}. \tag{2.2.18}$$

With the help of Figure 2.15, expressions for the surfaces bounding a small volume of incremental length can be easily found as follows:

$$d\vec{\ell} = \hat{r}dr + \hat{\theta}r\,d\theta + \hat{\phi}\,r\sin\theta d\phi, \tag{2.2.19}$$

$$d\vec{s}_1 = \pm\hat{r}r^2\sin\theta d\theta d\phi, \tag{2.2.20}$$

$$d\vec{s}_2 = \pm\hat{\theta}\,r\sin\theta drd\phi, \tag{2.2.21}$$

$$d\vec{s}_3 = \pm\hat{\phi}\,rdrd\theta, \tag{2.2.22}$$

and

$$dv = r^2 \sin\theta \, dr d\theta d\phi. \tag{2.2.23}$$

Occasionally a problem under consideration involves a combination of different coordinate systems. In other words, a part of the problem may be specified in one system of coordinates while the other is conveniently described in another system of coordinates. In such situations the description in one coordinate system can be transformed into the other with the help of Figure 2.16. These conversions are summarized in Tables 2.1 through 2.4.

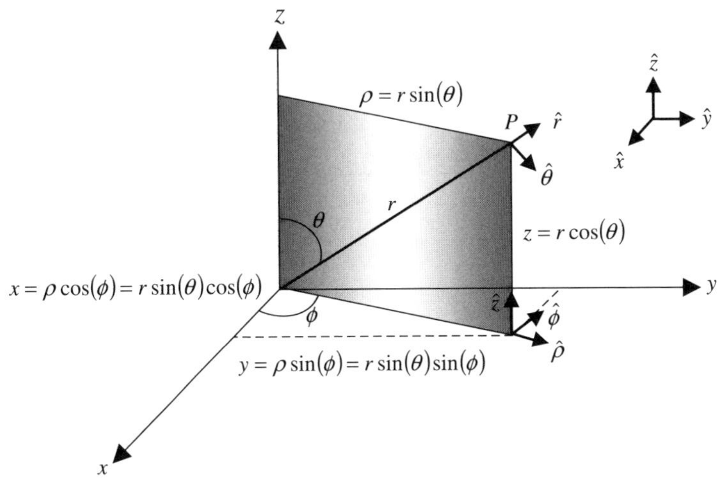

Figure 2.16 Conversions among the representations of coordinates.

TABLE 2.1 Conversions of the Coordinates

	From (x, y, z)	From (ρ, ϕ, z)	From (r, θ, ϕ)
To (x, y, z)	x	$x = \rho\cos(\phi)$	$x = r\sin(\theta)\cos(\phi)$
	y	$y = \rho\sin(\phi)$	$y = r\sin(\theta)\sin(\phi)$
	z	$z = z$	$z = r\cos(\theta)$
To (ρ, ϕ, z)	$\rho = \sqrt{x^2 + y^2}$	ρ	$\rho = r\sin(\theta)$
	$\phi = \tan^{-1}(y/x)$	ϕ	$\phi = \phi$
	$z = z$	z	$z = r\cos(\theta)$
To (r, θ, ϕ)	$r = \sqrt{x^2 + y^2 + z^2}$	$r = \sqrt{\rho^2 + z^2}$	r
	$\theta = \cos^{-1}\left(z/\sqrt{x^2 + y^2 + z^2}\right)$	$\theta = \tan^{-1}(\rho/z)$	θ
	$\phi = \tan^{-1}(y/x)$	$\phi = \phi$	ϕ

TABLE 2.2 Conversions of Unit Vectors to Rectangular Coordinates

	To $(\hat{x}, \hat{y}, \hat{z})$
From $(\hat{\rho}, \hat{\phi}, \hat{z})$	$\hat{\rho} = \hat{x}\cos(\phi) + \hat{y}\sin(\phi)$
	$\hat{\phi} = -\hat{x}\sin(\phi) + \hat{y}\cos(\phi)$
	$\hat{z} = \hat{z}$
From $(\hat{r}, \hat{\theta}, \hat{\phi})$	$\hat{r} = \hat{x}\sin(\theta)\cos(\phi) + \hat{y}\sin(\theta)\sin(\phi) + \hat{z}\cos(\theta)$
	$\hat{\theta} = \hat{x}\cos(\theta)\cos(\phi) + \hat{y}\cos(\theta)\sin(\phi) - \hat{z}\sin(\theta)$
	$\hat{\phi} = -\hat{x}\sin(\phi) + \hat{y}\cos(\phi)$

TABLE 2.3 Conversions of Unit Vectors to Cylindrical Coordinates

	To $(\hat{\rho}, \hat{\phi}, \hat{z})$
From $(\hat{x}, \hat{y}, \hat{z})$	$\hat{x} = \hat{\rho}\cos(\phi) - \hat{\phi}\sin(\phi)$
	$\hat{y} = \hat{\rho}\sin(\phi) + \hat{\phi}\cos(\phi)$
	$\hat{z} = \hat{z}$
From $(\hat{r}, \hat{\theta}, \hat{\phi})$	$\hat{r} = \hat{\rho}\sin(\theta) + \hat{z}\cos(\theta)$
	$\hat{\theta} = \hat{\rho}\cos(\theta) - \hat{z}\sin(\theta)$
	$\hat{\phi} = \hat{\phi}$

TABLE 2.4 Conversions of Unit Vectors to Spherical Coordinates

	To $(\hat{r}, \hat{\theta}, \hat{\phi})$
From $(\hat{x}, \hat{y}, \hat{z})$	$\hat{x} = \hat{r}\sin(\theta)\cos(\phi) + \hat{\theta}\cos(\theta)\cos(\phi) - \hat{\phi}\sin(\phi)$
	$\hat{y} = \hat{r}\sin(\theta)\sin(\phi) + \hat{\theta}\cos(\theta)\sin(\phi) + \hat{\phi}\cos(\phi)$
	$\hat{z} = \hat{r}\cos(\theta) - \hat{\theta}\sin(\theta)$
From $(\hat{\rho}, \hat{\phi}, \hat{z})$	$\hat{\rho} = \hat{r}\sin(\theta) + \hat{\theta}\cos(\theta)$
	$\hat{\phi} = \hat{\phi}$
	$\hat{z} = \hat{r}\cos(\theta) - \hat{\theta}\sin(\theta)$

Example 2.9

A receiver antenna produces an electromotive force (voltage) proportional to the y component of the incident electric field intensity. Assume that the proportionality constant is 0.25 m. If the electric field intensity is given by the following expression, then find the voltage induced at the receiver:

$$\vec{E} = \hat{\phi}\frac{1+\cos^2(\phi)}{1+\rho^2}\ \text{V/m}.$$

In this case the electric field intensity is given in cylindrical coordinates, whereas the antenna responds to y component. Therefore the given electric field intensity must be converted into rectangular coordinates. This can be easily done via the Tables 2.1 and 2.2. Hence

$$\vec{E} = (-\hat{x}\sin\phi + \hat{y}\cos\phi)\frac{1+\cos^2(\phi)}{1+x^2+y^2}\ \text{V/m}.$$

Since $\phi = \tan^{-1}(y/x)$,

$$\sin\phi = \frac{y}{\sqrt{x^2+y^2}},$$

$$\cos\phi = \frac{x}{\sqrt{x^2+y^2}},$$

and

$$\vec{E} = (-\hat{x}\sin\phi + \hat{y}\cos\phi)\frac{1+\cos^2(\phi)}{1+x^2+y^2} = (-x\hat{x} + y\hat{y})\frac{2x^2+y^2}{(x^2+y^2)(1+x^2+y^2)}\ \text{V/m}.$$

The induced voltage V_{induced} is found to be

$$V_{\text{induced}} = 0.25\vec{E}\cdot\hat{y} = 0.25y\frac{2x^2+y^2}{(x^2+y^2)(1+x^2+y^2)}\ \text{V}.$$

Example 2.10

A receiver antenna produces an electromotive force (voltage) proportional to the x component of the incident electric field intensity. Assume that the proportionality constant is 0.25 m. If the electric field intensity is given by the following expression, then find the induced voltage at the receiver:

$$\vec{E} = 4(\hat{\theta}\cos\phi - \hat{\phi}\sin\phi\cos\theta)\ \text{V/m}.$$

In this case, the electric field intensity is given in spherical coordinates, whereas the induced voltage is proportional to x component. Therefore the given electric field intensity must be converted into rectangular coordinates. This is easily done via the Tables 2.1 and 2.2. Hence

$$\vec{E} = 4[(\hat{x}\cos\theta\cos\phi + \hat{y}\cos\theta\sin\phi - \hat{z}\sin\theta)\cos\phi$$
$$- (-\hat{x}\sin\phi + \hat{y}\cos\phi)\sin\phi\cos\theta].$$

The induced voltage can be found as follows:

$$V_{\text{induced}} = 0.25\hat{x}\cdot\vec{E} = \cos\theta\cos^2\phi + \sin^2\phi\cos\theta = \cos\theta\left[\cos^2\phi + \sin^2\phi\right]$$
$$= \cos\theta = \frac{z}{\sqrt{x^2+y^2+z^2}}\ \text{V}.$$

Volume Integrals

Figure 2.17 shows a scalar field $f(\vec{r})$ in a volume V. The volume integral of this field over V (a scalar quantity) is expressed mathematically as $\int_V f(\vec{r})dv$. Since the volume region V can be considered as $V = \sum_{i=1}^{N}\Delta v_i$, the volume integral may be defined mathematically as

$$\int_V f(\vec{r})dv = \lim_{\substack{N\to\infty \\ \Delta v_i \to 0}} \sum_{i=1}^{N} f(\vec{r}_i)\Delta v_i. \tag{2.2.23}$$

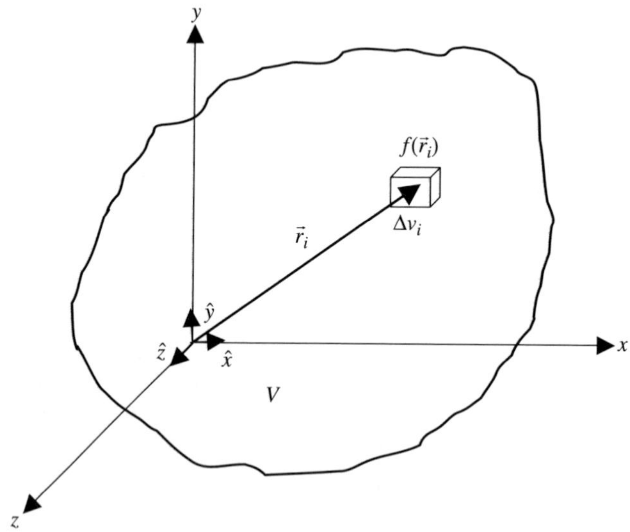

Figure 2.17 A scalar field in volume V and its volume integral.

Thus the volume integral in rectangular coordinates can be written as follows:

$$\int_V f(\vec{r})dv = \iiint_V f(x, y, z)dx\,dy\,dz \tag{2.2.24}$$

The volume integral of mass density $m(\vec{r})$ over a region V gives the total mass in V. Similarly the volume integral of a charge density gives the total charge contained in region V.

Example 2.11

The electrical charge density in a volume is given by

$$\rho_V = 1 - 0.5(x^2 + y^2 + z^2)\,\mathrm{C/m^{-3}}.$$

Determine the total charge enclosed in a cube defined by $-1\ \mathrm{m} \le x \le 1\ \mathrm{m}$, $-1\ \mathrm{m} \le y \le 1\ \mathrm{m}$, and $-1\ \mathrm{m} \le z \le 1\ \mathrm{m}$.

$$\begin{aligned}
Q &= \int_V \rho_V\,dv = \int_{-1}^{1}\int_{-1}^{1}\int_{-1}^{1}\left[1 - 0.5(x^2 + y^2 + z^2)\right]dx\,dy\,dz \\
&= \int_{-1}^{1}\int_{-1}^{1}\int_{-1}^{1} dx\,dy\,dz - 0.5\int_{-1}^{1}\int_{-1}^{1}\int_{-1}^{1}(x^2 + y^2 + z^2)dx\,dy\,dz \\
&= x|_{-1}^{1}y|_{-1}^{1}z|_{-1}^{1} - 0.5\left[\frac{x^3}{3}\bigg|_{-1}^{1} y|_{-1}^{1}z|_{-1}^{1} + x|_{-1}^{1}\frac{y^3}{3}\bigg|_{-1}^{1} z|_{-1}^{1} + x|_{-1}^{1}y|_{-1}^{1}\frac{z^3}{3}\bigg|_{-1}^{1}\right] = 4\,\mathrm{C}.
\end{aligned}$$

Example 2.12

Electrical charge density in a volume is given by

$$\rho_V = 1 - 0.5(x^2 + y^2 + z^2)\,\mathrm{C/m^{-3}}$$

Determine the total charge enclosed in a volume defined by $0 \le \rho \le 1\ \mathrm{m}$, $(\pi/4) \le \phi \le (\pi/2)$, and $-1\ \mathrm{m} \le z \le 1\ \mathrm{m}$.

$$\begin{aligned}
Q &= \int_V \rho_V\,dv = \int_{0}^{1}\int_{\pi/4}^{\pi/2}\int_{-1}^{1}\left[1 - 0.5(\rho^2 + z^2)\right]\rho d\rho d\phi dz \\
&= \int_{0}^{1}\int_{\pi/4}^{\pi/2}\int_{-1}^{1}\rho d\rho d\phi dz - 0.5\int_{0}^{1}\int_{\pi/4}^{\pi/2}\int_{-1}^{1}(\rho^2 + z^2)\rho d\rho d\phi dz \\
&= \frac{\rho^2}{2}\bigg|_{0}^{1}\phi|_{\pi/4}^{\pi/2}z|_{-1}^{1} - 0.5\left[\frac{\rho^4}{4}\bigg|_{0}^{1}\phi|_{\pi/4}^{\pi/2}z|_{-1}^{1} + \frac{\rho^2}{2}\bigg|_{0}^{1}\phi|_{\pi/4}^{\pi/2}\frac{z^3}{3}\bigg|_{-1}^{1}\right] = \frac{7\pi}{48}\,\mathrm{C}.
\end{aligned}$$

Example 2.13

Electrical charge density in a volume is given by

$$\rho_V = 1 - 0.5(x^2 + y^2 + z^2)\,\mathrm{C/m^{-3}}.$$

Determine the total charge enclosed in a wedge defined by $0 \leq r \leq 1$ m, $(\pi/4) \leq \theta \leq (\pi/2)$, and $(\pi/4) \leq \phi \leq (\pi/2)$.

$$Q = \int_V \rho_V\, dv = \int_0^1 \int_{\pi/4}^{\pi/2} \int_{\pi/4}^{\pi/2} [1 - 0.5r^2] r^2 \sin\theta dr d\theta d\phi$$

$$= \int_0^1 \int_{\pi/4}^{\pi/2} \int_{\pi/4}^{\pi/2} r^2 \sin\theta\, dr d\theta d\phi - 0.5 \int_0^1 \int_{\pi/4}^{\pi/2} \int_{\pi/4}^{\pi/2} (\rho^2 + z^2) r^2 \sin\theta dr d\theta d\phi$$

$$= \frac{r^3}{3}\Big|_0^1 (-\cos\theta)|_{\pi/4}^{\pi/2} \phi|_{\pi/4}^{\pi/2} - 0.5\left[\frac{r^5}{5}\Big|_0^1 (-\cos\theta)|_{\pi/4}^{\pi/2} \phi|_{\pi/4}^{\pi/2}\right] = \frac{7\pi}{120\sqrt{2}}\ \mathrm{C}.$$

2.3 DIFFERENTIATION AND INTEGRATION OF VECTORS

Differentiation of Vectors

The partial derivative of a vector field $\vec{A}(\vec{r}) = \vec{A}(x, y, z)$ is defined as follows:

$$\frac{\partial}{\partial x}\vec{A}(x, y, z) = \lim_{\Delta x \to 0} \frac{\vec{A}(x + \Delta x, y, z) - \vec{A}(x, y, z)}{\Delta x}. \tag{2.3.1}$$

Since the unit basis vectors are constant in the rectangular coordinates system, this equation may be expressed as

$$\begin{aligned}\frac{\partial}{\partial x}\vec{A}(x, y, z) = \hat{x} \lim_{\Delta x \to 0} \frac{A_x(x + \Delta x, y, z) - A_x(x, y, z)}{\Delta x} \\ + \hat{y} \lim_{\Delta x \to 0} \frac{A_y(x + \Delta x, y, z) - A_y(x, y, z)}{\Delta x} \\ + \hat{z} \lim_{\Delta x \to 0} \frac{A_z(x + \Delta x, y, z) - A_z(x, y, z)}{\Delta x}.\end{aligned}$$

Therefore

$$\frac{\partial}{\partial x}\vec{A}(x, y, z) = \hat{x}\frac{\partial}{\partial x}A_x(x, y, z) + \hat{y}\frac{\partial}{\partial x}A_y(x, y, z) + \hat{z}\frac{\partial}{\partial z}A_z(x, y, z). \tag{2.3.2}$$

In the product of a vector field $\vec{A}(x, y, z)$ with a scalar field $f(x, y, z)$, the following property can be easily verified:

$$\frac{\partial}{\partial x}(f\vec{A}) = \vec{A}\frac{\partial f}{\partial x} + f\frac{\partial \vec{A}}{\partial x}. \tag{2.3.3}$$

Similarly the following relations hold for two vector fields $\vec{A}(x, y, z)$ and $\vec{B}(x, y, z)$:

$$\frac{\partial}{\partial x}(\vec{A} + \vec{B}) = \frac{\partial \vec{A}}{\partial x} + \frac{\partial \vec{B}}{\partial x}, \tag{2.3.4}$$

$$\frac{\partial}{\partial x}(\vec{A} \cdot \vec{B}) = \frac{\partial \vec{A}}{\partial x} \cdot \vec{B} + \vec{A} \cdot \frac{\partial \vec{B}}{\partial x}, \tag{2.3.5}$$

$$\frac{\partial}{\partial x}(\vec{A} \times \vec{B}) = \frac{\partial \vec{A}}{\partial x} \times \vec{B} + \vec{A} \times \frac{\partial \vec{B}}{\partial x}. \tag{2.3.6}$$

Note that these derivatives require extra care in the cylindrical and spherical coordinates systems. If a unit vector is not constant, then the effect of its variation must be included when taking derivatives. For example, if $\vec{A}(x, y, z) = \hat{u}(x, y, z)\, A(x, y, z)$, then

$$\frac{\partial}{\partial x}\vec{A}(x, y, z) = A(x, y, z)\frac{\partial}{\partial x}\hat{u}(x, y, z) + \hat{u}(x, y, z)\frac{\partial}{\partial x}A(x, y, z). \tag{2.3.7}$$

Integration of Vectors

There are two significantly useful integration operations involving vectors. Depending on the quantities involved, the line integral of a vector field can lead to the work done by a force or the electromotive force induced by an electromagnetic field. Similarly the surface integral of a vector field over a closed surface can provide information about the enclosed source (or sink). These two operations of vector fields are described below.

Line Integral of a Vector Field Consider a vector field $\vec{A}(\vec{r})$ and a path Γ (a scalar quantity) as shown in Figure 2.18. The line integral of $\vec{A}(\vec{r})$ along the contour Γ is

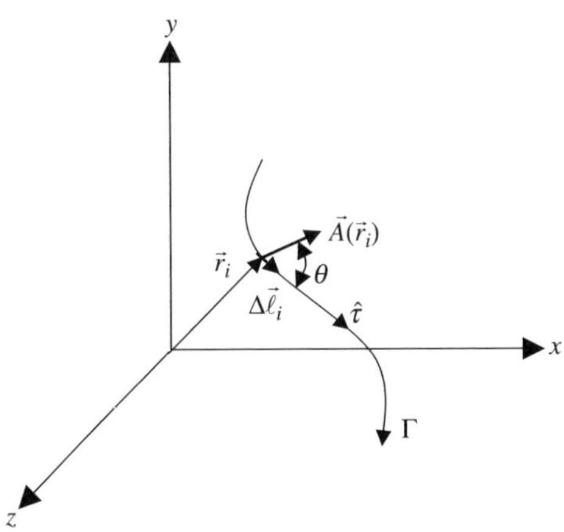

Figure 2.18 A vector field and its line integral along the path Γ.

expressed as $\int_\Gamma \vec{A}(\vec{r}) \cdot d\vec{\ell}$, where $d\vec{\ell}$ is a differential displacement along and tangent to the contour Γ.

Suppose that path Γ is made up of differential lengths $\Delta\ell_i$ and $\hat{\tau}$ is tangent at every point on this path. The line integral of $\vec{A}(\vec{r})$ is then interpreted geometrically as

$$\Gamma = \sum_{i=1}^{N} \Delta\vec{\ell}_i, \tag{2.3.8}$$

where

$$\Delta\vec{\ell}_i = \hat{\tau}\Delta\ell_i \tag{2.3.9}$$

and

$$\theta = \angle(\hat{\tau}, \vec{A}(\vec{r}_i)). \tag{2.3.10}$$

The line integral can be expressed as

$$\int_\Gamma \vec{A}(\vec{r}) \cdot d\vec{\ell} = \lim_{\substack{N\to\infty \\ \Delta\ell_i\to 0}} \sum_{i=1}^{N} \vec{A}(\vec{r}_i) \cdot \Delta\vec{\ell}_i \tag{2.3.11}$$

provided that the limit exists and it is finite.

Further

$$\vec{A}(\vec{r}_i) \cdot \Delta\vec{\ell}_i = \vec{A}(\vec{r}_i) \cdot \hat{\tau}\Delta\ell_i = |\vec{A}(\vec{r}_i)| \cos(\theta)\Delta\ell_i. \tag{2.3.12}$$

Therefore the line integral is an integration of the component of $\vec{A}(\vec{r})$ along and tangent to the contour Γ with respect to the arc length along Γ.

For example, consider a force field $\vec{F}(\vec{r})$. At a given point the force field can be resolved into two components as shown in Figure 2.19. Component $\vec{F}_n(\vec{r})$ is normal

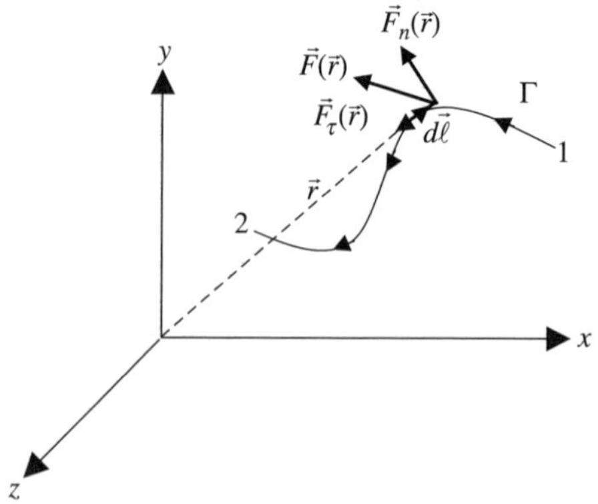

Figure 2.19 Physical interpretation of the line integral.

and $\vec{F}_\tau(\vec{r})$ is tangential to the path Γ. The work required to move a particle along Γ from point 1 to point 2 against this force field $\vec{F}(\vec{r})$ is found to be

$$W_{12} = -\int_1^2 \vec{F}(\vec{r}) \cdot d\vec{\ell}. \tag{2.3.13}$$

Note that if the contour Γ is a closed curve, then sometimes it is called a "circulation" integral (the closed line integral), and it is denoted as follows:

$$\oint \vec{A}(\vec{r}) \cdot d\vec{\ell} = \oint \hat{\tau} \cdot \vec{A}(\vec{r}) d\ell. \tag{2.3.14}$$

Example 2.14

Find the work done by the following force in moving a particle from P_1 to P_2 shown in Figure 2.20: (a) over the straight line connecting (0, 0, 2) and (1, 3, 0); (b) over the straight line path connecting (0, 0, 2) → (0, 0, 0) → (1, 0, 0) → (1, 3, 0); and (c) over the straight line path connecting (0, 0, 2) → (0, 0, 0) → (1, 3, 0). Assume that all distances are in meters.

$$\vec{F} = \hat{x}12xy^2 + \hat{y}15yz + \hat{z}9z^2 \text{ N}.$$

Since $d\vec{\ell} = \hat{x}dx + \hat{y}dy + \hat{z}dz$ in the rectangular coordinates,

$$\vec{F} \cdot d\vec{\ell} = 12xy^2\, dx + 15yz\, dy + 9z^2 dz \text{ Nm}.$$

The work done can be determined by evaluating the line integrals as follows:

(a) When moving on the straight line that connects points P_1 and P_2 directly, the relations among x, y, and z are found to be

$$y = 3x \rightarrow dy = 3dx$$

and

$$z = -\tfrac{2}{3}y + 2 \rightarrow dz = -\tfrac{2}{3}dy.$$

Therefore

$$\int_{P_1}^{P_2} \vec{F} \cdot d\vec{\ell} = \int_0^3 12\left(\frac{y}{3}\right)y^2 \frac{dy}{3} + \int_0^3 15y\left(-\frac{2}{3}y + 2\right)dy + \int_2^0 9z^2 dz = 48 \text{ Nm}.$$

(b) When moving on the path that follows the points (0, 0, 2) → (0, 0, 0) → (1, 0, 0) → (1, 3, 0), the line integral is evaluated in three parts. Note that x and y both remain zero for (0, 0, 2) → (0, 0, 0), y and z both remain

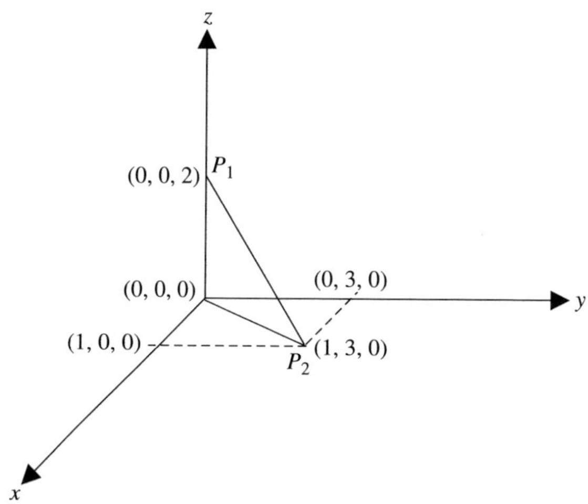

Figure 2.20 Geometry and integration paths for Example 2.14.

zero for (0, 0, 0) $\rightarrow$ (1, 0, 0), and z remains zero and x at 1 for (1, 0, 0) $\rightarrow$ (1, 3, 0). Therefore the line integral simplifies to

$$\int_{P_1}^{P_2} \vec{F} \cdot d\vec{\ell} = \int_2^0 9z^2\, dz = -24\,\text{Nm}.$$

(c) In this case the line integral splits into two parts. First segment is same as in (b) above and $y = 3x$ for $(0, 0, 0) \rightarrow (1, 3, 0)$. Therefore the line integral simplifies to

$$\int_{P_1}^{P_2} \vec{F} \cdot d\vec{\ell} = \int_2^0 9z^2 dz + \int_0^1 12x(3x)^2 dx = -24 + 27 = 3\,\text{Nm}.$$

Example 2.15

Find the circulation of the electric field intensity $\vec{E} = \hat{\rho}150\rho\cos\phi + \hat{\phi}200\sin\phi$ V/m over a semicircular path of radius 1.5 m with its center at (1.5 m, 0°, 0), as shown in Figure 2.21.

Since this problem calls for integrating over a circular path, use of cylindrical coordinates is appropriate. The electric field intensity is already given in cylindrical coordinates, and for the incremental length we can use the following expression:

$$d\vec{\ell} = \hat{\rho}d\rho + \hat{\phi}\rho d\phi + \hat{z}dz.$$

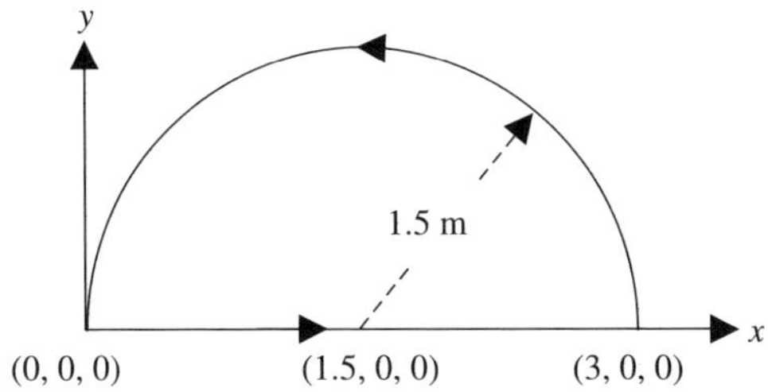

Figure 2.21 Geometry and integration path for Example 2.15.

Therefore

$$\vec{E} \cdot d\vec{\ell} = 150\rho \cos \phi \, d\rho + 200\rho \sin \phi \, d\phi.$$

Note that $d\phi$ is zero over the path (0, 0, 0) to (3, 0, 0). However, both $d\rho$ and $d\phi$ are nonzero over the circular arc. Therefore a relation between ρ and ϕ is found as follows:

The equation of a circle whose center is at ($x_1 = 1.5$ m, $y_1 = 0$) and radius is 1.5 m can be written as

$$(x - 1.5)^2 + y^2 = 1.5^2 \rightarrow x^2 + y^2 = 3x.$$

Using Table 2.1, we can transform this equation into cylindrical coordinates as follows:

$$\rho = 3 \cos \phi.$$

Therefore

$$\oint \vec{E} \cdot d\vec{\ell} = \int_0^3 150\rho \cos\phi \, d\rho \Big|_{\phi=0} + \int_3^0 150\rho \cos \phi \, d\rho \Big|_{\cos \phi = \rho/3} + \int_0^{\pi/2} 200\rho \sin \phi \, d\phi \Big|_{\rho = 3 \cos \phi}$$

$$= \int_0^3 150\rho \, d\rho + \int_3^0 150\rho \left(\frac{\rho}{3}\right) d\rho + \int_0^{\pi/2} 200(3 \cos \phi) \sin \phi \, d\phi = 525 \text{ V}.$$

Example 2.16

Find the circulation of the following magnetic field intensity around the contour *ABCDA*, as shown in Figure 2.22:

$$\vec{H} = \hat{\rho} \rho \sin \phi + \hat{\phi} 2\rho^2 \text{ A/m}.$$

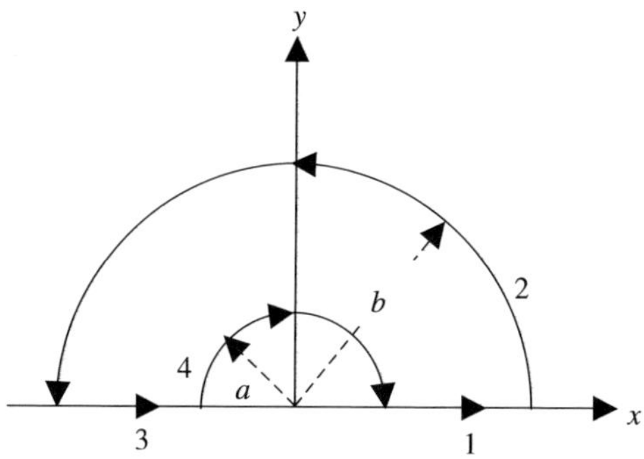

Figure 2.22 Geometry and integration path for Example 2.16.

The geometry indicates that the cylindrical coordinates are appropriate for this problem. Therefore

$$\vec{H} \cdot d\vec{\ell} = \rho \sin\phi \, d\rho + 2\rho^3 d\phi \text{ A}.$$

Dividing the integration path as indicated in the figure, the circulation integral can be written as follows:

$$\oint \vec{H} \cdot d\vec{\ell} = \int_1 \vec{H} \cdot d\vec{\ell}\bigg|_{\substack{\rho=a\to b \\ \phi=0}} + \int_2 \vec{H} \cdot d\vec{\ell}\bigg|_{\substack{\rho=b \\ \phi=0\to\pi}} + \int_3 \vec{H} \cdot d\vec{\ell}\bigg|_{\substack{\rho=b\to a \\ \phi=\pi}} + \int_4 \vec{H} \cdot d\vec{\ell}\bigg|_{\substack{\rho=a \\ \phi=\pi\to 0}}.$$

Therefore

$$\oint \vec{H} \cdot d\vec{\ell} = \int_a^b \rho \sin\phi \, d\rho\bigg|_{\phi=0} + \int_0^\pi 2\rho^3 d\phi\bigg|_{\rho=b} + \int_b^a \rho \sin\phi \, d\rho\bigg|_{\phi=\pi} + \int_\pi^0 2\rho^3 d\phi\bigg|_{\rho=a}$$

$$= 2\pi(b^3 - a^3)\text{A}.$$

Surface or Flux Integral of a Vector Field A surface (or flux) integral of the vector field $\vec{A}(\vec{r})$ over the surface S (a scalar quantity) is defined as follows:

$$\int_S \vec{A}(\vec{r}) \cdot d\vec{s} = \int_S \hat{n} \cdot \vec{A}(\vec{r}) ds, \tag{2.3.15}$$

where $d\vec{s} = \hat{n}ds$ and $\hat{n}$ is a unit outward normal vector at each point on the surface S.

Consider a vector field $\vec{A}(\vec{r})$ that passes through an open surface S bound by a closed contour C. This surface may be considered as composed of N small areas Δs. As shown in Figure 2.23, $\hat{n}$ is a unit vector normal to the surface at point $\vec{r}_1$ and the vector field $\vec{A}(\vec{r}_i)$ makes an angle θ with the normal at this point.

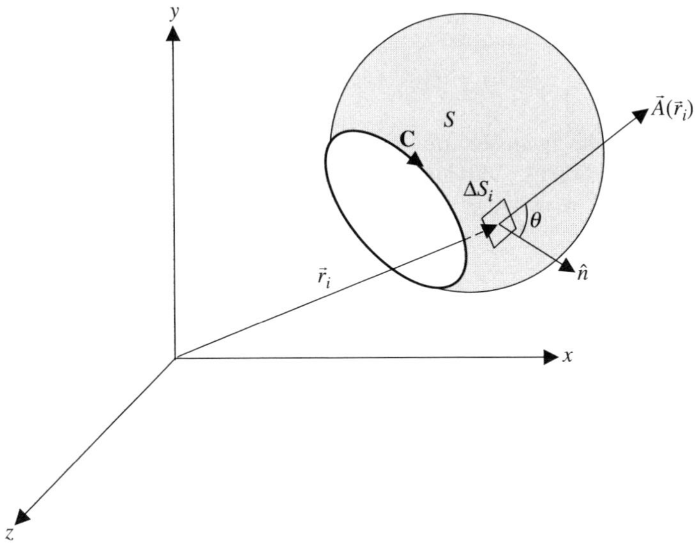

Figure 2.23 Surface integral of a vector field.

Mathematically

$$S = \sum_{i=1}^{N} \Delta s_i, \tag{2.3.16}$$

$$\Delta\vec{s}_i = \hat{n}\Delta s_i. \tag{2.3.17}$$

The surface integral is expressed as

$$\int_S \vec{A}(\vec{r}) \cdot d\vec{s} = \lim_{\substack{N\to\infty \\ \Delta s_i \to 0}} \sum_{i=1}^{N} \hat{n} \cdot \vec{A}(\vec{r}_i)\Delta s_i \tag{2.3.18}$$

provided that the limit exists and it is finite.

Physically the surface integral can be interpreted as follows:

$$\hat{n} \cdot \vec{A}(\vec{r}_i)\Delta s_i = |\vec{A}(\vec{r}_i)| \cos(\theta)\Delta s_i. \tag{2.3.19}$$

Hence the surface (or flux) integral expresses the "flux" of the normal component of $\vec{A}(\vec{r})$ over the surface S. For example, the surface integral $\int_S \hat{n} \cdot \vec{J}(\vec{r})ds$ of the current density $\vec{J}(\vec{r})$ over a surface S gives the total current I that flows across S.

Note that if S is a closed surface, then the notation becomes

$$\oint_S \vec{A}(\vec{r}) \cdot d\vec{s} = \oint_S \hat{n} \cdot \vec{A}(\vec{r})\, ds. \tag{2.3.20}$$

Example 2.17

For an electric flux density given as $\vec{D} = \hat{\rho}(5/\rho)\,\mathrm{C/m^2}$, find the total electric flux emanating from a closed surface that bounds (a) a cylindrical volume of 3 m in radius, with $0 \le \phi \le 2\pi$, and -1 m $\le z \le 1$ m, (b) a cube with -1 m $\le x \le 1$ m, -1 m $\le y \le 1$ m, -1 m $\le z \le 1$ m, and (c) a sphere of 1 m radius centered at the origin.

(a) Note that the given electric flux density has only radial component in cylindrical coordinates, and therefore the flux emanates only from the lateral surface of the cylinder:

$$\oint_S \vec{D} \cdot d\vec{s} = \int_{-1}^{1} \int_{0}^{2\pi} \frac{5}{\rho} \rho \, d\phi dz \bigg|_{\rho=3} = 5\phi|_0^{2\pi} z|_{-1}^{1} = 20\pi \text{ C}.$$

(b) Since rectangular coordinates are appropriate for a cube, electric flux density given in cylindrical coordinates is converted into rectangular coordinates as well. This can be done easily via Tables 2.1 and 2.3 as follows:

$$\vec{D} = \hat{\rho}\frac{5}{\rho} = (\hat{x}\cos\phi + \hat{y}\sin\phi)\frac{5}{\sqrt{x^2+y^2}} = \frac{5(\hat{x}x + \hat{y}y)}{x^2+y^2}.$$

Note that the given vector has no component in z direction. Therefore only four lateral surfaces will contribute to surface integral:

$$\oint_S \vec{D} \cdot d\vec{s} = \int_{-1}^{1}\int_{-1}^{1} \frac{5y\,dx\,dz}{x^2+y^2}\bigg|_{y=1} - \int_{-1}^{1}\int_{-1}^{1} \frac{5y\,dx\,dz}{x^2+y^2}\bigg|_{y=-1}$$
$$+ \int_{-1}^{1}\int_{-1}^{1} \frac{5x\,dy\,dz}{x^2+y^2}\bigg|_{x=1} - \int_{-1}^{1}\int_{-1}^{1} \frac{5x\,dy\,dz}{x^2+y^2}\bigg|_{x=-1}$$
$$= 20\int_{-1}^{1} \frac{dx}{1+x^2} + 20\int_{-1}^{1} \frac{dy}{1+y^2} = 40\tan^{-1}x\bigg|_0^1 + 40\tan^{-1}y\bigg|_0^1 = 20\pi \text{ C}.$$

(c) In this case the given vector field need to be converted into spherical coordinates. This can be done as follows by using Tables 2.1 and 2.4:

$$\vec{D} = \hat{\rho}\frac{5}{\rho} = (\hat{r}\sin\theta + \hat{\theta}\cos\theta)\frac{5}{r\sin\theta}.$$

Since $d\vec{s} = \hat{r} r \sin\theta d\theta d\phi$,

$$\oint_S \vec{D} \cdot d\vec{s} = \int_0^{2\pi}\int_0^{\pi} \frac{5(\hat{r}\sin\theta + \hat{\theta}\cos\theta)}{r\sin\theta} \cdot \hat{r} r \sin\theta d\theta d\phi \bigg|_{r=1}$$
$$= 5\int_0^{2\pi}\int_0^{\pi} \sin\theta d\theta d\phi = 20\pi \text{ C}.$$

2.4 GRADIENT OF THE SCALAR FIELD

The gradient of a scalar field $f(\vec{r})$ is a vector field that points in the direction of its maximum rate of change at each point $\vec{r}$. The magnitude of this vector field is equal to maximum directional derivative $\partial f/\partial \ell|_{\max}$. It is denoted by ∇f or grad(f), where ∇ is a vector differential operator called "del." In rectangular coordinates, it is defined as follows:

$$\nabla = \hat{x}\frac{\partial}{\partial x} + \hat{y}\frac{\partial}{\partial y} + \hat{z}\frac{\partial}{\partial z}. \tag{2.4.1}$$

Consider the contours of constant $f(\vec{r})$ as illustrated in Figure 2.24. Assume that the scalar field on the contour S_2 is $f(\vec{r}) = f(x, y, z)$, and let there be a point $P_1(\vec{r})$ on this contour. The field changes to $f(\vec{r} + d\vec{\ell}) = f(x + dx, y + dy, z + dz)$ at a nearby point $P_2(\vec{r} + d\vec{\ell})$. Since

$$d\vec{\ell} = \hat{x}dx + \hat{y}dy + \hat{z}dz \tag{2.4.2}$$

and from calculus (the total differential), we find that

$$\begin{aligned} df &= f(\vec{r} + d\vec{\ell}) - f(\vec{r}) = f(x + dx,\ y + dy,\ z + dz) \\ &= \frac{\partial f}{\partial x}dx + \frac{\partial f}{\partial y}dy + \frac{\partial f}{\partial z}dz; \end{aligned} \tag{2.4.3}$$

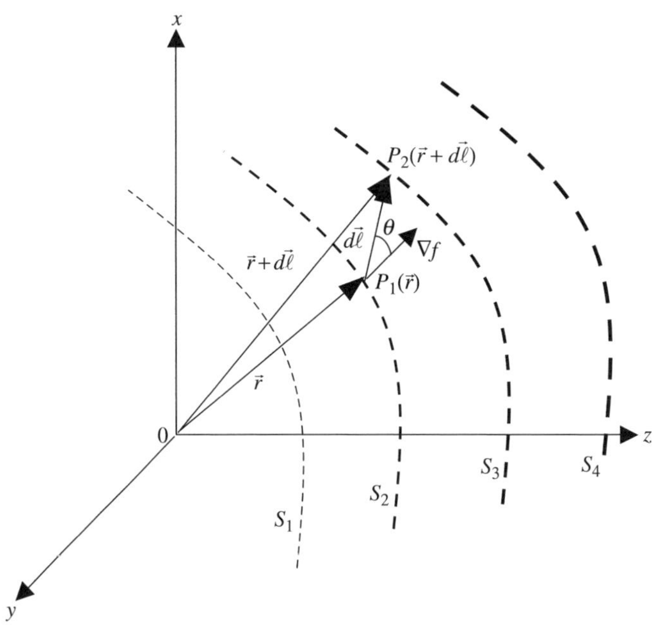

Figure 2.24 Contours of constant $f(\vec{r})$ and the definition of its gradient.

the gradient of $f(\vec{r})$ is defined as

$$\nabla f = \text{grad}(f) = \vec{G} = \hat{x}\frac{\partial f}{\partial x} + \hat{y}\frac{\partial f}{\partial y} + \hat{z}\frac{\partial f}{\partial z}. \tag{2.4.4}$$

Therefore, by the properties of scalar (dot) product of two vectors,

$$df = \vec{G} \cdot d\vec{\ell}. \tag{2.4.5}$$

If $\hat{u}$ is a unit vector along $d\vec{\ell}$, then $d\vec{\ell} = \hat{u}\,d\ell$ and

$$df = \vec{G} \cdot d\vec{\ell} = (\vec{G} \cdot \hat{u})d\ell. \tag{2.4.6}$$

Therefore

$$\frac{df}{d\ell} = (\vec{G} \cdot \hat{u}) = \left|\vec{G}\right| \cos\theta. \tag{2.4.7}$$

Note that this directional derivative has its maximum value when $\hat{u}$ is parallel to $\vec{G}$. Further, suppose that P_2 lies on the same equipotential surface S_2 as shown in Figure 2.25. Therefore

$$df = \vec{G} \cdot d\vec{\ell} = 0. \tag{2.4.8}$$

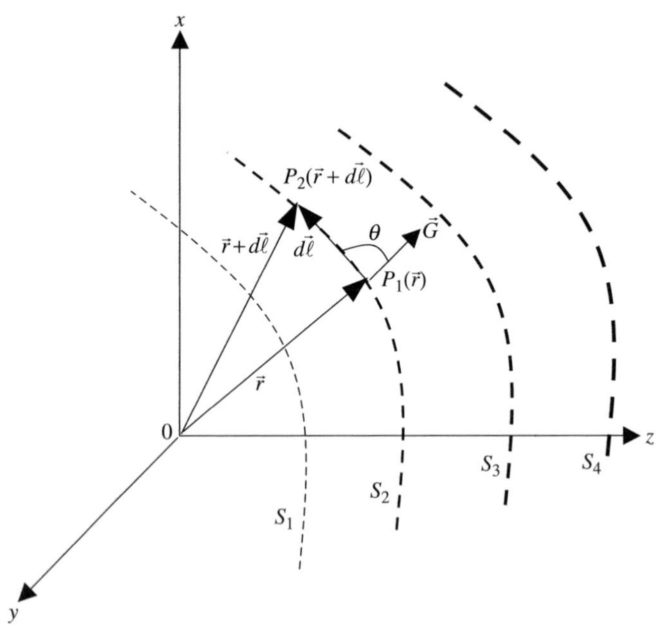

Figure 2.25 Interpretation of the gradient of a scalar field.

Since both points P_1 and P_2 are on the same surface, $d\vec{\ell}$ is tangent to S_2. Note that the magnitudes $|\vec{G}|$ and $|d\vec{\ell}|$ are not zero. For (2.4.8) to be true, $\vec{G}$ and $d\vec{\ell}$ got to be orthogonal to each other. Therefore $\vec{G}$ is normal to contour S_2.

Example 2.18

Find the unit vector normal to a surface described by $xy + yz + zx - 5 = 0$.

Since the gradient of a scalar field is normal to the surface, we can find a unit vector in that direction as follows.

$$\vec{U} = \nabla f = \hat{x}(y+z) + \hat{y}(x+z) + \hat{z}(y+x).$$

Therefore a unit vector normal to the surface is found to be

$$\hat{u} = \frac{\vec{U}}{|\vec{U}|} = \frac{\hat{x}(y+z) + \hat{y}(x+z) + \hat{z}(y+x)}{\sqrt{(y+z)^2 + (x+z)^2 + (y+x)^2}}.$$

Example 2.19

Determine the rate of change of a scalar field $f(\vec{r}) = x^3 + y^3 + z^3$ at (1, 2, 3) in the direction of $\vec{A} = 2\hat{x} + 4\hat{y} + 4\hat{z}$.

A unit vector $\hat{a}$ is the direction of $\vec{A}$ is found to be

$$\hat{a} = \frac{\vec{A}}{|\vec{A}|} = \frac{2\hat{x} + 4\hat{y} + 4\hat{z}}{\sqrt{2^2 + 4^2 + 4^2}} = \frac{2\hat{x} + 4\hat{y} + 4\hat{z}}{6} = \frac{\hat{x} + 2\hat{y} + 2\hat{z}}{3},$$

and the directional derivative of the scalar field is

$$\nabla f(\vec{r}) = 3x^2\hat{x} + 3y^2\hat{y} + 3z^2\hat{z} = 3(\hat{x}x^2 + \hat{y}y^2 + \hat{z}z^2).$$

Therefore

$$\begin{aligned}\nabla f(\vec{r}) \cdot \hat{a}|_{(1,2,3)} &= 3(\hat{x}x^2 + \hat{y}y^2 + \hat{z}z^2) \cdot \left(\frac{\hat{x} + 2\hat{y} + 2\hat{z}}{3}\right)\bigg|_{(1,2,3)} \\ &= (x^2 + 2y^2 + 2z^2)|_{(1,2,3)} = 1 + 8 + 18 = 27.\end{aligned}$$

Example 2.20

Source and field points are generally specified via primed and unprimed coordinates respectively:

$$\vec{r}' = \hat{x}x' + \hat{y}y' + \hat{z}z'$$

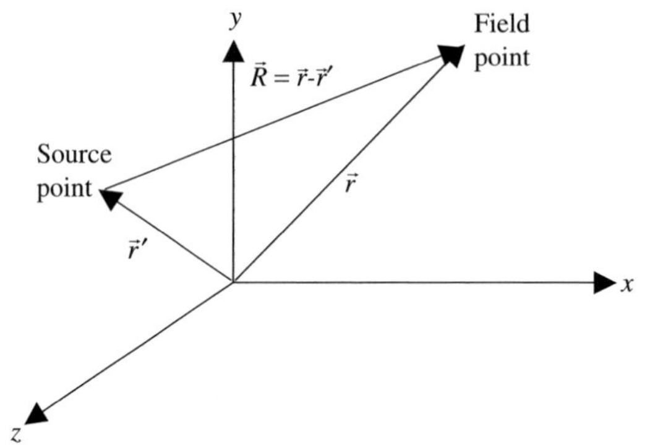

Figure 2.26 Geometry of the problem described in Example 2.20.

and

$$\vec{r} = \hat{x}x + \hat{y}y + \hat{z}z.$$

For $R = \left|\vec{r} - \vec{r}'\right|$, Determine the gradient of $1/R$ with respect to unprimed coordinates. What will be its gradient in primed coordinates?

Figure 2.26 shows the geometry that is used in this example. Since

$$\vec{r} = x\hat{x} + y\hat{y} + z\hat{z},$$
$$\vec{r}' = x'\hat{x} + y'\hat{y} + z'\hat{z},$$

and

$$\vec{R} = \vec{r} - \vec{r}' = (x - x')\hat{x} + (y - y')\hat{y} + (z - z')\hat{z}.$$

Therefore

$$R = \left|\vec{r} - \vec{r}'\right| = \sqrt{(x - x')^2 + (y - y')^2 + (z - z')^2}.$$

Hence the scalar field for this case is

$$f(\vec{r}) = \frac{1}{\sqrt{(x - x')^2 + (y - y')^2 + (z - z')^2}}.$$

The gradient of this field with respect to unprimed coordinates is found as follows:

$$\nabla\left(\frac{1}{R}\right) = \left(\hat{x}\frac{\partial}{\partial x} + \hat{y}\frac{\partial}{\partial y} + \hat{z}\frac{\partial}{\partial z}\right)\left(\frac{1}{\sqrt{(x - x')^2 + (y - y')^2 + (z - z')^2}}\right).$$

The first term simplifies to

$$\begin{aligned}\frac{\partial}{\partial x}\left(\frac{1}{R}\right) &= \frac{\partial}{\partial x}\left(\frac{1}{\sqrt{(x-x^2)+(y-y')^2+(z-z')^2}}\right)\\ &= -\frac{1}{2}\left\{(x-x^2)+(y-y')^2+(z-z')^2\right\}^{-3/2}2(x-x')\\ &= -\frac{(x-x')}{R^3}.\end{aligned}$$

Similarly

$$\frac{\partial}{\partial y}\left(\frac{1}{R}\right) = -\frac{(y-y')}{R^3}$$

and

$$\frac{\partial}{\partial z}\left(\frac{1}{R}\right) = -\frac{(z-z')}{R^3}.$$

Therefore

$$\nabla\left(\frac{1}{R}\right) = -\frac{(x-x')\hat{x}+(y-y')\hat{y}+(z-z')\hat{z}}{R^3} = -\frac{\vec{R}}{R^3} = -\frac{\hat{R}}{R^2},$$

where $\hat{R}$ is a unit vector along $\vec{R}$.

Similarly it can be proved that

$$\nabla'\left(\frac{1}{R}\right) = \frac{\hat{R}}{R^2},$$

where

$$\nabla' = \hat{x}\frac{\partial}{\partial x'} + \hat{y}\frac{\partial}{\partial y'} + \hat{z}\frac{\partial}{\partial z'}.$$

Gradient in Other Common Coordinates

The gradient of a scalar field in cylindrical coordinates (ρ, ϕ, z) is given by

$$\nabla f = \hat{\rho}\frac{\partial f}{\partial \rho} + \hat{\phi}\frac{\partial f}{\rho\partial\phi} + \hat{z}\frac{\partial f}{\partial z}. \tag{2.4.9}$$

In spherical coordinates (r, θ, ϕ) it is given as follows:

$$\nabla f = \hat{r}\frac{\partial f}{\partial r} + \hat{\theta}\frac{\partial f}{r\partial\theta} + \hat{\phi}\frac{\partial f}{r\sin(\theta)\partial\phi}. \tag{2.4.10}$$

Note that the notation ∇f is used for the gradient of f in every coordinate system. ∇f should not be confused with the operator ∇, which is defined by (2.4.1) only in rectangular coordinates.

Example 2.21

Find the gradient of the following scalar fields:

(a) $f(x, y, z) = x(a + bz - cy) + d$
(b) $f(\rho, \phi, z) = a \sin \phi - b\rho z + d$
(c) $f(r, \theta, \phi) = a \sin \theta \cos \phi / r^3$

Assume that a, b, c, and d do not change with space coordinates.

(a) Since the gradient in rectangular coordinates is defined as

$$\nabla f = \hat{x}\frac{\partial f}{\partial x} + \hat{y}\frac{\partial f}{\partial y} + \hat{z}\frac{\partial f}{\partial z},$$

then

$$\nabla f = \hat{x}(a + bz - cy) + \hat{y}(-cx) + \hat{z}(bx).$$

(b) Similarly, using the expression for the gradient in cylindrical coordinates,

$$\nabla f = \hat{\rho}\frac{\partial f}{\partial \rho} + \hat{\phi}\frac{1}{\rho}\frac{\partial f}{\partial \phi} + \hat{z}\frac{\partial f}{\partial z},$$

we find that

$$\nabla f = \hat{\rho}(-bz) + \hat{\phi}\frac{1}{\rho}(a \cos \phi) + \hat{z}(-b\rho) = -\hat{\rho}bz + \hat{\phi}\frac{a}{\rho}\cos \phi - \hat{z}b\rho.$$

(c) Using the expression for the gradient in the spherical coordinates,

$$\nabla f = \vec{r}\frac{\partial f}{\partial r} + \vec{\theta}\frac{1}{r}\frac{\partial f}{\partial \theta} + \vec{\phi}\frac{1}{r \sin \theta}\frac{\partial f}{\partial \phi},$$

we find that

$$\nabla f = \hat{r}\left(-\frac{3a \sin \theta \cos \phi}{r^4}\right) + \hat{\theta}\frac{1}{r}\left(\frac{a \cos \theta \cos \phi}{r^3}\right)$$
$$+ \hat{\phi}\frac{1}{r \sin \theta}\left(-\frac{a \sin \theta \sin \phi}{r^3}\right)$$
$$= -\hat{r}\frac{3a \sin \theta \cos \phi}{r^4} + \hat{\theta}\frac{a \cos \theta \cos \phi}{r^4} - \hat{\phi}\frac{a \sin \phi}{r^4}.$$

2.5 DIVERGENCE OF THE VECTOR FIELD

The divergence of a vector field $\vec{A}(\vec{r})$ is generally denoted as $\nabla \cdot \vec{A}(\vec{r})$ or $\operatorname{div} \vec{A}(\vec{r})$. It is a scalar field that is defined mathematically as follows:

$$\nabla \cdot \vec{A}(\vec{r}) = \lim_{\Delta V(\vec{r}) \to 0} \frac{\oint_S \hat{n} \cdot \vec{A}(\vec{r}\,')\, ds'}{\Delta V(\vec{r})}, \tag{2.5.1}$$

where S is the closed surface that bounds the incremental volume region $\Delta V(\vec{r})$ located by the position vector $\vec{r}$. Primed coordinates are used for as integration variables over this surface.

By its definition, the divergence of a vector at a point is the net outflux of that vector per unit volume. Thus it gives a measure of the strength of the sources that produce the vector field. If the divergence of a vector field $\vec{A}(\vec{r})$ is zero at every point $\vec{r}$, then it is called to be a solenoidal field.

Consider an incremental volume $\Delta V(\vec{r})$ and the vector $\vec{A}(\vec{r})$ in the rectangular coordinates as shown in Figure 2.27. Equation (2.5.1) can be evaluated after integrating the closed-surface integral over each of six faces as follows. On the right-hand-side face, it gives

$$\begin{aligned}\int_{S_{x1}} \hat{n} \cdot \vec{A}(\vec{r}\,')ds' &= \int_{S_{x1}} \vec{A}\left(x + \frac{dx}{2}, y, z\right) \cdot \hat{x}dydz \approx A_x\left(x + \frac{dx}{2}, y, z\right)dydz \\ &= A_x(x, y, z)dydz + \frac{\partial}{\partial x}A_x(x, y, z)\frac{dx}{2}dydz.\end{aligned}$$

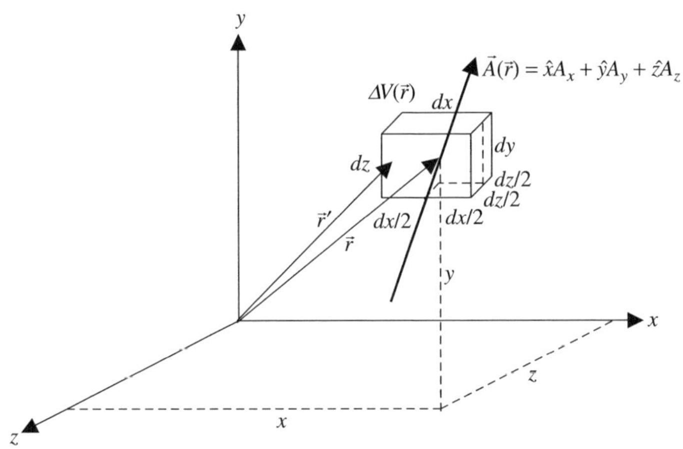

Figure 2.27 Incremental volume $\Delta V(\vec{r})$ and the vector $\vec{A}(\vec{r})$.

On the left face, it is found to be

$$\int_{S_{x2}} \hat{n} \cdot \vec{A}(\vec{r}')ds' = \int_{S_{x2}} \vec{A}\left(x - \frac{dx}{2}, y, z\right) \cdot (-\hat{x})dydz \approx A_x\left(x - \frac{dx}{2}, y, z\right)dydz$$

$$= -A_x(x, y, z)dydz - \frac{\partial}{\partial x}A_x(x, y, z)\left(-\frac{dx}{2}\right)dydz.$$

Adding the results of these two side-faces, we have

$$\int_{S_{x_1}+S_{x_2}} \hat{n} \cdot \vec{A}(\vec{r}')ds' = \frac{\partial}{\partial x}A_x(x, y, z)dxdydz.$$

Similarly the top and bottom surfaces of incremental volume produce

$$\int_{S_{y_1}+S_{y_2}} \hat{n} \cdot \vec{A}(\vec{r}')ds' = \frac{\partial}{\partial y}A_y(x, y, z)dxdydz.$$

The remaining two surfaces (front and back) give

$$\int_{S_{z_1}+S_{z_2}} \hat{n} \cdot \vec{A}(\vec{r}')ds = \frac{\partial}{\partial z}A_z(x, y, z)dxdydz.$$

Adding the results from all six faces of $\Delta V(\vec{r})$, we have

$$\oint_S \hat{n} \cdot \vec{A}(\vec{r}')ds = \left(\frac{\partial A_x}{\partial x} + \frac{\partial A_y}{\partial y} + \frac{\partial A_z}{\partial z}\right)dxdydz.$$

Therefore

$$\begin{aligned} \nabla \cdot \vec{A}(\vec{r}) &= \lim_{\Delta V(\vec{r}) \to 0} \frac{\oint_S \hat{n} \cdot \vec{A}(r')ds}{\Delta V(\vec{r})} = \frac{\partial A_x}{\partial x} + \frac{\partial A_y}{\partial y} + \frac{\partial A_z}{\partial z} \\ &= \left(\hat{x}\frac{\partial}{\partial x} + \hat{y}\frac{\partial}{\partial y} + \hat{z}\frac{\partial}{\partial z}\right) \cdot (\hat{x}A_x + \hat{y}A_y + \hat{z}A_z). \end{aligned} \tag{2.5.2}$$

Divergence in Other Common Coordinates

The divergence of a vector field in cylindrical coordinates (ρ, ϕ, z) is given by

$$\nabla \cdot \vec{A} = \frac{1}{\rho}\frac{\partial}{\partial \rho}(\rho A_\rho) + \frac{1}{\rho}\frac{\partial A_\phi}{\partial \phi} + \frac{\partial A_z}{\partial z}. \tag{2.5.3}$$

In spherical coordinates (r, θ, ϕ) it is given as follows:

$$\nabla \cdot \vec{A} = \frac{1}{r^2}\frac{\partial}{\partial r}(r^2 A_r) + \frac{1}{r \sin\theta}\frac{\partial}{\partial \theta}(\sin\theta A_\theta) + \frac{1}{r \sin\theta}\frac{\partial A_\phi}{\partial \phi}. \tag{2.5.4}$$

Example 2.22

Find the divergence of the following vector fields:

(a) $\vec{A} = \hat{x}x^2 + \hat{y}y + \hat{z}z$
(b) $\vec{B} = -\hat{\rho}bz + \hat{\phi}(a/\rho)\cos\phi - \hat{z}b\rho$, where a and b are constant.
(c) $\vec{E} = \hat{r}4r\cos\theta - \hat{\theta}6r\sin\theta + \hat{\phi}\sin\theta$

(a) From (2.5.2),

$$\nabla \cdot \vec{A} = \frac{\partial A_x}{\partial x} + \frac{\partial A_y}{\partial y} + \frac{\partial A_z}{\partial z} = 2x + 1 + 1 = 2 + 2x.$$

(b) From (2.5.3),

$$\nabla \cdot \vec{B} = \frac{1}{\rho}\frac{\partial(\rho B_\rho)}{\partial \rho} + \frac{1}{\rho}\frac{\partial B_\phi}{\partial \phi} + \frac{\partial B_z}{\partial z} = \frac{1}{\rho}\frac{\partial(-\rho bz)}{\partial \rho} + \frac{1}{\rho}\frac{\partial}{\partial \phi}\left(\frac{a}{\rho}\cos\phi\right) + \frac{\partial(-b\rho)}{\partial z}$$
$$= -\frac{bz}{\rho} - \frac{a}{\rho^2}\sin\phi.$$

(c) From (2.5.4),

$$\nabla \cdot \vec{E} = \frac{1}{r^2\sin\theta}\left[\sin\theta\frac{\partial}{\partial r}(r^2 E_r) + r\frac{\partial}{\partial \theta}(\sin\theta E_\theta) + r\frac{\partial E_\phi}{\partial \phi}\right]$$
$$= \frac{1}{r^2\sin\theta}\left[\sin\theta\frac{\partial}{\partial r}(r^2 4r\cos\theta) + r\frac{\partial}{\partial \theta}\{\sin\theta(-6r\sin\theta)\} + r\frac{\partial}{\partial \phi}(\sin\theta)\right]$$
$$= \frac{1}{r^2\sin\theta}\left[\sin\theta(12r^2\cos\theta) + r\{-12r\sin\theta\cos\theta\}\right] = 0.$$

2.6 CURL OF THE VECTOR FIELD

The curl of a vector field $\vec{A}(\vec{r})$ is generally denoted by $\nabla \times \vec{A}(\vec{r})$ or curl $\vec{A}(\vec{r})$. It is a vector field whose scalar component along a particular coordinate i is given as follows:

$$\left\{\nabla \times \vec{A}(\vec{r})\right\}_i = \lim_{\Delta S_i(\vec{r})\to 0} \frac{\oint_\Gamma \vec{A}(\vec{r}\,') \cdot d\vec{\ell}\,'}{\Delta S_i(\vec{r})}, \tag{2.6.1}$$

where $\Delta S_i(\vec{r})$ is an elemental surface area perpendicular to the direction i and located at $\vec{r}$. The closed contour Γ represents the boundary of elemental area $\Delta S_i(\vec{r})$. It is taken in the right-hand sense relative to the external normal vector to $\Delta S_i(\vec{r})$.

By its definition, $\nabla \times \vec{A}(\vec{r})$ at a point is the net circulation of $\vec{A}(\vec{r})$ per unit area at point $\vec{r}$. Thus it gives a measure of the strength of the sources producing $\vec{A}(\vec{r})$ at each point. If the curl of $\vec{A}(\vec{r})$ is zero at every point $\vec{r}$, then it is said to be irrotational or conservative vector field.

Consider an elemental area $\Delta S_x(\vec{r})$ and the vector $\vec{A}(\vec{r})$ in the rectangular coordinates as shown in Figure 2.28. Equation (2.6.1) can be evaluated after integrating the circulation integral over the edge 1 to 4 as follows:

For

$$\vec{A}(\vec{r}) = \hat{x}A_x(\vec{r}) + \hat{y}A_y(\vec{r}) + \hat{z}A_z(\vec{r}),$$

$$\begin{aligned}\oint_\Gamma \vec{A}(\vec{r}\,') \cdot d\vec{\ell} &= \oint_\Gamma (\hat{x}A_x + \hat{y}A_y + \hat{z}A_z) \cdot (\hat{x}dx + \hat{y}dy + \hat{z}dz) \\ &= \int_1 A_y dy + \int_2 A_z dz + \int_3 A_y dy + \int_4 A_z dz,\end{aligned}$$

where

$$\int_1 A_y dy \approx A_y\left(x, y, z - \frac{dz}{2}\right)dy = A_y(x, y, z) + \frac{\partial}{\partial z}A_y(x, y, z)\left(-\frac{dz}{2}\right)dy,$$

$$\int_2 A_z dz \approx A_z\left(x, y + \frac{dy}{2}, z\right)dz = A_z(x, y, z) + \frac{\partial}{\partial y}A_z(x, y, z)\left(\frac{dy}{2}\right)dz,$$

$$\int_3 A_y dy \approx -A_y\left(x, y, z + \frac{dz}{2}\right)dy = -A_y(x, y, z) - \frac{\partial}{\partial z}A_y(x, y, z)\left(\frac{dz}{2}\right)dy,$$

and

$$\int_4 A_z dz \approx -A_z\left(x, y - \frac{dy}{2}, z\right)dz = -A_z(x, y, z) - \frac{\partial}{\partial y}A_z(x, y, z)\left(-\frac{dy}{2}\right)dz.$$

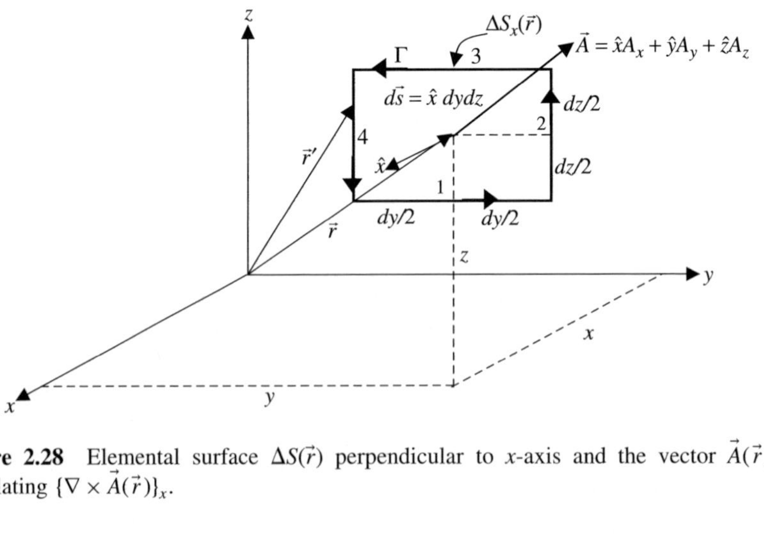

Figure 2.28 Elemental surface $\Delta S(\vec{r})$ perpendicular to x-axis and the vector $\vec{A}(\vec{r})$ for calculating $\{\nabla \times \vec{A}(\vec{r})\}_x$.

Substituting these results back, we get

$$\oint_{\Gamma} \vec{A}(\vec{r}\,') \cdot d\vec{\ell} = \left(\frac{\partial A_z}{\partial y} - \frac{\partial A_y}{\partial z}\right) dydz.$$

Therefore

$$\left\{\nabla \times \vec{A}(\vec{r})\right\}_x = \lim_{\Delta S_x(\vec{r}) \to 0} \frac{\oint_{\Gamma} \vec{A}(\vec{r}\,') \cdot d\vec{\ell}}{\Delta S_x(\vec{r})} = \frac{\oint_{\Gamma} \vec{A}(\vec{r}\,') \cdot d\vec{\ell}}{dydz} = \frac{\partial A_z}{\partial y} - \frac{\partial A_y}{\partial z}.$$

Similarly y and z components are found to be

$$\left\{\nabla \times \vec{A}(\vec{r})\right\}_y = \lim_{\Delta S_y(\vec{r}) \to 0} \frac{\oint_{\Gamma} \vec{A}(\vec{r}\,') \cdot d\vec{\ell}}{\Delta S_y(\vec{r})} = \frac{\oint_{\Gamma} \vec{A}(\vec{r}\,') \cdot d\vec{\ell}}{dxdz} = \frac{\partial A_x}{\partial z} - \frac{\partial A_z}{\partial x}$$

and

$$\left\{\nabla \times \vec{A}(\vec{r})\right\}_z = \lim_{\Delta S_z(\vec{r}) \to 0} \frac{\oint_{\Gamma} \vec{A}(\vec{r}\,') \cdot d\vec{\ell}}{\Delta S_z(\vec{r})} = \frac{s\oint_{\Gamma} \vec{A}(\vec{r}\,') \cdot d\vec{\ell}}{dxdy} = \frac{\partial A_y}{\partial x} - \frac{\partial A_x}{\partial y}.$$

On combining the results for all three components, we have

$$\nabla \times \vec{A}(\vec{r}) = \hat{x}\left(\frac{\partial A_z}{\partial y} - \frac{\partial A_y}{\partial z}\right) + \hat{y}\left(\frac{\partial A_x}{\partial z} - \frac{\partial A_z}{\partial x}\right) + \hat{z}\left(\frac{\partial A_y}{\partial x} - \frac{\partial A_x}{\partial y}\right). \tag{2.6.2}$$

This can also be written in determinant form as follows:

$$\nabla \times \vec{A}(\vec{r}) = \begin{vmatrix} \hat{x} & \hat{y} & \hat{z} \\ \dfrac{\partial}{\partial x} & \dfrac{\partial}{\partial y} & \dfrac{\partial}{\partial z} \\ A_x & A_y & A_z \end{vmatrix}. \tag{2.6.3}$$

Curl in Other Common Coordinates

The curl of a vector field in cylindrical coordinates (ρ, ϕ, z) is given by

$$\nabla \times \vec{A} = \hat{\rho}\left(\frac{1}{\rho}\frac{\partial A_z}{\partial \phi} - \frac{\partial A_\phi}{\partial z}\right) + \hat{\phi}\left(\frac{\partial A_\rho}{\partial z} - \frac{\partial A_z}{\partial \rho}\right) + \hat{z}\frac{1}{\rho}\left[\frac{\partial}{\partial \rho}(\rho A_\phi) - \frac{\partial A_\rho}{\partial \phi}\right]. \tag{2.6.4}$$

Alternatively,

$$\nabla \times \vec{A} = \frac{1}{\rho}\begin{vmatrix} \hat{\rho} & \rho\hat{\phi} & \hat{z} \\ \dfrac{\partial}{\partial \rho} & \dfrac{\partial}{\partial \phi} & \dfrac{\partial}{\partial z} \\ A_\rho & \rho A_\phi & A_z \end{vmatrix}. \tag{2.6.5}$$

In spherical coordinates $(r,\ \theta,\ \phi)$ it is given as follows:

$$\nabla \times \vec{A} = \hat{r}\frac{1}{r\sin\theta}\left[\frac{\partial}{\partial\theta}(A_\phi \sin\theta) - \frac{\partial A_\theta}{\partial\phi}\right] + \hat{\theta}\frac{1}{r}\left[\frac{1}{\sin\theta}\frac{\partial A_r}{\partial\phi} - \frac{\partial}{\partial r}(rA_\phi)\right]$$
$$+ \hat{\phi}\frac{1}{r}\left[\frac{\partial}{\partial r}(rA_\theta) - \frac{\partial A_r}{\partial\theta}\right]. \tag{2.6.6}$$

This can also be written in a determinant form as follows:

$$\nabla \times \vec{A} = \frac{1}{r^2\sin(\theta)}\begin{vmatrix} \hat{r} & r\hat{\theta} & r\sin(\theta)\hat{\phi} \\ \dfrac{\partial}{\partial r} & \dfrac{\partial}{\partial\theta} & \dfrac{\partial}{\partial\phi} \\ A_r & rA_\theta & r\sin(\theta)A_\phi \end{vmatrix}. \tag{2.6.7}$$

Note that the notation $\nabla \times \vec{A}(\vec{r})$ is used for the curl of $\vec{A}(\vec{r})$ in every coordinate system; it should not be confused with the operator ∇ in rectangular coordinates.

Example 2.23

Find the curl of the following vector fields:

(a) $\vec{U} = \hat{x}(y+z) + \hat{y}(x+z) + \hat{z}(y+x)$

(b) $\vec{E} = \hat{\rho}150\rho\cos\phi + \hat{\phi}200\sin\phi$

(c) $\vec{H} = \hat{r}r\cos\theta - \hat{\theta}(\sin\theta/r) + \hat{\phi}2r^2\sin\theta$

(a) From (2.6.3), we have

$$\nabla \times \vec{U} = \begin{vmatrix} \hat{x} & \hat{y} & \hat{z} \\ \dfrac{\partial}{\partial x} & \dfrac{\partial}{\partial y} & \dfrac{\partial}{\partial z} \\ U_x & U_y & U_z \end{vmatrix} = \begin{vmatrix} \hat{x} & \hat{y} & \hat{z} \\ \dfrac{\partial}{\partial x} & \dfrac{\partial}{\partial y} & \dfrac{\partial}{\partial z} \\ (y+z) & (x+z) & (y+x) \end{vmatrix}$$
$$= \hat{x}(1-1) + \hat{y}(1-1) + \hat{z}(1-1) = 0.$$

(b) From (2.6.5), we get

$$\nabla \times \vec{E} = \frac{1}{\rho}\begin{vmatrix} \hat{\rho} & \rho\hat{\phi} & \hat{z} \\ \dfrac{\partial}{\partial\rho} & \dfrac{\partial}{\partial\phi} & \dfrac{\partial}{\partial z} \\ E_\rho & \rho E_\phi & E_z \end{vmatrix} = \frac{1}{\rho}\begin{vmatrix} \hat{\rho} & \rho\hat{\phi} & \hat{z} \\ \dfrac{\partial}{\partial\rho} & \dfrac{\partial}{\partial\phi} & \dfrac{\partial}{\partial z} \\ 150\rho\cos\phi & 200\rho\sin\phi & 0 \end{vmatrix}$$
$$= \hat{z}\left(\frac{200}{\rho} + 150\right)\sin\phi.$$

(c) From (2.6.7), we get

$$\nabla \times \vec{H} = \frac{1}{r^2 \sin(\theta)} \begin{vmatrix} \hat{r} & r\hat{\theta} & r\sin\theta\hat{\phi} \\ \dfrac{\partial}{\partial r} & \dfrac{\partial}{\partial \theta} & \dfrac{\partial}{\partial \phi} \\ H_r & rH_\theta & r\sin\theta H_\phi \end{vmatrix}$$

$$= \frac{1}{r^2 \sin(\theta)} \begin{vmatrix} \hat{r} & r\hat{\theta} & r\sin\theta\hat{\phi} \\ \dfrac{\partial}{\partial r} & \dfrac{\partial}{\partial \theta} & \dfrac{\partial}{\partial \phi} \\ r\cos\theta & r\left(-\frac{\sin\theta}{r}\right) & r\sin\theta(2r^2\sin\theta) \end{vmatrix}$$

$$= \frac{1}{r^2 \sin(\theta)} \left[\hat{r}(4r^3 \sin\theta\cos\theta) + r\hat{\theta}(-6r^2\sin^2\theta) + r\sin\theta\hat{\phi}\, r\sin\theta \right]$$

$$= \hat{r}4r\cos\theta - \hat{\theta}6r\sin\theta + \hat{\phi}\sin\theta.$$

2.7 THE DIVERGENCE THEOREM

Consider a vector field $\vec{A}(\vec{r})$ defined throughout a volume region V that is bound by the surface S. According to the divergence theorem (also known as the Gauss's theorem), the integral of the divergence of $\vec{A}(\vec{r})$ in V is equivalent to integrating the normal component of $\vec{A}(\vec{r})$ over S. Mathematically

$$\int_V \nabla \cdot \vec{A}(\vec{r}) dv = \oint_S \hat{n} \cdot \vec{A}(\vec{r}) ds. \tag{2.7.1}$$

This theorem can be proved easily using (2.5.1), which defines $\nabla \cdot \vec{A}(\vec{r})$. Consider a vector field $\vec{A}(\vec{r})$ in a volume V bound by the surface S, as shown in Figure 2.29.

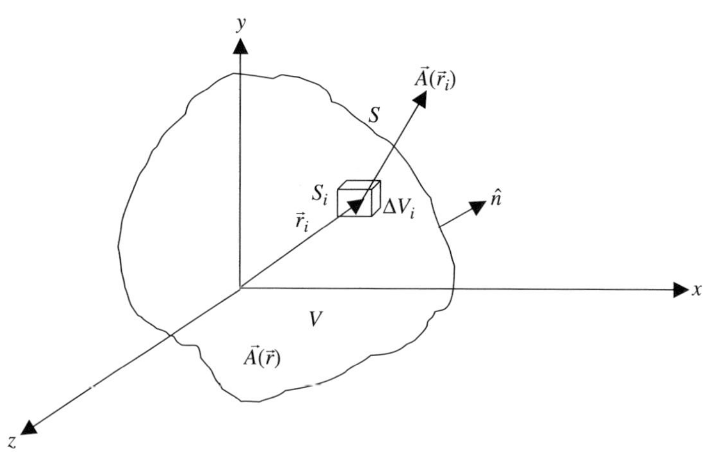

Figure 2.29 Geometry of the subdivided volume for establishing the divergence theorem.

Imagine that the volume V is made up of N volume-elements ΔV_i that is bound by a closed surface S_i. Therefore

$$V = \sum_{i=1}^{N} \Delta V_i.$$

Using (2.5.1) for $\vec{A}(\vec{r}_i)$ in this volume element, we get

$$\nabla \cdot \vec{A}(\vec{r}_i) = \lim_{\Delta V(\vec{r}_i)\to 0} \frac{\oint_{S_i} \hat{n} \cdot \vec{A}(\vec{r}_i) ds}{\Delta V(\vec{r}_i)}.$$

As $\Delta V_i \to 0$, this equation can be written as follows:

$$\nabla \cdot \vec{A}(\vec{r}_i)\Delta V_i = \oint_{S_i} \hat{n} \cdot \vec{A} ds.$$

Writing similar expressions for each volume-element and adding, we find that

$$\sum_{i=1}^{N} \nabla \cdot \vec{A}(\vec{r}_i)\Delta V_i = \sum_{i=1}^{N} \oint_{S_i} \hat{n} \cdot \vec{A} ds.$$

As illustrated in Figure 2.30, integrations over adjacent surfaces of neighboring volume-elements cancel each other, leaving only the integral over boundary surface S. Therefore this expression reduces to

$$\lim_{\substack{N\to\infty \\ \Delta V_i\to 0}} \sum_{i=1}^{N} \nabla \cdot \vec{A}(\vec{r}_i)\Delta V_i = \oint_{S} \hat{n} \cdot \vec{A} ds.$$

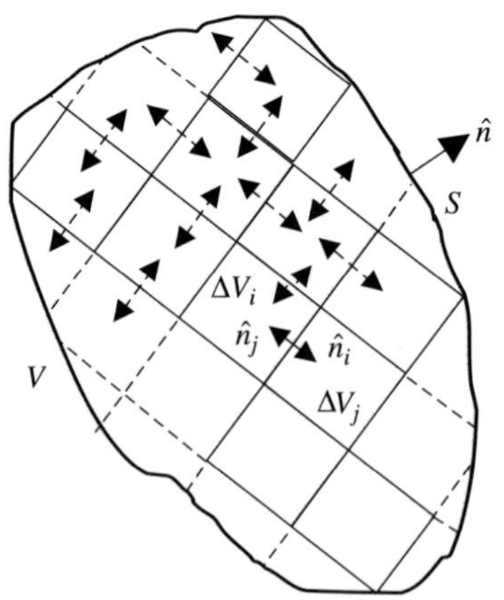

Figure 2.30 Cancellation of surface integrals over internal volume-elements.

Now using the definition of volume integral for the left-hand side of this, we have

$$\int_V \nabla \cdot \vec{A}(\vec{r})dv = \oint_S \hat{n} \cdot \vec{A}ds.$$

Example 2.24

Verify the divergence theorem for a vector $\vec{A} = \hat{r}5r$ when a closed surface is (a) a cylinder with $\rho = 1$ m, and $0 \leq z \leq 1$ m, (b) a sphere of $r = 1$ m, and (c) a cube with $0 \leq x \leq 1$ m, $0 \leq y \leq 1$ m, and $0 \leq z \leq 1$ m.

(a) Since both, the given vector as well as the closed surface, are in spherical coordinates, the corresponding formulas can be easily applied as follows:

$$\oint_S \vec{A} \cdot d\vec{s} = \int_{\phi=0}^{2\pi} \int_{\theta=0}^{\pi} (\hat{r}5r) \cdot \hat{r}r^2 \sin\theta d\theta d\phi\Big|_{r=1} = 5 \times 2 \times 2\pi = 20\pi.$$

Since

$$\nabla \cdot \vec{A} = \frac{1}{r^2 \sin\theta} \sin\theta \frac{\partial}{\partial r}(r^2 5r) = 15,$$

$$\int_V \nabla \cdot \vec{A}\, dv = 15 \int_{\phi=0}^{2\pi} \int_{\theta=0}^{\pi} \int_{r=0}^{1} r^2 \sin\theta\, dr d\theta d\phi$$

$$= 15 \times \frac{r^3}{3}\Big|_0^1 \times (-\cos\theta|_0^{\pi}) \times \phi\Big|_0^{2\pi} = 20\pi.$$

(b) In this case the closed surface is in cylindrical coordinates whereas the given vector is in spherical coordinates. Therefore it needs to be converted into cylindrical coordinates before integrations. From the Tables 2.1 and 2.3 we get

$$\vec{A} = \hat{r}5r = (\hat{\rho}\sin\theta + \hat{z}\cos\theta)5\sqrt{\rho^2 + z^2}.$$

Since

$$\sin\theta = \frac{\rho}{\sqrt{\rho^2 + z^2}} \quad \text{and} \quad \cos\theta = \frac{z}{\sqrt{\rho^2 + z^2}},$$

this simplifies to

$$\vec{A} = \hat{r}5r = (\hat{\rho}\rho + \hat{z}z)5.$$

Therefore

$$\nabla \cdot \vec{A} = \frac{1}{\rho}\frac{\partial}{\partial \rho}(5\rho^2) + 0 + \frac{\partial}{\partial z}(5z) = 15$$

and

$$\int_V \nabla \cdot \vec{A} dv = \int_{z=0}^{1} \int_{\phi=0}^{2\pi} \int_{\rho=0}^{1} 15\rho d\rho d\phi dz = 15\pi.$$

Since there are three surfaces in a cylinder to form a closed surface, we evaluate the integrals individually as follows:

For the side surface of the cylinder,

$$\int_{S_1} \vec{A} \cdot ds = \int_{z=0}^{1} \int_{\phi=0}^{2\pi} \vec{A} \cdot \hat{\rho} \rho d\phi dz \Big|_{\rho=1} = 5 \int_{z=0}^{1} \int_{\phi=0}^{2\pi} d\phi dz = 10\pi.$$

For the top surface of the cylinder,

$$\int_{S_2} \vec{A} \cdot ds = \int_{\phi=0}^{2\pi} \int_{\rho=0}^{1} \vec{A} \cdot \hat{z} \rho d\rho d\phi \Big|_{z=1} = 5 \int_{\phi=0}^{2\pi} \int_{\rho=0}^{1} \rho d\rho d\phi = 5\pi.$$

For the bottom surface,

$$\int_{S_3} \vec{A} \cdot ds = \int_{\phi=0}^{2\pi} \int_{\rho=0}^{1} \vec{A} \cdot \hat{z} \rho d\rho d\phi \Big|_{z=0} = 0.$$

Therefore

$$\oint_S \vec{A} \cdot d\vec{s} = \oint_{S_1} \vec{A} \cdot d\vec{s} + \oint_{S_2} \vec{A} \cdot d\vec{s} + \oint_{S_3} \vec{A} \cdot d\vec{s} = 10\pi + 5\pi = 15\pi$$

$$= \int_V \nabla \cdot \vec{A} dv.$$

(c) In this case the closed surface is in rectangular coordinates whereas the given vector is in spherical coordinates. Therefore it needs to be converted into rectangular coordinates before integrations. From the Tables 2.1 and 2.2 we get

$$\vec{A} = \hat{r} 5r = (\hat{x} \sin\theta \cos\phi + \hat{y} \sin\theta \sin\phi + \hat{z} \cos\theta) 5\sqrt{x^2 + y^2 + z^2}.$$

Further, since

$$\cos\theta = \frac{z}{\sqrt{x^2 + y^2 + z^2}},$$

$$\sin\theta = \frac{\sqrt{x^2 + y^2}}{\sqrt{x^2 + y^2 + z^2}},$$

$$\cos\phi = \frac{x}{\sqrt{x^2 + y^2}},$$

and

$$\sin\phi = \frac{y}{\sqrt{x^2+y^2}},$$

we get

$$\vec{A} = 5(\hat{x}x + \hat{y}y + \hat{z}z).$$

Therefore

$$\nabla \cdot \vec{A} = \frac{\partial}{\partial x}(5x) + \frac{\partial}{\partial y}(5y) + \frac{\partial}{\partial z}(5z) = 15$$

and

$$\int_V \nabla \cdot \vec{A} dv = \int_{z=0}^{1}\int_{y=0}^{1}\int_{x=0}^{1} 15 dxdydz = 15.$$

Since there are six surfaces on a cube, we evaluate the integrals individually as follows:

For the surface at $x = 1$,

$$\int_{S_1} \vec{A} \cdot ds_1 = \int_{z=0}^{1}\int_{y=0}^{1} 5xdydz\big|_{x=1} = 5.$$

For the surface at $x = 0$,

$$\int_{S_2} \vec{A} \cdot ds_2 = -\int_{z=0}^{1}\int_{y=0}^{1} 5xdydz\big|_{x=0} = 0.$$

For the surface at $y = 1$,

$$\int_{S_3} \vec{A} \cdot ds_3 = \int_{z=0}^{1}\int_{x=0}^{1} 5ydxdz\big|_{y=1} = 5.$$

For the surface at $y = 0$,

$$\int_{S_4} \vec{A} \cdot ds_1 = -\int_{z=0}^{1}\int_{x=0}^{1} 5ydxdz\big|_{y=0} = 0.$$

For the surface at $z = 1$,

$$\int_{S_5} \vec{A} \cdot ds_5 = \int_{y=0}^{1}\int_{x=0}^{1} 5zdxdy\big|_{z=1} = 5.$$

For the surface at $z = 0$,

$$\int_{S_6} \vec{A} \cdot ds_6 = -\int_{y=0}^{1} \int_{x=0}^{1} 5z dx dy\Big|_{z=0} = 0.$$

Therefore

$$\oint_S \vec{A} \cdot ds = \int_{S_1} \vec{A} \cdot ds_1 + \int_{S_2} \vec{A} \cdot ds_2 + \cdots + \int_{S_6} \vec{A} \cdot ds_6 = 15 = \int_V \nabla \cdot \vec{A} dv.$$

2.8 STOKES THEOREM

Consider a vector field $\vec{A}(\vec{r})$ defined over the surface S bound by a closed boundary contour Γ. Stokes's theorem states that total flux of the curl of $\vec{A}(\vec{r})$ over S is equal to circulation of $\vec{A}(\vec{r})$ along the contour Γ. Mathematically

$$\int_S \hat{n} \cdot \left[\nabla \times \vec{A}(\vec{r}) \right] ds = \oint_\Gamma \vec{A}(\vec{r}) \cdot d\vec{\ell}. \tag{2.8.1}$$

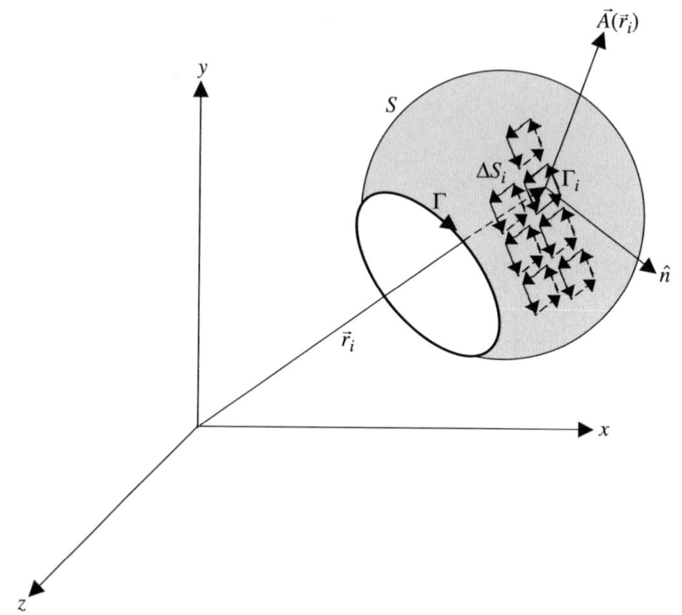

Figure 2.31 Geometry of the subdivided surface for establishing Stokes's theorem.

This theorem can be easily proved using (2.6.1), which defines $\nabla \times \vec{A}(\vec{r})$. Consider a vector field $\vec{A}(\vec{r})$ over the surface S bound by a boundary contour Γ, as shown in Figure 2.31. Imagine that the surface S is made up of N area-elements ΔS_i that is bound by the closed boundary Γ_i. Therefore

$$S = \sum_{i=1}^{N} \Delta S_i.$$

Using (2.6.1) for $\vec{A}(\vec{r}_i)$ in this area-element, we get

$$\nabla \times \vec{A}(\vec{r}_i)\Big|_n = \hat{n} \cdot \left[\nabla \times \vec{A}(\vec{r}_i)\right] = \lim_{\Delta S_i \to 0} \frac{\oint_{\Gamma_i} \vec{A}(\vec{r}_i) \cdot d\vec{\ell}}{\Delta S_i}$$

or

$$\lim_{\Delta S_i \to 0} \hat{n} \cdot \left[\nabla \times \vec{A}(\vec{r}_i)\right]\Delta S_i = \oint_{\Gamma_i} \vec{A}(\vec{r}_i) \cdot d\vec{\ell}$$

or

$$\sum_{i=1}^{N} \hat{n} \cdot \left[\nabla \times \vec{A}(\vec{r}_i)\right]\Delta S_i = \sum_{i=1}^{N} \oint_{\Gamma_i} \vec{A}(\vec{r}_i) \cdot d\vec{\ell}.$$

As illustrated, integrals over adjacent contours cancel out except the integral around boundary contour Γ of S. Hence the expression reduces to

$$\lim_{\substack{N \to \infty \\ \Delta S_i \to 0}} \sum_{i=1}^{N} \hat{n} \cdot \left[\nabla \times \vec{A}(\vec{r}_i)\right]\Delta S_i = \oint_{\Gamma} \vec{A}(\vec{r}) \cdot d\vec{\ell}.$$

Using the definition of a surface integral on the right-hand side of this expression, we get

$$\int_S \hat{n} \cdot \left[\nabla \times \vec{A}(\vec{r})\right] ds = \oint_{\Gamma} \vec{A}(\vec{r}) \cdot d\vec{\ell}.$$

Example 2.25

Verify Stokes's theorem for the case of Example 2.16.

For $\vec{H} = \hat{\rho}\rho \sin \phi + \hat{\phi} 2\rho^2$ A/m,

$$\nabla \times \vec{H} = \frac{1}{\rho} \begin{vmatrix} \hat{\rho} & \rho\hat{\phi} & \hat{z} \\ \dfrac{\partial}{\partial \rho} & \dfrac{\partial}{\partial \phi} & \dfrac{\partial}{\partial z} \\ \rho \sin \phi & 2\rho^3 & 0 \end{vmatrix} = \hat{z}(6\rho - \cos \phi)\ \mathrm{A/m^2}.$$

Therefore

$$\int_S (\nabla \times \vec{H}) \cdot d\vec{s} = \int_{\phi=0}^{\pi} \int_{\rho=a}^{b} \hat{z}(6\rho - \cos\phi) \cdot \hat{z}\rho d\rho d\phi$$

$$= \int_{\phi=0}^{\pi} \left(6\frac{\rho^3}{3} - \cos\phi \frac{\rho^2}{2}\right)\bigg|_a^b d\phi,$$

or

$$\int_S (\nabla \times \vec{H}) \cdot d\vec{s} = 2(b^3 - a^3)\phi\big|_0^{\pi} - \tfrac{1}{2}(b^2 - a^2)\sin\phi\big|_0^{\pi} = 2\pi(b^3 - a^3)\,\text{A}.$$

Now applying the result from Example 2.16, we find that

$$\int_S (\nabla \times \vec{H}) \cdot d\vec{s} = 2\pi(b^3 - a^3)\text{A} = \oint_\Gamma \vec{H} \cdot d\vec{\ell}.$$

Example 2.26

Verify Stokes's theorem for the case of Example 2.15.

In this case $\vec{E} = \hat{\rho}150\rho\cos\phi + \hat{\phi}200\sin\phi$ V/m, and from the results obtained in Example 2.15,

$$\oint_\Gamma \vec{E} \cdot d\vec{\ell} = 525\,\text{V}.$$

Note that ρ is dependent on ϕ, which was found to be $\rho = 3\cos\phi$. The relation and curl of the vector was evaluated in Example 2.23 as follows:

$$\nabla \times \vec{E} = \hat{z}\left(\frac{200}{\rho} + 150\right)\sin\phi.$$

Therefore

$$\int_S (\nabla \times \vec{E}) \cdot d\vec{s} = \int_{\phi=0}^{\pi/2} \int_{\rho=0}^{3\cos\phi} \hat{z}\left(\frac{200}{\rho} + 150\right)\sin\phi \cdot \hat{z}\rho d\rho d\phi$$

$$= \int_{\phi=0}^{\pi/2} \int_{\rho=0}^{3\cos\phi} (200 + 150\rho)\sin\phi d\rho d\phi,$$

or

$$\int_S (\nabla \times \vec{E}) \cdot d\vec{s} = \int_{\phi=0}^{\pi/2} \left(200\rho + 150\frac{\rho^2}{2}\right)\bigg|_0^{3\cos\phi} \sin\phi d\phi$$

$$= \int_0^{\pi/2} (600\cos\phi + 675\cos^2\phi)\sin\phi d\phi.$$

The integral can be evaluated by changing the integration variable (and the associated integration limits) as follows:

$$\cos\phi = t \;\rightarrow\; -\sin\phi d\phi = dt,$$
$$\phi = 0 \;\rightarrow\; t = 1,$$

and

$$\phi = \frac{\pi}{2} \;\rightarrow\; t = 0$$

Therefore

$$\int_S (\nabla \times \vec{E}) \cdot d\vec{s} = -\int_1^0 (600t - 675t^2)dt = 525\,\text{V} = \oint_\Gamma \vec{E} \cdot d\vec{\ell}.$$

2.9 OTHER OPERATIONS INVOLVING ∇

Laplacian of the Scalar Field $f(\vec{r})$

The Laplacian of a scalar field is defined in rectangular coordinates as follows:

$$\nabla \cdot \nabla f = \nabla^2 f = \nabla \cdot \left(\hat{x}\frac{\partial f}{\partial x} + \hat{y}\frac{\partial f}{\partial y} + \hat{z}\frac{\partial f}{\partial z}\right) = \frac{\partial^2 f}{\partial x^2} + \frac{\partial^2 f}{\partial y^2} + \frac{\partial^2 f}{\partial z^2}. \tag{2.9.1}$$

In cylindrical coordinates, it is given by

$$\nabla^2 f = \frac{1}{\rho}\frac{\partial}{\partial \rho}\left(\rho\frac{\partial f}{\partial \rho}\right) + \frac{1}{\rho^2}\frac{\partial^2 f}{\partial \phi^2} + \frac{\partial^2 f}{\partial z^2}. \tag{2.9.2}$$

The Laplacian of a scalar field is given in spherical coordinates as follows:

$$\nabla^2 f = \frac{1}{r^2}\frac{\partial}{\partial r}\left(r^2\frac{\partial f}{\partial r}\right) + \frac{1}{r^2 \sin\theta}\frac{\partial}{\partial \theta}\left(\sin\theta\frac{\partial f}{\partial \theta}\right) + \frac{1}{r^2 \sin^2\theta}\frac{\partial^2 f}{\partial \phi^2}. \tag{2.9.3}$$

Laplacian of the Vector Field $\vec{A}(\vec{r})$

From $\nabla \times \nabla \times \vec{A} = \nabla(\nabla \cdot \vec{A}) - \nabla^2\vec{A}$, we get

$$\nabla^2\vec{A}(\vec{r}) = \nabla\left[\nabla \cdot \vec{A}(\vec{r})\right] - \nabla \times \nabla \times \vec{A}(\vec{r}). \tag{2.9.4}$$

Note that the Laplacian of a vector is separable only in the rectangular coordinates, as follows:

$$\nabla^2\vec{A} = \hat{x}\nabla^2 A_x + \hat{y}\nabla^2 A_y + \hat{z}\nabla^2 A_z. \tag{2.9.5}$$

Two Important Differential Identities

The curl of the gradient of any scalar field is always zero. In other words, the gradient of a scalar field is always irrotational. Mathematically

$$\nabla \times \nabla f = 0. \tag{2.9.6}$$

The divergence of the curl of a vector field is always zero. In other words, the curl of a vector field is always solenoidal. Mathematically

$$\nabla \cdot \nabla \times \vec{A} = 0. \tag{2.9.7}$$

Several Often Used Identities

$$\nabla \cdot (f\vec{A}) = f\nabla \cdot \vec{A} + \vec{A} \cdot \nabla f, \tag{2.9.8}$$

$$\nabla \times (f\vec{A}) = f\nabla \times \vec{A} + \nabla f \times \vec{A}, \tag{2.9.9}$$

$$\nabla(f_1 f_2) = f_1 \nabla f_2 + f_2 \nabla f_1, \tag{2.9.10}$$

$$\nabla \cdot (\vec{A}_1 \times \vec{A}_2) = \vec{A}_2 \cdot (\nabla \times \vec{A}_1) - \vec{A}_1 \cdot (\nabla \times \vec{A}_2). \tag{2.9.11}$$

2.10 HELMHOLTZ THEOREM

As was mentioned earlier, a vector field is solenoidal if its divergence vanishes everywhere in the spatial domain. If the curl of a vector field vanishes everywhere, then it is called an irrotational field. Physically the former situation arises in magnetostatics in response to a current density, and the latter one is associated with the electrostatics created by a charge density. If both the divergence and the curl of a vector field are zero, then the field has no source. A general vector field $\vec{A}(\vec{r})$ is uniquely specified if both its divergence and its curl are given within a region, and its normal component is specified over the boundary. Furthermore such a field may always be expressed as follows:

$$\vec{A}(\vec{r}) = \nabla f(\vec{r}) + \nabla \times \vec{F}(\vec{r}), \tag{2.10.1}$$

where $f(\vec{r})$ and $\vec{F}(\vec{r})$ are called scalar and vector potential fields, respectively.

Note that according to the Helmholtz theorem, the divergence and curl of a vector field cannot simultaneously vanish at every point in space if the field $\vec{A}(\vec{r})$ is to be nonvanishing; that is,

$$\nabla \times \vec{A}(\vec{r}) = 0 \rightarrow \nabla \cdot \vec{A}(\vec{r}) \neq 0, \tag{2.10.2}$$

$$\nabla \cdot \vec{A}(\vec{r}) = 0 \rightarrow \nabla \times \vec{A}(\vec{r}) \neq 0. \tag{2.10.3}$$

In other words, a vector field $\vec{A}(\vec{r})$ is completely specified if both its divergence and curl are given at each point $\vec{r}$ in its domain V.

For $\nabla \cdot \vec{A}(r) = g(\vec{r})$ and $\nabla \times \vec{A}(\vec{r}) = \vec{J}(\vec{r})$, if we take divergence of (2.10.1) and use (2.9.7), then we find that

$$\nabla \cdot \vec{A}(r) = \nabla \cdot \nabla f(\vec{r}) + \nabla \cdot \left[\nabla \times \vec{F}(\vec{r})\right] \longrightarrow \nabla \cdot \nabla f(\vec{r}) = g(\vec{r}),$$

or

$$\nabla^2 f(r) = g(\vec{r}). \tag{2.10.4}$$

This is Poisson's equation as found in electrostatics. It is called a Laplace equation when its right-hand side is zero.

On the other hand, if we take curl of (2.10.1) and use (2.9.6), then we find that

$$\nabla \times \vec{A}(r) = \nabla \times \nabla f(\vec{r}) + \nabla \times \left[\nabla \times \vec{F}(\vec{r})\right] = \nabla \times \left[\nabla \times \vec{F}(\vec{r})\right] = \vec{J}(\vec{r}).$$

Therefore

$$\nabla \times \nabla \times \vec{F}(\vec{r}) = \vec{J}(\vec{r}). \tag{2.10.5}$$

PROBLEMS

2.1. For $\vec{A} = -2\hat{x} + 3\hat{y} - \hat{z}$, find the magnitude $\left|\vec{A}\right|$ and the unit vector in the direction of the vector.

2.2. A vector $\vec{A}$ is defined in the rectangular coordinate system as being directed from (1, 2, 3) to (3, 4, 5). Find (a) a vector expression for $\vec{A}$, (b) the magnitude of $\vec{A}$, and (c) a unit vector pointing in the direction of $\vec{A}$.

2.3. Find the position vectors for points $P_1(1, 3, 5)$ and $P_2(2, 4, -1)$. If a parallelogram is formed by these two position vectors as its two adjacent sides, then verify if the diagonals of this parallelogram bisect each other.

2.4. Find a vector $\vec{B}$ that is perpendicular to $\vec{A} = 3\hat{x} - 2\hat{y} + \hat{z}$, has no z component, and has a magnitude of unity.

2.5. If $\vec{A} = 2\hat{x} - \hat{y} + 4\hat{z}$ and $\vec{B} = -\hat{x} + \hat{y} - 2\hat{z}$, then find (a) the component of $\vec{B}$ in direction of $\vec{A}$, (b) the smallest angle between the two vectors, and (c) a unit vector perpendicular to the plane that contains two vectors.

2.6. A triangle ABC is found by connecting the points A (1, 2, 0), B (0, 0, 1), and C (−2, 1, 0). Find the point D that gives a plane parallelogram $ABDC$.

2.7. Verify (2.1.36) for $\vec{A} = \hat{x} + 4\hat{y} + 3\hat{z}$, $\vec{B} = 3\hat{y} + \hat{z}$, and $\vec{C} = 4\hat{x}$.

2.8. Verify (2.1.37) for $\vec{A} = \hat{x} + 4\hat{y} + 3\hat{z}$, $\vec{B} = 3\hat{y} + \hat{z}$, and $\vec{C} = 4\hat{x}$.

2.9. A receiver antenna produces an electromotive force (voltage) proportional to the z component of the incident electric field intensity. Assume that the proportionality constant is 0.3 m. If the electric field intensity is given by the following expression, then find the voltage induced at the receiver:

$$\vec{E} = \hat{\phi}\,\frac{2 + \cos(\phi)}{\sqrt{\rho}}\ \text{V/m}$$

2.10. A receiver antenna produces an electromotive force (voltage) proportional to the z component of the incident electric field intensity. Assume that the proportionality constant is 0.2 m. If the electric field intensity is given by the following expression, then find the induced voltage at the receiver:

$$\vec{E} = 3(\hat{\theta}\sin\theta + \hat{\phi}\sin\phi\cos\theta)\ \text{V/m}$$

2.11. The electrical charge density in a volume is given by

$$\rho_V = 2 - 0.2r^3\ \text{C/m}^{-3}.$$

Determine the total charge enclosed in a cube defined by $-1 \le x \le 1$ m, $-1 \le y \le 1$ m, and $-1 \le z \le 1$ m.

2.12. The electrical charge density in a volume is given by

$$\rho_V = 2 - 0.2r^3\ \text{C/m}^{-3}.$$

Determine the total charge enclosed in a volume defined by $0 \le \rho \le 1$ m, $(\pi/4) \le \phi \le (\pi/2)$, and $1 \le z \le 1$ m.

2.13. The electrical charge density in a volume is given by

$$\rho_V = 2 - 0.2r^3\ \text{C/m}^{-3}.$$

Determine the total charge enclosed in a wedge defined by $0 \le r \le 1$ m, $(\pi/4) \le \theta \le (\pi/2)$, and $(\pi/4) \le 0 \le (\pi/2)$.

2.14. Find the work done by the following force in moving a particle from P to P_2 (a) over the straight line connecting (0, 0, 2) and (1, 3, 0), (b) over the straight-line path connecting (0, 0, 2) → (0, 0, 0) → (1, 0, 0) → (1, 3, 0), and (c) (0, 0, 2) → (0, 0, 0) → (1, 3, 0). Assume that all distances are in meters:

$$\vec{F} = \hat{x}2x^2y + \hat{y}y^2 + \hat{z}yz\ \text{N}.$$

2.15. Find the circulation of the electric field intensity $\vec{E} = \hat{\rho}\rho^2\cos\phi + \hat{\phi}5\sin^2\phi\,\text{V/m}$ over a semicircular path of radius 1.5 m with its center at (1.5 m, 0°, 0) as shown in Figure 2.21.

2.16. Find the circulation of the following magnetic field intensity around the contour $ABCDA$, as shown in Figure 2.22.

$$\vec{H} = \hat{\rho}\rho^2\sin\phi + \hat{\phi}4\rho\ \text{A/m}.$$

2.17. Find the circulation of the following magnetic field intensity around the contour *ABCDA*, as shown in Figure 2.22.

$$\vec{H} = \hat{\rho}\rho \sin^2\phi + \hat{\phi}4\rho \,\text{A/m}.$$

2.18. For an electric flux density given as $\vec{D} = \hat{\rho}(3/\rho^2)\,\text{C/m}^2$, find the total electric flux emanating from a closed surface that bounds (a) a cylindrical volume of 2 m in radius, with $0 \leq \phi \leq \pi$, and $-1 \leq z \leq 1$ m, (b) a cube with $-1 \leq x \leq 1$ m, $-1 \leq y \leq 1$ m, and $-1 \leq z \leq 1$ m, and (c) a sphere of 1 m radius centered at the origin.

2.19. Determine the rate of change of a scalar field $f(\vec{r}) = x^2 + yz$ at (1, 2, 3) in the direction of $\vec{A} = \hat{x} + 2\hat{y} + 3\hat{z}$.

2.20. Find the gradient of the following scalar fields:

(a) $f(x, y, z) = z(5x + 3y + 2)$

(b) $f(\rho, \phi, z) = 5\cos^2\phi + 2\rho z$

(c) $f(r, \theta, \phi) = 2(\sin^2\theta \cos\phi/r^2)$

2.21. Find the divergence of the following vector fields:

(a) $\vec{A} = -\hat{x}x + \hat{y}4y^2 - \hat{z}9z$

(b) $\vec{B} = -\hat{\rho}\cos\phi - \hat{\phi}(5/\rho^4) + \hat{z}z^3$

(c) $\vec{E} = \hat{r}9r^2 - \hat{\theta}\sin\phi + \hat{\phi}\cos^2\theta$

2.22. Find the curl of the following vector fields:

(a) $\vec{U} = \hat{x}(yz) + \hat{y}(xz) + \hat{z}(yx)$

(b) $\vec{E} = \hat{\rho}5\rho^2 + \hat{\phi}2\rho\sin\phi$

(c) $\vec{H} = \hat{r}(\cos\theta/r) - \hat{\theta}r\sin\theta + \hat{\phi}\cos\phi$

2.23. Verify the divergence theorem for a vector $\vec{A} = \hat{r}r^3$ when a closed surface is (a) a cylinder with $\rho = 1$ m, and $0 \leq z \leq 1$ m, (b) a sphere of $r = 1$ m, and (c) a cube with $0 \leq x \leq 1$ m, $0 \leq y \leq 1$ m, and $0 \leq z \leq 1$ m.

2.24. Verify Stokes's theorem for the case of Problem 2.16.

2.25. Verify Stokes's theorem for the case of Problem 2.15.

2.26. Points $P_1(1, 0, 2)$, $P_2(-3, 1, 5)$, and $P_3(3, -4, 6)$ form a right triangle in space. Find the point where the angle is a right angle.

2.27. Determine the work done in carrying a 5 μC charge from point $P_1(1, 2, -4)$ to point $P_2(-2, 8, -4)$ in the field $\vec{E} = y\hat{x} + x\hat{y}\,\text{V/m}$ along the parabola $y = 2x^2$.

2.28. Points $P_1(-2, 0, 3)$ and $P_2(0, 4, -1)$ are given in space. Find (a) length of the line that joins P_1 and P_2. (b) How much is the perpendicular distance from point $P_3(3, 1, 3)$ to the line.

2.29. Find component of the vector $\vec{A} = 2\hat{x} - 5\hat{y} + 3\hat{z}$ that is perpendicular to the vector $\vec{B} = -\hat{x} + 4\hat{y}$.

2.30. If $\vec{A} = \hat{\rho}\rho \cos\phi + \hat{\phi}\rho^3 \sin(\phi)$, then evaluate $\oint \vec{A} \cdot d\vec{\ell}$, going counterclockwise around the contour bound by circles of radii 1 and 2 in the first quadrant.

2.31. Three vectors are given as follows:

$$\vec{A} = \hat{x} + 2\hat{y} - \hat{z},$$

$$\vec{B} = -3\hat{x} + \hat{y} - 2\hat{z},$$

and

$$\vec{C} = 9\hat{x} + \hat{z}.$$

Find (a) $\vec{A} \times \vec{C}$, and (b) $(\vec{C} \times \vec{B}) \cdot \vec{A}$.

2.32. For points $P_1(3, 2, -1)$, $P_2(3, 8, -5)$, and $P_3(1, 3, 0)$ given, find (a) the vector drawn from P_1 to P_2, (b) the straight line distance from P_2 to P_3, and (c) the unit vector along the line from P_1 to P_3.

2.33. Three vectors are given as follows:

$$\vec{A} = 6\hat{x} + 2\hat{y} - 3\hat{z},$$

$$\vec{B} = 4\hat{x} - 6\hat{y} + 12\hat{z},$$

and

$$\vec{C} = 5\hat{x} - 2\hat{z}.$$

Find (a) $\vec{C} \times \vec{B}$, and (b) $\vec{A} \cdot (\vec{C} \times \vec{B})$.

2.34. Given a vector field $\vec{A} = \rho^2\hat{\rho} + 2z\hat{z}$, compute the total outward flux $(\oint_S \vec{A} \cdot d\vec{s})$ from the surface of a 10 m high cylinder of radius 2 m.

2.35. For a vector function $\vec{A} = \hat{\rho}\rho^3 + \hat{z}\,z$, verify the divergence theorem over a circular cylindrical region bound by $\rho = 3$, $z = 0$, and $z = 2$.

2.36. For the vector field $\vec{A} = \hat{x}3x^2y^2 - \hat{y}x^3y^2$, verify the Stokes's theorem over a square region bound by $1 \le x \le 2$ and $1 \le y \le 2$.

2.37. Find the unit vector normal to the surface defined by

$$2 = \sqrt{x^2 + y^2 + z^2}.$$

2.38. Evaluate the curl of

$$\vec{A} = \hat{r}\frac{r}{1+r} + \hat{\theta}\, 2r\sin(\phi)$$

at the point $(3, \pi/4, 2\pi/7)$.

2.39. For $\vec{A} = (xy - 1)\hat{x} - y^2\hat{y} + (yz + 2)\hat{z}$, find $\oint_S \vec{A} \cdot d\vec{s}$, where S is the surface of a rectangular box bounded by the planes $x = 0$, $x = 2$, $y = 0$, $y = 1$, $z = 0$, and $z = 0.5$.

3

BASIC LAWS OF ELECTROMAGNETICS

This chapter begins with an introduction to fundamental laws of electromagnetic fields. These laws were formulated on the basis of various experimental observations. Historically static electric and magnetic fields were studied long before time-varying fields. Hence a number of laws (Coulomb's law, Biot-Savart law, Ampere's law, etc.) were formulated for static fields. Faraday's law for time-varying fields set the foundation for electrical generators. Later on Maxwell not only summarized elegantly the previous works through his celebrated equations but also introduced a displacement current term in the Ampere's law. This was a significant contribution that established relations between the electric and the magnetic fields, predicted the propagation of electromagnetic waves in the space, and set the foundation for modern wireless communication. Earlier results of static fields become mostly a special case of Maxwell's equations. In this chapter Maxwell's equations in integral (or large-scale) form are introduced initially because these are relatively easier to correlate with the experimental observations. The corresponding differential expressions (or point forms) are obtained via the Stokes and Gauss theorems. After a brief discussion on time-harmonic fields and constitutive relations, the boundary conditions are summarized next along with a few of its application. The chapter ends with sections on the Lorentz force equation and the Poynting vector.

Practical Electromagnetics: From Biomedical Sciences to Wireless Communication.
By Devendra K. Misra
Copyright © 2007 John Wiley & Sons, Inc.

3.1 MAXWELL'S EQUATIONS IN LARGE-SCALE OR INTEGRAL FORM

In general, electromagnetic fields and sources vary with space-coordinates and the time. In vector notation these may be represented as follows:

$\vec{\mathscr{E}}(\vec{r}, t)$ Electric field intensity in V/m
$\vec{\mathscr{H}}(\vec{r}, t)$ Magnetic field intensity in A/m
$\vec{\mathscr{D}}(\vec{r}, t)$ Electric flux density in C/m^2
$\vec{\mathscr{B}}(\vec{r}, t)$ Magnetic flux density in Tesla
$\vec{\mathscr{J}}(\vec{r}, t)$ Electric current density in A/m^2
$\rho(\vec{r}, t)$ Electric charge density in C/m^3

Laws using these notations are presented here, the same as they appear in the Maxwell's equations. For simplicity, the space and time dependences of these field quantities are not shown explicitly but are implied.

Faraday's Law of Induction

Faraday discovered that if a conducting wire loop is exposed to magnetic flux, it can induce an electromotive force (emf). The induced emf depends on the time-rate of change of the magnetic flux that leaves the surface bound by the loop. Induction is possible only when the magnetic flux is changing with time or the loop is moving through a nonuniform magnetic field. Consider a loop c that bounds the surface s, as illustrated in Figure 3.1. The magnetic flux leaving the surface can be evaluated by integrating the normal component of magnetic flux density at every point on it.

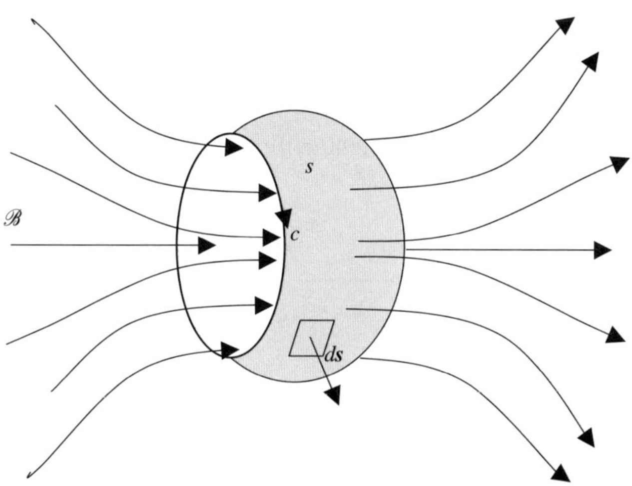

Figure 3.1 Magnetic flux density $\mathscr{B}$ passing through an area s bound by curve c.

Mathematically

$$\oint_c \vec{\mathscr{E}} \cdot d\vec{\ell} = -\frac{\partial}{\partial t}\int_s \vec{\mathscr{B}} \cdot d\vec{s}. \tag{3.1.1}$$

Example 3.1

As shown in Figure 3.2, a metal bar oscillates over a pair of conducting rails, with its position given by $x = 2 - \cos(20\pi t)$ m. The system lies on $z = 0$ plane, and the magnetic flux with a density of 5 mT is in the z direction. Find the current i due to the induced emf. What happens if the magnetic flux varies with time as follows?

$$\vec{B} = \hat{z}5\,\cos(20\pi t)\,\text{mT}.$$

The voltage induced across R is found to be

$$\oint_c \vec{\mathscr{E}} \cdot d\vec{\ell} = -\frac{\partial}{\partial t}\int_{x'=0}^{x}\int_{y=0}^{2} \vec{\mathscr{B}} \cdot \hat{z}dx'\,dy = -\frac{\partial}{\partial t}(5 \times 2 \times x) = -10\frac{dx}{dt}$$
$$= -200\pi\,\sin(20\pi t)\,\text{mV}$$

and

$$i = \frac{200\pi}{0.1}\sin(20\pi t)\,\text{mA} = 2\pi\sin(20\pi t)\,\text{A}.$$

For the case of magnetic field changing with time,

$$\oint_c \vec{\mathscr{E}} \cdot d\vec{\ell} = -\frac{\partial}{\partial t}\int_{x'=0}^{x}\int_{y=0}^{2} \vec{\mathscr{B}} \cdot \hat{z}dx'dy = -\frac{\partial}{\partial t}[2 \times x \times 5\cos(20\pi t)]\,\text{mV}$$
$$= -10\cos(20\pi t)\frac{dx}{dt} + 200\pi x\sin(20\pi t)\,\text{mV}$$
$$= 200\pi[2 - \cos(20\pi t)]\sin(20\pi t) - 200\pi\sin(20\pi t)\cos(20\pi t)\,\text{mV}$$
$$= 200\pi[2\sin(20\pi t) - \sin(40\pi t)]\,\text{mV} = 0.2\pi[2\sin(20\pi t) - \sin(40\pi t)]\,\text{V}.$$

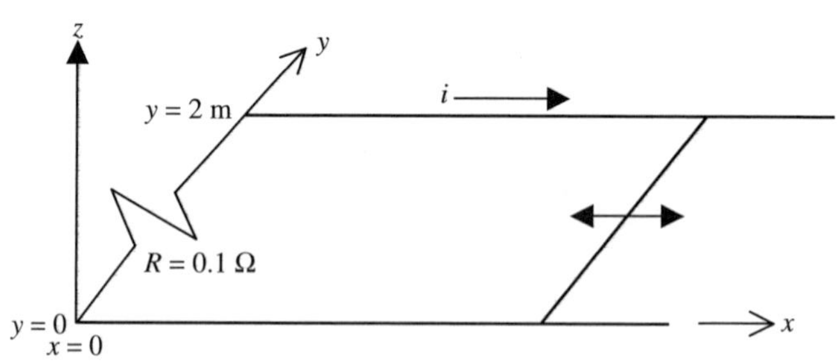

Figure 3.2 System geometry for Example 3.1.

Therefore

$$i = -2\pi[2\sin(20\pi t) - \sin(40\pi t)]\,\text{A}.$$

Example 3.2

A rigid rectangular loop is lying on the xy-plane with its vertices at $(1, y, 0)$, $(3, y, 0)$, $(3, y+2, 0)$, and $(1, y+2, 0)$, as shown in Figure 3.3. It is immersed in a magnetic field $\vec{B}(\vec{r}, t) = \hat{z}5\cos[(\pi y/2) - 250\pi t]\,\text{mT}$. (a) For a stationary loop find the electromotive force induced counterclockwise around it. (b) How will the induced emf change if the loop is moving with velocity $\vec{v} = \hat{y}500\,\text{m/s}$.

The magnetic flux passing through the loop area is found to be

$$\begin{aligned}\varphi(\vec{r}, t) &= \int_S \vec{\mathscr{B}}(\vec{r}, t) \cdot d\vec{s} \\ &= \int_{y'=y}^{y+2} \int_{x=1}^{3} \hat{z}\,5\cos\left(\frac{\pi y'}{2} - 250\pi t\right) \cdot \hat{z}dxdy' \\ &= 10 \int_{y'=y}^{y+2} \cos\left(\frac{\pi y'}{2} - 250\pi t\right) dy',\end{aligned}$$

or

$$\begin{aligned}\varphi(\vec{r}, t) &= 10\left.\frac{\sin[(\pi y'/2) - 250\pi t]}{\pi/2}\right|_{y'=y}^{y+2} \\ &= -\frac{40}{\pi}\sin\left(\frac{\pi y}{2} - 250\pi t\right)\,\text{mWb} \\ &= -\frac{0.04}{\pi}\sin\left(\frac{\pi y}{2} - 250\pi t\right)\,\text{Wb}.\end{aligned}$$

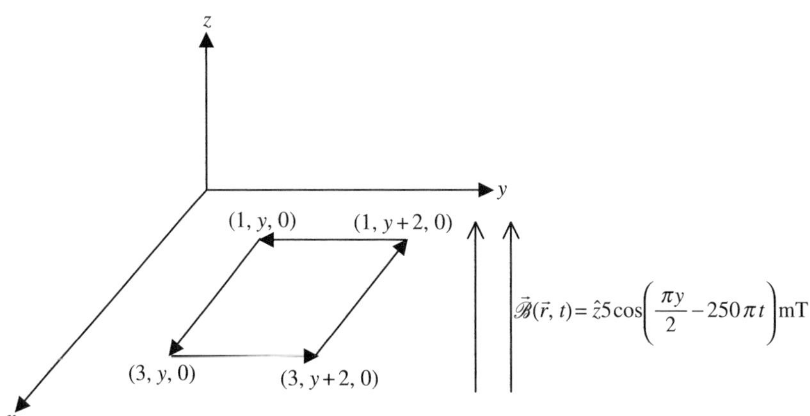

Figure 3.3 System geometry for Example 3.2.

(a) When the loop is stationary, the time derivative of y is zero. Therefore

$$\oint_c \vec{\mathscr{E}} \cdot d\vec{\ell} = \text{induced emf} = -\frac{d\varphi}{dt} = -10\cos\left(\frac{\pi y}{2} - 250\pi t\right)\text{V}.$$

(b) Because the loop is moving in the y direction at a velocity of 500 m/s,

$$\oint_c \vec{\mathscr{E}} \cdot d\vec{\ell} = \text{induced emf} = -\frac{d\varphi}{dt}$$

$$= -\frac{40}{\pi}\left[\cos\left(\frac{\pi y}{2} - 250\pi t\right)\frac{\pi}{2}\frac{dy}{dt} - 250\pi\cos\left(\frac{\pi y}{2} - 250\pi t\right)\right] = 0.$$

Example 3.3

A rigid rectangular loop of 3×2 m is situated on the yz-plane, as shown in Figure 3.4. It is rotating symmetrically about the z-axis with an angular velocity of ω_1 rad/s. If there is a magnetic field with its flux density $\vec{\mathscr{B}}(\vec{r}, t) = \hat{x}5\cos(\omega_2 t)$ mT, find the open-circuit voltage V_o induced in the loop.

Small area on the surface rotating about z-axis can be expressed as follows:

$$d\vec{s} = \hat{\phi} dydz, \qquad -1.5 \le y \le 1.5\,\text{m}, \quad 0 \le z \le 2\,\text{m}.$$

Since the magnetic flux density is directed along x-axis, it can be transformed to cylindrical coordinates as well with the help of Table 2.3. Hence

$$\vec{B}(\vec{r}, t) = (\hat{\rho}\cos\phi - \hat{\phi}\sin\phi)5\cos(\omega_2 t)\,\text{mT}$$

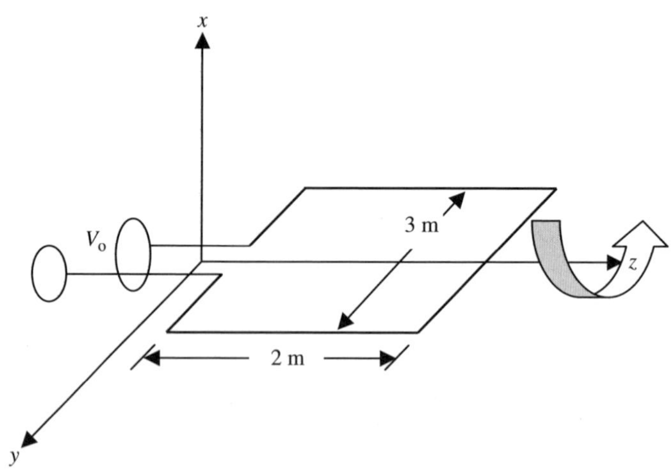

Figure 3.4 System geometry for Example 3.3.

and

$$\varphi(\vec{r}, t) = \int_S \vec{\mathscr{B}}(\vec{r}, t) \cdot d\vec{s} = -5\cos(\omega_2 t)\sin\phi \int_{z=0}^{2} \int_{y=-1.5}^{1.5} dydz$$
$$= -30\cos(\omega_2 t)\sin\phi \, \text{mWb},$$

or

$$\varphi(\vec{r}, t) = -30\cos(\omega_2 t)\sin(\omega_1 t)$$
$$= -15[\sin(\omega_1 + \omega_2)t + \sin(\omega_1 - \omega_2)t] \, \text{mWb}.$$

Therefore

$$\oint_c \vec{\mathscr{E}} \cdot d\vec{\ell} = -\frac{\partial \varphi(\vec{r}, t)}{\partial t}$$
$$= -0.015[(\omega_1 + \omega_2)\cos(\omega_1 + \omega_2)t + (\omega_1 - \omega_2)\cos(\omega_1 - \omega_2]\,\text{V}.$$

Note that ω_2 is zero for the static magnetic field, and therefore the expression for total magnetic flux simplifies to

$$\varphi(\vec{r}, t) = -30\sin(\omega_1 t)\,\text{mWb}.$$

In this case the induced electromotive force is found to be

$$\oint_c \vec{\mathscr{E}} \cdot d\vec{\ell} = -\frac{\partial \varphi(\vec{r}, t)}{\partial t} = 0.03\omega_1 \cos(\omega_1 t)\,\text{V}.$$

Example 3.4

A conducting fluid is flowing through a plastic tube of 3.5 cm in diameter, as shown in Figure 3.5. The flow rate is 0.24 L/s along the z-axis. It is subjected to a magnetic field with its flux density $\vec{\mathscr{B}}(\vec{r}, t) = \hat{x}45\,\text{mT}$. As Shown in the figure, there are two electrodes placed across its diameter along the y-axis. Determine the voltage induced across these electrodes.

For a constant uniform magnetic flux density of B_oT along the x-axis, tube radius a and diameter Dm, we can write

$$\oint_c \vec{\mathscr{E}} \cdot d\vec{\ell} = -\frac{\partial}{\partial t} \int_{y=-a}^{a} \int_{z'=0}^{z} B_o \hat{x} \cdot \hat{x} dydz' = -\frac{\partial}{\partial t}(B_o 2az) = -2aB_o \frac{dz}{dt} = -DB_o v_z.$$

If the fluid flow rate is Q m^3/s, then

$$Q = \frac{\pi D^2}{4} v_z \rightarrow v_z = \frac{4Q}{\pi D^2}.$$

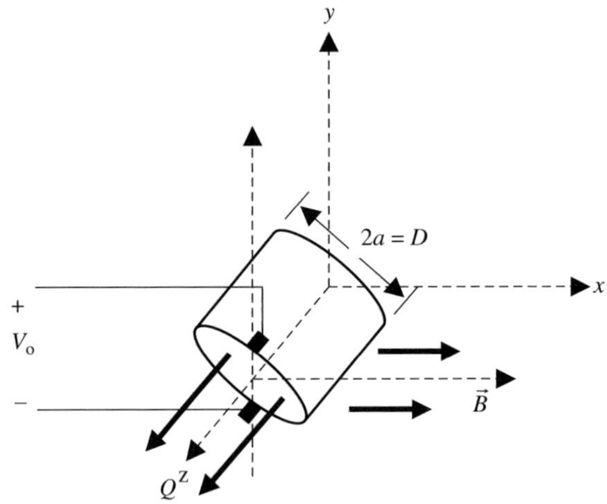

Figure 3.5 System geometry for Example 3.4.

On substituting it back, we find that

$$\oint_c \vec{\mathscr{E}} \cdot d\vec{\ell} = -DB_o \frac{4Q}{\pi D^2} = -\frac{4QB_o}{\pi D}.$$

Using the given data, we find that

$$V_o = -\frac{4 \times 0.24 \times 10^{-3} \times 0.045}{\pi 0.035} = 0.3929 \times 10^{-3}\,\text{V} = 392.9\,\mu\text{V}.$$

Note that the flow rate is given in L/s and the diameter of the tube in cm, whereas the formula derived is in SI units. Commercial flow meters are designed using this kind of arrangement.

Generalized Ampere's Law

The generalized Ampere's law is based on the experiments conducted by Oersted and Ampere and the theoretical reasoning of Maxwell. Oersted found that the electric current flowing through a conductor generates an encircling magnetic field. Maxwell introduced the displacement current term along with the conduction current. This significant contribution could explain the current flow through capacitors in a circuit as well as predict the propagation of electromagnetic waves in space.

Consider a closed path c that bounds the surface s, as illustrated in Figure 3.6. Integration of the tangential component of the magnetic field intensity along this closed path (the line integral of vector magnetic field intensity along c) gives the circulation of the magnetic field intensity, which is called the magnetomotive force (mmf). It is found that this mmf is equal to the net current enclosed by the closed

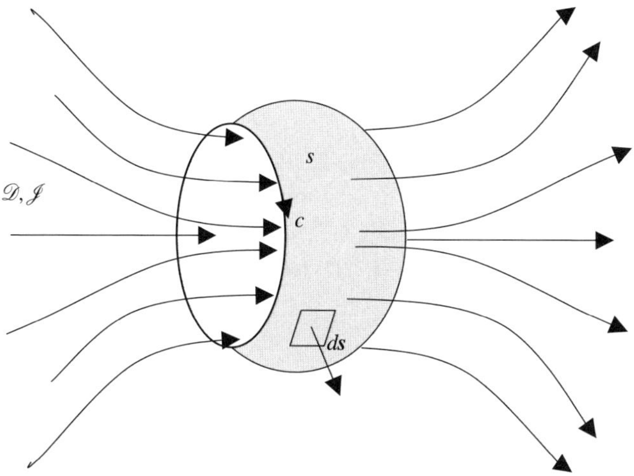

Figure 3.6 Electric flux density $\mathscr{D}$ and current density $\mathscr{J}$ passing through an area s bound by curve c.

path c. Mathematically

$$\oint_c \vec{\mathscr{H}} \cdot d\vec{\ell} = \int_s \vec{\mathscr{J}} \cdot d\vec{s} + \frac{\partial}{\partial t}\int_s \vec{\mathscr{D}} \cdot d\vec{s}. \tag{3.1.2}$$

Thus Ampere's law is analogous to the Faraday's law. Since the electric field intensity is expressed in volt per meter, its line integral (the emf) is in volts. Similarly the unit of magnetic field intensity is ampere per meter, and therefore the mmf has unit of ampere. The first term on right-hand side of (3.1.2) represents the conduction currents (net transfer of electric charge) and its second term is the time rate of change of electric flux that leaves the surface s (the displacement current). There is no term analogous to the conduction current in Faraday's law because in reality there is never a net transfer of magnetic charge. As mentioned earlier, this term is significant because it establishes the link between electric and magnetic fields. Further Ampere's law sets the foundation for the wireless communication.

Example 3.5

An infinitely long solid cylindrical wire with a radius of 2 m lies along the z-axis. It carries a current with density $\vec{\mathscr{J}}(\rho) = \mathscr{J}_o(1 - 0.5\rho)\hat{z}\ \mathrm{A/m^2}$. Find the magnetic field intensity inside as well as outside the wire.

From the given condition it is clear that only the encircling magnetic field H_ϕ exists. Hence Ampere's law for inside the wire gives

$$\oint_c \vec{\mathscr{H}} \cdot d\vec{\ell} = \int_s \vec{\mathscr{J}} \cdot d\vec{s} \longrightarrow \int_0^{2\pi} H_\phi \rho d\phi = \int_0^{2\pi}\int_0^{\rho} \mathscr{J}_o(1 - 0.5\rho')\rho' d\rho' d\phi,$$

or

$$2\pi\rho H_\phi = 2\pi\mathcal{J}_o\left(\frac{\rho^2}{2} - 0.5\frac{\rho^3}{3}\right) \longrightarrow H_\phi = \mathcal{J}_o\left(\frac{\rho}{2} - \frac{\rho^2}{6}\right) \text{ A/m}, \qquad 0 \le \rho \le 2.$$

Similarly, for outside the wire,

$$\oint_c \vec{\mathcal{H}} \cdot d\vec{\ell} = \int_s \vec{\mathcal{J}} \cdot d\vec{s} \longrightarrow \int_0^{2\pi} H_\phi \rho d\phi = \int_0^{2\pi}\int_0^{2} \mathcal{J}_o(1 - 0.5\rho')\rho' d\rho' d\phi,$$

or

$$2\pi\rho H_\phi = 2\pi\mathcal{J}_o\left(\frac{2^2}{2} - 0.5\frac{2^3}{3}\right) \longrightarrow H_\phi = \frac{2\mathcal{J}_o}{3\rho} \text{ A/m}, \qquad 0 \le \rho.$$

Example 3.6

A coaxial line with its inner and outer conductor radii a and b, respectively, carries a current constant with time. The outer conductor has a finite thickness, as shown in Figure 3.7. The current densities on the two conductors are given as follows:

$$\vec{J}_a = \hat{z}\frac{I}{\pi a^2} \text{ A/m}^2 \quad \text{and} \quad \vec{J}_b = -\hat{z}\frac{I}{\pi(c^2 - b^2)} \text{ A/m}^2.$$

Find the corresponding magnetic field intensity everywhere.

The geometry of this problem requires cylindrical coordinates. From the symmetry of the problem, only the ϕ component of the magnetic field exists.

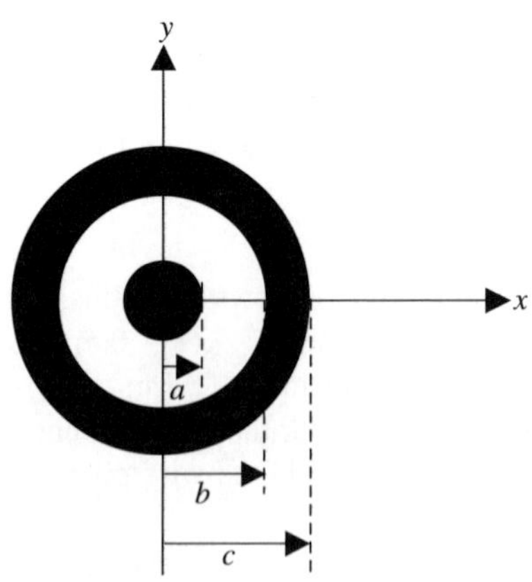

Figure 3.7 System geometry for Example 3.6.

Therefore Ampere's law can be used for four regions as follows:

For $0 \leq \rho \leq a$,

$$\oint_c \vec{\mathscr{H}} \cdot d\vec{\ell} = \int_S \vec{\mathscr{J}} \cdot d\vec{s} \rightarrow \int_0^{2\pi} H_\phi \hat{\phi} \cdot \hat{\phi} \rho d\phi$$
$$= \int_{\phi=0}^{2\pi} \int_{\rho'=0}^{\rho} \frac{I}{\pi a^2} \hat{z} \cdot \hat{z} \rho' d\rho' d\phi,$$

or

$$H_\phi 2\pi\rho = \frac{I}{\pi a^2} 2\pi \frac{\rho^2}{2} \rightarrow H_\phi = \frac{I\rho}{2\pi a^2} \text{ A/m}.$$

For $a \leq \rho \leq b$,

$$\oint_c \vec{\mathscr{H}} \cdot d\vec{\ell} = \int_S \vec{\mathscr{J}} \cdot d\vec{s} \rightarrow \int_0^{2\pi} H_\phi \hat{\phi} \cdot \hat{\phi} \rho d\phi$$
$$= \int_{\phi=0}^{2\pi} \int_{\rho'=0}^{a} \frac{I}{\pi a^2} \hat{z} \cdot \hat{z} \rho' d\rho' d\phi,$$

or

$$H_\phi 2\pi\rho = \frac{I}{\pi a^2} 2\pi \left. \frac{\rho'^2}{2} \right|_{\rho'=0}^{a} \rightarrow H_\phi = \frac{I}{2\pi\rho} \text{ A/m}.$$

For $b \leq \rho \leq c$,

$$\oint_c \vec{\mathscr{H}} \cdot d\vec{\ell} = \int_S \vec{\mathscr{J}} \cdot d\vec{s} \rightarrow \int_0^{2\pi} H_\phi \hat{\phi} \cdot \hat{\phi} \rho d\phi$$
$$= \int_{\phi=0}^{2\pi} \int_{\rho'=0}^{a} \frac{I}{\pi a^2} \hat{z} \cdot \hat{z} \rho' d\rho' d\phi - \int_{\phi=0}^{2\pi} \int_{\rho'=b}^{\rho} \frac{I}{\pi(c^2 - b^2)} \hat{z} \cdot \hat{z} \rho' d\rho' d\phi$$

or

$$H_\phi 2\pi\rho = \frac{I}{\pi a^2} 2\pi \left. \frac{\rho'^2}{2} \right|_{\rho'=0}^{a} - \frac{I}{\pi(c^2 - b^2)} 2\pi \left. \frac{\rho'^2}{2} \right|_{\rho'=b}^{\rho}$$

or

$$H_\phi = \frac{I}{2\pi\rho} \left[\frac{c^2 - \rho^2}{c^2 - b^2} \right] \text{ A/m}.$$

Finally, for $c \leq \rho$,

$$\oint_c \vec{\mathscr{H}} \cdot d\vec{\ell} = \int_S \vec{\mathscr{J}} \cdot d\vec{s} \longrightarrow \int_0^{2\pi} H_\phi \hat{\phi} \cdot \hat{\phi} \rho d\phi$$

$$= \int_{\phi=0}^{2\pi} \int_{\rho'=0}^{a} \frac{I}{\pi a^2} \hat{z} \cdot \hat{z} \rho' d\rho' d\phi - \int_{\phi=0}^{2\pi} \int_{\rho'=b}^{c} \frac{I}{\pi (c^2 - b^2)} \hat{z} \cdot \hat{z} \rho' d\rho' d\phi,$$

or

$$H_\phi 2\pi\rho = \frac{I}{\pi a^2} 2\pi \frac{\rho'^2}{2}\bigg|_{\rho'=0}^{a} - \frac{I}{\pi (c^2 - b^2)} 2\pi \frac{\rho'^2}{2}\bigg|_{\rho'=b}^{c} = 0 \longrightarrow H_\phi = 0.$$

Gauss's Law for the Electric Field

Gauss's law is based on his experimental observations and those of Faraday. Gauss's law for the electric field relates the enclosed charge with total displacement flux that emanates from the closed surface s. Consider a closed surface s that bounds a volume v, as shown in Figure 3.8. There is electric charge distributed in volume v, with a charge density of ρ. This electrical charge sets up a displacement flux with its density as $\mathscr{D}$.

According to Gauss's law, the electric displacement flux that emanates from a closed surface s is equal to the net charge contained within the volume v. Mathematically

$$\oint_s \vec{\mathscr{D}} \cdot d\vec{s} = \int_v \rho dv. \tag{3.1.3}$$

Example 3.7

A spherical volume of radius a contains electrical charge with $\rho = \rho_o\ \mathrm{C/m^3}$. Find the corresponding electric flux density everywhere. (Assume that ρ_o is a constant.)

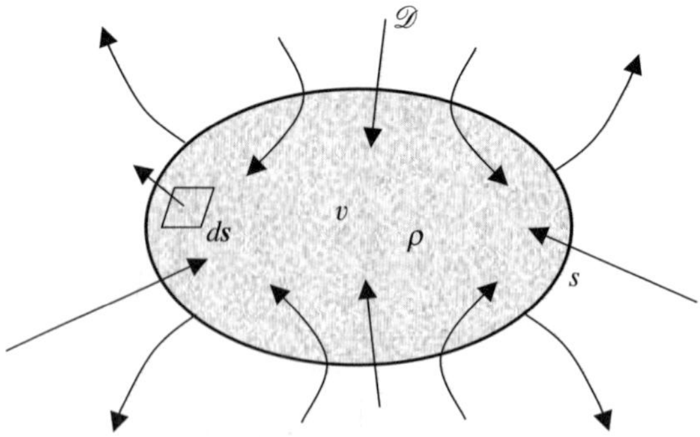

Figure 3.8 Electric flux density $\mathscr{D}$ emanating from a volume v bound by the surface s.

Using spherical symmetry, we find that only radial component of the flux density exists. Therefore Gauss's law gives

$$\oint_S \vec{\mathscr{D}} \cdot d\vec{s} = \int_V \rho_v dv \longrightarrow \int_{\phi=0}^{2\pi} \int_{\theta=0}^{\pi} \mathscr{D}_r \hat{r} \cdot \hat{r} r^2 \sin\theta \, d\theta d\phi \Big|_{r=r_o}$$

$$= \int_{\phi=0}^{2\pi} \int_{\theta=0}^{\pi} \int_{r=0}^{r_o} \rho_o r^2 \sin\theta \, dr d\theta d\phi.$$

For $0 \le r_o \le a$,

$$\mathscr{D}_r r_o^2 2\pi(-\cos\theta|_{\theta=0}^{\pi}) = \rho_o \frac{r^3}{3}\Big|_{r=0}^{r_o} 2\pi(-\cos\theta|_{\theta=0}^{\pi}) \longrightarrow \mathscr{D}_r = \rho_o \frac{r_o}{3} \text{ C/m}^2.$$

For $a \le r_o$,

$$\mathscr{D}_r r_o^2 2\pi(-\cos\theta|_{\theta=0}^{\pi}) = \rho_o \frac{r^3}{3}\Big|_{r=0}^{a} 2\pi(-\cos\theta|_{\theta=0}^{\pi}) \longrightarrow \mathscr{D}_r = \rho_o \frac{a^3}{3r_o^2} \text{ C/m}^2.$$

Gauss's Law for the Magnetic Field

This law is analogous to Gauss's law for the electric field. Since the magnetic flux lines are always closed and magnetic charges do not separate like positive or negative electric charges, net magnetic charge in a volume v has to be zero. As illustrated in Figure 3.9, the magnetic flux emanating from the closed surface s is therefore equal to zero. Mathematically

$$\oint_s \vec{\mathscr{B}} \cdot d\vec{s} = 0. \tag{3.1.4}$$

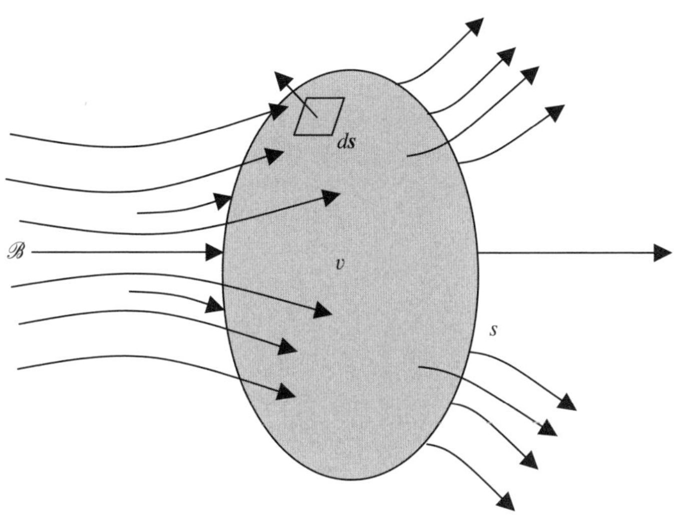

Figure 3.9 Magnetic flux density $\mathscr{B}$ emanating from a volume v bound by the surface s.

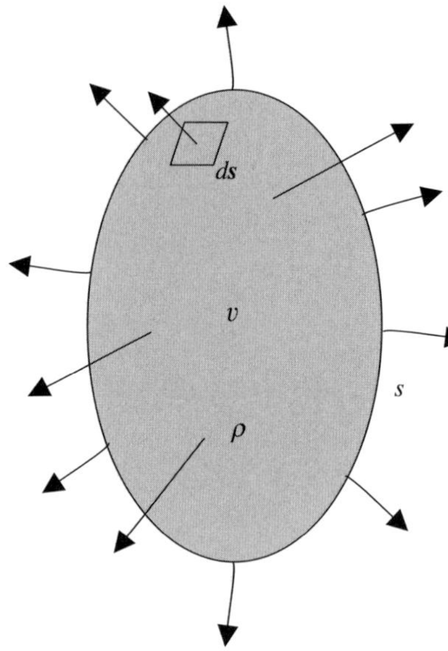

Figure 3.10 Electric charges leaving a volume v bound by the surface s.

Maxwell studied the preceding four laws that were formulated purely based on the experimental observations, and added the displacement current term to the original Ampere's law (hence, generalized it). Therefore Equations (3.1.1) through (3.1.4) are known as Maxwell's equations. Since these equations involve integrals, this formulation is known as the integral or large-scale form of the Maxwell's equations.

Equation of Continuity/Conservation of Charge

Consider a closed surface s that bounds a volume v, as shown in Figure 3.10. There is an electrical charge distributed in this volume that has a charge density ρ. If there is a current flow across the closed surface s, then there is an effect on the charge distribution. If the net current flowing through the surface s is zero, then there is no change in total charge. The net current emanating from the surface s is equal to the time rate of decrease in the charge enclosed. Mathematically this is stated as

$$\oint_s \vec{\mathscr{J}} \cdot d\vec{s} = -\frac{\partial}{\partial t}\int_v \rho\, dv. \tag{3.1.5}$$

3.2 MAXWELL'S EQUATIONS IN POINT OR DIFFERENTIAL FORM

As mentioned earlier, Equations (3.1.1) through (3.1.4) are known as the Maxwell equations in integral (or large-scale) form. A differential (or point) form of these equations can be obtained via the application of Stokes and Gauss's theorems as follows:

From (3.1.1), (3.1.2), and (2.8.1) we have

$$\oint_c \vec{\mathscr{E}} \cdot d\vec{\ell} = \int_s (\nabla \times \vec{\mathscr{E}}) \cdot d\vec{s} = -\frac{\partial}{\partial t}\int_s \vec{\mathscr{B}} \cdot d\vec{s} \rightarrow \nabla \times \vec{\mathscr{E}} = -\frac{\partial \vec{\mathscr{B}}}{\partial t}$$

and

$$\oint_c \vec{\mathscr{H}} \cdot d\vec{\ell} = \int_s (\nabla \times \vec{\mathscr{H}}) \cdot d\vec{s} = \int_s \vec{\mathscr{J}} \cdot d\vec{s} + \frac{\partial}{\partial t}\int_s \vec{\mathscr{D}} \cdot d\vec{s} \rightarrow \nabla \times \vec{\mathscr{H}} = \vec{\mathscr{J}} + \frac{\partial \vec{\mathscr{D}}}{\partial t}.$$

Similarly from (3.1.3), (3.1.4), and (2.7.1) we write

$$\oint_s \vec{\mathscr{D}} \cdot d\vec{s} = \int_v \nabla \cdot \vec{\mathscr{D}}\, dv = \int_v \rho dv \rightarrow \nabla \cdot \vec{\mathscr{D}} = \rho$$

and

$$\oint_s \vec{\mathscr{B}} \cdot d\vec{s} = \int_v \nabla \cdot \vec{\mathscr{B}}\, dv = 0 \rightarrow \nabla \cdot \vec{\mathscr{B}} = 0.$$

Application of the divergence theorem (2.7.1) to (3.1.5) gives

$$\oint_s \vec{\mathscr{J}} \cdot d\vec{s} = \int_v \nabla \cdot \vec{\mathscr{J}} dv = -\frac{\partial}{\partial t}\int_v \rho dv \rightarrow \nabla \cdot \vec{\mathscr{J}} = -\frac{\partial \rho}{\partial t}.$$

These results may be summarized as

$$\nabla \times \vec{\mathscr{E}} = -\frac{\partial \vec{\mathscr{B}}}{\partial t}, \tag{3.2.1}$$

$$\nabla \times \vec{\mathscr{H}} = \vec{\mathscr{J}} + \frac{\partial \vec{\mathscr{D}}}{\partial t}, \tag{3.2.2}$$

$$\nabla \cdot \vec{\mathscr{D}} = \rho, \tag{3.2.3}$$

$$\nabla \cdot \vec{\mathscr{B}} = 0, \tag{3.2.4}$$

and

$$\nabla \cdot \vec{\mathscr{J}} = -\frac{\partial \rho}{\partial t}. \tag{3.2.5}$$

Equations (3.2.1) through (3.2.4) are known as the Maxwell equations in point (or differential) form, whereas (3.2.5) is the equation of continuity in differential form.

Example 3.8

The electric flux density in a spherical volume is given as follows:

$$\vec{\mathscr{D}} = \hat{r}\frac{r\rho_o}{3}\ \mathrm{C/m^3}.$$

Find the associated electrical charge density. (Assume that ρ_o is a constant.)

Using (3.2.3), we have

$$\rho = \nabla \cdot \vec{\mathscr{D}} = \frac{1}{r^2 \sin\theta}\left[\sin\theta \frac{\partial}{\partial r}\left(r^2 \frac{\rho_o r}{3}\right) + 0 + 0\right] = \frac{1}{r^2}\frac{\rho_o}{3}\frac{dr^3}{dr} = \rho_o\ \mathrm{C/m^3}.$$

Note that this example verifies the result obtained in Example 3.7.

Example 3.9

The current density in a region is given as follows:

$$\vec{\mathscr{J}} = \hat{x}yz + \hat{y}yz + \hat{z}zx\ \mathrm{A/m^2}.$$

Find the current leaving the surface that bounds a unit cube in first quadrant, as shown in Figure 3.11.

This problem can be solved two different ways. One way is to evaluate the left-hand side of (3.1.5). That is, since there are six surfaces in a cube, the evaluation can be carried out individually on each, and then the final result found after summing. Hence the current I_1 leaving the front surface is written as

$$I_1 = \int_{z=0}^{1}\int_{y=0}^{1} \vec{\mathscr{J}} \cdot \hat{x}dydz\Big|_{x=1} = \int_{z=0}^{1}\int_{y=0}^{1} yzdydz\Big|_{x=1} = \tfrac{1}{4}\ \mathrm{A}.$$

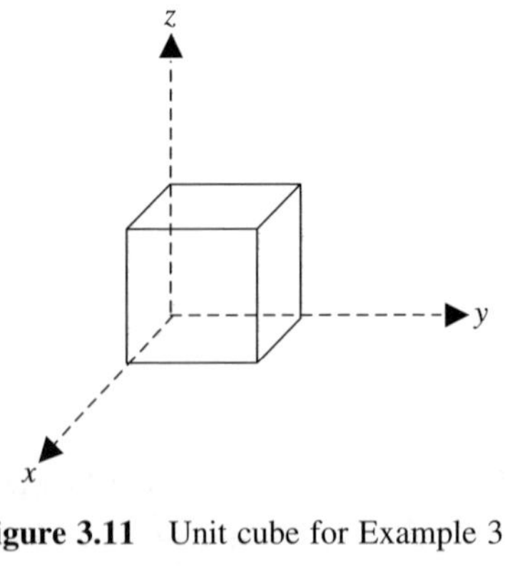

Figure 3.11 Unit cube for Example 3.9.

Current I_2 leaving back surface is

$$I_2 = \int_{z=0}^{1}\int_{y=0}^{1} \vec{\mathcal{J}} \cdot (-\hat{x})dydz\Big|_{x=0} = -\int_{z=0}^{1}\int_{y=0}^{1} yzdydz\big|_{x=1} = -\tfrac{1}{4}\text{ A}.$$

Current I_3 leaving right-side surface is

$$I_3 = \int_{z=0}^{1}\int_{x=0}^{1} \vec{\mathcal{J}} \cdot \hat{y}dxdz\Big|_{y=1} = \int_{z=0}^{1}\int_{x=0}^{1} yzdxdz\big|_{y=1} = \tfrac{1}{2}\text{ A}.$$

Current I_4 leaving left-side surface is

$$I_4 = \int_{z=0}^{1}\int_{x=0}^{1} \vec{\mathcal{J}} \cdot (-\hat{y})dxdz\Big|_{y=0} = -\int_{z=0}^{1}\int_{x=0}^{1} yzdxdz\big|_{y=0} = 0.$$

Current I_5 leaving the top surface is

$$I_5 = \int_{x=0}^{1}\int_{y=0}^{1} \vec{\mathcal{J}} \cdot \hat{z}dxdy\Big|_{z=1} = \int_{x=0}^{1}\int_{y=0}^{1} xzdxdy\big|_{z=1} = \tfrac{1}{2}\text{ A}.$$

Finally, current I_6 leaving the bottom surface is

$$I_6 = \int_{x=0}^{1}\int_{y=0}^{1} \vec{\mathcal{J}} \cdot (-\hat{z})dxdy\Big|_{z=0} = -\int_{x=0}^{1}\int_{y=0}^{1} xzdxdy\big|_{z=0} = 0.$$

Therefore the total current I leaving the closed surface is found to be

$$I = \tfrac{1}{4} - \tfrac{1}{4} + \tfrac{1}{2} + 0 + \tfrac{1}{2} + 0 = 1\text{ A}.$$

Alternatively, from (3.2.5),

$$\nabla \cdot \vec{\mathcal{J}} = \frac{\partial(yz)}{\partial x} + \frac{\partial(yz)}{\partial y} + \frac{\partial(xz)}{\partial z} = z + x\text{ A/m}^3 = -\frac{\partial \rho}{\partial t}.$$

Therefore the current leaving the unit cube is

$$I = \int_{z=0}^{1}\int_{y=0}^{1}\int_{x=0}^{1} (z + x)dxdydz = \frac{z^2}{2}\Big|_{z=0}^{1} + \frac{x^2}{2}\Big|_{x=0}^{1} = \frac{1}{2} + \frac{1}{2} = 1\text{ A}.$$

Time-Harmonic Fields

Electromagnetic fields and sources considered up to this point were assumed to be arbitrary function of time and space. The analysis can be simplified significantly with the assumption that these vary sinusoidally with time. It is to be noted that this assumption includes a large number of cases. Further this formulation can be extended to include nonsinusoidal cases via Fourier series or Fourier integrals, as necessary.

Consider a vector field $\vec{\alpha}(\vec{r}, t)$ that is sinusoidal with time. Using complex notations, it can be expressed as follows:

$$\vec{\alpha}(\vec{r}, t) = \vec{A}(\vec{r})\cos(\omega t + \theta) = \mathrm{Re}\Big[\{\vec{A}(\vec{r})e^{j\theta}\}e^{j\omega t}\Big]. \tag{3.2.6}$$

Therefore the time derivative of this vector can be written as

$$\frac{\partial}{\partial t}\vec{\alpha}(\vec{r}, t) = \mathrm{Re}\Big[\{j\omega\vec{A}(\vec{r})e^{j\theta}\}e^{j\omega t}\Big]. \tag{3.2.7}$$

This indicates that a complex vector $\vec{A}(\vec{r})e^{j\theta}$ (a phasor quantity) can be used in place of $\vec{\alpha}(\vec{r}, t)$, $j\omega$ can replace the time derivative, and $e^{j\omega t}$ can be suppressed during the field analysis. Further, it can be proved that a division by $j\omega$ will replace the time integral. The time dependence can be recovered after multiplying the given complex vector by $e^{j\omega t}$ and then extracting the real part of that. This is a familiar process used in ac circuit analysis to define the reactance of an inductor or a capacitor.

Using phasor representations for the field quantities, Equations (3.2.1) through (3.2.5) can be written as follows:

$$\nabla \times \vec{E}(\vec{r}) = -j\omega\vec{B}(\vec{r}), \tag{3.2.8}$$

$$\nabla \times \vec{H}(\vec{r}) = \vec{J}(\vec{r}) + j\omega\vec{D}(\vec{r}), \tag{3.2.9}$$

$$\nabla \cdot \vec{D}(\vec{r}) = \rho(\vec{r}), \tag{3.2.10}$$

$$\nabla \cdot \vec{B}(\vec{r}) = 0, \tag{3.2.11}$$

and

$$\nabla \cdot \vec{J}(\vec{r}) = -j\omega\rho(\vec{r}). \tag{3.2.12}$$

Upright print notation is now employed in place of script notation to distinguish the phasor field quantities. Further the space dependence of the field quantities will not be included explicitly from this point on in order to simplify the notation.

3.3 CONSTITUTIVE RELATIONS

A general analysis involves the evaluation of $\vec{E}$, $\vec{D}$, $\vec{H}$, and $\vec{B}$ for given sources (current and charge densities). Maxwell's equations represent only 8 scalar equations whereas unknown fields have 12 scalar components. Further the two Gauss laws can be found from the other two Maxwell equations and the equation of continuity. Therefore Maxwell's equations give only 6 independent scalar equations whereas 12 are needed to find 12 unknown field components. The constitutive relations provide the remaining 6 independent scalar equations as follows:

$$\vec{D} = \varepsilon_o\vec{E} + \vec{P}, \tag{3.3.1}$$

where

$$\vec{P} = \begin{cases} \overline{\overline{\chi_e}}\varepsilon_o\vec{E} & \text{for a linear medium} \\ \chi_e\varepsilon_o\vec{E} & \text{for a linear and isotropic medium.} \end{cases} \tag{3.3.2}$$

$\vec{P}$ is the polarization density vector, ε_o is permittivity of free space (8.854×10^{-12} F/m), $\overline{\overline{\chi_e}}$ is a dimensionless electrical susceptibility tensor, and χ_e is a dimensionless electrical susceptibility scalar.

Therefore, for a linear and isotropic medium,

$$\vec{D} = \varepsilon_o(1 + \chi_e)\vec{E} = \varepsilon\vec{E} = \varepsilon_o\varepsilon_r\vec{E}, \tag{3.3.3}$$

where ε_r is called relative permittivity (or dielectric constant) of the medium and ε is its permittivity in F/m. Similarly

$$\vec{B} = \mu_o(\vec{H} + \vec{M}), \tag{3.3.4}$$

where

$$\vec{M} = \begin{cases} \overline{\overline{\chi_m}}\vec{H} & \text{for a linear medium} \\ \chi_m\vec{H} & \text{for a linear and isotropic medium.} \end{cases} \tag{3.3.5}$$

$\vec{M}$ is called the magnetization density vector, μ_o is permeability of free space ($4\pi \times 10^{-7}$ H/m), $\overline{\overline{\chi_m}}$ is a dimensionless magnetic susceptibility tensor, and χ_m is a dimensionless magnetic susceptibility scalar.

Therefore, for a linear and isotropic medium,

$$\vec{B} = \mu_o(1 + \chi_m)\vec{H} = \mu\vec{H} = \mu_o\mu_r\vec{H}, \tag{3.3.6}$$

where μ is permeability of the medium and μ_r is its relative permeability. Also the current density term in (3.2.9) can have two parts, one due to an impressed source and the other due to conduction. Conduction current density can be found via Ohm's law as follows:

$$\vec{J} = \sigma\vec{E}, \tag{3.3.7}$$

where σ is conductivity of the medium in S/m.

Equation (3.2.9) can be also expressed as

$$\nabla \times \vec{H} = \vec{J}^e + \sigma\vec{E} + j\omega\varepsilon\,\vec{E} = \vec{J}^e + j\omega\varepsilon\left(1 - j\frac{\sigma}{\omega\varepsilon}\right)\vec{E} = \vec{J}^e + j\omega\varepsilon^*\vec{E}, \tag{3.3.8}$$

where

$$\varepsilon^* = \left(\varepsilon - j\frac{\sigma}{\omega}\right) = \varepsilon(1 - j\tan\delta) \longrightarrow \varepsilon_r^* = \frac{\varepsilon}{\varepsilon_o} - j\frac{\sigma}{\omega\varepsilon_o} = \varepsilon' - j\varepsilon'', \tag{3.3.9}$$

$$\tan\delta = \frac{\sigma}{\omega\varepsilon}. \tag{3.3.10}$$

$\vec{J}^e$ is the impressed current source density, ε^* is the complex permittivity of the material, ε_r^* is the complex relative permittivity, and $\tan\delta$ is what is known as a loss tangent. This loss tangent is a ratio of the magnitude of conduction current density to the magnitude of displacement current density in the medium.

The conduction current is defined here in a broad sense. When a dielectric material is subjected to time-varying fields, its bound electrons move back and forth following the Lorentz force. However, movement of the bound charges can lag behind the applied force at high frequencies. As a result there can be some power loss, which is represented by an equivalent conduction current as well. It is included here in the conduction current term. In other words, the conductivity is assumed to include a time-dependent term that is zero at dc but increases with the frequency.

Example 3.10

The electric field intensity in a source-free region is given as follows:

$$\vec{E} = \hat{z}4e^{-j(x-3y)}\ \mathrm{V/m}.$$

If $\varepsilon = \varepsilon_o$ and $\mu = \mu_o$ in that region, find the signal frequency.

Since all electromagnetic fields must satisfy Maxwell's equations, we first determine the corresponding magnetic field and then try to find the electric field back. We proceed as follows:

$$\nabla \times \vec{E} = -j\omega\mu_o\vec{H} \longrightarrow \vec{H} = -\frac{1}{j\omega\mu_o}\begin{vmatrix} \hat{x} & \hat{y} & \hat{z} \\ \dfrac{\partial}{\partial x} & \dfrac{\partial}{\partial y} & \dfrac{\partial}{\partial z} \\ 0 & 0 & 4e^{-j(x-3y)} \end{vmatrix} = -\frac{(12\hat{x}+4\hat{y})}{\omega\mu_o}e^{-j(x-3y)}$$

and

$$\nabla \times \vec{H} = j\omega\varepsilon_o\vec{E} \longrightarrow \vec{E} = -\frac{1}{j\omega\varepsilon_o}\begin{vmatrix} \hat{x} & \hat{y} & \hat{z} \\ \dfrac{\partial}{\partial x} & \dfrac{\partial}{\partial y} & \dfrac{\partial}{\partial z} \\ \dfrac{12}{\omega\mu_o}e^{-j(x-3y)} & \dfrac{4}{\omega\mu_o}e^{-j(x-3y)} & 0 \end{vmatrix}$$

$$= \hat{x}0 + \hat{y}0 + \hat{z}\frac{40}{\omega^2\mu_o\varepsilon_o}e^{-j(x-3y)}.$$

Therefore

$$\frac{40}{\omega^2\mu_o\varepsilon_o} = 4 \longrightarrow \omega = \frac{3.16}{\sqrt{\mu_o\varepsilon_o}} = 9.49 \times 10^8\ \mathrm{rad/s}.$$

3.4 BOUNDARY CONDITIONS

The electromagnetic field problems considered so far involved a single medium of infinite extent. However, there are numerous real-world problems that require analysis of electromagnetic fields in the presence of more than one medium. Therefore it is important to know the behavior of electromagnetic fields across the interface of two media.

Consider an interface of the two media, as illustrated in Figure 3.12. Electrical characteristics of these media are as indicated. When (3.1.2) is applied to an infinitesimal area across the interface and its side Δh is reduced to zero under the limit, the result is the relation

$$(\vec{\mathscr{H}}_2 - \vec{\mathscr{H}}_1) \times \hat{n} = \vec{\mathscr{J}}_s, \tag{3.4.1}$$

where subscripts 1 and 2 are used to indicate the magnetic field intensity in the corresponding medium, $\mathscr{J}_s$ represents the surface current density in A/m on the boundary, and the unit vector $\hat{n}$ is directed into medium 1.

Similarly, when (3.1.1) is applied to this infinitesimal area across the interface and its side Δh is reduced to zero under the limit, the result is the relation

$$\hat{n} \times (\vec{\mathscr{E}}_2 - \vec{\mathscr{E}}_1) = 0. \tag{3.4.2}$$

Again, subscripts 1 and 2 are used to indicate the electric field intensity in the corresponding medium. Now consider a pillbox volume across the interface of the two media, as shown in Figure 3.13. When (3.1.3) is applied to this volume and the side Δh is reduced to zero under the limit, the results is the relation

$$\hat{n} \cdot (\vec{\mathscr{D}}_1 - \vec{\mathscr{D}}_2) = \rho_s, \tag{3.4.3}$$

where subscripts 1 and 2 are used to indicate the electric flux density in the corresponding medium, ρ_s represents the surface charge density in C/m^2 on the boundary, and the unit vector $\hat{n}$ is directed into medium 1.

Similarly the following relation is found via (3.1.4) for the magnetic flux densities in the two media:

$$\hat{n} \cdot (\vec{\mathscr{B}}_1 - \vec{\mathscr{B}}_2) = 0. \tag{3.4.4}$$

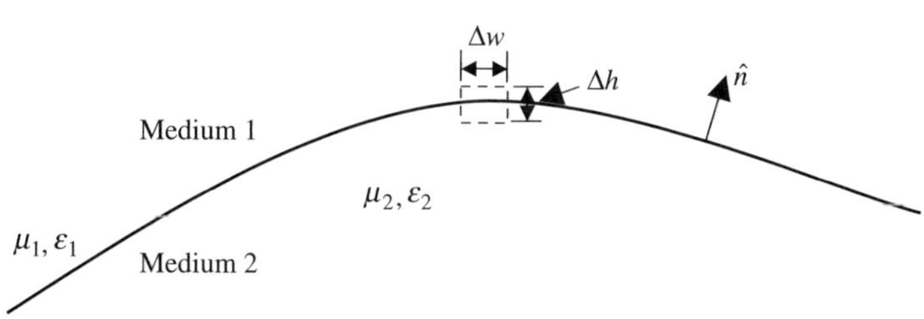

Figure 3.12 An infinitesimal area across the interface of medium 1 and medium 2.

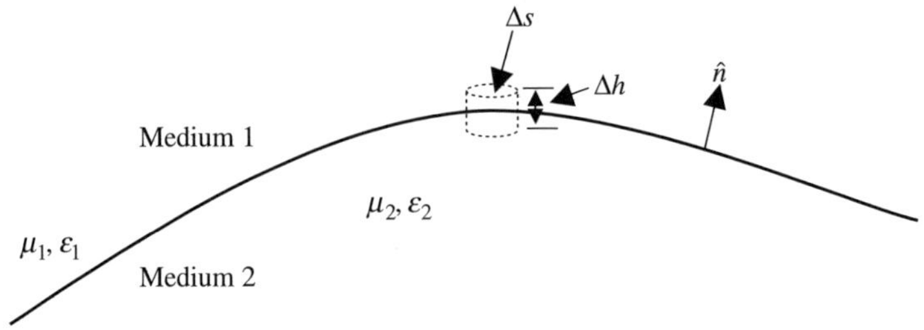

Figure 3.13 Pillbox volume across the interface of medium 1 and medium 2.

Equations (3.4.1) through (3.4.4) can be used to determine the electromagnetic fields from one medium to the other medium across the boundary. It is to be noted that (3.4.3) and (3.4.4) are formulated via Gauss's laws and are not independent. Therefore these conditions are automatically satisfied if (3.4.1) and (3.4.2) are met. Further, if medium 2 is a perfect conductor, then the fields inside it vanish. Therefore the boundary conditions (3.4.1) through (3.4.4) reduce to

$$\hat{n} \times \vec{\mathscr{H}}_1 = \vec{\mathscr{J}}_s, \tag{3.4.5}$$

$$\hat{n} \times \vec{\mathscr{E}}_1 = 0, \tag{3.4.6}$$

$$\hat{n} \cdot \vec{\mathscr{D}}_1 = \rho_s, \tag{3.4.7}$$

and

$$\hat{n} \cdot \vec{\mathscr{B}}_1 = 0. \tag{3.4.8}$$

If the second medium is a good conductor (finite conductivity), then the fields penetrate mostly up to its skin depth. In such a case the following Leontovich's impedance boundary condition can be used provided that the radius of curvature of the surface is significantly greater than the skin depth $(2/\omega\mu\sigma)^{1/2}$:

$$\vec{\mathscr{E}}_t = Z_s(\hat{n} \times \vec{\mathscr{H}}), \tag{3.4.9}$$

where

$$Z_s = \sqrt{\frac{\mu}{\varepsilon}} \approx \sqrt{\frac{j\mu\omega}{\sigma}}. \tag{3.4.10}$$

Example 3.11

The region $x > 0$ is a perfect dielectric with $\varepsilon_r = 2.25$ while the region $x < 0$ is a free space. At the interface, subscript 1 denotes the field components on the $+x$ side of the boundary and the subscript 2 on the $-x$ side. If $\vec{D}_1 = \hat{x} + 2\hat{y}\,\text{C/m}^2$, find $\vec{D}_2$, $\vec{E}_1$, and $\vec{E}_2$.

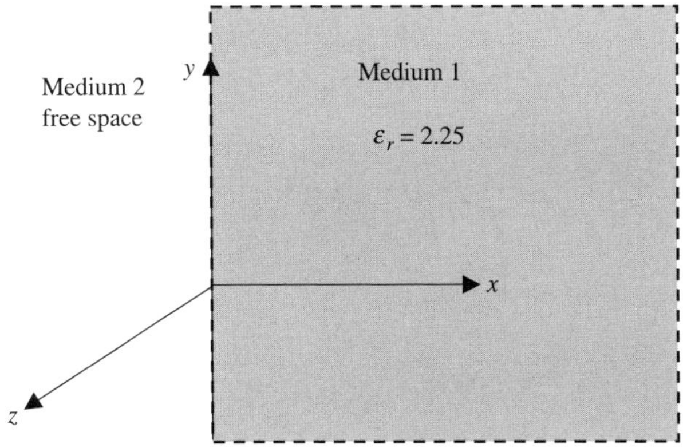

Figure 3.14 Geometry used for Example 3.11.

Conditions stated in the problem are illustrated in Figure 3.14. The interface is $x = 0$ plane. Therefore the x component will be normal while the y component tangential to this plane:

$$\vec{D}_1 = \varepsilon_1 \vec{E}_1 \rightarrow \vec{E}_1 = \frac{\vec{D}_1}{\varepsilon_1} = \frac{1}{2.25\varepsilon_o}\hat{x} + \frac{2}{2.25\varepsilon_o}\hat{y} \text{ V/m}.$$

From the boundary conditions, the x component of electric flux density and the y component of electric field intensity in medium 2 will be same as in medium 1. Hence

$$D_{2x} = 1 \quad \text{and} \quad E_{2y} = \frac{2}{2.25\varepsilon_o}.$$

Therefore

$$\vec{D}_2 = \hat{x} + \frac{2}{2.25}\hat{y} \text{ C/m}^2.$$

and

$$\vec{E}_2 = \frac{\vec{D}_2}{\varepsilon_o} = \frac{1}{\varepsilon_o}\hat{x} + \frac{2}{2.25\varepsilon_o}\hat{y} \text{ V/m}.$$

Example 3.12

A sphere of 2 m radius is made of a perfect dielectric material (medium 1). It is surrounded by free space (medium 2). The electric field intensities in the two media are given as follows:

$$\vec{E}_1 = E_{01}(\hat{r}\cos\theta - \hat{\theta}\sin\theta) \qquad \text{for} \quad r \leq 2\,\text{m}$$

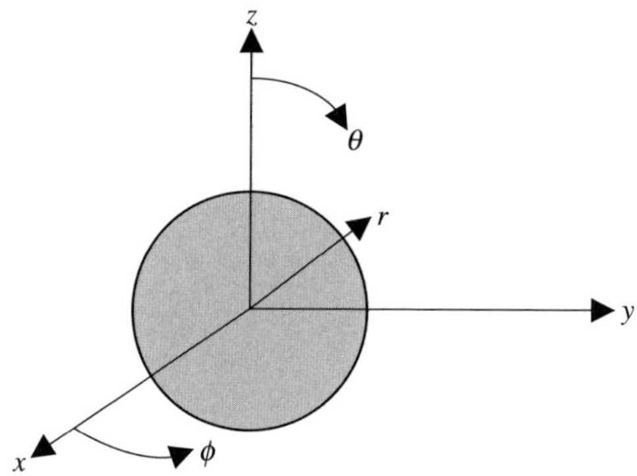

Figure 3.15 Geometry used for Example 3.12.

and

$$\vec{E}_2 = E_{02}\left(\hat{r}\left\{1+\frac{8}{r^3}\right\}\cos\theta - \hat{\theta}\left\{1-\frac{4}{r^3}\right\}\sin\theta\right) \qquad \text{for} \quad r \geq 2\,\text{m}.$$

Find the permittivity of the spherical medium.

As indicated in Figure 3.15, the r components of the electric fields are normal while the θ components are tangential at the interface. Therefore

$$-\sin\theta\, E_{01} = -\left(1-\tfrac{4}{8}\right)\sin\theta E_{02} \longrightarrow E_{02} = 2E_{01}$$

and

$$\varepsilon_1 E_{01}\cos\theta = \varepsilon_o E_{02} 2\cos\theta \longrightarrow \varepsilon_1 = 2\varepsilon_o \frac{E_{02}}{E_{01}} = 4\varepsilon_o\,\text{F/m}.$$

3.5 LORENTZ'S FORCE EQUATION

This equation facilitates determining the force experienced by a charged particle in presence of electromagnetic fields. According to the Lorentz equation, if an electrical charge q C is subjected to an electric field, then it experiences a force that is equal to the charge times the electric field's intensity $\vec{E}$. The direction of this force is same as that of the electric field's intensity. On the other hand, if the electrical change is subjected to a magnetic field with a flux density of $\vec{B}$, then the charge experiences a force only if it is moving. In this case the direction of the force is orthogonal to both the velocity of the charge and the direction of magnetic flux. These two results can be expressed mathematically as follows:

$$\vec{F} = q(\vec{E} + \vec{v} \times \vec{B}). \tag{3.5.1}$$

Equation (3.5.1) is called the Lorentz force equation. Thus the total force comprises of two parts. This equation is significant in the sense that it links electrical and mechanical quantities. A number of engineering concepts and applications are understandable via this force equation. Some of these applications are illustrated in this section.

Motion of a Charged Particle in a Uniform Electric Field

For simplicity, we consider the motion of a charged particle only in the presence of a uniform electric field. Therefore the second term (the one with the magnetic flux density) in (3.5.1) is assumed to be zero. As illustrated in Figure 3.16, there are two conducting plates on the y-z plane that have a separation of d m. A voltage V is applied to this set of parallel plates that produces an electric field as follows:

$$\vec{E} = \hat{x}E_x = -\hat{x}\frac{V}{d}\ \text{V/m}. \qquad (3.5.2)$$

As illustrated, a particle with an electrical charge of q C and velocity $\vec{v}_o$ enters this system. Since there is an electric field in the medium, the particle experiences a force that can be expressed using (3.5.1) as follows:

$$\vec{F} = -\hat{x}q\frac{V}{d}\ \text{N} \qquad (3.5.3)$$

By Newton's law, this force will accelerate the particle only in x direction. Therefore

$$\vec{F} = \hat{x}m\frac{dv_x}{dt} = -\hat{x}\frac{qV}{d} \quad \to \quad \frac{dv_x}{dt} = -\frac{qV}{md}.$$

Note that this formulation is valid only when the velocity v_x is very small in comparison to speed of light (so that the relativistic effects are negligible). Recall that

$$m = \frac{m_o}{\sqrt{1-(v_x/c)^2}},$$

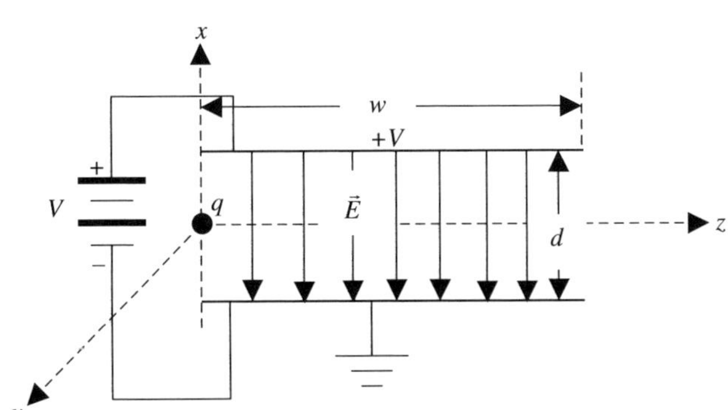

Figure 3.16 A charged particle moving through a uniform electric field.

where m_o is mass of the particle when it is stationary and c is speed of light in free space.

On integrating, we get

$$v_x = -\int \frac{qV}{md}\,dt = -\frac{qV}{md}t + C_1.$$

Because the initial velocity $\vec{v}(t=0) = \vec{v}_o = \hat{x}v_{xo} + \hat{y}v_{yo} + \hat{z}v_{zo}$, the integration constant C_1 can be easily evaluated. Hence

$$v_x = -\frac{qV}{md}t + v_{xo}. \tag{3.5.4}$$

Further

$$v_x = \frac{dx}{dt} = -\frac{qV}{md}t + v_{xo} \longrightarrow x = -\int \left(\frac{qV}{md}t + v_{xo}\right)dt = -\frac{qV}{md}t^2 + v_{xo}t + C_2.$$

Because the particle starts from (0, 0, 0), $C_2 = 0$. Therefore

$$x = -\frac{qV}{md}t^2 + v_{xo}t. \tag{3.5.5}$$

The total velocity and the location of the particle inside the parallel plates is then written as

$$\vec{v}(t) = \hat{x}\left(-\frac{qV}{md}t + v_{xo}\right) + \hat{y}v_{yo} + \hat{z}v_{zo} \tag{3.5.6}$$

and

$$\hat{r}(t) = \hat{x}\left(-\frac{qV}{md}t^2 + v_{xo}t\right) + \hat{y}v_{yo}t + \hat{z}v_{zo}t. \tag{3.5.7}$$

For v_{xo} and v_{yo} zero, $t = z/v_{zo}$. Therefore (3.5.5) gives

$$x = -\frac{qV}{md}\left(\frac{z}{v_{zo}}\right)^2. \tag{3.5.8}$$

Thus the particle follows a parabolic path up to $z = w$ (between the parallel plates) and then traces a straight line that is tangential to this parabola at the exit point.

Example 3.13

As shown in Figure 3.17, the anode accelerates an electron emitted by the cathode in an electron gun. If its initial velocity is zero and the anode is at 182 V with respect to the cathode, then find the electron's final velocity.

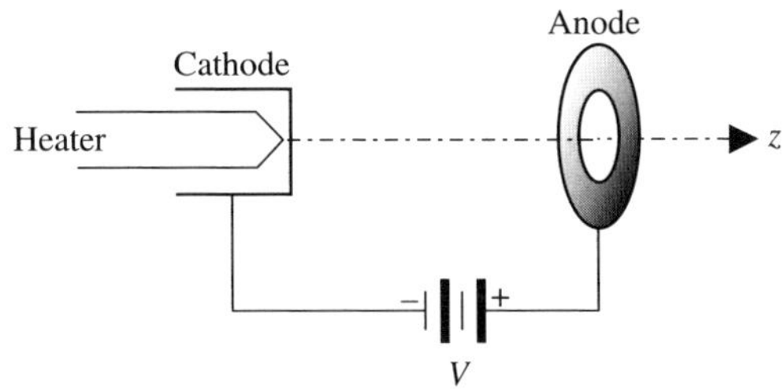

Figure 3.17 Simplified arrangement of an electron gun.

Since $\vec{F} = q\vec{E}$, the work done in moving a particle over a distance $d\vec{\ell}$ is written as

$$\vec{F} \cdot d\vec{\ell} = q\vec{E} \cdot d\vec{\ell}.$$

Using Newton's law, this expression can be rewritten as follows:

$$\vec{a} \cdot d\vec{\ell} = \frac{q}{m}\vec{E} \cdot d\vec{\ell} \longrightarrow \frac{d\vec{v}}{dt} \cdot d\vec{\ell} = v dv = \frac{q}{m}\vec{E} \cdot d\vec{\ell},$$

where $\vec{v}$ is the velocity and $\vec{a}$ is the acceleration of the particle.

As the particle moves from point a to point b,

$$\int_{v_a}^{v_b} v dv = \frac{q}{m}\int_a^b \vec{E} \cdot d\vec{\ell} \longrightarrow \frac{v_b^2 - v_a^2}{2} = -\frac{qV_{ba}}{m}.$$

Note that $-\int_a^b \vec{E} \cdot d\vec{\ell} = V_{ba}$ is the potential at point b with respect to point a. If the charge is negative (as in case of an electron), then the right-hand side of the equation becomes positive. The positive or negative sign can be traced back to whether the energy is supplied or released in moving the charge q. Assume zero initial velocity so that this relation can be simplified to

$$v_b = \sqrt{\frac{2qV_{ba}}{m}} \text{ m/s}.$$

Since the electronic charge and mass are 1.602×10^{-19} C and 9.11×10^{-31} kg,

$$v_b = \sqrt{\frac{2qV_{ba}}{m}} = \sqrt{\frac{2 \times 1.602 \times 10^{-19} \times 182}{9.11 \times 10^{-31}}} = 8.0006 \times 10^6 \text{ m/s}.$$

Example 3.14

Electron generated by the electron gun described in Example 3.13 enters between the parallel plates of Figure 3.16. There is a fluorescent screen placed at $z = 40$ cm. For $w = d = 1$ cm and $V = 5$ V, find the point where this electron hits the screen. Ignore the fringing fields of the deflecting plates.

For $v_{xo} = 0$, v_x at $z = w$ may be found from (3.5.4) as follows:

$$v_x\left(t = \frac{w}{v_{zo}}\right) = -\frac{qV}{md} \times \frac{w}{v_{zo}}.$$

Since the charged particle velocity does not change beyond this point, it traces a straight line that make an angle θ with z-axis. This angle may be found as

$$\theta = \tan^{-1}\left(\frac{v_x(z = w)}{v_{zo}}\right) = \tan^{-1}\left(-\frac{qVw}{mdv_{zo}^2}\right).$$

For the screen to be at $z_s = 40$ cm such that $w \ll z_s$, the x coordinate x_s of the point where the charge particle hits the screen can be approximated as

$$x_s \approx z_s \tan\theta = -\frac{qVwz_s}{mdv_{zo}^2}.$$

Therefore for the present case

$$x_s = \frac{1.602 \times 10^{-19} \times 5 \times 0.01 \times 0.4}{9.11 \times 10^{-31} \times 0.01 \times (8.0006 \times 10^6)^2}\ \mathrm{m} = 0.0055\ \mathrm{m} = 0.55\ \mathrm{cm}.$$

Motion of a Charged Particle in a Uniform Magnetic Field

Consider a particle moving along the z-axis with velocity v m/s. The particle enters a region that has a y-directed magnetic field of BT. On entering, it experiences a force due to the magnetic field that can be found from Lorentz's force equation as follows:

$$\vec{F}_o = q(\hat{z}v) \times (\hat{y}B) = -\hat{x}qvB. \tag{3.5.9}$$

Therefore the direction of the moving charge changes. Since $\vec{F} = q\vec{v} \times \vec{B}$ has to hold everywhere, the moving charge keeps changing its direction as illustrated in Figure 3.18. This movement traces a circle of radius R provided that the magnetic region is sufficiently large. The centrifugal force acting on the charged particle balances this radial inward force due to the magnetic field. Hence

$$\frac{mv^2}{R} = qvB \longrightarrow R = \frac{mv}{qB}. \tag{3.5.10}$$

The cyclotron frequency (also known as the gyro frequency) of the charged particle is defined as the number of revolution per second. Therefore

$$f_c = \frac{v}{2\pi R} = \frac{qB}{2\pi m},$$

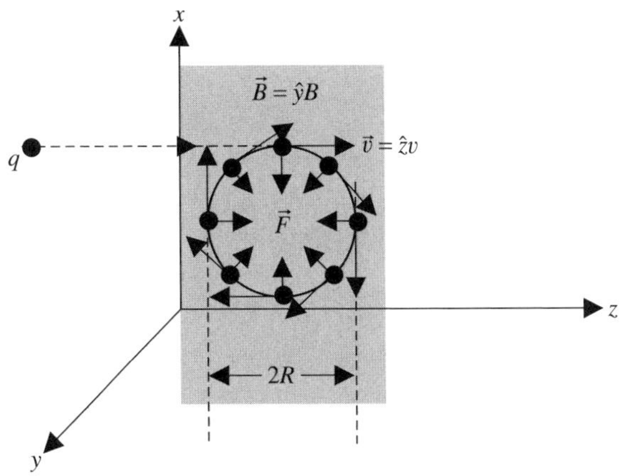

Figure 3.18 Charged particle moving through a uniform magnetic field.

or

$$\omega_c = 2\pi f_c = \frac{qB}{m}. \tag{3.5.11}$$

The force acting on the charged particle is always perpendicular to its velocity, and therefore there is no work done by the magnetic field.

Magnetic fields have been used in television picture tubes. There are a number of other applications including the separation of isotopes that have different velocities and the masses. A typical procedure used for such separation is as follows: An ionized compound of isotopes is accelerated through a constant voltage. If all particles have equal electrical charge but different masses, then they acquire different velocities. As these particles pass through a magnetic field in a perpendicular direction as shown in Figure 3.19, their paths take on different curvatures. This phenomenon is explained further in Example 3.16.

Example 3.15

A particle with a charge $q = 1.602 \times 10^{-19}$ C and a mass $m = 1.6725 \times 10^{-27}$ kg is at rest. It is accelerated for 2 μs by an electric field of 5 kV/m on the *xz*-plane. The particle then enters a magnetic field where $\vec{B} = \hat{y}5$ mT, as illustrated in Figure 3.18. Find the velocity before it enters the magnetic field region, and the radius of the circle that it will trace in the magnetic field.

Since

$$\vec{a} = \frac{q}{m}\vec{E} \longrightarrow \frac{d\vec{v}}{dt} = \frac{q}{m}\vec{E} \longrightarrow \int_{v_o}^{v_1} dv = \int_0^{t_1} \frac{q}{m}\vec{E}\,dt \longrightarrow \vec{v}_1 - \vec{v}_o = \frac{q}{m}\vec{E}t,$$

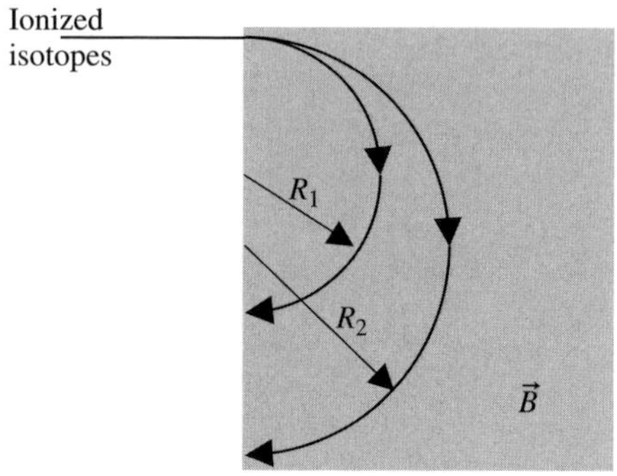

Figure 3.19 Ionized isotopes moving through a uniform magnetic field.

for $v_o = 0$,

$$\vec{v}_1 = \frac{q}{m}\vec{E}t = \frac{1.602 \times 10^{-19} \times 5000 \times 2 \times 10^{-6}}{1.6725 \times 10^{-27}}\hat{z} = \hat{z}\, 9.5785 \times 10^5 \text{ m/s}.$$

Now from (3.5.10),

$$R = \frac{mv}{qB} = \frac{1.6725 \times 10^{-27} \times 9.5785 \times 10^5}{1.602 \times 10^{-19} \times 5 \times 10^{-3}}\text{ m} = 2\text{ m}.$$

Example 3.16

A compound of isotopes is vaporized and then ionized. The charge on the ions is 1.602×10^{-19} C. A constant 1000 V accelerates these ions before they enter a region within a magnetic field of 5 mT in the perpendicular direction, as shown in Figure 3.19. If two of the charge particles have masses of $m_1 = 10^{-25}$ kg and $m_2 = 10^{-24}$ kg, determine the radii of the circles they trace.

Since $\frac{1}{2}mv^2 = qV$, both particles acquire the same energy, but they will have different velocities. These velocities can be found as follows:

$$v_1 = \sqrt{\frac{2qV}{m_1}} = \sqrt{\frac{2 \times 1.602 \times 10^{-19} \times 1000}{10^{-25}}} = 5.6604 \times 10^4 \text{ m/s}$$

and

$$v_2 = \sqrt{\frac{2qV}{m_2}} = \sqrt{\frac{2 \times 1.602 \times 10^{-19} \times 1000}{10^{-24}}} = 1.79 \times 10^4 \text{ m/s}.$$

For $B = 5$ mT, from (3.5.10) we find that

$$R_1 = \frac{m_1 v_1}{qB} = \frac{10^{-25} \times 5.6604 \times 10^4}{1.602 \times 10^{-19} \times 5 \times 10^{-3}}\text{ m} = 7.07\text{ m}$$

and

$$R_2 = \frac{m_2 v_2}{qB} = \frac{10^{-24} \times 1.79 \times 10^4}{1.602 \times 10^{-19} \times 5 \times 10^{-3}} \text{m} = 22.35 \text{ m}.$$

Motion of a Charged Particle in Uniform Electric and Magnetic Fields

We considered so far the cases when a charged particle moves in presence of only a static electric field or a magnetic field. Now we consider this in presence of both fields simultaneously. In order to keep the analysis simple, assume that $\vec{B} = \hat{y}B_o$ T and $\vec{E} = \hat{x}E_o$ V/m. Using the Lorentz force equation, we find that

$$\vec{F} = m\vec{a} = q[\hat{x}E_o + (\hat{x}v_x + \hat{y}v_y + \hat{z}v_z) \times \hat{y}\vec{B}] \rightarrow$$

$$\frac{d\vec{v}}{dt} = \frac{q}{m}[\hat{x}E_o + (\hat{z}v_x + 0 - \hat{x}v_z)B_o]. \tag{3.5.12}$$

Equating vector components on the two sides of this equation, we get

$$\frac{dv_x}{dt} = \frac{qE_o}{m} - \frac{qB_o}{m}v_z, \tag{3.5.13}$$

$$\frac{dv_y}{dt} = 0, \tag{3.5.14}$$

and

$$\frac{dv_z}{dt} = \frac{qB_o v_x}{m}. \tag{3.5.15}$$

According to (3.5.14) the particle does not accelerate in the y direction (as expected, since we assumed that the magnetic field is directed this way). Therefore the particle continues with the initial velocity v_{yo} (if not zero) in this direction. Equations (3.5.13) and (3.5.15) are solved, as follows, to determine the other two velocity components.

Differentiating (3.5.15) with respect to time, we get

$$\frac{d^2 v_z}{dt^2} = \frac{qB_o}{m}\frac{dv_x}{dt}.$$

Substituting (3.5.13) on its right-hand side, we have

$$\frac{m}{qB_o}\frac{d^2 v_z}{dt^2} = \frac{qE_o}{m} - \frac{qB_o}{m}v_z,$$

or

$$\frac{d^2 v_z}{dt^2} + \left(\frac{qB_o}{m}\right)^2 v_z = \left(\frac{q}{m}\right)^2 B_o E_o. \tag{3.5.16}$$

Since $qB_o/m = \omega_c$, (3.5.16) can be rewritten as follows:

$$\frac{d^2 v_z}{dt^2} + \omega_c^2 v_z = \left(\frac{q}{m}\right)^2 B_o E_o. \tag{3.5.17}$$

Equation (3.5.17) is a nonhomogeneous linear differential equation that can be solved using one of the standard techniques. Since its general solution has two parts—a solution to the corresponding homogeneous equation and its particular solution—we proceed with the homogeneous equation.

For the homogeneous equation $(d^2 v_z/dt^2) + \omega_c^2 v_z = 0$, two possible solutions can be found easily as $v_{z1} = \sin(\omega_c t)$ and $v_{z2} = \cos(\omega_c t)$. Hence the general solution of (3.5.17) may be written as

$$v_z = C_1 \sin(\omega_c t) + C_2 \cos(\omega_c t) + v_{zp}. \tag{3.5.18}$$

where C_1 and C_2 are the integration constants, and v_{zp} is the particular solution that can be found through the method of variation of parameters.

The method of finding variation of the parameters proceeds as follows: We take the Wronskian W to be

$$W = \begin{vmatrix} v_{z1} & v_{z2} \\ v'_{z1} & v'_{z2} \end{vmatrix} = \begin{vmatrix} \sin(\omega_c t) & \cos(\omega_c t) \\ \omega_c \cos(\omega_c t) & -\omega_c \sin(\omega_c t) \end{vmatrix} = -\omega_c[\sin^2(\omega_c t) + \cos^2(\omega_c t)] = -\omega_c.$$

From this the particular solution is found to be

$$v_{zp} = -\sin(\omega_c t)\int \frac{\cos(\omega_c t)(q/m)^2 B_o E_o}{-\omega_c} dt + \cos(\omega_c t)\int \frac{\sin(\omega_c t)(q/m)^2 B_o E_o}{-\omega_c} dt,$$

or

$$v_{zp} = \frac{\sin(\omega_c t)}{\omega_c}\left(\frac{q}{m}\right)^2 B_o E_o \frac{\sin(\omega_c t)}{\omega_c} - \frac{\cos(\omega_c t)}{\omega_c}\left(\frac{q}{m}\right)^2 B_o E_o \left(-\frac{\cos(\omega_c t)}{\omega_c}\right),$$

or

$$v_{zp} = \left(\frac{q}{m\omega_c}\right)^2 B_o E_o = \frac{E_o}{B_o}.$$

Therefore the general solution of (3.5.17) is found to be

$$v_z = \frac{E_o}{B_o} + C_1 \sin(\omega_c t) + C_2 \cos(\omega_c t). \tag{3.5.19}$$

Then, using (3.5.15), we get

$$v_x = \frac{m}{qB_o}\frac{dv_z}{dt} = \frac{m}{qB_o}[C_1 \omega_c \cos(\omega_c t) - C_2 \omega_c \sin(\omega_c t)]. \tag{3.5.20}$$

Next, the integration constants are determined from initial velocities (the initial conditions). For $v_x(t = 0) = v_{xo}$ and $v_z(t = 0) = v_{zo}$, (3.5.19) and (3.5.20) give

$$C_1 = v_{xo}$$

and

$$C_2 = v_{zo} - \frac{E_o}{B_o}.$$

Therefore

$$v_z = \frac{E_o}{B_o} + v_{xo} \sin(\omega_c t) + \left(v_{zo} - \frac{E_o}{B_o}\right) \cos(\omega_c t) \tag{3.5.21}$$

and

$$v_x = \frac{m}{qB_o}\left[v_{xo}\omega_c \cos(\omega_c t) - \left(v_{zo} - \frac{E_o}{B_o}\right)\omega_c \sin(\omega_c t)\right]. \tag{3.5.22}$$

Hall Effect

Consider a metallic bar of cross section $w \times h$ immersed in an x-directed magnetic field, as shown in Figure 3.20. If there is an electron moving through the bar with a velocity v_o in the $-z$ direction, then it will experience the Lorentz force in the $+y$ direction (due to $q = -e$, the negative charge on an electron). Mathematically

$$\vec{B} = \hat{x}B_o, \tag{3.5.23}$$

$$\vec{v} = -\hat{z}v_o, \tag{3.5.24}$$

$$\vec{F} = -e(-\hat{z}v_o) \times (\hat{x}B_o) = \hat{y}ev_oB_o. \tag{3.5.25}$$

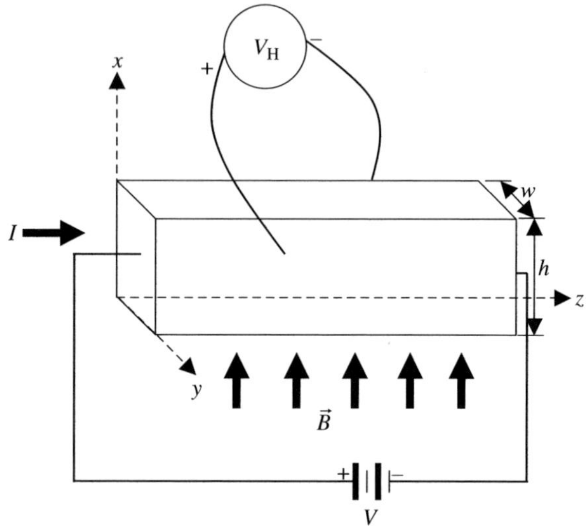

Figure 3.20 Simplified setup to demonstrate the Hall effect.

The electric field created by this effect is given as follows:

$$\vec{E} = \frac{\vec{F}}{e} = \hat{y} v_o B_o.$$

The corresponding voltage is found to be

$$V_H = -\int_a^b \vec{E} \cdot d\vec{\ell} = -v_o B_o w \text{ V}. \tag{3.5.26}$$

This is known as the Hall voltage after the scientist who first observed it. Note that the negative sign indicates the polarity of the voltage on the front side of the bar with respect to its back. In other words, the dc voltmeter shown in the figure will give a negative reading.

Since there is a current I flowing through the bar as shown in the figure, there is a stream of electrons flowing in $-z$ direction. If the electron density in the bar is n per m^3 then the current density J is found to be nev_o A/m^2. Since

$$I = Jwh = nev_o wh \longrightarrow v_o = \frac{I}{newh},$$

(3.5.26) may be rewritten as follows:

$$V_H = -\frac{IB_o}{neh} \text{ V}. \tag{3.5.27}$$

Note that if this current is due to the motion of positive charges (e.g., holes in a p-type semiconductor), then (3.5.27) will result in a positive Hall voltage. In case of intrinsic semiconductors, the Hall voltage will be zero because of equal concentrations of electrons and holes. Thus the Hall effect can be used to identify the type of a semiconductor as well as its charge density. Further there are Hall probes available commercially for sensing the unknown magnetic fields that are directly related to the Hall voltage, as (3.5.27) indicates.

Example 3.17

As shown in Figure 3.20, there is a 5 mA current flowing through the rectangular block of a semiconductor. It is subjected to an x-directed magnetic field of 5 T. If its cross section is a square of 1 cm on each side and the Hall voltage is found to be 1 mV, then find the concentration of charge (the charge density) and also the their polarity.

Since the meter reading is positive in this case, the semiconductor block is a p-type extrinsic semiconductor. Substituting a commonly used notation p for concentration of holes (in place of n for the electrons) into (3.5.27) gives

$$p = \frac{IB_o}{ehV_H} = \frac{5 \times 10^{-3} \times 5}{1.602 \times 10^{-19} \times 10^{-2} \times 10^{-3}} = 15.6055 \times 10^{21}\ \text{m}^{-3}.$$

dc Generators and Motors

Consider a conducting rigid rectangular loop with sides $a \times b$ and the commutator, as shown in Figure 3.21a. The loop is exposed to an x-directed uniform magnetic field with a flux density B_o T. The same kind of magnetic field can be generated by a permanent magnet, as illustrated in Figure 3.21b. However, for the time being, assume that the dc source is not connected to the commutator and the loop

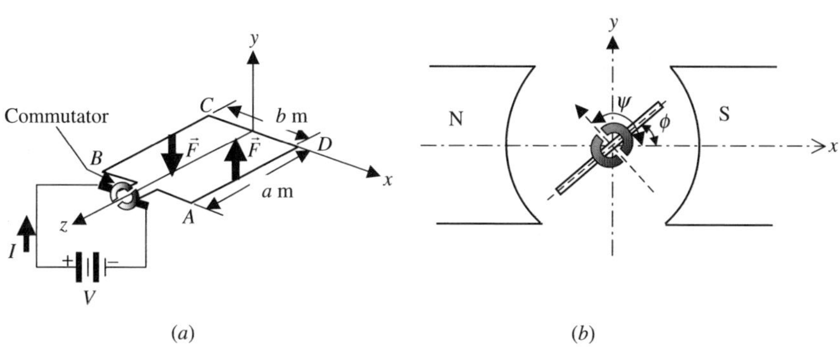

Figure 3.21 Rotating rectangular loop with a commutator to demonstrate the working principle of dc motors and generators.

is lying on the xz-plane. In this case the magnetic flux passing through the loop is zero (because the loop area, which is perpendicular to the magnetic flux, is zero). If instead the loop is repositioned to lie on the yz-plane, then the magnetic flux that passes through the loop becomes $a \times b \times B_o$ Wb. In other words, if the loop is rotated via an external force, the magnetic flux passing through the loop changes, and the emf induced as a result of this follows Faraday's law. We considered this situation earlier in Example 3.3. In that case there were two slip rings on the loop in place of the commutator. It can be easily verified that the commutator flips the negative half cycle of the sinusoidal voltage to the positive side, making the voltage unidirectional. The same result can be found via the Lorentz force equation as follows: A conductor is moving with a velocity $\vec{v}$ in a magnetic field $\vec{B}$, and the Lorentz force $q\vec{v} \times \vec{B}$ is applied to the charge in that loop. As a result the free electrons of the conductor move to one side leaving behind a positive charge. This charge distribution creates a potential difference (and hence an electric field) that exerts a $q\vec{E}$ force on the moving charges. In equilibrium, the two forces balance each other out:

$$\vec{F} = q\vec{v} \times \vec{B} \longrightarrow \frac{\vec{F}}{q} = \vec{E} = \vec{v} \times \vec{B} \longrightarrow V_{ba} = -\int_a^b \vec{E} \cdot d\vec{\ell} = -\int_a^b (\vec{v} \times \vec{B}) \cdot d\vec{\ell}.$$

In case of a closed circuit this expression can be written as follows:

$$V = -\oint_\Gamma (\vec{v} \times \vec{B}) \cdot d\vec{\ell}\,\text{V}. \tag{3.5.28}$$

This emf is called a motional or a flux-cutting electromotive force. Because the electric field intensity is directed from the positive to the negative potential, a negative sign is needed to indicate correct polarity of this emf.

Next, before discussing the basic principle of a dc motor, we analyze further the case of a conductor with its cross section A m^2 immersed in a uniform magnetic field. The conductor's electronic charge carriers have density n m^{-3} and are moving with velocity v m/s along its length $d\ell$. The force acting on this loop can be found by using the Lorentz force equation:

$$d\vec{F} = q\vec{v} \times \vec{B} = -nA|d\vec{\ell}|e\vec{v} \times \vec{B} = -nA|\vec{v}|ed\vec{\ell} \times \vec{B} = Id\vec{\ell} \times \vec{B}\,\text{N}, \tag{3.5.29}$$

where e is the charge of an electron and I is the current through the conductor. Note that the velocity $\vec{v}$ is in the direction of length $d\vec{\ell}$ and that $-nA|\vec{v}|e = I$.

To determine the total magnetic force on a closed circuit of contour Γ we integrate equation (3.5.29). Hence

$$\vec{F} = I\oint_\Gamma d\vec{\ell} \times \vec{B} = -I\oint_\Gamma \vec{B} \times d\vec{\ell}\,\text{N}. \tag{3.5.30}$$

For a conductor length b that makes an angle θ with the magnetic field, the magnitude of the force is

$$|\vec{F}| = Ib|\vec{B}|\sin\theta\,\text{N}.$$

If there are N such conductors with the current in the same direction, then the total magnetic force is added:

$$|\vec{F}| = NIb|\vec{B}|\sin\,\theta\,\mathrm{N}. \tag{3.5.31}$$

Now consider the case where the dc source is connected as shown in Figure 3.21*a* but the external force that was rotating the loop is removed. This time there is a current I that flows through the loop. It can be easily found that the magnetic field parallel to the loop (the surface bound by the loop) produces a torque about the axis of the loop that is perpendicular to it. The fields normal to the surface, however, produce forces that try to move the side conductors either inward or outward. The loop experiences no net force in any of these situations. Because the loop is lying on xz-plane, its arms BC and DA are perpendicular to the x-directed magnetic field while CD and AB are in the same direction. Therefore the latter two sides experience no force at all. Nevertheless, sides BC and DA experience a force that is equal in magnitude but opposite in direction, as shown in the figure. There is then no net force applied to the loop but the two forces create a torque that rotates the loop counterclockwise. Once the loop is rotated, components of the force responsible for the torque change by a factor of $\cos\phi$ because the forces only normal to the loop can produce it. Therefore the torque T is found to be

$$\vec{T} = 2\left(\frac{b}{2}\hat{x}\right) \times (\hat{y}IaB_o\cos\phi) = \hat{z}IabB_o\cos\phi = \hat{z}IabB_o\cos(\psi - 90^\circ)$$
$$= \hat{z}IabB_o\sin\,\psi,$$

or

$$\vec{T} = I\vec{A} \times \vec{B} = \vec{m} \times \vec{B}, \tag{3.5.32}$$

where A is the area bound by the loop and as usual, its direction is normal to the surface; $\vec{m} = I\vec{A}$ is called the magnetic dipole moment. Note that the torque is zero at $\psi = 0^\circ$ ($\phi = 90^\circ$). The commutator helps keep the torque going the same way by changing the direction of I. Also (3.5.32) stays the same for loops of any other shape, although it is formulated here for a rectangular loop.

Example 3.18

A conducting disk of radius b is rotating about its axis with an angular velocity of ω rad/s, as shown in Figure 3.22. It is immersed in a uniform magnetic field with flux density $\vec{B} = \hat{z}B_o$ T. A voltmeter is connected between the center and the edge of disk through sliding contacts. This arrangement is known as a Faraday disk generator. Find the reading of the voltmeter. *Hint*: $\vec{v} = \hat{\phi}\omega\rho$ and $d\vec{\ell} = \hat{\rho}d\rho$.

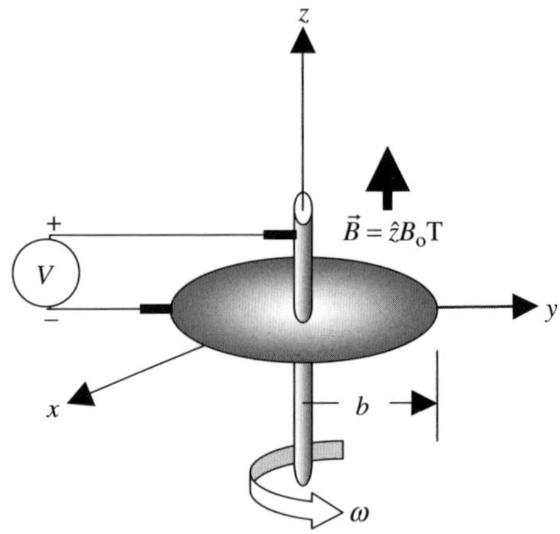

Figure 3.22 Circular rotating disk in the presence of a magnetic field.

Applying (3.5.28) gives

$$V = -\int_0^b \left\{ \left(\hat{\phi}\omega v \right) \times \hat{z}\,B_o \right\} \cdot \hat{\rho}d\rho = -\omega B_o \int_0^b \rho d\rho = -\frac{\omega B_o b^2}{2}\ \text{V}.$$

Example 3.19

A circular metallic loop of radius a lies on the x-y plane, as shown in Figure 3.23. The loop carries a current I in the ϕ direction. If there is a magnetic field with flux density $\vec{B} = (\hat{y} + \hat{z})B_o$ T, find the resulting torque on the loop.

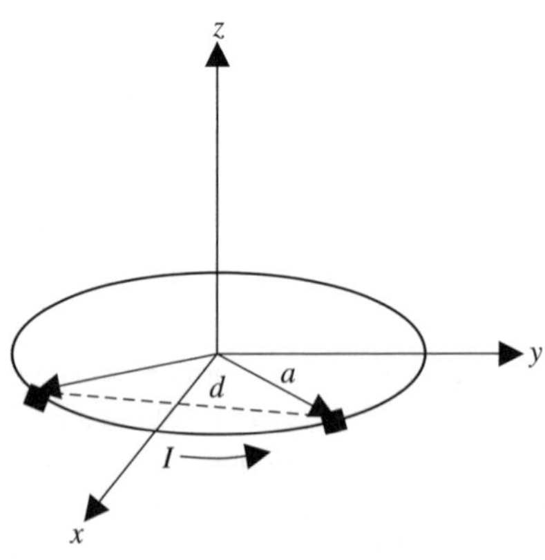

Figure 3.23 Circular metallic loop in the presence of a uniform magnetic field.

Using (3.5.29) for a small length $d\vec{\ell}$ at ϕ, we find that

$$d\vec{F} = Id\vec{\ell} \times \vec{B} = I(\hat{\phi} ad\phi) \times (\hat{y} + \hat{z})B_o = IB_o(\hat{\phi} ad\phi) \times (\hat{\rho} \sin \phi + \hat{\phi} \cos \phi + \hat{z}),$$

or

$$d\vec{F} = IB_o ad\phi(-\hat{z} \sin \phi + \hat{\rho}).$$

Note that a conversion of unit vectors was needed here, for which we used Table 2.3. Further the radial component of this force works outward uniformly everywhere, whereas its z component provides a torque about the x-axis because of a similar element at $-\phi$, as shown in the figure. The separation d between these two elements is found to be $2a \sin \phi$. Therefore

$$d\vec{T} = -\hat{x}(dF)2a \sin \phi = -\hat{x}(IB_o a \sin \phi d\phi)2a \sin \phi = -\hat{x}2IB_o a^2 \sin^2 \phi d\phi$$

and

$$\vec{T} = -\hat{x}2IB_o a^2 \int_0^{\pi} \sin^2 \phi \, d\phi = -\hat{x}2IB_o a^2 \int_0^{\pi} \frac{1 - \cos 2\phi}{2} d\phi = -\hat{x}I(\pi a^2)B_o \, \text{N} \cdot \text{m}.$$

Applying (3.5.32) gives the same result in one line, as follows:

$$\vec{T} = I\vec{A} \times \vec{B} = \vec{m} \times \vec{B} = I(\pi a^2 \hat{z}) \times (\hat{y} + \hat{z})B_o = -\hat{x}I(\pi a^2)B_o \, \text{N} \cdot \text{m}.$$

Example 3.20

Suppose that the rectangular loop of Figure 3.21 is 0.2 $\times$ 0.3 m and has 50 turns. If it carries a 10 A current and is immersed in a magnetic field of flux density $\vec{B} = \hat{x}\, 0.5\,\text{T}$. Find the resulting torque.

Since there are 50 turns in the loop, (3.5.31) and (3.5.32) give

$$\vec{T} = NI\vec{A} \times \vec{B} = \vec{m} \times \vec{B} = (50 \cdot 10 \cdot 0.2 \cdot 0.3\hat{y}) \times (\hat{x}0.5) = -\hat{z}15 \, \text{N} \cdot \text{m}.$$

Lenz's Law

We saw above that when a loop is rotated in the presence of a magnetic field, the emf induced is the same as in dc generators. If there is a current flowing through the loop, then it experiences torque, the same as in dc motors. In other words, we find that there is a current through the loop if the generator has a load. That means that the loop is experiencing a force as well following the working of the motor. This force is known to be opposing the force applied to rotate the loop. In a similar way a motor generates an emf that opposes an applied dc voltage. This phenomenon in generators of an induced emf always countering an applied voltage, as well as the counter-rotation generated in motors, well reminds us of the basic physical fact that energy is conserved. The tendency of systems to resist change is formally termed Lenz's law. The minus sign in Faraday's law of induction (3.1.1) can be used to explain this phenomenon as well.

Example 3.21

A large sheet of thickness t at 1 cm is falling with velocity v at 5 m/s through the magnetic field of B at 5 mT, as shown in Figure 3.24. The magnetic field is perpendicular to the velocity and the conductivity σ of the sheet is 5.8×10^7 S/m. Find the force per unit area of the sheet that resists its motion.

If there is a charge of ρ C per cubic meter (i.e., the charge density of sheet material), then the current density due to motion of charge is found to be

$$\vec{J} = \rho\vec{v} \rightarrow \vec{v} = \frac{\vec{J}}{\rho}.$$

Since $q = \rho \times \Delta V$ (where ΔV is small volume),

$$\vec{F} = q(\vec{v} \times \vec{B}) = \rho\Delta V(\vec{v} \times \vec{B}) = \rho\Delta V\left(\frac{\vec{J}}{\rho} \times \vec{B}\right) = \Delta V(\vec{J} \times \vec{B}).$$

For $\vec{v} = -\hat{y}v_o$ and $\vec{B} = \hat{x}B_o$,

$$\frac{\vec{F}}{q} = \vec{E} = \vec{v} \times \vec{B} = (-\hat{y}v_o) \times (\hat{x}B_o) = \hat{z}v_oB_o.$$

The motional emf produces a current in the sheet (known as the eddy current) with a current density of $\vec{J} = \sigma\vec{E} = \hat{z}\sigma v_oB_o$ that creates a secondary force. (Note that the primary force is along the $-y$-axis and moves the sheet with

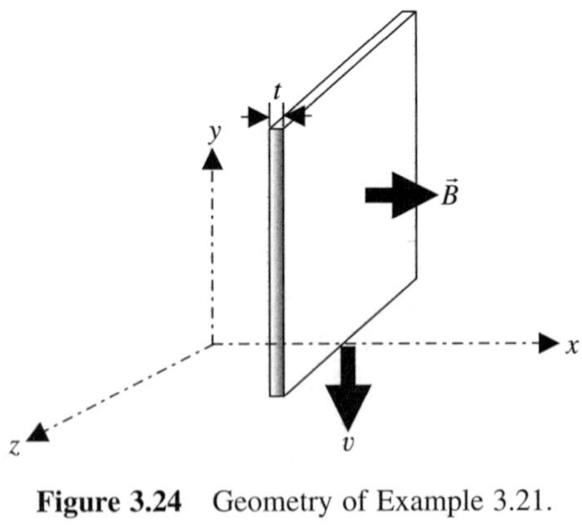

Figure 3.24 Geometry of Example 3.21.

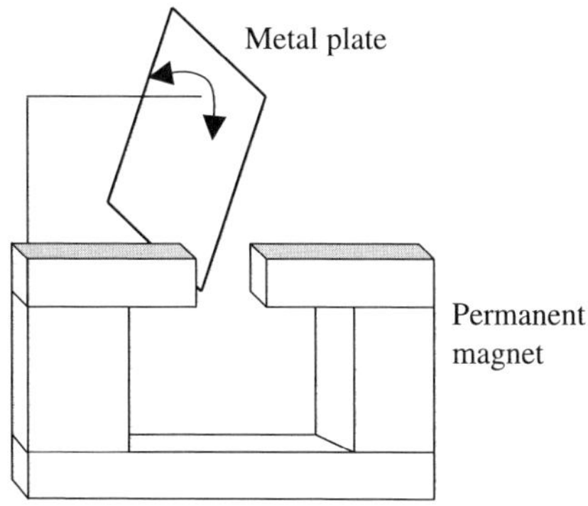

Figure 3.25 Oscillating metal plate near a magnetic field.

velocity v.) This secondary force is found to be

$$\vec{F} = \Delta V(\vec{J} \times \vec{B}) = \Delta V(\hat{z}\sigma v_o B_o) \times (\hat{x} B_o) = \hat{y}\Delta V \sigma v_o B_o^2.$$

Since $\Delta V = \Delta A \times t$, the force per unit area is found to be

$$\frac{\vec{F}}{\Delta A} = \hat{y} t \sigma v_o B_o^2 \,\mathrm{N/m^2}.$$

Since the primary force on the sheet is in the $-$y direction, this force opposes the motion, in accordance with Lenz's law. Substituting in the given numbers, we find that

$$\frac{\vec{F}}{\Delta A} = \hat{y} t \sigma v_o B_o^2 = \hat{y} \times 0.01 \times 5.8 \times 10^7 \times 5 \times (5 \times 10^{-3})^2 = 72.5 \,\mathrm{N/m^2}.$$

The idea illustrated in Example 3.21 has a number of applications. For instance, it can be used to design a magnetic brake, as illustrated in Figure 3.25. The metal plate can move freely if there is no magnetic field. When the plate enters the magnetic field of a permanent magnet (in this case), eddy currents are induced. These currents create a force that opposes the plate's motion.

3.6 POYNTING'S VECTOR AND POWER FLOW

The Poynting vector represents power flow through a unit area. In this section we start with the power associated with instantaneous electromagnetic fields and formulate Poynting's theorem. Note that sinusoidal fields are fairly common in practice, easy to analyze using phasors, and can be used as a basis for analyzing any other type of instantaneous field (via the Fourier series or Fourier integrals, as the case may be). Therefore the complex Poynting's theorem is considered and the relation for time-average power flow is formulated.

Instantaneous Poynting Vector

Consider the cross product of an instantaneous electric field intensity with a magnetic field intensity in a volume V that is bound by the closed surface S. Begin by taking the scalar product of (3.2.1) with instantaneous magnetic field intensity $\vec{\mathscr{H}}$ and of (3.2.2) with instantaneous electric field intensity $\vec{\mathscr{E}}$, to get

$$\vec{\mathscr{H}} \cdot \nabla \times \vec{\mathscr{E}} = -\vec{\mathscr{H}} \cdot \frac{\partial \vec{\mathscr{B}}}{\partial t} \tag{3.6.1}$$

and

$$\vec{\mathscr{E}} \cdot \nabla \times \vec{\mathscr{H}} = \vec{\mathscr{E}} \cdot \vec{\mathscr{J}} + \vec{\mathscr{E}} \cdot \frac{\partial \vec{\mathscr{D}}}{\partial t}. \tag{3.6.2}$$

Subtract (3.6.2) from (3.6.1), and noting by vector identity that $\vec{B} \cdot (\nabla \times \vec{A}) - \vec{A} \cdot (\nabla \times \vec{B}) = \nabla \cdot (\vec{A} \times \vec{B})$, obtain

$$\vec{\mathscr{H}} \cdot \nabla \times \vec{\mathscr{E}} - \vec{\mathscr{E}} \cdot \nabla \times \vec{\mathscr{H}} = \nabla \cdot (\vec{\mathscr{E}} \times \vec{\mathscr{H}}) = -\vec{\mathscr{H}} \cdot \frac{\partial \vec{\mathscr{B}}}{\partial t} - \vec{\mathscr{E}} \cdot \vec{\mathscr{J}} - \vec{\mathscr{E}} \cdot \frac{\partial \vec{\mathscr{D}}}{\partial t}.$$

Now assume that the medium is linear and isotropic such that $\vec{\mathscr{B}} = \mu \vec{\mathscr{H}}$ and $\vec{\mathscr{D}} = \varepsilon \vec{\mathscr{E}}$, and μ and ε do not change with time. Then find

$$\nabla \cdot (\vec{\mathscr{E}} \times \vec{\mathscr{H}}) = -\mu \vec{\mathscr{H}} \cdot \frac{\partial \vec{\mathscr{H}}}{\partial t} - \vec{\mathscr{E}} \cdot \vec{\mathscr{J}} - \varepsilon \vec{\mathscr{E}} \cdot \frac{\partial \vec{\mathscr{E}}}{\partial t}. \tag{3.6.3}$$

Since

$$\vec{\mathscr{H}} \cdot \frac{\partial \vec{\mathscr{H}}}{\partial t} = \frac{1}{2} \frac{\partial}{\partial t} (\vec{\mathscr{H}} \cdot \vec{\mathscr{H}})$$

and

$$\vec{\mathscr{E}} \cdot \frac{\partial \vec{\mathscr{E}}}{\partial t} = \frac{1}{2} \frac{\partial}{\partial t} (\vec{\mathscr{E}} \cdot \vec{\mathscr{E}}),$$

(3.6.3) may be rewritten as

$$\nabla \cdot (\vec{\mathscr{E}} \times \vec{\mathscr{H}}) = -\vec{\mathscr{E}} \cdot \vec{\mathscr{J}} - \frac{\varepsilon}{2} \frac{\partial}{\partial t} (\vec{\mathscr{E}} \cdot \vec{\mathscr{E}}) - \frac{\mu}{2} \frac{\partial}{\partial t} (\vec{\mathscr{H}} \cdot \vec{\mathscr{H}}). \tag{3.6.4}$$

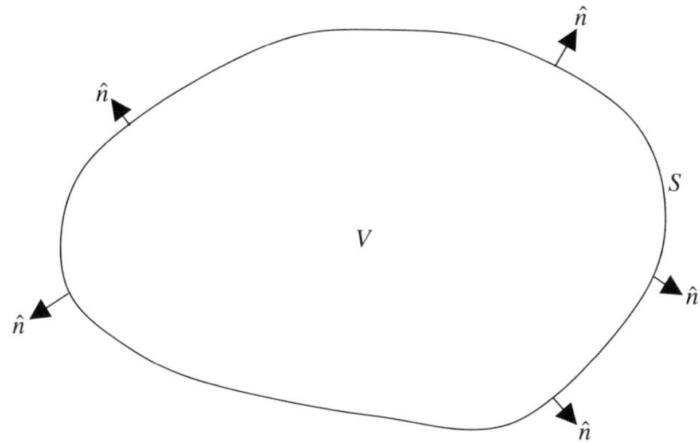

Figure 3.26 Volume V bounded by the surface S.

This equation is in point (differential) form. The large-scale (integral) form can be easily obtained by integrating both sides over the volume V, as shown is Figure 3.26. Further, using the divergence theorem, write

$$\int_V \nabla \cdot (\vec{\mathscr{E}} \times \vec{\mathscr{H}}) dv = -\int_V \vec{\mathscr{E}} \cdot \vec{\mathscr{J}} dv - \frac{\varepsilon}{2} \frac{\partial}{\partial t} \int_V (\vec{\mathscr{E}} \cdot \vec{\mathscr{E}}) dv - \frac{\mu}{2} \frac{\partial}{\partial t} \int_V (\vec{\mathscr{H}} \cdot \vec{\mathscr{H}}) dv,$$

or

$$-\oint_S (\vec{\mathscr{E}} \times \vec{\mathscr{H}}) \cdot ds = \frac{1}{2} \frac{\partial}{\partial t} \int_V (\vec{\mathscr{E}} \cdot \vec{\mathscr{D}} + \vec{\mathscr{H}} \cdot \vec{\mathscr{B}}) dv + \int_V \vec{\mathscr{E}} \cdot \vec{\mathscr{J}} dv. \tag{3.6.5}$$

The first term on the right-hand side of (3.6.5) is recognized as the time rate of change of energy stored in electric and magnetic fields, respectively, and the second term, in conjunction with (3.3.7), is identified as the power dissipated in this volume because of the conductivity of the medium. The left-hand side of (3.6.5) represents power entering the surface $\vec{\mathscr{E}} \times \vec{\mathscr{H}}$ has the unit of watt per square meter (volt per meter times ampere per meter). Thus the equation simply says that the power entering the closed surface is equal to the rate of energy stored in electric and magnetic fields, and the power dissipated in the medium bound by the surface. This effect is consistent with the conservation of energy. Equation (3.6.5) is generally referred to as Poynting's theorem, and $\vec{\mathscr{E}} \times \vec{\mathscr{H}}$ is regarded as the Poynting vector $\vec{\mathscr{S}}$.

Complex Poynting Vector

In case of sinusoidal signals we work with phasor fields. Multiplying (3.2.8) by a complex conjugate of magnetic field intensity and (3.2.9) by electric field intensity, we have, respectively,

$$\vec{H}^*(\vec{r}) \cdot \nabla \times \vec{E}(\vec{r}) = -j\omega\vec{H}^*(\vec{r}) \cdot \vec{B}(\vec{r}) \tag{3.6.6}$$

and

$$\vec{E}(\vec{r}) \cdot \nabla \times \vec{H}^*(\vec{r}) = \vec{E}(\vec{r}) \cdot \vec{J}^*(\vec{r}) - j\omega\vec{E}(\vec{r}) \cdot \vec{r}D^*(\vec{r}). \tag{3.6.7}$$

Subtracting (3.6.7) from (3.6.6), we get

$$\vec{H}^*(\vec{r}) \cdot \nabla \times \vec{E}(\vec{r}) - \vec{E}(\vec{r}) \cdot \nabla \times \vec{H}^*(\vec{r}) = -j\omega\vec{H}^*(\vec{r}) \cdot B(\vec{r}) - \vec{E}(\vec{r}) \cdot \vec{J}^*(\vec{r}) + j\omega\vec{E}(\vec{r}) \cdot \vec{D}^*(\vec{r}).$$

Since $\vec{H}^*(\vec{r}) \cdot \nabla \times \vec{E}(\vec{r}) - \vec{E}(\vec{r}) \cdot \nabla \times \vec{H}^*(\vec{r}) = \nabla \cdot \{\vec{E}(\vec{r}) \times \vec{H}^*(\vec{r})\}$, we can rewrite the expression above as

$$\nabla \cdot \{\vec{E}(\vec{r}) \times \vec{H}^*(\vec{r})\} = -j\omega\vec{H}^*(\vec{r}) \cdot B(\vec{r}) - \vec{E}(\vec{r}) \cdot \vec{J}^*(\vec{r}) + j\omega\vec{E}(\vec{r}) \cdot \vec{D}^*(\vec{r}).$$

To find its integral form we combine this point form equation with the divergence theorem as follows:

$$\begin{aligned} &-\frac{1}{2}\oint_S \left\{\vec{E}(\vec{r}) \times \vec{H}^*(\vec{r})\right\} \cdot d\vec{s} \\ &= \frac{1}{2}\int_V \vec{E}(\vec{r}) \cdot \vec{J}^*(\vec{r})dv + \frac{j\omega}{2}\int_V \{\vec{H}^*(\vec{r}) \cdot B(\vec{r}) - \vec{E}(\vec{r}) \cdot \vec{D}^*(\vec{r})\}dv. \end{aligned} \tag{3.6.8}$$

The result is Poynting's theorem for complex fields. As was noted earlier in relation to (3.3.8), $\vec{J}(\vec{r})$ can include the conduction current as well as the source current. Because the time average power is of most interest to us, we reformulate the expression for the time average Poynting vector in terms of phasor fields. We proceed by assuming that

$$\vec{E}(\vec{r}) = \vec{E}_{\text{re}}(\vec{r}) + j\vec{E}_{\text{im}}(\vec{r})$$

and

$$\vec{H}(\vec{r}) = \vec{H}_{\text{re}}(\vec{r}) + j\vec{H}_{\text{im}}(\vec{r}).$$

Therefore

$$\vec{\mathscr{E}}(\vec{r}, t) = \text{Re}[\vec{E}(\vec{r})e^{j\omega t}] = \text{Re}[\{\vec{E}_{\text{re}}(\vec{r}) + j\vec{E}_{\text{im}}(\vec{r})\}\{\cos(\omega t) + j\sin(\omega t)\}]$$

or

$$\vec{\mathscr{E}}(\vec{r}, t) = \text{Re}[\vec{E}_{\text{re}}(\vec{r}) \cos(\omega t) + j\vec{E}_{\text{im}}(\vec{r}) \cos(\omega t) + j\vec{E}_{\text{re}}(\vec{r}) \sin(\omega t) + j^2 \vec{E}_{\text{im}}(\vec{r}) \sin(\omega t)]$$

or

$$\vec{\mathscr{E}}(\vec{r}, t) = \vec{E}_{\text{re}}(\vec{r}) \cos(\omega t) - \vec{E}_{\text{im}}(\vec{r}) \sin(\omega t). \tag{3.6.9}$$

Similarly

$$\vec{\mathscr{H}}(\vec{r}, t) = \vec{H}_{\text{re}}(\vec{r}) \cos(\omega t) - \vec{H}_{\text{im}}(\vec{r}) \sin(\omega t). \tag{3.6.10}$$

Therefore

$$\begin{aligned} \vec{\mathscr{E}}(\vec{r}, t) \times \vec{\mathscr{H}}(\vec{r}, t) = {} & [\vec{E}_{\text{re}}(\vec{r}) \cos(\omega t) - \vec{E}_{\text{im}}(\vec{r}) \sin(\omega t)] \\ & \times [\vec{H}_{\text{re}}(\vec{r}) \cos(\omega t) - \vec{H}_{\text{im}}(\vec{r}) \sin(\omega t)] \end{aligned}$$

or

$$\begin{aligned} \vec{\mathscr{E}}(\vec{r}, t) \times \vec{\mathscr{H}}(\vec{r}, t) = {} & \vec{E}_{\text{re}}(\vec{r}) \times \vec{H}_{\text{re}}(\vec{r}) \cos^2(\omega t) - \vec{E}_{\text{im}}(\vec{r}) \times \vec{H}_{\text{re}}(\vec{r}) \sin(\omega t) \cos(\omega t) \\ & - E_{\text{re}}(\vec{r}) \times \vec{H}_{\text{im}}(\vec{r}) \cos(\omega t) \sin(\omega t) + \vec{E}_{\text{im}}(\vec{r}) \times \vec{H}_{\text{im}}(\vec{r}) \sin^2(\omega t) \end{aligned}$$

or

$$\begin{aligned} \vec{\mathscr{E}}(\vec{r}, t) \times \vec{\mathscr{H}}(\vec{r}, t) = {} & \vec{E}_{\text{re}}(\vec{r}) \times \vec{H}_{\text{re}}(\vec{r}) \cos^2(\omega t) + \vec{E}_{\text{im}}(\vec{r}) \times \vec{H}_{\text{im}}(\vec{r}) \sin^2(\omega t) \\ & - \{\vec{E}_{\text{im}}(\vec{r}) \times \vec{H}_{\text{re}}(\vec{r}) + \vec{E}_{\text{re}}(\vec{r}) \times \vec{H}_{\text{im}}(\vec{r})\} \cos(\omega t) \sin(\omega t) \end{aligned}$$

or

$$\begin{aligned} \vec{\mathscr{E}}(\vec{r}, t) \times \vec{\mathscr{H}}(\vec{r}, t) = {} & \vec{E}_{\text{re}}(\vec{r}) \times \vec{H}_{\text{re}}(\vec{r}) \left\{ \frac{1 + \cos(2\omega t)}{2} \right\} + \vec{E}_{\text{im}}(\vec{r}) \times \vec{H}_{\text{im}}(\vec{r}) \left\{ \frac{1 - \cos(2\omega t)}{2} \right\} \\ & - \frac{1}{2} \left\{ \vec{E}_{\text{im}}(\vec{r}) \times \vec{H}_{\text{re}}(\vec{r}) + \vec{E}_{\text{re}}(\vec{r}) \times \vec{H}_{\text{im}}(\vec{r}) \right\} \sin(2\omega t). \end{aligned}$$

Now, multiplying both sides by dt in order to integrate the last equation over the time period T and then dividing that by T (time averaging process), we have

$$\langle \vec{\mathscr{E}}(\vec{r}, t) \times \vec{\mathscr{H}}(\vec{r}, t) \rangle = \langle \vec{S} \rangle = \vec{S}_{\text{av}} = \tfrac{1}{2} \left\{ \vec{E}_{\text{re}}(\vec{r}) \times \vec{H}_{\text{re}}(\vec{r}) + \vec{E}_{\text{im}}(\vec{r}) \times \vec{H}_{\text{im}}(\vec{r}) \right\}. \tag{3.6.11}$$

If instead we take the cross product of phasor electric field intensity with complex conjugate of magnetic field intensity, we get

$$\vec{E}(\vec{r}) \times \vec{H}^*(\vec{r}) = \left\{\vec{E}_{\rm re}(\vec{r}) + j\vec{E}_{\rm im}(\vec{r})\right\} \times \{\vec{H}_{\rm re}(\vec{r}) - j\vec{H}_{\rm im}(\vec{r})\}$$

or

$$\vec{E}(\vec{r}) \times \vec{H}^*(\vec{r}) = \vec{E}_{\rm re}(\vec{r}) \times \vec{H}_{\rm re}(\vec{r}) + j\vec{E}_{\rm im}(\vec{r}) \times \vec{H}_{\rm re}(\vec{r}) - j\vec{E}_{\rm re}(\vec{r}) \times \vec{H}_{\rm im}(\vec{r}) + \vec{E}_{\rm im}(\vec{r}) \times \vec{H}_{\rm im}(\vec{r})$$

or

$$\vec{E}(\vec{r}) \times \vec{H}^*(\vec{r}) = \vec{E}_{\rm re}(\vec{r}) \times \vec{H}_{\rm re}(\vec{r}) + \vec{E}_{\rm im}(\vec{r}) \times \vec{H}_{\rm im}(\vec{r}) + j\left\{\vec{E}_{\rm im}(\vec{r}) \times \vec{H}_{\rm re}(\vec{r}) - \vec{E}_{\rm re}(\vec{r}) \times \vec{H}_{\rm im}(\vec{r})\right\}.$$

Therefore

$$\mathrm{Re}\left\{\vec{E}(\vec{r}) \times \vec{H}^*(\vec{r})\right\} = \vec{E}_{\rm re}(\vec{r}) \times \vec{H}_{\rm re}(\vec{r}) + \vec{E}_{\rm im}(\vec{r}) \times \vec{H}_{\rm im}(\vec{r}).$$

Combining it with (3.6.11), we have

$$\vec{S}_{\rm av} = \tfrac{1}{2}\left\{\vec{E}_{\rm re}(\vec{r}) \times \vec{H}_{\rm re}(\vec{r}) + \vec{E}_{\rm im}(\vec{r}) \times \vec{H}_{\rm im}(\vec{r})\right\} = \tfrac{1}{2}\mathrm{Re}\left\{\vec{E}(\vec{r}) \times \vec{H}^*(\vec{r})\right\}. \tag{3.6.12}$$

Example 3.22

The electromagnetic fields of an antenna at a large distance are found as follows:

$$\vec{E}(\vec{r}, \theta) = \hat{\theta} j \frac{120\pi}{r\sin(\theta)} \cos\left(\frac{\pi}{2}\cos(\theta)\right) e^{-jk_o r} \ \mathrm{V/m}$$

and

$$\vec{H}(\vec{r}, \theta) = \hat{\phi} j \frac{1}{r\sin(\theta)} \cos\left(\frac{\pi}{2}\cos(\theta)\right) e^{-jk_o r} \ \mathrm{A/m}.$$

Find the power radiated by this antenna.

$$\vec{S}_{\rm av} = \frac{1}{2}\mathrm{Re}\left(\vec{E} \times \vec{H}^*\right) = \hat{r} \frac{60\pi}{r^2 \sin^2(\theta)} \cos^2\left(\frac{\pi}{2}\cos(\theta)\right) \mathrm{W/m^2}$$

and

$$P_{\rm radiated} = \int_0^{\pi}\int_0^{2\pi} \vec{S}_{\rm av} \bullet \hat{r} r^2 \sin(\theta)\, d\theta d\phi = \int_0^{\pi}\int_0^{2\pi} \frac{60\pi}{\sin(\theta)} \cos^2\left(\frac{\pi}{2}\cos(\theta)\right) d\theta d\phi$$
$$= 1443.5\,\mathrm{W} = 1.4435\,\mathrm{kW}.$$

Example 3.23

There is a 1 A current flowing through a solid metallic wire of 5 cm in radius. If the conductivity σ of metal is 11×10^5 S/m, then verify the Poynting theorem for a unit length of the wire.

Using cylindrical coordinates, we write the expressions for the current density J and the electric field intensity in the wire as follows:

$$\vec{J} = \hat{z}\frac{I}{A} = \hat{z}\frac{I}{\pi \times 0.05^2} \text{ A/m}$$

and

$$\vec{E} = \frac{\vec{J}}{\sigma} = \hat{z}\frac{I}{A\sigma} = \hat{z}\frac{I}{(\pi \times 0.05^2)\sigma} \text{ V/m}.$$

Therefore

$$\int_V \vec{E} \cdot \vec{J} dv = \int_{z=0}^{1}\int_{\phi=0}^{2\pi}\int_{\rho=0}^{0.05} \frac{I^2}{(\pi \times 0.05^2)^2\sigma}\rho d\rho d\phi dz = \frac{I^2}{(\pi \times 0.05^2)^2\sigma}(\pi \times 0.05^2)$$
$$= 11.57 \times 10^{-5} \text{ W}.$$

Following the Example 3.6, we get

$$\vec{H} = \hat{\phi}\frac{I}{2\pi\rho} \text{ A/m}.$$

Therefore

$$\oint_S (\vec{E} \times \vec{H}) \cdot d\vec{s} = \int_{z=0}^{1}\int_{\phi=0}^{2\pi} \left(-\hat{\rho}\frac{I^2}{2\pi\rho\sigma(\pi \times 0.05^2)}\right)_{\rho=0.05} \cdot \hat{\rho}(0.05\, d\phi dz)$$
$$= -\frac{I^2}{(\pi \times 0.05^2)\sigma} = -11.57 \times 10^{-5} \text{ W}.$$

This verifies the Poynting's theorem.

Induction Heating

We found that a time-varying magnetic field induces an electromotive force in conducting materials, which in turn produces eddy currents. These currents produce localized heat in conducting material due to the ohmic loss of electrical power. It is highly desired that the currents be minimized when designing electrical devices (motors, generators, transformers, etc.). The loss of electrical power, however, can be maximized when designing a furnace for melting metals. This is the working principle of commercially available induction furnaces. In order to get a better understanding, consider a thin metallic cylindrical shell shown in Figure 3.27. Its radius and wall thickness are a and d, respectively.

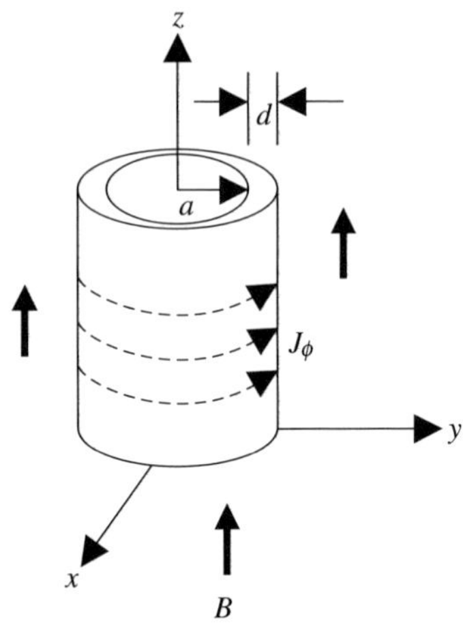

Figure 3.27 Metallic cylindrical shell in the presence of a time-varying magnetic field.

If a magnetic field with flux density $\vec{\mathscr{B}} = \hat{z}B_o \cos(\omega t)\,\text{T}$ is applied to the shell of Figure 3.27, then a ϕ-directed voltage $\mathscr{E}_\phi$ is induced, following Faradays' law. This voltage, in turn, generates the eddy currents $J_\phi = \sigma\mathscr{E}_\phi$, as shown in the figure. Following Lenz's law, these currents produce a magnetic field that opposes the applied field. The flux density of this opposing magnetic field can be found through Ampere's law as $\mu_o J_\phi d$. Therefore the effective magnetic flux density at the shell may be written as follows:

$$\vec{\mathscr{B}} = \hat{z}(B_o \cos \omega t + \sigma\mu_o d\mathscr{E}_\phi)\,\text{T}. \tag{3.6.13}$$

Using Faraday's law, we write

$$\oint_c \vec{\mathscr{E}} \cdot d\vec{\ell} = -\frac{\partial}{\partial t}\int_s \vec{\mathscr{B}} \cdot d\vec{s} \rightarrow \mathscr{E}_\phi 2\pi a = -\frac{\partial}{\partial t}[\pi a^2(B_o \cos \omega t + \sigma\mu_o d\mathscr{E}_\phi)].$$

This relation can be simplified as follows:

$$\frac{\partial\mathscr{E}_\phi}{\partial t} + \frac{2}{\sigma\mu_o da}\mathscr{E}_\phi = \frac{B_o\omega}{\sigma\mu_o d}\sin \omega t. \tag{3.6.14}$$

Instead of looking for the solution to this differential equation, we switch to the corresponding phasor equation (we could have used phasor fields from the beginning of this formulation). Alternatively, we can apply the conversion rules discussed in Section 3.2 to (3.6.14). This way we find that

$$j\omega E_\phi + \frac{2}{\sigma\mu_o da}E_\phi = -j\frac{B_o\omega}{\sigma\mu_o d} \rightarrow E_\phi = \frac{-B_o/\sigma\mu_o d}{1 + 2/j\omega\sigma\mu_o da}$$

$$E_\phi = \frac{-j\omega B_o a/2}{1 + j\omega\sigma\mu_o da/2}. \tag{3.6.15}$$

Therefore the average power dissipated per unit length in the shell is found to be

$$P_{\text{diss}} = \frac{1}{2}\text{Re}\int_V J_\phi^* E_\phi dv = \frac{1}{2}(2\pi ad)\sigma|E_\phi|^2$$

$$= \pi ad\sigma \frac{(\omega B_o a/2)^2}{1+(\omega\mu_o\sigma da/2)^2}\ \text{W/m}. \tag{3.6.16}$$

Since the energy U required to raise the temperature of a mass m of the conducting material by ΔT is $m \times$ specific heat $\times\ \Delta T$, the time t required for the magnetic field (and the eddy currents) to do that may be found as

$$t = \frac{m \cdot s \cdot \Delta T}{P_{\text{diss}}}. \tag{3.6.17}$$

Example 3.24

A 0.5 m long cylindrical shell of copper is at the room temperature (20°C), as shown in Figure 3.27. Its radius and wall thickness are 1 cm and 0.1 cm, respectively. If the shell is placed in a 60 Hz z-directed magnetic field with a flux density of 2 T, find the time required to melt the shell. For copper, the electrical conductivity $\sigma = 5.813 \times 10^7$ S/m, specific heat = 387 J/kg · °C, density = 8.92×10^3 kg/m^3, and melting point = 1083.4°C.

From (3.6.16), we can find the total power dissipated (for h = 0.5 m) as follows:

$$P_{\text{diss}} = \pi ad\sigma \frac{(\omega B_o a/2)^2}{1+(\omega\mu_o\sigma da/2)^2} h = 1.2736 \cdot 10^4\ \text{W}.$$

The total mass m of the material is found to be

$$m = 8.92 \cdot 10^3 \cdot \pi \lfloor (0.01+0.001)^2 - 0.01^2 \rfloor \cdot 0.5 = 0.2942\ \text{kg}.$$

Therefore (3.6.17) gives

$$t = \frac{m \cdot s \cdot \Delta T}{P_{\text{diss}}} = \frac{0.2942 \cdot 387 \cdot (1083.4 - 20)}{1.2736 \cdot 10^4} = 9.51\ \text{s}.$$

PROBLEMS

3.1. A metal bar oscillates over a pair of conducting rails, with its position given by $x = 5 - \sin(\pi\ t)$ m, as shown in Figure 3.2. The system lies on the $z = 0$ plane, and there is a magnetic flux density of 3 mT in the z direction. Find the current i due to the induced emf. What happens if the magnetic flux varies with time as follows?

$$\vec{\mathscr{B}} = \hat{z}\cos(\pi t)\,\text{mT}.$$

3.2. A rigid rectangular loop lies on the xy-plane with its vertices at $(1, 0, z)$, $(3, 0, z)$, $(3, 0, z+2)$, and $(1, 0, z+2)$. It is immersed in a magnetic field $\vec{B}(\vec{r}, t) = \hat{y}5\cos((\pi z/2) - 120\pi t)\,\text{mT}$. (a) For the loop being stationary, find the electromotive force induced around it in the counterclock sense. (b) How will the induced emf change if the loop is moving with velocity $\vec{v} = \hat{z}2\,\text{m/s}$.

3.3. A rigid rectangular loop of 1×4 m is situated on the yz-plane, as shown in Figure 3.4. It is rotating symmetrically about the z-axis with an angular velocity of 500 rad/s. If there is a magnetic field with a flux density $\vec{\mathscr{B}}(\vec{r}, t) = \hat{x}2\cos(t + 0.15)\,\text{mT}$, find the open-circuit voltage V_o induced in the loop.

3.4. A conducting fluid is flowing through a plastic tube of 2 cm in diameter, as shown in Figure 3.5. The flow rate is 0.6 L/s along the z-axis. It is subjected to a magnetic field with a flux density $\vec{\mathscr{B}}(\vec{r}, t) = \hat{x}80\,\text{mT}$. As shown in the figure, there are two electrodes placed across the tube's diameter along the y-axis. Determine the voltage induced across these electrodes.

3.5. An infinitely long solid cylindrical wire with its radius at 1.5 m is laying along the z-axis. It carries a current with density given as $\vec{\mathscr{J}}(\rho) = 2(1 + 0.2\rho^2)\hat{z}\,\text{A/m}^2$. Find the magnetic field intensity inside as well as outside the wire.

3.6. A coaxial line with inner and outer conductor radii of 2 cm and 5 cm, respectively, carries a dc of 5 A. The outer conductor is 2 mm thick. Find the magnetic field intensity everywhere.

3.7. A spherical volume of 5 cm radius contains an electrical charge of $\rho = 3\,\text{C/m}^3$. Find the electric flux density everywhere.

3.8. The electric flux density in a spherical volume is given as follows:

$$\vec{\mathscr{D}} = \hat{r}2r^2\,\text{C/m}^3.$$

Find the associated electrical charge density.

3.9. The current density in a region is given as follows:

$$\vec{\mathscr{J}} = \hat{x}y^2z + \hat{y}x^2y + \hat{z}zy\,\text{A/m}^2.$$

Find the current leaving the surface that bounds a unit cube in the first quadrant.

3.10. Electric field intensity in a source-free region is given as follows:

$$\vec{E} = \hat{y}2.5e^{+j2z}\,\text{V/m}.$$

If $\varepsilon = \varepsilon_o$ and $\mu = \mu_o$ in the region, find the signal frequency.

3.11. The region $x > 0$ is a perfect dielectric with $\varepsilon_r = 6.25$ while the region $x < 0$ is a perfect dielectric with $\varepsilon_r = 21.25$. At the interface, subscript 1 denotes field components on the $+x$ side of the boundary while the subscript 2 on the $-x$ side. If $\vec{D}_1 = 6\hat{x} + 12\hat{y}\,\mathrm{C/m^2}$, find $\vec{D}_2$, $\vec{E}_1$, and $\vec{E}_2$.

3.12. A sphere with its radius as 1 m is made of a perfect dielectric material (medium 1). It is surrounded by a perfect dielectric medium of $\varepsilon_r = 2.25$ (medium 2). The electric field intensities in the two media are given as follows:

$$\vec{E}_1 = E_{01}(\hat{r}\cos\theta - \hat{\theta}\sin\theta) \qquad \text{for} \quad r \leq 1\,\mathrm{m}$$

and

$$\vec{E}_2 = E_{02}\left(\hat{r}\left\{1 + \frac{1}{r^4}\right\}\cos\theta - \hat{\theta}\left\{1 - \frac{1}{2r^2}\right\}\sin\theta\right) \qquad \text{for } r \geq 1\,\mathrm{m}.$$

Find the permittivity of the spherical medium.

3.13. As shown in Figure 3.17, an anode accelerates an electron emitted by a cathode in an electron gun. If its initial velocity is zero and the anode is at 250 V with respect to the cathode, find its final velocity.

3.14. An electron generated by the electron gun described in Problem 3.13 enters between the parallel plates of Figure 3.16. There is a fluorescent screen placed at $z = 60$ cm. For $w = d = 1$ cm and $V = 2$ V, find the point where this electron hits the screen. Ignore the fringing fields of the deflecting plates.

3.15. A particle with $q = 3.2 \times 10^{-19}$ C and $m = 3.35 \times 10^{-27}$ kg is at rest. It is accelerated for 3 μs using an electric field of 2 kV/m on the x-z plane. This particle then enters a magnetic field with $\vec{B} = \hat{y}10\,\mathrm{mT}$, as shown in Figure 3.18. Find the velocity before it enters the magnetic field and the radius of the circle that it will trace due to the magnetic field.

3.16. A compound of isotopes is vaporized and then ionized. The charge on the ions is 3.2×10^{-19} C. A constant 2 kV accelerates these ions before they enter a region with a magnetic field of 10 mT in the perpendicular direction, as shown in Figure 3.19. If mass of these charge particles are $m_1 = 10^{-25}$ kg and $m_2 = 10^{-24}$ kg, determine the radii of the circles that are traced.

3.17. As shown in Figure 3.20, there is a 10 mA current flowing through a rectangular block of a semiconductor. The semiconductor is subjected to an x-directed magnetic field of 8 T. If its cross section is a square of 5 cm on each side and the Hall voltage is found to be 0.5 mV, find the concentration of charge (the charge density) and also the charge particles' polarity.

3.18. In a Faraday disk generator, the conducting disk of radius b is rotating about its axis with an angular velocity of 150 rad/s. The disk is immersed in a uniform magnetic field with flux density $\vec{B} = \hat{z}10$ T. A voltmeter is connected between the center and the edge of the disk through sliding contacts. Find the reading of the voltmeter.

3.19. A circular metallic loop of 9 cm in radius lies on x-y plane. The loop carries a current of 2 A in the ϕ direction. If there is a magnetic field with flux density $\vec{B} = (\hat{y} + \hat{z})10\,\text{T}$, find the resulting torque on the loop.

3.20. Suppose that the rectangular loop of Figure 3.21 is 50 cm $\times$ 40 m and has 100 turns. If the loop carries a 5 A current and is immersed in a magnetic field of flux density $\vec{B} = \hat{x}\,0.2\,\text{T}$, find the resulting torque.

3.21. A large sheet of thickness t at 2 mm is falling with velocity v at 2 m/s through a magnetic field of B at 20 mT. The magnetic field is perpendicular to the velocity, and the conductivity σ of the sheet is 5.8×10^7 S/m. Find the force per unit area of the sheet that resists its motion.

3.22. The electromagnetic fields of an antenna at a large distance are found as follows:

$$\vec{E}(\vec{r}, \theta) = \hat{\theta} j \frac{400}{r} \cos\left(2\cos(\theta)\right) \cos\phi e^{-jk_o r}\ \text{V/m}$$

and

$$\vec{H}(\vec{r}, \theta) = \hat{\phi} j \frac{1}{r} \cos(2\cos(\theta)) \cos\phi e^{-jk_o r}\ \text{A/m}.$$

Find the power radiated by this antenna.

3.23. There is a 2 A current flowing through a solid metallic wire of an 8 cm radius. If the conductivity σ of the metal is 7×10^6 S/m, verify the Poynting theorem for a unit length of the wire.

3.24. A 20 cm long cylindrical shell of copper is at the room temperature (20°C). Its radius and wall thickness are 5 cm and 0.2 cm, respectively. If it is placed in a 60 Hz z-directed magnetic field with a flux density of 5 T, find the time required to melt the shell. For copper, electrical conductivity $\sigma = 5.813 \times 10^7$ S/m, specific heat = 387 J/kg°C, density = 8.92×10^3 kg/m^3, and melting point = 1083.4°C.

3.25. For the following charge distribution, find the displacement flux emanating from a cubical surface bounded by $x = \pm 0.5$, $y = \pm 0.5$, and $z = \pm 0.5$:

$$\rho(x, y, z) = 3 - x^3 - y^3 - z^3\ \text{C/m}^3.$$

3.26. If $\vec{\mathscr{B}} = \hat{z}5\cos[10^5 t - (0.5\pi x) - (\pi y)]\,\text{T}$, find the emf induced around a closed loop formed by connecting successively the points (0, 0, 0), (2, 0, 0), (2, 1, 0), (0, 1, 0), and (0, 0, 0).

3.27. Find the charge densities that produce the following electric flux densities: (a) $\vec{D} = xy\hat{x} + yz\hat{y} + zx\hat{z}\,\text{C/m}^2$ and (b) $\vec{D} = \rho\sin(\phi)\hat{\phi}\,\text{C/m}^2$.

3.28. If $\vec{J} = x\hat{x} + y\hat{y} + z\hat{z}\,\mathrm{A/m^2}$, find the time rate of decrease of the charge contained within a volume bounded by the planes $x = \pm 0.5$ m, $y = \pm 0.5$ m, and $z = \pm 0.5$ m.

3.29. If an electric field $\vec{E} = E_o \cos(6\pi \cdot 10^8 t - x + ky)\hat{z}$ V/m in free space, find the value(s) of k for which the field satisfies Maxwell's equations.

3.30. The electric flux density in a medium is given as follows:

$$\vec{D} = xy^2\hat{x} + y^2 z\hat{y} + zx\hat{z}\ \mathrm{C/m^2}.$$

Find the associated electric charge density ρ_v.

3.31. The region $x > 0$ is a perfect dielectric of $\varepsilon_r = 2$, and the region $x < 0$ is a perfect dielectric of $\varepsilon_r = 4.2$. Consider the field components at point 1 on the $+x$ side of the boundary, to be denoted by subscript 1, and the field components at the adjacent point 2 on the $-x$ side of the boundary, to be denoted by subscript 2. If $\vec{D}_1 = 4\hat{x} + 2\hat{y}\,\mathrm{C/m^2}$, find $\vec{D}_2$, $\vec{E}_1$, and $\vec{E}_2$.

3.32. A wire of circular cross section with its radius as 1 mm carries a conduction current of 0.4 cos $(6\pi \cdot 10^8 t)$ μA. If its conductivity and dielectric constant are $2 \cdot 10^5$ S/m and 4.5, respectively, determine the magnitude of the displacement current density.

4

UNIFORM PLANE WAVES

This chapter begins with formulation of the Helmholtz equation and its solutions for uniform plane waves. Propagation characteristics of uniform plane waves in linear, isotropic, and homogeneous media are studied. Phase and group velocities of the wave are defined along with its polarization behaviors. The chapter concludes with sections on the wave reflections and transmission for normal as well for oblique incidence. A number of engineering applications are included throughout the chapter.

4.1 WAVE EQUATION AND UNIFORM PLANE WAVE SOLUTIONS

In a source-free, linear, isotropic, and homogeneous region, Maxwell's curl equations (3.2.8) and (3.2.9) combined with (3.3.6) and (3.3.8) can be written in phasor form as follows:

$$\nabla \times \vec{E} = -j\omega\mu\vec{H}, \tag{4.1.1}$$

$$\nabla \times \vec{H} = -j\omega\varepsilon^{*}\vec{E}. \tag{4.1.2}$$

Practical Electromagnetics: From Biomedical Sciences to Wireless Communication.
By Devendra K. Misra
Copyright © 2007 John Wiley & Sons, Inc.

After taking the curl of (4.1.1) and substituting (4.1.2) on its right-hand side, we get

$$\nabla \times \nabla \times \vec{E} = -j\omega\mu\nabla \times \vec{H} = -j\omega\mu(j\omega\varepsilon^* \vec{E}) = k^2\vec{E}, \tag{4.1.3}$$

where

$$k = \omega\sqrt{\mu\varepsilon^*} = \beta - j\alpha, \tag{4.1.4}$$

and k, α, and β are the wave number, the attenuation constant, and the phase constant, respectively.

Further

$$\nabla \times \nabla \times \vec{E} = \nabla(\nabla \cdot \vec{E}) - \nabla^2\vec{E} = k^2\vec{E} \rightarrow \nabla^2\vec{E} + k^2\vec{E} = 0. \tag{4.1.5}$$

Similarly, starting with (4.1.2), we find that

$$\nabla^2\vec{H} + k^2\vec{H} = 0. \tag{4.1.6}$$

Equations (4.1.5) and (4.1.6) are known as Helmholtz equations or as vector wave equations for the electric field and magnetic field, respectively. As noted earlier in Section 2.9, these equations are separable only in rectangular coordinates.

Consider an electric field that is directed along the x-axis. The field is uniform in x and y directions (i.e., with no variations in x or y). Therefore the wave equation (4.1.5) for this case reduces to

$$\frac{d^2E_x(z)}{dz^2} + k^2E_x(z) = 0. \tag{4.1.7}$$

This is an ordinary differential equation of the second order that can be solved as follows: Assume that $E_x(z) = e^{Cz}$, and substitute it back into (4.1.7) to find that

$$(C^2 + k^2)e^{Cz} = 0 \rightarrow C = \pm jk.$$

Then write the two solutions $E_{x1}(z)$ and $E_{x2}(z)$ of (4.1.7) as

$$E_{x1}(z) = E^+e^{-jkz} \tag{4.1.8}$$

and

$$E_{x2}(z) = E^-e^{jkz}, \tag{4.1.9}$$

where E^+ and E^- arc the integration constants.

The complete solution to (4.1.7) is found to be

$$E_x(z) = E^+e^{-jkz} + E^-e^{jkz} = E^+e^{-\alpha z}e^{-j\beta z} + E^-e^{\alpha z}e^{j\beta z}. \tag{4.1.10}$$

Similarly (4.1.6) can be solved to find the corresponding magnetic fields. Alternatively, (4.1.1) can be used to find the two solutions of (4.1.6) as follows: First,

$$\vec{H}_1 = -\frac{1}{j\omega\mu}\nabla \times (\hat{x}E_{x1}) = -\frac{1}{j\omega\mu}\begin{vmatrix} \hat{x} & \hat{y} & \hat{z} \\ \dfrac{\partial}{\partial x} & \dfrac{\partial}{\partial y} & \dfrac{\partial}{\partial z} \\ E^+e^{-jkz} & 0 & 0 \end{vmatrix} = \hat{y}\left(\frac{k}{\omega\mu}\right)E^+e^{-jkz},$$

or

$$\vec{H}_1 = \hat{y}\left(\frac{k}{\omega\mu}\right)E_{x1}. \tag{4.1.11}$$

Next,

$$\vec{H}_2 = -\frac{1}{j\omega\mu}\nabla \times (\hat{x}E_{x2}) = -\frac{1}{j\omega\mu}\begin{vmatrix} \hat{x} & \hat{y} & \hat{z} \\ \dfrac{\partial}{\partial x} & \dfrac{\partial}{\partial y} & \dfrac{\partial}{\partial z} \\ E^-e^{jkz} & 0 & 0 \end{vmatrix} = -\hat{y}\left(\frac{k}{\omega\mu}\right)E^-e^{jkz},$$

or

$$\vec{H}_2 = -\hat{y}\left(\frac{k}{\omega\mu}\right)E_{x2}. \tag{4.1.12}$$

We proceed to determine the Poynting vector and hence the power flow due to these fields. Using (4.1.8) and (4.1.11) with the assumption that μ is real, we get

$$\vec{S}^+_{\text{av}} = \frac{1}{2}\text{Re}\,(\vec{E} \times \vec{H}^*) = \frac{1}{2}\text{Re}\left[(\hat{x}E_{x1}) \times \left(\hat{y}\frac{k}{\omega\mu}E_{x1}\right)^*\right] = \hat{z}\frac{1}{2}\text{Re}\left(\frac{k^*}{\omega\mu}|E_{x1}|^2\right),$$

or

$$\vec{S}^+_{\text{av}} = \hat{z}\frac{1}{2}\text{Re}\left(\frac{\beta + j\alpha}{\omega\mu}|E^+|^2\right) = \hat{z}\frac{\beta}{2\omega\mu}|E^+|^2. \tag{4.1.13}$$

Similarly, using (4.1.9) and (4.1.12) in (3.6.12), we find that

$$\vec{S}^-_{\text{av}} = \frac{1}{2}\text{Re}(\vec{E} \times \vec{H}^*) = \frac{1}{2}\text{Re}\left[(\hat{x}E_{x2}) \times \left(\hat{y}\frac{k}{\omega\mu}E_{x2}\right)^*\right] = -\hat{z}\frac{1}{2}\text{Re}\left(\frac{k^*}{\omega\mu}|E_{x2}|^2\right),$$

or

$$\vec{S}^-_{\text{av}} = -\hat{z}\frac{1}{2}\text{Re}\left(\frac{\beta + j\alpha}{\omega\mu}|E^-|^2\right) = -\hat{z}\frac{\beta}{2\omega\mu}|E^-|^2. \tag{4.1.14}$$

Equation (4.1.13) says that the first solution of the wave equation (the one with e^{-jkz}) represents the power flow in $+z$ direction. Equation (4.1.14) says that the second solution of the wave equation (the one with e^{jkz}) represents the power flow in the $-z$ direction. In other words, the first solution represents a wave traveling along $+z$ (as an outgoing wave), and the second solution represents a wave traveling along $-z$ (as an incoming wave). Therefore the electric field, the magnetic field, and the direction of wave-propagation are orthogonal to each other. Since the electric and magnetic field intensities of the wave are transverse to the direction of propagation, the wave is called a transverse electromagnetic (TEM) wave. To analyze this wave further, we need to determine the time-dependent expression of the electric field intensity from (4.1.10), which is written as follows:

$$\begin{aligned}\vec{\mathscr{E}}(\vec{r}, t) = \mathrm{Re}[\hat{x}E_x e^{j\omega t}] = \hat{x}\lfloor |E^+|e^{-\alpha z}\cos(\omega t - \beta z + \theta_1) \\ + |E^-|e^{\alpha z}\cos(\omega t + \beta z + \theta_2)\rfloor,\end{aligned} \tag{4.1.15}$$

where

$$E^+ = |E^+|e^{j\theta_1}$$

and

$$E^- = |E^-|e^{j\theta_2}.$$

Each of the two terms on the right-hand side of (4.1.15) can be graphed as a function of distance z, while keeping other parameters same. Figure 4.1 shows the typical characteristics of these terms. As the figure shows, the first sinusoidal term is exponentially decreasing in amplitude as z increases, and the second sinusoidal term is exponentially increasing with z. As is easily intuited, the wave's amplitude in a lossy medium ($\alpha \neq 0$) decreases as the wave moves further away from the source. Thus the first term of (4.1.15) represents a wave that propagates away

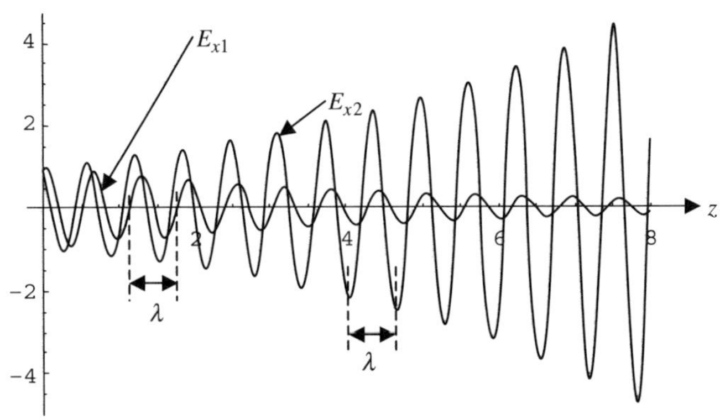

Figure 4.1 Graphical representation of the two solutions of (4.1.7) at a given time.

from a source, or origin. The second term represents a wave moving toward the origin (moving in the $-z$ direction). The distance over which a wave traverses in a 360°-phase change is called a wavelength. A wavelength is generally denoted by λ, as is shown in the figure. λ is defined as

$$\lambda = \frac{2\pi}{\beta}. \tag{4.1.16}$$

There is still another way to interpret the two solutions of (4.1.7). Equation (4.1.15) can be re-arranged as follows:

$$\vec{\mathscr{E}}(\vec{r}, t) = \hat{x}\left[|E^+|e^{-\alpha z}\cos\left\{\omega\left(t - \frac{\beta z}{\omega}\right) + \theta_1\right\} + |E^-|e^{\alpha z}\cos\left\{\omega\left(t + \frac{\beta z}{\omega}\right) + \theta_2\right\}\right].$$

Both terms have corrections in their time variables. The first cosine function needs the correction term to be subtracted, whereas the second one needs it to be added in order to get the correct time. Thus the first term represents a phenomenon that occurred earlier, whereas the second term represents one that will be occurring in future. This is possible only when E_{x1} represents an outgoing wave and E_{x2} an incoming wave. This correction in time (and hence the phase of the wave) has distance z in its numerator, and therefore it should be divided by the velocity in order to obtain the correction. This velocity is known as the phase velocity, v_p, of the wave. Hence

$$v_p = \frac{\omega}{\beta} = \frac{2\pi f}{2\pi/\lambda} = f\lambda. \tag{4.1.17}$$

The ratio of electric and magnetic field intensities of a wave is known as the wave impedance. Hence from (4.1.11), or (4.1.12), and (3.3.9) we find that

$$Z_{TEM} = \frac{E_{x1}}{H_{y1}} = -\frac{E_{x2}}{H_y} = \frac{\omega\mu}{k} = \frac{\omega\mu}{\omega\sqrt{\mu\varepsilon^*}} = \sqrt{\frac{\mu}{\varepsilon^*}}\ \Omega = \eta, \tag{4.1.18}$$

where η is the intrinsic impedance of the medium.

For $\alpha = 0$ in (4.1.4), and hence $\sigma = 0$ in (3.3.9), k is real, and (4.1.18) reduces to

$$\eta = \sqrt{\frac{\mu}{\varepsilon}}\ \Omega. \tag{4.1.19}$$

Note that in the case of free space, $\sigma = 0$ (true for any lossless medium), and therefore $k = \omega\sqrt{\mu\varepsilon} = \beta$ and $\alpha = 0$. Since $\mu = \mu_o = 4\pi \cdot 10^{-7}$ H/m and $\varepsilon = \varepsilon_o = 8.854 \cdot 10^{-12}$ F/m for free space, we find that

$$v_p = \frac{\omega}{\beta} = \frac{\omega}{\omega\sqrt{\mu\varepsilon}} = \frac{1}{\sqrt{\mu\varepsilon}} = \frac{1}{\sqrt{4\pi \cdot 10^{-7} \cdot 8.854 \cdot 10^{-12}}} = 2.998 \cdot 10^8 \text{ m/s}$$
$$= c \approx 3 \cdot 10^8 \text{ m/s}$$

and

$$\eta = \sqrt{\frac{\mu}{\varepsilon}} = \sqrt{\frac{\mu_o}{\varepsilon_o}} = \sqrt{\frac{4\pi \cdot 10^{-7}}{8.854 \cdot 10^{-12}}} = 376.73 \approx 377\,\Omega.$$

Example 4.1

The electric field intensity of a uniform plane wave propagating through a nonmagnetic medium is given as follows:

$$\vec{\mathscr{E}}(z,t) = \hat{y}10\cos(10^9 t - 5z)\,\mathrm{mV/m}.$$

Find (a) the direction of wave travel, (b) the phase velocity of the wave, (c) the wavelength, and (d) the associated magnetic field intensity.

(a) On the basis of $-5z$ term in the argument, we conclude that the wave is propagating in the $+z$ direction.
(b) $v_p = \omega/\beta = 10^9/5 = 2 \cdot 10^8$ m/s.
(c) $\lambda = 2\pi/\beta = 2\pi/5 = 0.4\pi$ m $= 1.2566$ m.
(d) There are a number of ways to determine its magnetic field intensity. One is to use (3.2.1) and (3.3.6) with $\mu_r = 1$ (because the medium is nonmagnetic). Perhaps the easiest way is as follows! Since the wave is propagating in the $+z$ direction and it has an electric field in the $+y$ direction, with the help of right-hand rule (note that $\vec{S} = \vec{\mathscr{E}} \times \vec{\mathscr{H}}$) we find that its magnetic field has to be in the $-x$ direction. Hence

$$\vec{H}(z) = -\hat{x}\frac{E_y}{Z_{\mathrm{TEM}}} = -\hat{x}\frac{10e^{-5z}}{\eta}\,\mathrm{mA/m}.$$

Note that we are using phasor fields with wave impedances. It is essential when the wave impedance is a complex number (and therefore the medium conductivity is not zero). From the given field we find that the attenuation constant α is zero, and therefore the medium is lossless. Hence

$$Z_{\mathrm{TEM}} = \frac{\omega\mu}{k} = \frac{\omega\mu_o}{\beta} = \frac{10^9 \cdot 4\pi \cdot 10^{-7}}{5} = 80\pi\,\Omega = 251.3274\,\Omega,$$

$$\vec{H}(z) = -\hat{x}\frac{10e^{-5z}}{251.3274}\,\mathrm{mA/m} = -\hat{x}\,0.0398e^{-5z} \rightarrow$$

$$\vec{\mathscr{H}}(z,t) = -\hat{x}39.8\cos(10^9 t - 5z)\,\mu\mathrm{A/m}.$$

Example 4.2

The time-harmonic electric field intensity of a uniform plane wave propagating in free space is given as follows:

$$\vec{E} = \hat{z}(1 - j)e^{-j(4\pi x)}\,\mathrm{mV/m}.$$

Find (a) the direction of propagation, (b) the wavelength, (c) the frequency, and (d) the associated magnetic field intensity.

(a) Comparing the given expression with (4.1.8) and (4.1.9), we find that this wave is propagating in the $+x$ direction.

(b) $\lambda = 2\pi/\beta = 2\pi/4\pi = 0.5$ m.

(c) $f = (3 \times 10^8)/\lambda = 6 \times 10^8$ Hz $= 600$ MHz.

(d)

$$\vec{H} = -\frac{1}{j\omega\mu_o}\nabla \times \vec{E} = -\frac{1}{j\omega\mu_o}\begin{vmatrix} \hat{x} & \hat{y} & \hat{z} \\ \frac{\partial}{\partial x} & \frac{\partial}{\partial y} & \frac{\partial}{\partial z} \\ 0 & 0 & E_z \end{vmatrix} = -\hat{y}\frac{4\pi}{\omega\mu_o}(1-j)e^{-j4\pi x} \text{ mA/m}$$

or

$$\vec{H} = -\hat{y}\frac{(1-j)}{376.9911}e^{-j4\pi x} \text{ mA/m} = -\hat{y}2.6526(1-j)e^{-j4\pi x} \text{ μA/m}.$$

Alternatively, following the procedure used in previous example, we find that its magnetic field should be in the $-y$ direction. Therefore

$$\eta = \frac{E_z}{-H_y} = \frac{\omega\mu_o}{k} = \frac{2\pi \times 6 \times 10^8 \times 4\pi \times 10^{-7}}{4\pi} = 376.9911\,\Omega \rightarrow$$

$$H_y = -\frac{E_z}{376.9911}.$$

Example 4.3

A safe radio frequency and microwave exposure standard for the humans is set at 5 mW/cm^2. Find the corresponding electric and magnetic field intensities in air. How these fields compare with 1.5 kW/m^2 radiation arriving typically from the sun (even though not at a single frequency)?

Since $|\vec{S}_{av}| = |\vec{E}|^2/2\eta = \frac{1}{2}\eta|\vec{H}|^2$ and 5 mW/cm^2 = 50 W/m^2,

$$|\vec{E}| = \sqrt{2\eta|\vec{S}_{av}|} = \sqrt{2 \cdot 377 \cdot 50} = 194.16 \text{ V/m}$$

and

$$|\vec{H}| = \sqrt{\frac{2|\vec{S}_{av}|}{\eta}} = \sqrt{\frac{2 \cdot 50}{377}} = 0.5150 \text{ A/m}.$$

Similarly for 1.5 kW/m^2 we get

$$|\vec{E}| = \sqrt{2\eta\,|\vec{S}_{\mathrm{av}}|} = \sqrt{2 \cdot 377 \cdot 1500} = 1063.48\ \mathrm{V/m}$$

and

$$|\vec{H}| = \sqrt{\frac{2|\vec{S}_{\mathrm{av}}|}{\eta}} = \sqrt{\frac{2 \cdot 1500}{377}} = 2.8209\ \mathrm{A/m}.$$

Example 4.4

The electric and magnetic field intensities of a uniform plane wave are given as follows:

$$\vec{\mathscr{E}}(\vec{r}, t) = \hat{x}100\cos(10^9 t - \beta z)\ \mathrm{V/m}$$

and

$$\vec{\mathscr{H}}(\vec{r}, t) = \hat{y}0.2\cos(10^9 t - \beta z)\ \mathrm{A/m}.$$

If the wave's phase velocity is 10^8 m/s, find (a) the wavelength λ, (b) the intrinsic impedance η, (c) the relative permeability μ_r, and (d) the dielectric constant ε_r of the medium.

Note that there is no exponential decay (or growth) term in the field amplitude. Therefore the medium is lossless, meaning $\alpha = 0$. Further the wave is propagating in the $+z$ direction. The corresponding phasor fields can be found as follows:

$$\vec{E} = \hat{x}100e^{-j\beta z}\ \mathrm{V/m}$$

and

$$\vec{H} = \hat{y}0.2e^{-j\beta z}\ \mathrm{A/m}.$$

(a) $\beta = \omega/v_{\mathrm{p}} = 10^9/10^8 = 10\ \mathrm{rad/m} \rightarrow \lambda = 2\pi/\beta = 2\pi/10 = 0.2\pi = 0.6283$ m.

(b) $\eta = E_x/H_y = 100/0.2 = 500\ \Omega$.

(c) Since $\eta = \sqrt{\mu/\varepsilon}$ and $v_{\mathrm{p}} = 1/\sqrt{\mu\varepsilon}$,

$$\mu = \frac{\eta}{v_{\mathrm{p}}} = \frac{500}{10^8} = 5 \cdot 10^{-6}\ \mathrm{H/m} \rightarrow \mu_r = \frac{\mu}{\mu_{\mathrm{o}}} = \frac{5 \cdot 10^{-6}}{4\pi \cdot 10^{-7}} = 3.9789.$$

(d) $\varepsilon = 1/v_{\mathrm{p}}\eta = 1/(10^8 \cdot 500) = 2 \cdot 10^{-11}\ \mathrm{F/m} \rightarrow \varepsilon_{\mathrm{r}} = \varepsilon/\varepsilon_{\mathrm{o}}$
$\varepsilon_{\mathrm{r}} = (2 \cdot 10^{-11})/(8.854 \cdot 10^{-12}) = 2.2589$.

Uniform Plane Wave Propagating at an Arbitrary Angle

The preceding analysis can be easily extended to the case of a wave that propagates at an angle from a given axis. Consider a uniform plane wave propagating along r at an angle θ from the z-axis, as illustrated in Figure 4.2. This wave has an x component of the electric field intensity, which can be written as

$$\vec{E} = \hat{x}E_o e^{-jk_r r}. \tag{4.1.20}$$

Since the magnetic field intensity must be orthogonal to both its electric field and the direction of propagation, it will have a y as well as a z component.

We select a point B at r on the propagation path and drop a perpendicular on the y-axis from that point. Further we drop a perpendicular from point C to point A in the direction of the propagation. As indicated in this figure, the angle ABC is equal to θ as well. Assume that the Cartesian coordinates of point B are y and z. Hence

$$r = \text{OA} + \text{AB} = \text{OC}\sin(\theta) + \text{BC}\cos(\theta) = y\sin(\theta) + z\cos(\theta) \tag{4.1.21}$$

and

$$k_r r = k_r \sin(\theta) y + k_r \cos(\theta) z = k_y y + k_z z, \tag{4.1.22}$$

where

$$k_y = k_r \sin(\theta)$$

and

$$k_z = k_r \cos(\theta).$$

Therefore (4.1.20) can be written as follows:

$$\vec{E} = \hat{x}E_o e^{-j(k_y y + k_z z)}. \tag{4.1.23}$$

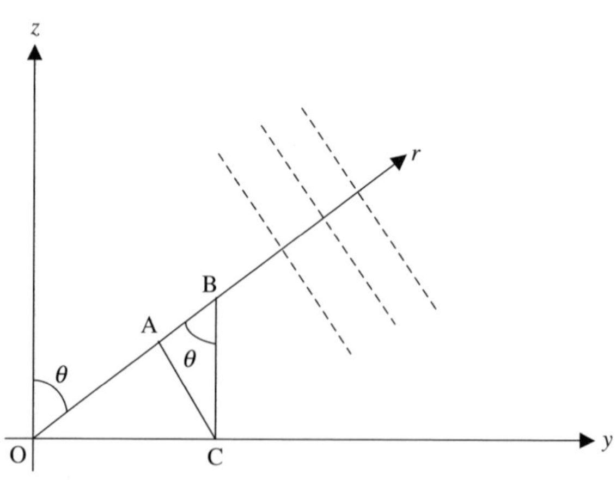

Figure 4.2 Geometry for a uniform plane wave propagating at an angle θ from the z-axis.

The phase velocities $v_{\rm py}$ and $v_{\rm pz}$ of this wave with respect to the y- and z-axes can be found as follows:

$$v_{\rm py} = \frac{\omega}{\mathrm{Re}(k_y)} \tag{4.1.24}$$

and

$$v_{\rm pz} = \frac{\omega}{\mathrm{Re}(k_z)}. \tag{4.1.25}$$

Equation (4.1.23) can be generalized for a plane wave propagating in an arbitrary direction $\hat{u}$ as follows:

$$\vec{E}(\vec{r}) = \vec{A}e^{-j\vec{k}\cdot\vec{r}}, \tag{4.1.26}$$

where $\vec{k} = k\hat{u} = k_x\hat{x} + k_y\hat{y} + k_z\hat{z}$ and $\vec{r} = x\hat{x} + y\hat{y} + z\hat{z}$. The associated magnetic field intensity is

$$\vec{H}(\vec{r}) = \frac{\hat{u}\times\vec{E}(\vec{r})}{\eta}. \tag{4.1.27}$$

Example 4.5

The electric field intensity of a time-harmonic wave traveling in a source-free free space is given as follows:

$$\vec{E} = (4\hat{y} + 3\hat{z})e^{-j(6y-8z)}\ \mathrm{mV/m}$$

Assuming that y and z represent their respective distances in meters, determine (a) the angle of the propagation direction relative to the z-axis, (b) the wavelengths of the wave along the r, y, and z directions, (c) the phase velocities along the r, y, and z directions, (d) the energy velocities along the r, y, and z directions, (e) the frequency of the wave, and (f) the associated magnetic field intensity.

From the given information, $k_y = k\sin\theta = 6$, $k_z = k\cos\theta = 8$. Therefore $k_r = \sqrt{k_y^2 + k_z^2} = \sqrt{6^2+8^2} = 10\ \mathrm{m}^{-1}$.

(a) $\sin\theta = k_y/k_r = 0.6 \rightarrow \theta = 36.87^\circ$ An analysis of the given electric field indicates that this angle is from the $-z$ direction, and therefore it should be subtracted from 180° to obtain the angle measured from $+z$. Hence $\theta = 180^\circ - 36.87^\circ = 143.13^\circ$.

(b) $\lambda_r = 2\pi/k_r = 2\pi/10 = 0.6283\ \mathrm{m}$, $\lambda_y = 2\pi/k_y = 2\pi/6 = 1.0472\ \mathrm{m}$, and $\lambda_z = 2\pi/k_z = 2\pi/8 = 0.7854\ \mathrm{m}$.

(c) Since the wave is propagating in free space, its phase velocity along r is 3×10^8 m/s. The frequency of this wave may be found as follows:

$$\omega = v_r \cdot k_r = 3\times10^9\ \mathrm{rad/s}, \quad \text{or} \quad f = \frac{3\cdot10^8}{\lambda_r}\mathrm{Hz} = 477.47\ \mathrm{MHz},$$

$$v_{\text{py}} = \frac{\omega}{k_y} = \frac{3 \cdot 10^9}{6} = 5 \times 10^8 \text{ m/s},$$

and

$$v_{\text{pz}} = \frac{\omega}{k_z} = \frac{3 \cdot 10^9}{8} = 3.75 \times 10^8 \text{ m/s}.$$

(d) $v_{er} = 3 \times 10^8$ m/s, $v_{ey} = v_{er} \sin\theta = 1.8 \times 10^8$ m/s, and $v_{ez} = v_{er} \cos\theta = 2.4 \times 10^8$ m/s.

(e) The frequency is 477.47 MHz, as found in (c) above.

(f)

$$\vec{H} = -\frac{1}{j\omega\mu_o} \nabla \times \vec{E} = -\frac{1}{j\omega\mu_o} \begin{vmatrix} \hat{x} & \hat{y} & \hat{z} \\ \frac{\partial}{\partial x} & \frac{\partial}{\partial y} & \frac{\partial}{\partial z} \\ 0 & E_y & E_z \end{vmatrix}$$

$$= \hat{x} 13.2629 \times 10^{-6} e^{-j(6y - 8z)} \text{ A/m}.$$

Example 4.6

Phase velocities of a plane wave propagating in a lossless medium are measured in three different directions with unit vectors $\hat{y}$, $(3\hat{x} + 4\hat{y})/5$, and $(2\hat{x} + 2\hat{y} - \hat{z})/3$ at $4 \cdot 10^8$ m/s, $2 \cdot 10^8$ m/s, and $3 \cdot 10^8$ m/s, respectively. Find the direction of propagation and phase velocity of the wave along $(4\hat{y} + 3\hat{z})/5$.

For $\vec{k} = k_x\hat{x} + k_y\hat{y} + k_z\hat{z}$, $k_y = \omega/(4 \cdot 10^8)\,\text{m}^{-1}$,

$$(k_x\hat{x} + k_y\hat{y} + k_z\hat{z}) \cdot \left(\frac{3\hat{x} + 4\hat{y}}{5}\right) = \frac{\omega}{2 \cdot 10^8} \rightarrow \frac{3k_x + 4k_y}{5} = \frac{\omega}{2 \cdot 10^8} \rightarrow$$

$$k_x = \frac{\omega}{2 \cdot 10^8} \text{ m}^{-1}$$

and

$$(k_x\hat{x} + k_y\hat{y} + k_z\hat{z}) \cdot \left(\frac{2\hat{x} + 2\hat{y} - \hat{z}}{3}\right) = \frac{\omega}{3 \cdot 10^8} \rightarrow \frac{2k_x + 2k_y - k_z}{3} = \frac{\omega}{3 \cdot 10^8} \rightarrow$$

$$k_z = \frac{\omega}{2 \cdot 10^8} \text{ m}^{-1}.$$

Therefore the direction of wave propagation is

$$\hat{u} = \frac{\vec{k}}{|\vec{k}|} = \frac{1}{3}(2\hat{x} + \hat{y} + 2\hat{z}).$$

Since

$$(k_x\hat{x} + k_y\hat{y} + k_z\hat{z}) \cdot \left(\frac{4\hat{y} + 3\hat{z}}{5}\right) = \frac{\omega}{4 \cdot 10^8}(2\hat{x} + \hat{y} + 2\hat{z}) \cdot \left(\frac{4\hat{y} + 3\hat{z}}{5}\right) = \frac{\omega}{2 \cdot 10^8}.$$

The velocity along the vector is $2 \cdot 10^8$ m/s.

Phase and Group Velocities

Single-frequency (monochromatic) waves are ideal but not practically realizable. According to the Fourier theorem, any information-carrying wave may be considered as a superposition of ideal waves of different frequencies. In other words, a real wave will have a finite bandwidth. If this bandwidth is narrow, then the wave may be considered as almost single frequency. For a bandwidth of $2\Delta\omega$ about the central frequency ω, we consider two waves as follows:

$$E_1 = A\cos\{(\omega + \Delta\omega)t - (\beta + \Delta\beta)z\} \tag{4.1.28}$$

and

$$E_2 = A\cos\{(\omega - \Delta\omega)t - (\beta - \Delta\beta)z\}. \tag{4.1.29}$$

Note that the propagation constant β also changes with ω. The sum of these two waves is found to be

$$E = E_1 + E_2 = A[\cos\{(\omega + \Delta\omega)t - (\beta + \Delta\beta)z\} + \cos\{(\omega + \Delta\omega)t - (\beta + \Delta\beta)z\}],$$

or

$$E = 2A\cos(\Delta\omega t - \Delta\beta z)\cos(\omega t - \beta z). \tag{4.1.30}$$

Equation (4.1.30) is graphically displayed in Figure 4.3 for a selected set of parameters. The amplitude shown is that of a typical modulated sinusoidal signal. Note that the wave with frequency ω has an amplitude modulated by a wave of frequency $\Delta\omega$. Thus, for a narrow $\Delta\omega$, the wave may be considered as a single frequency ω. The phase of this central frequency signal is moving with a phase velocity v_p, as was defined earlier. In the case where a wave group (or wave packet) moves with a different velocity, the velocity is called group velocity.

Thus the group velocity v_g is defined as

$$v_g = \frac{\Delta\omega}{\Delta\beta}. \tag{4.1.31}$$

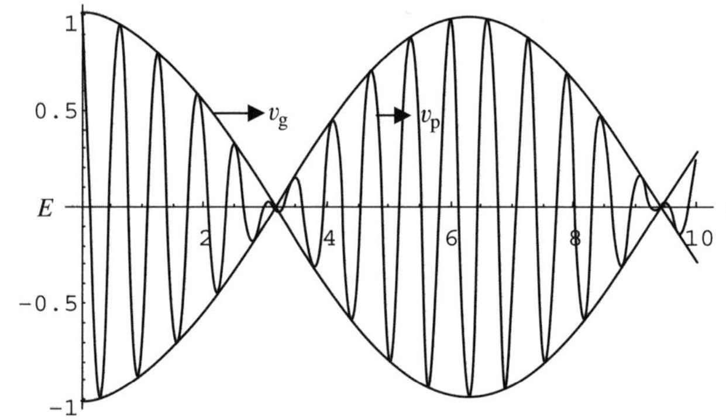

Figure 4.3 Graphical display of (4.1.30) showing the wave group.

For plane waves propagating in a lossless medium, the phase constant β is linearly related with ω if its permittivity and permeability parameters are frequency independent. However, this relation may be nonlinear for certain other media (e.g., media with finite loss or with a permittivity/permeability change with frequency). In this case the phase velocity will vary with frequency. Figure 4.4 shows the typical characteristics of such media. Since an information-carrying signal has a group of frequencies, each signal will travel at a different velocity. This behavior of the medium will distort the information. Such media are called dispersive and the phenomenon is called dispersion.

For a dispersive medium, v_p and v_g are

$$v_p\Big|_{\beta_o} = \frac{\omega_o}{\beta_o} \tag{4.1.32}$$

and

$$v_g\Big|_{\beta_o} = \frac{\delta\omega}{\delta\beta}\Big|_{\beta_o}. \tag{4.1.33}$$

Thus the group velocity will be equal to the phase velocity if ω is linearly related with β (i.e., ω versus β is a straight line). Such materials are called nondispersive. If $v_g < v_p$, then the medium is said to be exhibiting normal dispersion. It exhibits anomalous dispersion instead if $v_g > v_p$.

Example 4.7

A uniform plane wave of 1 GHz is propagating through a lossless medium that exhibits the following relation between the phase velocity v_p and the wavelength λ.

$$v_p = \xi\lambda^{(1/3)},$$

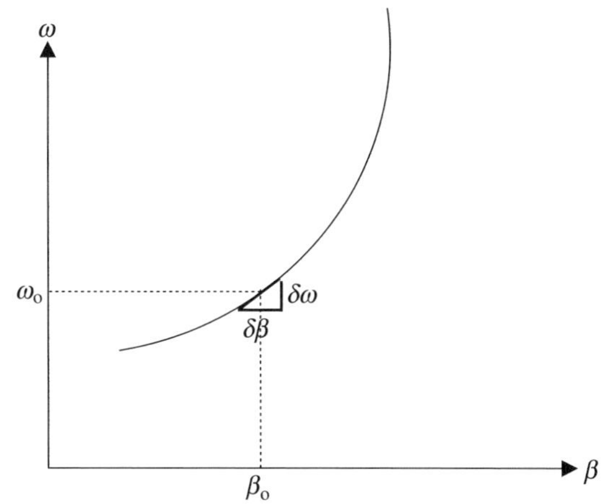

Figure 4.4 The ω–β diagram.

where ξ is a constant. The phase velocity of this signal is $1.5 \cdot 10^8$ m/s. Find the group velocity.

$$v_{\mathrm{p}} = \frac{\omega}{\beta} \rightarrow \omega = \beta v_{\mathrm{p}} = \beta \xi \lambda^{(1/3)} = \beta \xi \left(\frac{2\pi}{\beta} \right)^{(1/3)} = \xi \, (2\pi)^{(1/3)} \beta^{(2/3)}.$$

Therefore

$$\begin{aligned} v_{\mathrm{g}} &= \frac{\delta\omega}{\delta\beta} = \xi \, (2\pi)^{(1/3)} \frac{2}{3} \beta^{(-1/3)} = \frac{2}{3} \xi \left(\frac{2\pi}{\beta} \right)^{(1/3)} \\ &= \frac{2}{3} \xi \lambda^{(1/3)} = \frac{2}{3} v_{\mathrm{p}} = \frac{2}{3} \cdot 1.5 \cdot 10^8 = 10^8 \text{ m/s}. \end{aligned}$$

The medium exhibits normal dispersion.

Polarization

Consider the general case of a uniform plane wave that propagates along the $+z$ axis. It has two electric field components, one along the $+x$-axis and the other along the $+y$-axis. The corresponding magnetic field components are along the $+y$-axis and the $-x$-axis, respectively. The resulting electric and magnetic fields maintain the usual characteristics (i.e., electric and magnetic fields are orthogonal to each other and also to the direction of propagation). The wave is assumed to be propagating in a lossless medium (i.e., $\alpha = 0$). Hence the electric field intensity in time domain can be expressed as follows:

$$\vec{\mathscr{E}}(\vec{r}, t) = \hat{x}\mathscr{E}_x + \hat{y}\mathscr{E}_y = \hat{x}A_x \cos(\omega t - \beta z + \theta_1) + \hat{y}A_y \cos(\omega t - \beta z + \theta_2). \tag{4.1.34}$$

The corresponding phasor expression is found to be

$$\vec{E}(\vec{r}) = \hat{x}E_x + \hat{y}E_y = (\hat{x}A_x + \hat{y}A_y e^{j(\theta_2-\theta_1)})e^{-j(\beta z-\theta_1)}. \tag{4.1.35}$$

Note that the phase of each term in (4.1.34) is changing with time t as well as with distance z. If we track the characteristics of the total electric field on a fixed plane (i.e., at constant z), we find that the field vector rotates with time and traces an ellipse or a circle provided that $\theta_1 \neq \theta_2$. The field vector moves in a straight line when $\theta_1 = \theta_2$. The polarization of the wave is specified by the curve this field traces. The polarization behavior can be found from the magnetic field as well because the wave is always orthogonal to it. This information has many practical applications. For example, if a receiving antenna is not polarization matched, then the power input to the receiver may not be maximized. This characteristic is also used in wireless communication area to maximize usage of the available frequency spectrum. Some more details of each example are included below.

Linear Polarization Consider the case where $\theta_1 = \theta_2 = \theta$. Equation (4.1.34) simplifies to

$$\vec{\mathscr{E}}(\vec{r}, t) = (\hat{x}A_x + \hat{y}A_y)\cos(\omega t - \beta z + \theta). \tag{4.1.36}$$

Over time this field traces a straight line on the $z = z_1$ plane, as shown in Figure 4.5a. The angle φ may be found as follows:

$$\varphi = \tan^{-1}\left(\frac{A_y}{A_x}\right). \tag{4.1.37}$$

Thus the wave is horizontally polarized (along the y-axis) if A_y is zero, and it is vertically polarized (along the x-axis) for A_x zero.

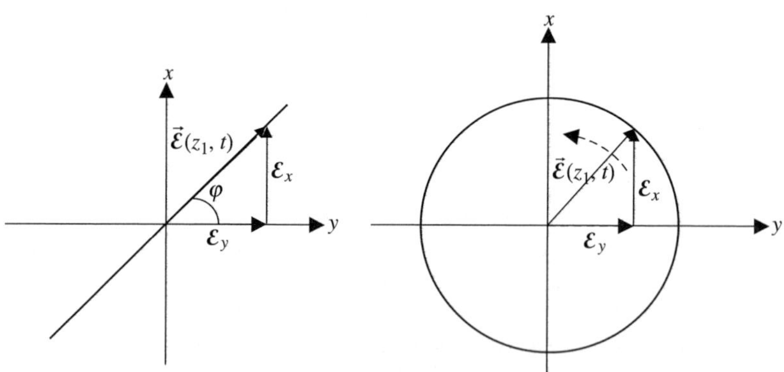

(a) Linear polarization (b) Left-handed circular polarization

Figure 4.5 Linear and circular polarizations.

Circular Polarization Consider now the situation where $A_x = A_y = A$ and $\theta_2 = (\pi/2) + \theta_1$. In this case E_y leads E_x by 90°, and (4.1.34) simplifies to

$$\vec{\mathscr{E}}(\vec{r}, t) = \hat{x}A\cos(\omega t - \beta z + \theta_1) - \hat{y}A\sin(\omega t - \beta z + \theta_1). \tag{4.1.38}$$

Thus the magnitude of the total electric field intensity with time remains constant on a given plane $z = z_o$. However, the vector rotates counterclockwise as time lapses. This can be found by analyzing the right-hand side of (4.1.38). To begin with, assume that the argument $\omega t - \beta z + \theta_1$ is zero. This gives the y component of the field zero, whereas the x component is A. As the argument increases with time, the magnitude of the x component decreases whereas the y component's increases to a negative value. Thus the vector in Figure 4.5*b* seems to rotate counterclockwise. Since the wave is propagating along $+z$ (into the page), this wave is left-handed circularly polarized (LHCP).

Similarly, for $A_x = A_y = A$, and $\theta_2 = -(\pi/2) + \theta_1$, E_y lags behind E_x by 90°. In this case two components of the electric field are found to be

$$\mathscr{E}_x = A_x\cos(\omega t - \beta z + \theta_1) \tag{4.1.39}$$

and

$$\mathscr{E}_y = A_y\cos\left(\omega t - \beta z + \theta_1 - \frac{\pi}{2}\right) = A_y\sin(\omega t - \beta z + \theta). \tag{4.1.40}$$

Looking along the $+z$-axis from the origin, the electric field vector in this case is found to rotate with time in clockwise direction. This kind of wave is called the right-handed circularly polarized (RHCP).

Elliptical Polarization In a most general case, two components of the electric field can be expressed as follows:

$$\frac{\mathscr{E}_x}{A_x} = \cos(\omega t - \beta z + \theta_1) = \cos\chi \tag{4.1.41}$$

and

$$\frac{\mathscr{E}_y}{A_y} = \cos(\omega t - \beta z + \theta_2) = \cos(\chi + \theta) = \cos\chi\cos\theta - \sin\chi\sin\theta, \tag{4.1.42}$$

where $\omega t - \beta z + \theta_1 = \chi$ and $\theta = \theta_2 - \theta_1$. Therefore

$$\left(\frac{\mathscr{E}_y}{A_y} - \cos\chi\cos\theta\right)^2 = (-\sin\chi\sin\theta)^2.$$

After simplifying it, we have

$$\left(\frac{\mathscr{E}_y}{A_y}\right)^2 + \left(\frac{\mathscr{E}_x}{A_x}\right)^2 - 2\left(\frac{\mathscr{E}_y}{A_y}\right)\left(\frac{\mathscr{E}_x}{A_x}\right)\cos\theta = \sin^2\theta. \tag{4.1.43}$$

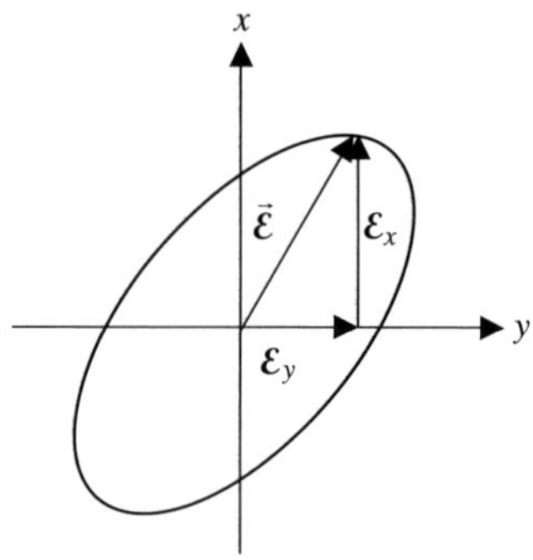

Figure 4.6 Elliptical polarization.

As shown in Figure 4.6, equation (4.1.43) represents equation of an ellipse. Therefore the electric field vector traces an ellipse in this case, and the wave is elliptically polarized. Further it is left-handed elliptically polarized (LHEP) for E_y leading the E_x component. It is right-handed elliptically polarized (RHEP) if E_y lags behind E_x. It can be proved that circular and linear polarizations are special cases of the elliptical polarization.

Example 4.8

A uniform plane wave propagating in free space has the following magnetic field intensity:

$$\vec{H} = (\hat{y}e^{j(\pi/2)} + \hat{z})e^{j\beta x} \text{ mA/m}.$$

Find the instantaneous electric field intensity, time-average power per unit area, and the polarization of the wave. The signal frequency is 1 GHz.

$$\nabla \times \vec{H} = j\omega\varepsilon_o \vec{E} \rightarrow \vec{E} = \frac{1}{j\omega\varepsilon_o}\begin{vmatrix} \hat{x} & \hat{y} & \hat{z} \\ \frac{\partial}{\partial x} & \frac{\partial}{\partial y} & \frac{\partial}{\partial z} \\ 0 & e^{j(\beta x + \pi/2)} & e^{j\beta x} \end{vmatrix}$$

$$= \frac{\beta}{\omega\varepsilon_o}(-\hat{y}e^{j\beta x} + \hat{z}\, je^{j\beta x})\,\text{mV/m},$$

where $\beta/\omega\varepsilon_o = \sqrt{\mu_o/\varepsilon_o} \approx 120\pi = 377\ \Omega$.

Therefore the instantaneous electric field intensity is found to be

$$\vec{\mathscr{E}}(x, t) = \text{Re}(\vec{E}e^{j\omega t}) = -377\{\hat{y}\cos(\omega t + \beta x) + \hat{z}\sin(\omega t + \beta x)\},$$

where $\omega = 2\pi \cdot 10^9$ rad/s and $\beta = 20\pi/3 = 20.944$ rad/m.

The time average power per unit area, S_{av}, is found to be

$$S_{av} = \tfrac{1}{2}\mathrm{Re}(\vec{E} \times \vec{H}^*) = \tfrac{1}{2}\mathrm{Re}\Big[377\{-\hat{y}e^{j\beta x} + \hat{z}e^{j(\beta x+\pi/2)}\}$$
$$\times \{\hat{y}e^{-j(\beta x+\pi/2)} + \hat{z}e^{-j\beta x}\}\Big] \cdot 10^{-6}\ \mathrm{W/m^2}$$

or

$$\vec{S}_{av} = -\hat{x}\,377 \cdot 10^{-6}\ \mathrm{W/m^2} = -\hat{x}\,0.377\ \mathrm{mW/m^2}.$$

For $(\omega t + \beta x) = 0$ the electric field is along the $-y$-axis. As this argument increases over time, the z component of the field increases on the negative side while the y component starts decreasing. The resultant vector of the total electric field rotates counterclockwise (left hand) when moving in the direction of the wave's propagation. Therefore it is a LHCP wave.

4.2 PLANE ELECTROMAGNETIC WAVES IN LOSSY MEDIA

From (4.1.4) and (3.3.9), we have

$$k = \beta - j\alpha = \omega\sqrt{\mu\varepsilon(1 - j\tan\delta)} \rightarrow \beta^2 - \alpha^2 - 2j\beta\alpha = \omega^2\mu\varepsilon(1 - j\tan\delta).$$

Therefore

$$\beta^2 - \alpha^2 = \omega^2\mu\varepsilon \tag{4.2.1}$$

and

$$\beta^2 + \alpha^2 = (\beta - j\alpha)(\beta - j\alpha)^* = \omega^2\mu\varepsilon\sqrt{(1 + \tan^2\delta)}. \tag{4.2.2}$$

From (4.2.1) and (4.2.2), we have

$$\beta = \omega\sqrt{\frac{\mu\varepsilon}{2}(\sqrt{1 + \tan^2\delta} + 1)} \tag{4.2.3}$$

and

$$\alpha = \omega\sqrt{\frac{\mu\varepsilon}{2}(\sqrt{1 + \tan^2\delta} - 1)}. \tag{4.2.4}$$

Low-Loss Dielectric Media

For a low-loss dielectric medium, $\tan \delta \approx \delta$, and therefore (4.2.3) and (4.2.4) simplify to

$$\beta \approx \omega\sqrt{\frac{\mu\varepsilon}{2}\left(1+\frac{\delta^2}{2}+1\right)} = \omega\sqrt{\mu\varepsilon\left(1+\frac{\delta^2}{4}\right)} \approx \omega\sqrt{\mu\varepsilon}\left(1+\frac{\delta^2}{8}\right)$$
$$= \omega\sqrt{\mu\varepsilon}\left(1+\frac{1}{8}\left(\frac{\sigma}{\omega\varepsilon}\right)^2\right) \tag{4.2.5}$$

and

$$\alpha \approx \omega\sqrt{\frac{\mu\varepsilon}{2}\left(1+\frac{\delta^2}{2}-1\right)} = \frac{\omega\delta}{2}\sqrt{\mu\varepsilon} = \frac{\omega}{2}\cdot\frac{\sigma}{\omega\varepsilon}\sqrt{\mu\varepsilon} = \frac{\sigma}{2}\sqrt{\frac{\mu}{\varepsilon}}. \tag{4.2.6}$$

The penetration depth of the wave is defined as the distance d_{p} over which the amplitude of electric field decays to $1/e$ of its value at $z = 0$. Hence

$$d_{\text{p}} = \frac{1}{\alpha}. \tag{4.2.7}$$

Conducting Media

For good conductors, $\sigma \gg \omega\varepsilon$ and $\tan \delta \gg 1$. Therefore

$$\sqrt{1+\tan^2\delta} \approx \tan \delta.$$

In this case (4.2.3) and (4.2.4) can be approximated as follows:

$$\beta \approx \omega\sqrt{\frac{\mu\varepsilon}{2}}(\tan\delta+1) \approx \omega\sqrt{\frac{\mu\varepsilon}{2}}\tan\delta = \sqrt{\frac{\omega\mu\sigma}{2}}\ \text{rad/m} \tag{4.2.8}$$

and

$$\alpha \approx \omega\sqrt{\frac{\mu\varepsilon}{2}}(\tan\delta-1) \approx \omega\sqrt{\frac{\mu\varepsilon}{2}}\tan\delta = \sqrt{\frac{\omega\mu\sigma}{2}}\ \text{np/m}. \tag{4.2.9}$$

Note that $\alpha = \beta$ in this case. Therefore

$$\eta = \frac{\omega\mu}{\beta - j\alpha} \approx \frac{\omega\mu}{1-j}\cdot\sqrt{\frac{2}{\omega\mu\sigma}} = \sqrt{\frac{\omega\mu}{2\sigma}}(1+j)\ \Omega. \tag{4.2.10}$$

When the medium is highly conductive, the depth of penetration is called the skin depth, δ. It is found to be

$$\delta = \sqrt{\frac{2}{\omega\mu\sigma}}. \tag{4.2.11}$$

Example 4.9

A uniform plane wave of 1 MHz propagates in a certain material. Its phase decreases by 90° over a distance of 33 m and the fields are attenuated by a factor of e^{-1} for every 120 m. Further the ratio of its electric and magnetic field intensities at a point in the medium is 163.39 Ω. Find the wave number k, the intrinsic impedance η of the medium, the conductivity σ, and the permittivity ε of the medium, and the permeability μ of the medium.

Since the phase of the wave decreases by 90° over a distance of 33 m, its wavelength is 132 m. Also the depth of penetration is given at 120 m. Therefore

$$\alpha = \frac{1}{120} = 8.3333 \cdot 10^{-3}\,\text{Np/m}$$

and

$$\beta = \frac{2\pi}{\lambda} = \frac{2\pi}{132} = 0.0476\,\text{rad/m}.$$

Hence

$$k = \beta - j\alpha = 0.0476 - j8.3333 \cdot 10^{-3}\,\text{m}^{-1}.$$

Since

$$k = \beta - j\alpha = \omega\sqrt{\mu\varepsilon\left(1 - j\frac{\sigma}{\omega\varepsilon}\right)}$$

and

$$\eta = \frac{\omega\mu}{k} = \frac{\omega\mu}{\beta - j\alpha} = \frac{\mu}{\sqrt{\mu\varepsilon[1 - j(\sigma/\omega\varepsilon)]}} = \sqrt{\frac{\mu}{\varepsilon[1 - j(\sigma/\omega\varepsilon)]}},$$

$$\frac{k}{\eta} = \frac{k^2}{\omega\mu} = \frac{\omega^2\mu\varepsilon[1 - j(\sigma/\omega\varepsilon)]}{\omega\mu} = \omega\varepsilon - j\sigma.$$

The magnitude of η is given at 163.39. Its phase angle ϕ is found to be

$$\phi = \tan^{-1}\left(\frac{\alpha}{\beta}\right) = \tan^{-1}\left(\frac{8.3333 \cdot 10^{-3}}{0.0476}\right) = 0.1733\,\text{rad}.$$

Therefore, $\eta = 163.39\ \Omega\ \angle 0.1733$ rad

$$\sigma = -\text{Im}\left(\frac{k}{\eta}\right) = 1.0048 \cdot 10^{-4}\ \text{S/m},$$

$$\varepsilon_r = \frac{\varepsilon}{\varepsilon_o} = \frac{1}{\omega\varepsilon_o}\text{Re}\left(\frac{k}{\eta}\right) = 5.0002,$$

and

$$\mu_r = \frac{k\eta}{\omega\mu_o} = 1.$$

Therefore the wave is traveling through a nonmagnetic material.

Example 4.10

Two submerged submarines are using a 10 kHz plane electromagnetic wave for their communication. The magnitude of the electric field at the transmitter is 100 mV/m, whereas the receiver requires at least 1 mV/m (peak value) for a reliable communication. Assuming that the conductivity and the dielectric constant of the seawater are 4 S/m and 81, respectively, find (a) the wavelength, (b) the attenuation constant, (c) the phase velocity, (d) the skin depth of the wave, and (e) the maximum range over which a reliable communication is possible.

Since the seawater is nonmagnetic, $\mu = \mu_o$,

$$k = \omega\sqrt{\mu_o\varepsilon_o\varepsilon^*} = \omega\sqrt{\mu_o\varepsilon_o\left(\varepsilon_r - j\frac{\sigma}{\omega\varepsilon_o}\right)}.$$

Further

$$\frac{\sigma}{\omega\varepsilon_o} = \frac{4}{2\pi \times 10^4 \times 8.854 \times 10^{-12}} = 7190.2 \gg 81.$$

Therefore

$$k \approx \omega\sqrt{\mu_o\varepsilon_o\left(-j\frac{\sigma}{\omega\varepsilon_o}\right)} = \sqrt{\frac{\omega\mu_o\sigma}{2}} - j\sqrt{\frac{\omega\mu_o\sigma}{2}} = 0.3974 - j0.3974\ \text{m}^{-1} = \beta - j\alpha.$$

Hence

(a)
$$\lambda = \frac{2\pi}{\beta} = \frac{2\pi}{0.3974} = 15.8107\ \text{m}.$$

(b)
$$\alpha = 0.3974\ \text{Np/m}.$$

(c)
$$v_p = \frac{\omega}{\beta} = \frac{2\pi \times 10^4}{0.3974} = 15.8107 \times 10^4 \text{ m/s}.$$

(d)
$$\delta = \frac{1}{\alpha} = \frac{1}{0.3974} = 2.5163 \text{ m}.$$

(e) Since $\left|\vec{E}(r)\right| = 100 \cdot e^{-0.3974 \cdot r}$ mV/m, the distance d over which the magnitude reduces to 1 mV/m is found to be

$$1 = 100 \cdot e^{-0.3974 \cdot d} \rightarrow d = -\frac{\ln(0.01)}{0.3974} = 11.59 \text{ m}.$$

Example 4.11

An airplane wants to communicate with a submerged submarine using a 10 MHz transmitter. Transmitted uniform plane wave travels downward in the $+z$ direction and produces a magnetic field at the ocean surface ($z = 0$), which is given as follows:

$$\vec{\mathscr{H}}(z = 0, t) = \hat{y}300 \cos(2\pi \cdot 10^7 t) \text{ mA/m}.$$

(a) Express the magnetic field intensity in the ocean. (b) If the submarine requires at least 3 mA/m (peak value) for a reliable communication, how deep it can go without losing the link? (c) How deep the submarine can go if the transmitter frequency is switched to 10 kHz? Assume that the ocean water is nonmagnetic with $\varepsilon_r = 72$, and $\sigma = 4$ S/m.

Since

$$k = \omega\sqrt{\mu_o\varepsilon_o\left(\varepsilon_r - j\frac{\sigma}{\omega\varepsilon_o}\right)}$$

$$= 2\pi \cdot 10^7 \sqrt{4\pi \cdot 10^{-7} \cdot 8.854 \cdot 10^{-12}\left(72 - j\frac{4}{2\pi \cdot 10^7 \cdot 8.854 \cdot 10^{-12}}\right)},$$

or

$$k = 12.6294 - j12.5036 \text{ m}^{-1} = \beta - j\alpha.$$

(a)
$$\vec{\mathscr{H}}(z, t) = \hat{y}300e^{-12.5036 \cdot z} \cos(2\pi \cdot 10^7 t - 12.6294 \cdot z) \text{ mA/m}.$$

(b) Assume that the peak magnetic field reduces to 3 mA/m at $z = d$. Therefore

$$300e^{-12.5036 \cdot d} = 3 \rightarrow d = -\frac{\ln(0.01)}{12.5036} = 0.3683 \text{ m}.$$

(c) For $f = 10$ kHz,

$$k = 2\pi \cdot 10^4 \sqrt{4\pi \cdot 10^{-7} \cdot 8.854 \cdot 10^{-12}\left(72 - j\frac{4}{2\pi \cdot 10^4 \cdot 8.854 \cdot 10^{-12}}\right)}$$

$$= 0.3974 - j0.3974\,\text{m}^{-1}.$$

Therefore the attenuation rate α is smaller at this frequency and

$$300e^{-0.3974 \cdot d} = 3 \rightarrow d = -\frac{\ln(0.01)}{0.3974} = 11.5888\,\text{m}.$$

Example 4.12

An electronic circuit is designed to operate at 1 MHz. To minimize the electromagnetic interference with nearby systems, it is desirable to enclose the circuit in an aluminum box. If the radiation set by the circuit produces an electric field of 100 mV/m on the inner surface of the box and no more than 1 mV/m is permissible outside, then determine the minimum thickness of aluminum required. Can this shield your circuit from electromagnetic radiations above 1 kHz? Determine the minimum aluminum thickness required to shield the circuit. Assume $\sigma = 3.54 \cdot 10^7$ S/m.

For aluminum, $\mu = \mu_o$ and $\varepsilon = \varepsilon_o$. Therefore at 1 MHz,

$$k = \omega\sqrt{\mu_o\varepsilon_o\left(\varepsilon_r - j\frac{\sigma}{\omega\varepsilon_o}\right)}$$

$$= 2\pi \cdot 10^7 \sqrt{4\pi \cdot 10^{-7} \cdot 8.854 \cdot 10^{-12}\left(1 - j\frac{3.54 \cdot 10^7}{2\pi \cdot 10^7 \cdot 8.854 \cdot 10^{-12}}\right)}$$

$$= 1.1822 \cdot 10^4(1 - j) = \beta - j\alpha.$$

If the required thickness of the aluminum sheet is t, then

$$100e^{-1.1822 \cdot 10^4 \cdot t} = 1 \rightarrow t = -\frac{\ln(0.01)}{1.1822 \cdot 10^4} = 3.9 \cdot 10^{-4}\,\text{m}.$$

At 1 kHz,

$$k = \omega\sqrt{\mu_o\varepsilon_o\left(\varepsilon_r - j\frac{\sigma}{\omega\varepsilon_o}\right)}$$

$$= 2\pi \cdot 10^3 \sqrt{4\pi \cdot 10^{-7} \cdot 8.854 \cdot 10^{-12}\left(1 - j\frac{3.54 \cdot 10^7}{2\pi \cdot 10^3 \cdot 8.854 \cdot 10^{-12}}\right)}$$

$$= 373.8363(1 - j) = \beta - j\alpha.$$

Now the attenuation constant is smaller than its earlier value. Therefore it will require a thicker sheet, which can be determined as follows:

$$100e^{-373.8363 \cdot t} = 1 \to t = -\frac{\ln(0.01)}{373.8363} = 0.0123\,\text{m}.$$

Thus 0.4 mm thick sheet is good at 1 MHz but has problem at 1 kHz. A much thicker sheet of about 1.2 cm is needed to shield radiation above 1 kHz.

4.3 UNIFORM PLANE WAVE INCIDENT NORMALLY ON AN INTERFACE

Figure 4.7 shows a uniform plane electromagnetic wave that is propagating along $+z$ in medium 1. Its electric field $\vec{E}^i$ is along the y-axis and the magnetic field $\vec{H}^i$ is in the $-x$ direction. The wave strikes normally a plane interface of medium 2 with medium 1. As shown in the figure, the conductivity, permittivity, and permeability of medium 1 are σ_1, ε_1, and μ_1, respectively. The corresponding parameters for medium 2 are σ_2, ε_2, and μ_2. Since the wave experiences a change in electrical characteristics in its path, a part of incident signal is reflected back as an echo signal while the rest of the power is transmitted into medium 2. Thus the medium 1 has two waves propagating in opposite directions, whereas there is only one forward-moving wave in medium 2.

The time-harmonic electromagnetic fields associated with the incident wave can be expressed as follows:

$$\vec{E}^i = \hat{y}E_o e^{-jk_1 z} \tag{4.3.1}$$

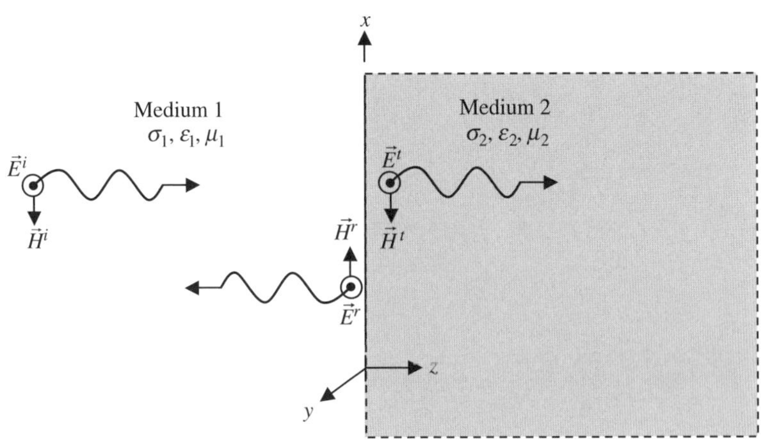

Figure 4.7 Normal incident uniform plane electromagnetic wave.

and

$$\vec{H}^i = -\hat{x}H_o e^{-jk_1 z} = -\hat{x}\frac{E_o}{\eta_1}e^{-jk_1 z}, \tag{4.3.2}$$

where E_o and H_o are arbitrary constants, k_1 is the complex wave number, and η_1 is the complex intrinsic impedance of medium 1. If the medium 1 is a perfect dielectric, then these parameters will be real.

Similarly the electromagnetic fields associated with the reflected wave ($\vec{E}^r$ and $\vec{H}^r$) as well as the transmitted wave ($\vec{E}^t$ and $\vec{H}^t$) can be expressed as follows:

$$\vec{E}^r = \hat{y}RE_o e^{+jk_1 z}, \tag{4.3.3}$$

$$\vec{H}^r = \hat{x}RH_o e^{-jk_1 z} = \hat{x}R\frac{E_o}{\eta_1}e^{+jk_1 z}, \tag{4.3.4}$$

$$\vec{E}^t = \hat{y}TE_o e^{-jk_2 z}, \tag{4.3.5}$$

and

$$\vec{H}^t = -\hat{x}TH_o e^{-jk_2 z} = -\hat{x}T\frac{E_o}{\eta_2}e^{-jk_2 z}, \tag{4.3.6}$$

where R and T are reflection and transmission coefficients, respectively, k_2 is the wave number in medium 2, and η_2 is intrinsic impedance of medium 2. These parameters are complex because $\sigma_2 \neq 0$. Note that the ratio of reflected-to-incident electric field is $+R$. On the other hand, the corresponding ratio of the magnetic field intensity is $-R$. This happens everywhere because of their fundamental relation. Similar behavior may be observed in case of transmission lines where we deal with voltage and current reflection coefficients.

Now enforcing the boundary conditions (tangential components of the fields are continuous across the boundary) at $z = 0$, we get

$$E_o + RE_o = TE_o \rightarrow 1 + R = T \tag{4.3.7}$$

and

$$-\frac{E_o}{\eta_1} + R\frac{E_o}{\eta_1} = -T\frac{E_o}{\eta_2} \rightarrow 1 - R = \frac{\eta_1}{\eta_2}T. \tag{4.3.8}$$

Equations (4.3.7) and (4.3.8) can be solved for R and T as follows:

$$T = \frac{2\eta_2}{\eta_1 + \eta_2} \tag{4.3.9}$$

and

$$R = \frac{\eta_2 - \eta_1}{\eta_1 + \eta_2}. \tag{4.3.10}$$

Further the total electric field at any point in medium 1 is found to be

$$\vec{E}_1 = \hat{y}E_o e^{-jk_1 z} + \hat{y}RE_o e^{+jk_1 z}.$$

Therefore the magnitude of the total electric field intensity in medium 1 is

$$|\vec{E}_1|^2 = \vec{E}_1 \cdot \vec{E}_1^* = |E_o|^2\{e^{-2\alpha_1 z} + \rho^2 e^{2\alpha_1 z} + 2\rho\cos(2\beta_1 z + \theta)\}, \qquad (4.3.11)$$

where

$$R = \rho e^{j\theta}, \qquad (4.3.12)$$

$$k_1 = \beta_1 - j\alpha_1. \qquad (4.3.13)$$

As defined earlier, β_1 is the phase constant in rad/s and α_1 is the attenuation constant in Np/m for medium 1. If medium 1 is lossless, then (4.3.11) simplifies to

$$|\vec{E}_1|^2 = |E_o|^2\{1 + \rho^2 + 2\rho\cos(2\beta_1 z + \theta)\}. \qquad (4.3.14)$$

This relation indicates that the magnitude of total electric field in medium 1 varies with z. Its maximum and minimum values are found as

$$|\vec{E}_1|_{max} = |E_o|\sqrt{\{1 + \rho^2 + 2\rho\}} = |E_o|\{1 + \rho\} \qquad (4.3.15)$$

and

$$|\vec{E}_1|_{min} = |E_o|\sqrt{\{1 + \rho^2 - 2\rho\}} = |E_o|\{1 - \rho\}. \qquad (4.3.16)$$

At the maximum magnitude,

$$\cos(2\beta_1 z + \theta) = 1 \rightarrow 2\beta_1 z + \theta = 2n\pi, \qquad n = 0, 1, 2 \ldots. \qquad (4.3.17)$$

At the minimum magnitude,

$$\cos(2\beta_1 z + \theta) = -1 \rightarrow 2\beta_1 z + \theta = (2m + 1)\pi, \qquad m = 0, 1, 2 \ldots. \qquad (4.3.18)$$

Equation (4.3.17) indicates that the magnitude maxima are separated by $\lambda/2$. Similar results are found from (4.3.18) for minima as well. Further maxima and minima are $\lambda/4$ apart and alternate. The maximum occurs where the incident and reflected waves are in phase, and the two wave types are out of phase at the minimum. The standing wave ratio (SWR) is defined as the ratio of the maximum to the minimum of the total electric field. Hence

$$\mathrm{SWR} = \frac{|\vec{E}_1|_{max}}{|\vec{E}_1|_{max}} = \frac{1 + \rho}{1 - \rho}. \qquad (4.3.19)$$

These characteristics are similar to those of a wave propagating on a transmission line.

Example 4.13

A uniform plane electromagnetic wave propagating in free space impinges normally on a lossless nonmagnetic medium of $\varepsilon_r = 2.25$ in the $z \geq 0$ region. The electric field intensity of the incident wave is given as follows:

$$\vec{\mathscr{E}}^i(z, t) = \hat{y}4.8\cos(10^9 t - k_o z)\,\mathrm{V/m}.$$

Find k_o, k_2, R, T, SWR, and the average power transmitted into the medium.

Wave numbers and the intrinsic impedance of the wave in the two media are found as follows:

$$k_o = \omega\sqrt{\mu_o\varepsilon_o} = 10^9\sqrt{4\pi \times 10^{-7} \times 8.854 \times 10^{-12}} = 3.3356\,\mathrm{rad/m},$$

$$k_2 = \omega\sqrt{\mu_o\varepsilon_o\varepsilon_r} = k_o\sqrt{2.25} = 5.0034\,\mathrm{rad/m},$$

$$\eta_1 = \sqrt{\frac{\mu_o}{\varepsilon_o}} = \sqrt{\frac{4\pi \times 10^{-7}}{8.854 \times 10^{-12}}} = 376.7343\,\Omega,$$

and

$$\eta_2 = \sqrt{\frac{\mu_o}{\varepsilon_o\varepsilon_r}} = \sqrt{\frac{4\pi \times 10^{-7}}{8.854 \times 10^{-12} \times 2.25}} = 251.1562\,\Omega.$$

From (4.3.9) and (4.3.10) we get

$$R = \frac{\eta_2 - \eta_1}{\eta_1 + \eta_2} \approx -0.2$$

and

$$T = \frac{2\eta_2}{\eta_1 + \eta_2} \approx 0.8.$$

The SWR in medium 1 is found from (4.3.19) as follows:

$$\mathrm{VSWR} = \frac{1+\rho}{1-\rho} = \frac{1+0.2}{1-0.2} = 1.5.$$

The electric field intensity in medium 2 is

$$\vec{\mathscr{E}}^t(z, t) = \hat{y}\,3.84\cos(10^9 t - 5.0034z)\,\mathrm{V/m}.$$

Therefore the time average power transmitted into medium 2 is found as follows:

$$P_{av} = \frac{1}{2}\mathrm{Re}(\vec{E} \times \vec{H}^*) = \frac{1}{2}\frac{|\vec{E}^t|^2}{\eta_2} = \frac{1}{2}\frac{3.84^2}{251.1562} = 0.0294\,\mathrm{W/m^2}.$$

Example 4.14

A uniform plane wave of 500 MHz propagating in air is incident normally on the interface with a semi-infinite dielectric slab. If it shows a SWR of 1.8 and the electric field has a minimum right at the interface, find the intrinsic impedance and the dielectric constant of this slab. Further find the average power density of the wave in the slab if the incident wave has a peak electric field of 100 V/m.

From (4.3.19),

$$\mathrm{SWR} = \frac{1+\rho}{1-\rho} \rightarrow \rho = \frac{\mathrm{SWR}-1}{\mathrm{SWR}+1} = \frac{1.8-1}{1.8+1} = 0.2857.$$

Since minimum occurs right at the interface ($z = 0$), $\theta = \pi$.

For free space, $\eta_1 = \sqrt{\mu_o/\varepsilon_o} = 376.7343\,\Omega$. From (4.3.10),

$$R = 0.2857e^{j\pi} = \frac{\eta_2 - \eta_1}{\eta_1 + \eta_2} \rightarrow \eta_2 = \frac{1+(-0.2857)}{1-(-0.2857)} \cdot 376.7343 = 209.3033\,\Omega.$$

Since $\eta_2 = \sqrt{\mu_o/\varepsilon_o\varepsilon_r} = \eta_1/\sqrt{\varepsilon_r}$,

$$\varepsilon_r = \left(\frac{\eta_1}{\eta_2}\right)^2 = 3.2398.$$

Further

$$1 + R = T \rightarrow T = 1 - 0.2857 = 0.7143.$$

The power density S_t in the slab is found to be

$$S_t = \frac{1}{2}\frac{(TE_o)^2}{\eta_2} = \frac{1}{2}\frac{(0.7143 \cdot 100)^2}{209.3033} = 12.1886\ \mathrm{W/m^2}.$$

The power densities associated with incident and reflected waves are

$$S_{in} = \frac{1}{2}\frac{(E_o)^2}{\eta_1} = \frac{1}{2}\frac{(100)^2}{376.7343} = 13.272\ \mathrm{W/m^2}$$

and

$$S_r = \frac{1}{2}\frac{|RE_o|^2}{\eta_1} = \frac{1}{2}\frac{(0.2857 \cdot 100)^2}{376.7343} = 1.0833\ \mathrm{W/m^2}.$$

As expected, $S_t = S_{in} - S_r$.

Example 4.15

A uniform plane wave of 100 MHz propagating in air is incident normally on the interface with a semi-infinite medium. If it shows a reflection coefficient of $0.25\angle 15^\circ$, find the intrinsic impedance of the material and the standing wave ratio in air. Further find the total electric field at the boundary if the incident wave has a peak electric field of 100 V/m.

For air $\eta_1 = 376.7343\ \Omega$,

$$R = \frac{\eta_2 - \eta_1}{\eta_1 + \eta_2} \rightarrow \eta_2 = \frac{1+R}{1-R} \cdot \eta_1 = 615.2107e^{j0.1372}\ \Omega = 609.4319 + j84.1241\ \Omega$$

and

$$\text{SWR} = \frac{1+\rho}{1-\rho} = \frac{1+0.25}{1-0.25} = 1.6667.$$

The total electric field $|E_1|$ at the boundary is found as

$$|E_1| = E_o\sqrt{1+\rho^2+2\rho\cos\theta} = 100\sqrt{1+0.25^2+2\cdot 0.25\cdot\cos(15^\circ)} = 124.32\ \text{V}.$$

Since the tangential electric field is continuous across the boundary, the transmitted field at the boundary should give same results as well. This can be verified as follows:

$$|E_1| = 100\cdot|T| = 100\cdot|1+R| = 100\cdot|1.2415+j0.0647| = 124.32\ \text{V}.$$

Example 4.16

An airplane communicates with a submerged submarine using a uniform plane wave of 100 MHz. The wave propagating along $+z$ (downward) in air is incident normally on the seawater (interface at $z=0$) with a power density of 20 W/m^2. Find the electric and magnetic fields in the seawater. If the submarine requires at least 1 μW/m^2 for a reliable communication, find the depth up to which it can go without losing the signal. Assume $\varepsilon_r = 80$ and $\sigma = 4.5$ S/m for the seawater.

For air, $\eta_1 = 376.7343\ \Omega$. For seawater,

$$\varepsilon^* = \varepsilon_o\left(\varepsilon_r - j\frac{\sigma}{\omega\varepsilon_o}\right) = \varepsilon_o\left(80 - j\frac{4.5}{2\pi\cdot 10^8\cdot 8.854\cdot 10^{-12}}\right)$$

$$= \varepsilon_o(80 - j808.8968)\ \text{F/m}.$$

Therefore

$$\eta_2 = \sqrt{\frac{\mu_o}{\varepsilon^*}} = 9.7927 + j8.8719\ \Omega = 13.2139e^{j0.7361}\ \Omega.$$

Using the relation (4.1.13) for the power density, we find that

$$S_{\text{in}} = \frac{1}{2}\frac{(E_o)^2}{\eta_1} = \frac{1}{2}\eta_1 (H_o)^2 \rightarrow E_o = \sqrt{2S_{\text{in}}\eta_1}$$
$$= \sqrt{2 \cdot 20 \cdot 376.7343} = 122.7574 \text{ V/m}$$

and

$$H_o = 0.3258 \text{ A/m}.$$

The reflection and transmission coefficients are found to be

$$R = \frac{\eta_2 - \eta_1}{\eta_2 + \eta_1} = 0.9494e^{j3.0945}$$

and

$$T = 1 + R = 0.0684e^{j0.7132}.$$

Since $|E_oT| = 8.391$ V/m, the electric and magnetic fields in seawater can be expressed as follows:

$$\vec{E} = \hat{x}8.391e^{-j(k_2 z - 0.7132)} \text{ V/m}$$

and

$$\vec{H} = \hat{y}0.635e^{-j(k_2 z + 0.0229)} \text{ A/m}.$$

Therefore the average power density in seawater can be written as follows:

$$S_{\text{avg}} = \tfrac{1}{2}\text{Re}\left[\vec{E} \times \vec{H}^*\right] = \tfrac{1}{2}\text{Re}\left[8.391 \cdot 0.635 \cdot e^{j(0.7132+0.0229)} \cdot e^{-2\alpha_2 z}\right]$$
$$= 1.9744e^{-2\alpha_2 z} \text{ W/m}^2.$$

Assume that 1 μW/m^2 is available at a depth of d. Therefore

$$10^{-6} = 1.9744e^{-2\alpha_2 d} \rightarrow d = \frac{6}{2 \cdot 40.1185}[\ln(10) - \ln(1.9744)] = 0.1213 \text{ m}.$$

Note that the skin depth for medium 2 (seawater) is only 0.0249 m.

Example 4.17

A uniform plane wave of 5 MHz is incident normally on a thick copper sheet. If its incident electric field is 1000 V/m, find the electric and magnetic fields inside the sheet. What is the wave-velocity inside the sheet? What happens if the signal frequency decreases to 5 kHz? For copper, $\sigma = 5.813 \cdot 10^7$ S/m.

For air, $\eta_1 = 376.7343\ \Omega$, whereas η_2 and k_2 for copper are found to be

$$\eta_2 = \sqrt{\frac{\mu_o}{\varepsilon_o - (j\sigma/\omega)}} = 8.24 \cdot 10^{-4} e^{j0.7854} = 5.8273 \cdot 10^{-4}(1 + j)\,\Omega$$

and

$$k_2 = \omega\sqrt{\mu_o\left(\varepsilon_o - j\frac{\sigma}{\omega}\right)} = 3.3874 \cdot 10^4(1 - j)\ \mathrm{m}^{-1}.$$

Therefore the reflection and transmission coefficient at the air–copper interface are

$$R = \frac{\eta_2 - \eta_1}{\eta_2 + \eta_1} = 1e^{j3.1416}$$

and

$$T = 1 + R = 4.375 \cdot 10^{-6} e^{j0.7854}.$$

The magnitude of the transmitted electric field is found to be $1000|T| = 4.375 \cdot 10^{-3}$ V/m, and the corresponding magnetic field is $(1000 \cdot T)/\eta_2 = 5.3088 \cdot e^{-j1.5468 \cdot 10^{-6}}$ A/m. The skin depth $= 2.9521 \cdot 10^{-6}$ m, and the phase velocity in copper $v_p = \omega/\beta = 927.4379$ m/s.

When frequency changes to 5 kHz, η_2 and k_2 change as follows:

$$\eta_2 = \sqrt{\frac{\mu_o}{\varepsilon_o - (\sigma/\omega)}} = 2.606 \cdot 10^{-5} e^{j0.7854} = 1.8427 \cdot 10^{-5}(1 + j)\,\Omega$$

and

$$k_2 = \omega\sqrt{\mu_o\left(\varepsilon_o - j\frac{\sigma}{\omega}\right)} = 1.0712 \cdot 10^3(1 - j)\,\mathrm{m}^{-1}.$$

Therefore the new reflection and transmission coefficients are

$$R = \frac{\eta_2 - \eta_1}{\eta_2 + \eta_1} = 1e^{j3.1416}$$

and

$$T = 1 + R = 1.3835 \cdot 10^{-7} e^{j0.7854}.$$

Now the magnitudes of the electric and magnetic fields are found to be $1000|T| = 1.3835 \cdot 10^{-4}$ V/m and $(1000 \cdot T)/\eta_2 = 5.3088 \cdot e^{-j4.8914 \cdot 10^{-8}}$ A/m, respectively. The skin depth is increased to $9.3354 \cdot 10^{-4}$ m, whereas the phase velocity is reduced to $v_p = \omega/\beta = 29.3282$ m/s.

Example 4.18

A wireless communication network is allowed to use a 1 V/m radiation at 2.45 GHz. Find the magnetic field intensity in humans working there if the wave is incident normally and also the depth over which fields decrease by $1/e$. Assume that the human body is a semi-infinite plane medium with $\varepsilon_r = 47$ and $\sigma = 2.21$ S/m and that the radiation is in the form of a uniform plane wave. How do these results compare if the radiation frequency decreases to 40 MHz, with $\varepsilon_r = 97$ and $\sigma = 0.7$ S/m at this frequency?

For $f = 2.45$ GHz, $\varepsilon_r = 47$, and $\sigma = 2.21$ S/m,

$$\eta_2 = \sqrt{\frac{\mu_o}{\varepsilon_o\varepsilon_r - j(\sigma/\omega)}} = 53.429e^{j0.1661} = 52.6936 + j8.834\ \Omega$$

and

$$k_2 = \omega\sqrt{\mu_o\left(\varepsilon_o\varepsilon_r - j\frac{\sigma}{\omega}\right)} = 357.0755 - j59.8629\,\text{m}^{-1}.$$

Therefore the reflection and transmission coefficients are found to be

$$R = \frac{\eta_2 - \eta_1}{\eta_2 + \eta_1} = 0.7547e^{j3.0938}$$

and

$$T = 1 + R = 0.2488e^{j0.1455}.$$

The magnitude of transmitted magnetic field is found to be $T/\eta_2 = 4.6564 \cdot 10^{-3} \cdot e^{-j0.0206}$ A/m. The skin depth = 0.0167 m, and the phase velocity $v_p = \omega/\beta = 4.3111 \cdot 10^7$ m/s.

When the frequency changes to 40 MHz, $\varepsilon_r = 97$ and $\sigma = 0.7$ S/m, η_2 and k_2 change to

$$\eta_2 = \sqrt{\frac{\mu_o}{\varepsilon_o\varepsilon_r - j(\sigma/\omega)}} = 20.7641e^{j0.6358} = 16.7062 + j12.331\ \Omega$$

and

$$k_2 = \omega\sqrt{\mu_o\left(\varepsilon_o\varepsilon_r - j\frac{\sigma}{\omega}\right)} = 12.2377 - j9.0327\,\text{m}^{-1}.$$

Therefore the new reflection and transmission coefficients are found to be

$$R = \frac{\eta_2 - \eta_1}{\eta_2 + \eta_1} = 0.9152e^{j3.076}$$

and

$$T = 1 + R = 0.1055e^{j0.6045}.$$

Hence the magnitude of the transmitted magnetic field is found as follows:

$$\frac{T}{\eta_2} = 5.0809 \cdot 10^{-3} \cdot e^{-j0.0313}\ \text{A/m}.$$

The skin depth at this frequency is found to be 0.1107 m, and the phase velocity is

$$v_{\text{p}} = \frac{\omega}{\beta} = 2.0537 \cdot 10^{7}\ \text{m/s}.$$

Reflection with Multiple Interfaces

Consider a medium of thickness d sandwiched between two semi-infinite media, as illustrated in Figure 4.8. A part of the wave incident from medium 1 is transmitted into medium 2 and the remaining power is reflected back from the interface. The wave transmitted into medium 2 splits again into two parts at the interface. One of these parts is reflected back into medium 2, and the other propagates forward into medium 3.

On the basis of the preceding analysis, the total electric and magnetic fields in medium 2 can be expressed as follows:

$$\vec{E}(z) = \hat{y}E_{\text{o}}(e^{-jk_2z} + R_2e^{jk_2z}) = \hat{y}E_y(z) \tag{4.3.20}$$

and

$$\vec{H}(z) = -\hat{x}\frac{E_{\text{o}}}{\eta_2}(e^{-jk_2z} - R_2e^{jk_2z}) = -\hat{x}H_x(z), \tag{4.3.21}$$

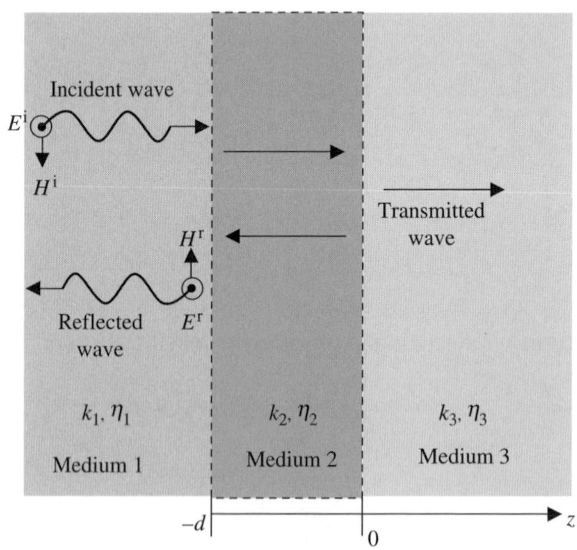

Figure 4.8 Geometry of three media with interfaces at $z = 0$ and $z = -d$.

where R_2 is the reflection coefficient at the interface between medium 2 and medium 3. Therefore the wave impedance $Z(z)$ in medium 2 is found to be

$$Z(z) = \frac{E_y(z)}{-H_x(z)} = \eta_2 \frac{e^{-jk_2 z} + R_2 e^{jk_2 z}}{e^{-jk_2 z} - R_2 e^{jk_2 z}} = \eta_2 \frac{e^{-jk_2 z} + [(\eta_3 - \eta_2)/(\eta_3 + \eta_2)]e^{jk_2 z}}{e^{-jk_2 z} - [(\eta_3 - \eta_2)/(\eta_3 + \eta_2)]e^{jk_2 z}}.$$

This relation can be simplified to

$$Z(z) = \eta_2 \frac{\eta_3 - j\eta_2 \tan(k_2 z)}{\eta_2 - j\eta_3 \tan(k_2 z)}. \tag{4.3.22}$$

The wave impedance at $z = -d$ in medium 2 is

$$Z(z = -d) = \eta_2 \frac{\eta_3 + j\eta_2 \tan(k_2 d)}{\eta_2 + j\eta_3 \tan(k_2 d)}. \tag{4.3.23}$$

Note that (4.3.23) represents the wave impedance on the medium 2 side of the interface. If this is equal to the wave impedance η_1 of medium 1, then there will be no reflected wave in medium 1. In other words, medium 1 will be the impedance matched at the interface. For the lossless medium, $k_2 = \beta_2$, and (4.3.23) may be written as follows:

$$Z(z = -d) = \eta_2 \frac{\eta_3 + j\eta_2 \tan(\beta_2 d)}{\eta_2 + j\eta_3 \tan(\beta_2 d)}. \tag{4.3.24}$$

In fact there are two cases of practical interests that can be studied at this point. Consider the case of $\tan(\beta_2 d) = 0$ that occurs when $\beta_2 d = n\pi \rightarrow d = n\lambda/2$, $n = 0, 1, 2.\ldots$ Then (4.3.24) reduces to

$$Z(z = -d) = \eta_3. \tag{4.3.25}$$

This says that there will be no reflection in medium 1 if $\eta_1 = \eta_3$. This relation is useful to observe in designing protective enclosures (known as the radome) for outdoor antennas. There are many other possible design applications of this case.

Now consider the case where medium 1 is different from medium 3 (i.e., $\eta_1 \neq \eta_3$). If $d = (2n + 1)(\lambda/4)$, $n = 0, 1, 2, \ldots$, then $\tan(\beta_2 d) = \infty$, and (4.3.24) reduces to

$$Z(z = -d) = \frac{\eta_2^2}{\eta_3}.$$

Since

$$R_1 = \frac{Z(-d) - \eta_1}{Z(-d) + \eta_1},$$

medium 1 will be matched (no reflection in medium 1) if

$$\eta_1 = \frac{\eta_2^2}{\eta_3} \rightarrow \eta_2 = \sqrt{\eta_1 \eta_3}. \tag{4.3.26}$$

This case is known as the quarter-wave plate matching technique.

Example 4.19

An LED light source operating at 0.9 μm is generating uniform plane waves that are incident at a semi-infinite glass ($\varepsilon_r = 4.82$). Determine its reflection coefficient. Suggest a mechanism to reduce this reflection to zero.

The reflection coefficient is found as follows:

$$R_1 = \frac{\sqrt{\mu_o/\varepsilon_o\varepsilon_r} - \sqrt{\mu_o/\varepsilon_o}}{\sqrt{\mu_o/\varepsilon_o\varepsilon_r} + \sqrt{\mu_o/\varepsilon_o}} = -\frac{1.1954}{3.1954} = -0.3741.$$

We need a quarter-wave plate to reduce this reflection to zero. Therefore

$$\eta_2 = \sqrt{\eta_1 \eta_3} \rightarrow \sqrt{\frac{\mu_o}{\varepsilon_o \varepsilon_{r2}}} = \sqrt{\sqrt{\frac{\mu_o}{\varepsilon_o \varepsilon_{r1}}} \cdot \sqrt{\frac{\mu_o}{\varepsilon_o \varepsilon_{r3}}}} \rightarrow \varepsilon_{r2} = \sqrt{\varepsilon_{r1} \cdot \varepsilon_{r3}}$$

$$\varepsilon_{r2} = \sqrt{1 \cdot 4.82} = 2.1954$$

and

$$d = \frac{\lambda_2}{4} = \frac{\lambda_{\text{air}}}{4 \cdot \sqrt{\varepsilon_{r2}}} = \frac{0.9}{4 \cdot \sqrt{2.1954}} = 0.1519\,\mu\text{m}$$

Example 4.20

A wireless network using a 1.9 GHz signal is established in an office. A few cubicals are needed for the office staff, and these are to be made from wooden boards of maple ($\varepsilon_r = 2.1$). Find the appropriate thickness of the boards so that the partitions will not affect the signal strength. Assume that the network uses the uniform plane waves.

In this case there is air on two sides of the board. Therefore we need half-wavelength thick boards to reduce the reflections. Since $\lambda_{\text{air}} = (3 \cdot 10^8)/(1.9 \cdot 10^9)$ m = 15.7895 cm,

$$d = \frac{\lambda_2}{2} = \frac{\lambda_{\text{air}}}{2 \cdot \sqrt{\varepsilon_{r2}}} = \frac{15.7895}{2 \cdot \sqrt{2.1}} = 5.4479\,\text{cm}.$$

4.4 UNIFORM PLANE WAVE INCIDENT OBLIQUELY ON AN INTERFACE

When a uniform plane wave is obliquely incident on an interface, the reflected wave does not trace back the path. The plane normal to the interface that contains incident and reflected wave paths is called the plane of incidence. The electric field of the incident wave may be perpendicular or parallel to this plane. The former situation is referred as the perpendicularly polarized or the TE wave and the latter as the parallel polarized or the TM wave. These two cases are separately analyzed below.

Electric Field Perpendicular to the Plane of Incidence

Consider a perpendicularly polarized uniform plane wave incident on an interface of two lossless dielectric media, as shown in Figure 4.9. The plane of incidence is y-z plane (the same as the plane of this sheet of paper) and the electric field is x directed (coming out of the paper). As indicated, the angle of incidence is θ_i. A part of the incident signal is transmitted into medium 2 at an angle θ_t while the remaining signal is reflected into medium 1 at θ_r.

From (4.1.21) and (4.1.22) the electric field intensities associated with incident, reflected, and transmitted signals can be found to be

$$\vec{E}_i(\vec{r}) = \hat{x}E_o e^{-jk_1(y\sin\theta_i + z\cos\theta_i)}, \tag{4.4.1}$$

$$\vec{E}_r(\vec{r}) = \hat{x}R_\perp E_o e^{-jk_1(y\sin\theta_r - z\cos\theta_r)}, \tag{4.4.2}$$

and

$$\vec{E}_t(\vec{r}) = \hat{x}T_\perp E_o e^{-jk_2(y\sin\theta_t + z\cos\theta_t)}, \tag{4.4.3}$$

where subscripts i, r, and t are used to represent incident, reflected, and transmitted (also known as refracted) signals, and $R_\perp$ and $T_\perp$ are the reflection and transmission coefficients at the interface. From

$$\nabla \times \vec{E}(\vec{r}) = -j\omega\mu\vec{H}(\vec{r}) \rightarrow \vec{H}(\vec{r}) = \frac{j}{\omega\mu}\left(\hat{y}\frac{\partial E_x}{\partial z} - \hat{z}\frac{\partial E_x}{\partial y}\right) = \hat{y}H_y - \hat{z}H_z.$$

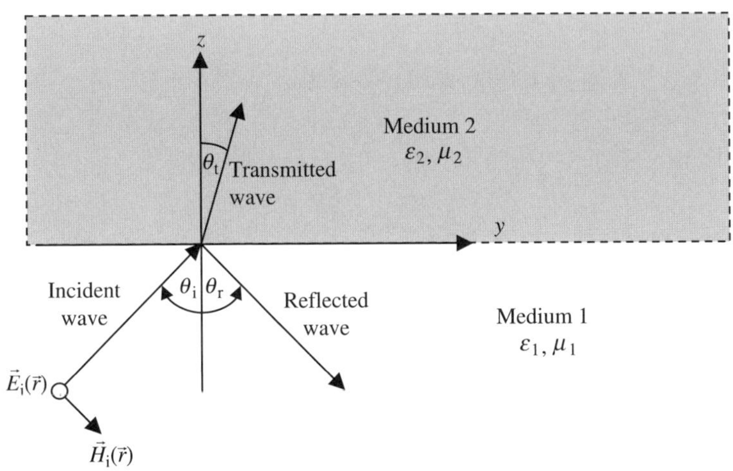

Figure 4.9 TE wave incident on a planar interface.

Therefore

$$H_{y1} = \frac{k_1 E_o}{\omega\mu_1}\left[\cos\theta_i e^{-jk_1(y\sin\theta_i + z\cos\theta_i)} - R_\perp \cos\theta_r e^{-jk_1(y\sin\theta_r - z\cos\theta_r)}\right] \tag{4.4.4}$$

and

$$H_{y2} = \frac{k_2}{\omega\mu_2} T_\perp E_o \cos\theta_t e^{-jk_2(y\sin\theta_t + z\cos\theta_t)}. \tag{4.4.5}$$

Application of the boundary condition that the tangential electric intensity must be continuous across the interface yields

$$\hat{x} \cdot \left[\vec{E}_i(y, z=0) + \vec{E}_r(y, z=0)\right] = \hat{x} \cdot \vec{E}_t(y, z=0).$$

Next we find that

$$e^{-jk_1 y\sin\theta_i} + R_\perp e^{-jk_1 y\sin\theta_r} = T_\perp e^{-jk_2 y\sin\theta_t}. \tag{4.4.6}$$

Since (4.4.6) should hold for all y, the phase terms must be identical for each term. Therefore

$$k_1 \sin\theta_i = k_1 \sin\theta_r = k_2 \sin\theta_t. \tag{4.4.7}$$

This equation is known as the phase-matching condition. The following two relations, known as the Snell laws for reflection and refraction, respectively, are direct consequence of this requirement:

$$k_1 \sin\theta_i = k_1 \sin\theta_r \rightarrow \theta_i = \theta_r$$

and

$$k_1 \sin\theta_i = k_2 \sin\theta_t \rightarrow \sin\theta_t = \sqrt{\frac{\mu_1\varepsilon_1}{\mu_2\varepsilon_2}}\sin\theta_i. \tag{4.4.8}$$

For nonmagnetic dielectric materials, $\mu = \mu_o$, and therefore (4.4.8) simplifies to

$$\sin\theta_t = \sqrt{\frac{\varepsilon_1}{\varepsilon_2}}\sin\theta_i = \frac{n_1}{n_2}\sin\theta_i, \tag{4.4.9}$$

where $n = \sqrt{\varepsilon/\varepsilon_o} = \sqrt{\varepsilon_r}$ is refractive index of the material.

With the phase-matching condition satisfied, (4.4.6) reduces to

$$1 + R_\perp = T_\perp. \tag{4.4.10}$$

From the continuity of tangential magnetic fields at the interface, we have

$$H_{y1}|_{z=0} = H_{y2}|_{z=0}.$$

From (4.4.4), (4.4.5), and the phase-matching condition (Snell's laws satisfied) this boundary condition gives

$$1 - R_{\perp} = \frac{\eta_1}{\eta_2} \cdot \frac{\cos\theta_t}{\cos\theta_i} T_{\perp}. \tag{4.4.11}$$

Equations (4.4.10) and (4.4.11) can be solved to find $R_{\perp}$ and $T_{\perp}$ as follows:

$$R_{\perp} = \frac{\eta_2 \cos\theta_i - \eta_1 \cos\theta_t}{\eta_2 \cos\theta_i + \eta_1 \cos\theta_t} \tag{4.4.12}$$

and

$$T_{\perp} = \frac{2\eta_2 \cos\theta_i}{\eta_2 \cos\theta_i + \eta_1 \cos\theta_t}. \tag{4.4.13}$$

Electric Field Parallel to the Plane of Incidence

Now consider the case when a parallel-polarized uniform plane wave is incident on an interface of two lossless dielectric media, as shown in Figure 4.10. The plane of incidence is y-z plane (plane of the paper) and the magnetic field is x directed (coming out of paper). As indicated, the angle of incidence is θ_i. A part of incident signal is transmitted into medium 2 at an angle θ_t while the remaining is reflected into medium 1 at θ_r.

Following (4.1.21) and (4.1.22), expressions for the magnetic field intensities associated with incident, reflected, and transmitted signals can be found to be

$$\vec{H}_i(\vec{r}) = \hat{x} H_o e^{-jk_1(y \sin\theta_i + z\cos\theta_i)}, \tag{4.4.14}$$

$$\vec{H}_r(\vec{r}) = \hat{x} R_{\parallel} H_o e^{-jk_1(y \sin\theta_r - z\cos\theta_r)}, \tag{4.4.15}$$

and

$$\vec{H}_t(\vec{r}) = \hat{x} T_{\parallel} H_o e^{-jk_2(y \sin\theta_t + z\cos\theta_t)}, \tag{4.4.16}$$

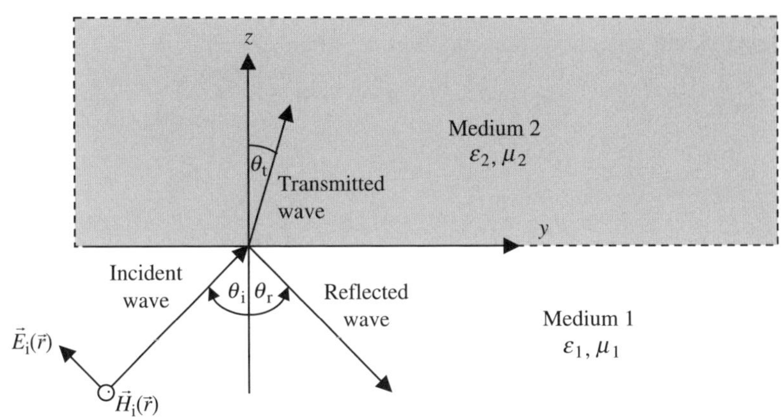

Figure 4.10 TM wave incident on a planar interface.

where $R_{\|}$ and $T_{\|}$ are the reflection and the transmission coefficients, respectively. From

$$\nabla \times \vec{H}(\vec{r}) = j\omega\varepsilon\vec{E}(\vec{r}) \rightarrow \vec{E}(\vec{r}) = -\frac{j}{\omega\varepsilon}\left(\hat{y}\frac{\partial H_x}{\partial z} - \hat{z}\frac{\partial H_x}{\partial y}\right) = \hat{y}E_y + \hat{z}E_z.$$

Therefore the associated electric field intensities are

$$\vec{E}_{\mathrm{i}} = \frac{k_1 H_{\mathrm{o}}}{\omega\varepsilon_1}[-\hat{y}\cos\theta_{\mathrm{i}} + \hat{z}\sin\theta_{\mathrm{i}}]e^{-jk_1(y\sin\theta_{\mathrm{i}} + z\cos\theta_{\mathrm{i}})}, \tag{4.4.17}$$

$$\vec{E}_{\mathrm{r}} = \frac{k_1 R_{\|} H_{\mathrm{o}}}{\omega\varepsilon_1}[\hat{y}\cos\theta_{\mathrm{r}} - \hat{z}\sin\theta_{\mathrm{r}}]e^{-jk_1(y\sin\theta_{\mathrm{i}} - z\cos\theta_{\mathrm{i}})}, \tag{4.4.18}$$

and

$$\vec{E}_{\mathrm{t}} = \frac{k_2 T_{\|} H_{\mathrm{o}}}{\omega\varepsilon_2}[-\hat{y}\cos\theta_{\mathrm{t}} + \hat{z}\sin\theta_{\mathrm{t}}]e^{-jk_2(y\sin\theta_{\mathrm{t}} + z\cos\theta_{\mathrm{t}})} \tag{4.4.19}$$

The boundary conditions require that the tangential electric and magnetic fields be continuous across the boundary. Enforcing the continuity of magnetic fields, we have

$$\hat{x} \bullet \{\vec{H}_{\mathrm{i}}(\vec{r}) + \vec{H}_{\mathrm{r}}(\vec{r}) - \vec{H}_{\mathrm{t}}(\vec{r})\}|_{z=0} = 0.$$

Therefore

$$e^{-jk_1 y\sin\theta_{\mathrm{i}}} + R_{\|}e^{-jk_1 y\sin\theta_{\mathrm{r}}} = T_{\|}e^{-jk_2 y\sin\theta_{\mathrm{t}}}. \tag{4.4.20}$$

As in the previous case, (4.4.20) should hold for all y, and therefore the phase terms must be identical for each term:

$$k_1\sin\theta_{\mathrm{i}} = k_1\sin\theta_{\mathrm{r}} = k_2\sin\theta_{\mathrm{t}}. \tag{4.4.21}$$

Note that it is same phase-matching condition as for the TE case. Therefore the laws of reflection and refraction (transmission) are same as well:

$$k_1\sin\theta_{\mathrm{i}} = k_1\sin\theta_{\mathrm{r}} \rightarrow \theta_{\mathrm{i}} = \theta_{\mathrm{r}} \tag{4.4.22}$$

and

$$k_1\sin\theta_{\mathrm{i}} = k_2\sin\theta_{\mathrm{t}} \rightarrow \sin\theta_{\mathrm{t}} = \sqrt{\frac{\mu_1\varepsilon_1}{\mu_2\varepsilon_2}}\sin\theta_{\mathrm{i}}. \tag{4.4.23}$$

For the wave traveling from a nonmagnetic dielectric material to air, relation (4.4.23) reduces to

$$\sin\theta_{\mathrm{t}} = \sqrt{\frac{\varepsilon_1}{\varepsilon_2}}\sin\theta_{\mathrm{i}} = \frac{n_1}{n_2}\sin\theta_{\mathrm{i}}.$$

where $n = \sqrt{\varepsilon/\varepsilon_o} = \sqrt{\varepsilon_r}$ is refractive index of the material. With the phase-matching condition satisfied, (4.4.20) reduces to

$$1 + R_{\|} = T_{\|}. \tag{4.4.24}$$

The continuity of tangential electric field across the boundary requires that

$$E_{y1}\big|_{z=0} = E_{y2}\big|_{z=0}.$$

Therefore, from (4.4.17) through (4.4.19) along with (4.4.21), we have

$$1 - R_{\|} = \frac{\eta_2}{\eta_1} \cdot \frac{\cos\theta_t}{\cos\theta_i} T_{\|}. \tag{4.4.25}$$

Equations (4.4.24) and (4.4.25) can be solved for $R_{\|}$ and $T_{\|}$ as follows:

$$R_{\|} = \frac{\eta_1 \cos\theta_i - \eta_2 \cos\theta_t}{\eta_1 \cos\theta_i + \eta_2 \cos\theta_t} \tag{4.4.26}$$

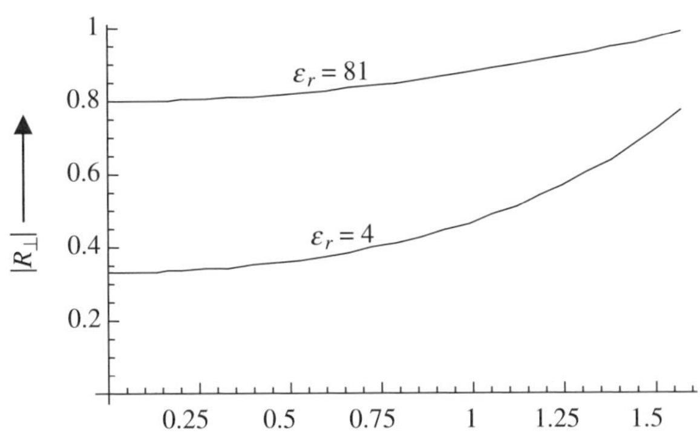

(a) Perpendicular polarization

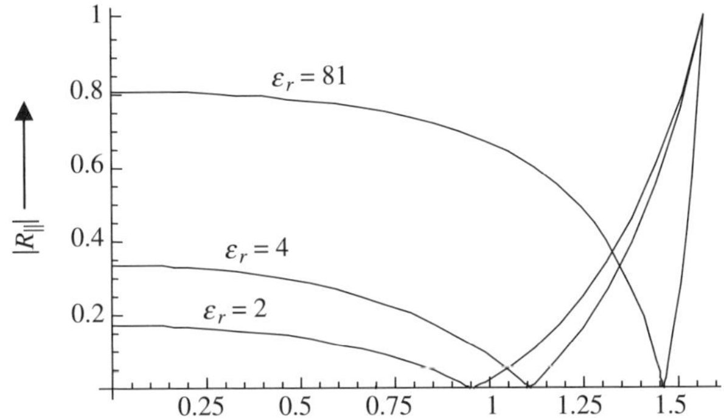

(b) Parallel polarization

Figure 4.11 Reflection coefficient versus the incident angle for TE (a) and TM waves in air to selected dielectrics (b).

and

$$T_{\parallel} = \frac{2\eta_1 \cos\theta_i}{\eta_1 \cos\theta_i + \eta_2 \cos\theta_t}. \tag{4.4.27}$$

Typical characteristics of the reflection coefficients for the two cases are illustrated in Figure 4.11.

Total Transmission

Note that $R_{\parallel} = 0$ for $\eta_2 \cos\theta_t = \eta_1 \cos\theta_i$. Hence the signal is transmitted totally into medium 2 if the incident angle satisfies the following condition:

$$\cos\theta_i = \frac{\eta_2}{\eta_1}\cos\theta_t = \sqrt{\frac{\mu_2\varepsilon_1}{\varepsilon_2\mu_1}}\cos\theta_t. \tag{4.4.28}$$

If both media are nonmagnetic, then $\mu_2 = \mu_1 = \mu_o$ and we denote this incident angle as θ_{iB}. Therefore (4.4.28) can be written as follows:

$$\cos\theta_{iB} = \sqrt{\frac{\varepsilon_1}{\varepsilon_2}}\cos\theta_t. \tag{4.4.29}$$

Since $\sin\theta_i = (k_2/k_1)\sin\theta_t$, we find that for $\mu_2 = \mu_1 = \mu_o$,

$$\sin\theta_{iB} = \sqrt{\frac{\varepsilon_2}{\varepsilon_1}}\sin\theta_t$$

or

$$\sin\theta_t = \sqrt{\frac{\varepsilon_1}{\varepsilon_2}}\sin\theta_{iB}. \tag{4.4.30}$$

Equation (4.4.30) is used to draw the right-angle triangle of Figure 4.12*a*. From this figure we find that

$$\cos\theta_t = \sqrt{1 - \left(\frac{\varepsilon_1}{\varepsilon_2}\right)\sin^2\theta_i}. \tag{4.4.31}$$

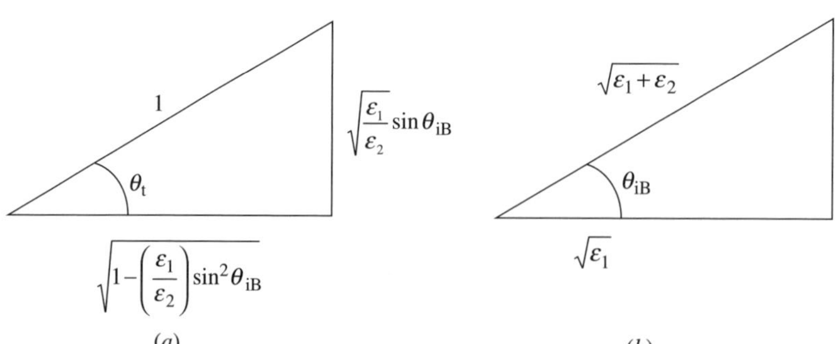

Figure 4.12 Trigonometric relations of (4.4.30) and (4.4.32).

Using (4.4.31), (4.4.29) can be written as follows:

$$\cos\theta_{iB} = \sqrt{\frac{\varepsilon_1}{\varepsilon_2}} \cdot \sqrt{1 - \left(\frac{\varepsilon_1}{\varepsilon_2}\right)\sin^2\theta_{iB}} \rightarrow \cos^2\theta_{iB} = 1 - \sin^2\theta_{iB}$$

$$\cos^2\theta_{iB} = \left(\frac{\varepsilon_1}{\varepsilon_2}\right)\left[1 - \left(\frac{\varepsilon_1}{\varepsilon_2}\right)\sin^2\theta_{iB}\right]$$

or

$$\sin^2\theta_{iB} = \frac{1}{1 + (\varepsilon_1/\varepsilon_2)} \rightarrow \sin\theta_{iB} = \sqrt{\frac{\varepsilon_2}{\varepsilon_1 + \varepsilon_2}}. \tag{4.4.32}$$

Equation (4.4.32) is used to draw the right-angle triangle of Figure 4.12*b*. From the figure we find that

$$\tan\theta_{iB} = \sqrt{\frac{\varepsilon_2}{\varepsilon_1}}. \tag{4.4.33}$$

The incident angle for which total transmission takes place is known as the Brewster angle. Note that Brewster's angle θ_{iB} exists only for TM waves unless the media involved are magnetic, meaning $\mu_1 \neq \mu_2 \neq \mu_o$.

Since total transmission of the incident signal in a nonmagnetic dielectric medium can occur only for TM waves, the Brewster's window can be employed to filter out TE waves. As illustrated in Figure 4.13, Brewer's windows are used to generate a linearly polarized gas laser source. The gas discharge tube has Brewster's windows, one on each side. As a mixture of TE and TM waves propagates along the z-axis, the TM waves, which are y-polarized, are transmitted through the windows because they are set such that the incident angle is equal to the Brewster angle. However, some of the TE waves, which are x-polarized are reflected in the opposite direction, as shown. If quartz ($\varepsilon_r = 2.25$ at optical frequencies) is used for the windows, the Brewster angle is about 56.31°. Then a parallel-polarized (TM) wave will pass through the windows without reflection and a perpendicularly polarized (TE) wave will be reflected at an angle of $2\theta_{iB}$ from the z-axis (axis of the discharge tube). The reflection coefficient for this case is found to be -0.3846. The

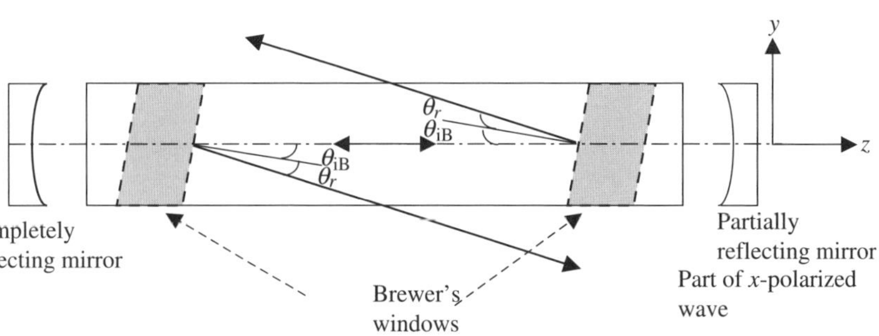

Figure 4.13 Brewster's windows in a typical gas laser source.

wave traveling along $-z$ is reflected back completely from the mirror set on the left, and the wave traveling along $+z$ is partially reflected back. The rest of the signal is available as the output of the laser source. Note that the output is y-polarized.

Total Internal Reflection

Consider the situation where a planar electromagnetic wave is incident upon an interface with a nonmagnetic material of a lower dielectric constant (i.e., $\varepsilon_1 > \varepsilon_2$). Since the phase-matching conditions for TE and TM both require that $k_1 \sin\theta_i = k_2 \sin\theta_t$, as given by (4.4.8) and (4.4.23), respectively, $\sin\theta_t$ can exceed unity for certain θ_i. The incident angle, for which θ_t is 90°, is known as the critical angle. Hence

$$k_1 \sin\theta_i = k_2 \sin\theta_t \rightarrow \sin\theta_i = \frac{\omega\sqrt{\mu_o \varepsilon_2}}{\omega\sqrt{\mu_o \varepsilon_1}} \sin\theta_t \rightarrow \sin\theta_{ic} = \sqrt{\frac{\varepsilon_2}{\varepsilon_1}}. \tag{4.4.34}$$

If the incident angle is increased beyond θ_{ic}, then in order to satisfy the phase matching, θ_t becomes an imaginary number. Using the analysis of a uniform plane wave that propagates at an arbitrary angle, and (4.1.19) through (4.1.22), we find that

$$k_{z2} = \sqrt{k_2^2 - k_2^2 \sin\theta_t} = \sqrt{k_2^2 - k_1^2 \sin^2\theta_i} = \pm j\sqrt{k_1^2 \sin^2\theta_i - k_2^2}. \tag{4.4.35}$$

In this case the wave is propagating along the y-axis without attenuation, whereas it is attenuated in the z direction. Note that a negative sign is physically appropriate in (4.4.35) for such TE and TM wave propagations because of the resulting exponential decay of the fields. Further (4.4.35) gives

$$\cos\theta_t = \pm j\sqrt{\left(\frac{k_1}{k_2}\right)^2 \sin^2\theta_i - 1} = \pm j\sqrt{\left(\frac{\varepsilon_1}{\varepsilon_2}\right) \sin^2\theta_i - 1}. \tag{4.4.36}$$

Therefore (4.4.12) and (4.4.26) simplify in this case to

$$R_\perp = \frac{\cos\theta_i + j\sqrt{\sin^2\theta_i - (\varepsilon_2/\varepsilon_1)}}{\cos\theta_i - j\sqrt{\sin^2\theta_i - (\varepsilon_2/\varepsilon_1)}} = 1\ \angle 2\tan^{-1}\left[\frac{1}{\cos\theta_i}\sqrt{\sin^2\theta_i - \left(\frac{\varepsilon_2}{\varepsilon_1}\right)}\right] \tag{4.4.37}$$

and

$$\begin{aligned} R_\parallel &= \frac{\sqrt{(\varepsilon_2/\varepsilon_1)}\cos\theta_i + j\sqrt{(\varepsilon_1/\varepsilon_2)\sin^2\theta_i - 1}}{\sqrt{(\varepsilon_2/\varepsilon_1)}\cos\theta_i - j\sqrt{(\varepsilon_1/\varepsilon_2)\sin^2\theta_i - 1}} \\ &= 1\ \angle 2\tan^{-1}\left[\frac{\sqrt{(\varepsilon_1/\varepsilon_2)\sin^2\theta_i - 1}}{\sqrt{(\varepsilon_2/\varepsilon_1)}\cos\theta_i}\right]. \end{aligned} \tag{4.4.38}$$

Hence the magnitude of the reflection coefficient is unity for both polarizations. Since the wave is totally reflected back for $\theta_i > \theta_{ic}$, this situation is known as total internal reflection.

Another way to analyze this phenomenon is as follows: For a perpendicularly polarized wave the transmitted electric and magnetic fields in medium 2 are

$$\vec{E}_t = \hat{x} T_\perp E_o e^{-jk_2(y \sin\theta_t + z\cos\theta_t)}$$

and

$$\vec{H}_t = T_\perp E_o \frac{k_2}{\omega\mu_o}[\hat{y}\cos\theta_t - \hat{z}\sin\theta_t] e^{-jk_2(y\sin\theta_t + z\cos\theta_t)}.$$

Then the average power flow in this medium is found to be

$$S_{av} = \frac{1}{2}\mathrm{Re}\left(\vec{E}_t \times \vec{H}_t^*\right) = \frac{1}{2}\frac{k_2}{\omega\mu_o}|T_\perp E_o|^2\ \mathrm{Re}[\hat{y}\sin\theta_t + \hat{z}\cos\theta_t]. \tag{4.4.39}$$

Since $\cos\theta_t$ is imaginary if the incident angle is greater than the critical angle, (4.4.39) says that there is no power flow in the z direction. However, there is power transfer in the y direction because $\sin\theta_t$ is a real number. Similar results can be found for TM waves. Note that the transmitted waves no longer exhibit the uniform plane wave characteristics. These waves are called surface waves because they are transmitted only along the boundary surface. It is by this principle that optical signals are transmitted through fibers. A dielectric rod can be used to guide any electromagnetic signal.

Example 4.21

A 2 mm thick quartz panel ($\varepsilon_r = 2.25$) lies on the x-y plane, as shown in Figure 4.14. An optical uniform plane wave is incident upon its surface at $y = 0$ with an incident angle of 89°. Determine if the panel will guide this wave. What happens if incident angle reduces to 30°?

For the quartz-to-air signal, the critical angle is found to be

$$\theta_{ic} = \sin^{-1}\left(\frac{1}{\sqrt{\varepsilon_r}}\right) = \sin^{-1}\left(\frac{1}{1.5}\right) = 41.8103°.$$

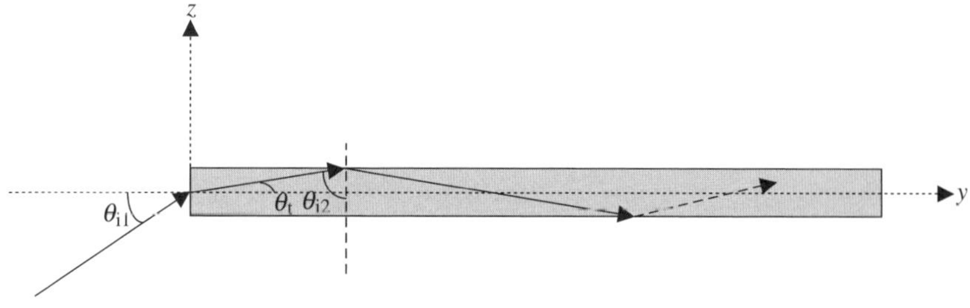

Figure 4.14 Geometry of a quartz panel and an incident optical signal.

Note that $\theta_{i2} = 90^\circ - \theta_t$. Further

$$\sin\theta_t = \sqrt{\frac{\mu_1\varepsilon_1}{\mu_2\varepsilon_2}}\sin\theta_i = \sqrt{\frac{1}{\varepsilon_r}}\sin 89^\circ = 0.6666 \rightarrow \theta_t = 41.8025^\circ.$$

Therefore

$$\theta_{i2} = 90^\circ - \theta_t = 48.1975^\circ.$$

Since θ_{i2} is larger than the critical angle, this wave will be totally reflected back and strike the other side of the panel as shown. Thus this wave will be guided along the panel.

If incident angle reduces to 30°, then

$$\sin\theta_t = \sqrt{\frac{\mu_1\varepsilon_1}{\mu_2\varepsilon_2}}\sin\theta_i = \sqrt{\frac{1}{\varepsilon_r}}\sin 30^\circ = 0.3333 \rightarrow \theta_t = 19.4712^\circ$$

and

$$\theta_{i2} = 90^\circ - \theta_t = 70.5288^\circ.$$

This also meets the requirement of total internal reflection, and it too will be guided.

Example 4.22

An LED is submerged in a swimming pool. If the light seen on the surface is a circle of radius 2 m, find its depth d in the pool. Assume that the water is lossless and its dielectric constant is 1.77 in the optical range.

Note that the signal travels from water to air, and therefore there will be total internal reflection after the critical angle is reached. The light is visible on the surface up to this angle that forms a circle. Further

$$\theta_{ic} = \sin^{-1}\left(\frac{1}{\sqrt{\varepsilon_r}}\right) = \sin^{-1}\left(\frac{1}{1.3304}\right) = 48.7332^\circ$$

Therefore

$$\tan\theta_{ic} = \frac{a}{d} \rightarrow d = \frac{a}{\tan\theta_{ic}} = \frac{2}{\tan(48.7332^\circ)} = 1.775\,\text{m}.$$

Example 4.23

A TM polarized uniform plane wave traveling in water ($\varepsilon_r = 81$) is incident upon its boundary with air. The angle of incidence is 30°. Find the reflection and transmission coefficients, and the transmitted electric and magnetic fields into the air.

Taking the wave number in air as k_o, we find that

$$k_1\sin\theta_i = \sqrt{81}k_o\sin 30^\circ = 4.5k_o = k_o\sin\theta_t.$$

The angle of transmission must be imaginary in this case. To verify that further, the critical angle is obtained as

$$\theta_{ic} = \sin^{-1}\sqrt{\frac{1}{81}} = 6.3794^\circ$$

Since the incident angle is larger than the critical angle, the transmission angle will be an imaginary number. From (4.4.38) we have

$$R_\parallel = 1\angle 2\tan^{-1}\left[\frac{\sqrt{(\varepsilon_1/\varepsilon_2)\sin^2\theta_i - 1}}{\sqrt{(\varepsilon_2/\varepsilon_1)}\cos\theta_i}\right] = 1\angle 177.4872^\circ$$

and

$$T_\parallel = 1 + R_\parallel = 1 + 1\angle 177.4872^\circ = 0.0438\angle 88.6934^\circ.$$

Since $k_2 = k_o$, $\sin\theta_t = 4.5$, and $\cos\theta_t = -j4.3875$, the electric and magnetic fields transmitted into air are found from (4.4.16) and (4.4.19) as follows:

$$\vec{H}_t(\vec{r}) = \hat{x}T_\parallel H_o e^{-jk_2(y\sin\theta_t + z\cos\theta_t)} = \hat{x}\,0.0438\,H_o e^{-4.3875k_oz}\,e^{-j(4.5k_oy-1.548)}\ \text{A/m}$$

and

$$\vec{E}_t = \frac{k_2T_\parallel H_o}{\omega\varepsilon_2}[-\hat{y}\cos\theta_t + \hat{z}\sin\theta_t]e^{-jk_2(y\sin\theta_t + z\cos\theta_t)}$$

$$= \sqrt{\frac{\mu_o}{\varepsilon_o}}[\hat{y}\,j4.3875 + \hat{z}4.5]0.0438H_o e^{-4.3875k_oz}\,e^{-j(4.5k_oy-1.548)}\ \text{V/m}.$$

Note that the fields quickly attenuate along the z-axis while propagating in the y direction. The field attenuation rate in the z direction is 38.1 dB for each wavelength, and the power density is 76.2 dB for each wavelength.

PROBLEMS

4.1. The electric field intensity of a uniform plane wave propagating through a nonmagnetic medium is given as follows:

$$\vec{\mathscr{E}}(x,t) = \hat{y}5\cos(10^9t + 5x)\ \text{mV/m}.$$

Find (a) the direction of wave travel, (b) the phase velocity of the wave, (c) the wavelength, and (d) the associated magnetic field intensity.

4.2. The time-harmonic electric field intensity of a uniform plane wave propagating in free space is given as follows:

$$\vec{E} = \hat{z}(1+j)e^{+j(4\pi y)}\ \text{mV/m}.$$

Find (a) the direction of propagation, (b) the wavelength, (c) the frequency, and (d) the associated magnetic field intensity.

4.3. Assume that a safe radio frequency and microwave exposure standard for the humans is set at 2 μW/cm^2. Find the corresponding electric and magnetic field intensities in air.

4.4. The electric and magnetic field intensities of a uniform plane wave are given as follows:

$$\vec{\mathscr{E}}(\vec{r}, t) = -\hat{y}150\cos(10^9 t - \beta z)\ \text{V/m}$$

and

$$\vec{\mathscr{H}}(\vec{r}, t) = \hat{x}0.15\cos(10^9 t - \beta z)\ \text{A/m}.$$

If its phase velocity is $2 \cdot 10^8$ m/s, find (a) the wavelength λ, (b) the intrinsic impedance η, (c) the relative permeability μ_r, and (d) the dielectric constant ε_r of the medium.

4.5. The electric field intensity of a time-harmonic wave traveling in a source-free free space is given as follows:

$$\vec{E} = (7\hat{y} + 5\hat{z})e^{-j(10y-14z)}\ \text{mV/m}.$$

Assuming y and z represent their respective distances in meters, determine (a) the angle of the propagation direction relative to the z-axis, (b) the wavelengths of the wave along the r, y, and z directions, (c) the phase velocities along the r, y, and z directions, (d) the energy velocities along the r, y, and z directions, (e) the frequency of the wave, and (f) the associated magnetic field intensity.

4.6. The phase velocities of a plane wave propagating in a lossless medium are measured in three different directions with unit vectors $\hat{x}$, $(4\hat{x} + 3\hat{z})/5$ and $(2\hat{x} - 2\hat{y} + 2\hat{z})/3$ at $4 \cdot 10^8$ m/s, 10^8 m/s, and $3 \cdot 10^8$ m/s, respectively. Find the direction of propagation and phase velocity of the wave along $(3\hat{y} + 4\hat{z})/5$.

4.7. A uniform plane wave of 500 MHz is propagating through a lossless medium that exhibits the following relation between the phase velocity v_p and the wavelength λ.

$$v_p = \xi\lambda^{1/3},$$

where ξ is a constant. The phase velocity of this signal is 10^8 m/s. Find the group velocity.

4.8. A uniform plane wave propagating in free space has the following magnetic field intensity:

$$\vec{H} = (\hat{y} + \hat{z}e^{j(\pi/2)})e^{-j\beta x}\ \text{mA/m}.$$

Find the instantaneous electric field intensity, the time-average power per unit area, and the polarization of the wave. The signal frequency is 9.375 GHz.

4.9. When a uniform plane wave of 1 MHz propagates in a certain material, it is found that its phase decreases by 90° over a distance of 50 m and the fields are attenuated by a factor of e^{-1} for every 170 m. Further the ratio of the wave's electric and magnetic field intensities at a point in the medium is 163.39 Ω. Find the wave number k, the intrinsic impedance η of the medium, the conductivity σ, the permittivity ε of the medium, and the permeability μ of the medium.

4.10. Two submerged submarines are using a 10 kHz plane electromagnetic wave for their communication. The magnitude of the electric field at the transmitter is 200 mV/m, whereas the receiver requires at least 50 μV/m (peak value) for a reliable communication. Assuming that the conductivity and the dielectric constant of the seawater are 4 S/m and 81, respectively, find (a) the wavelength, (b) the attenuation constant, (c) the phase velocity, (d) the skin depth of the wave, and (e) the maximum range over which a reliable communication is possible.

4.11. An airplane wants to communicate with a submerged submarine using a 100 MHz transmitter. The transmitted uniform plane wave travels downward in the $+z$ direction and produces a magnetic field at the ocean surface ($z = 0$) that is given as follows:

$$\vec{\mathscr{H}}(z = 0, t) = \hat{y}\, 500 \cos(2\pi \cdot 10^8 t)\ \mathrm{mA/m}.$$

(a) Express the magnetic field intensity in the ocean. (b) If the submarine requires at least 500 nA/m (peak value) for a reliable communication, how deep it can go without losing the link? (c) How deep the submarine can go if the transmitter frequency is switched to 100 kHz? Assume that the ocean water is nonmagnetic with $\varepsilon_r = 72$, and $\sigma = 4$ S/m.

4.12. An electronic circuit is designed to operate at 800 MHz. To minimize the electromagnetic interference with nearly systems, it is desired to enclose the circuit in an aluminum box. If the radiation set by the circuit produces an electric field of 100 mV/m on the inner surface of the box and no more than 100 μV/m is permissible outside the box, determine the minimum thickness of aluminum required. Can this shield your circuit from electromagnetic radiations above 60 Hz? Determine the minimum aluminum thickness required to shield the circuit. Assume $\sigma = 3.54 \cdot 10^7$ S/m.

4.13. A uniform plane electromagnetic wave propagating in free space impinges normally on a lossless nonmagnetic medium of $\varepsilon_r = 5.2$ in a $z \geq 0$ region. The electric field intensity of the incident wave is given as follows:

$$\vec{\mathscr{E}}^i(z, t) = \hat{y} \cos(2\pi \cdot 10^9 t - k_o z)\ \mathrm{V/m}.$$

Find k_o, k_2, R, T, SWR, and the average power transmitted into the medium.

4.14. A uniform plane wave of 1 GHz propagating in air is incident normally on an interface with a semi-infinite dielectric slab. If the wave shows an SWR of 2.3 and the electric field has a minimum right at the interface, find the intrinsic impedance and the dielectric constant of this slab. Further find the average power density of the wave in the slab if the incident wave has a peak electric field of 5 V/m.

4.15. A uniform plane wave of 1 GHz propagating in air is incident normally on the interface with a semi-infinite medium. If the wave shows a reflection coefficient of $0.3\angle -30^\circ$, find the intrinsic impedance of the material and the standing wave ratio in air. Further find the total electric field at the boundary if the incident wave has a peak electric field of 5 nV/m.

4.16. An airplane communicates with a submerged submarine using a uniform plane wave of 5 GHz. The wave propagating along $+z$ (downward) in air is incident normally on the seawater (interface at $z = 0$) with a power density of 1 W/m^2. Find the electric and magnetic fields in the seawater. If the submarine requires at least 1500 pW/m^2 for a reliable communication, find the depth up to which it can go without losing the signal. Assume $\varepsilon_r = 80$ and $\sigma = 4.5$ S/m for the seawater.

4.17. A uniform plane wave of 800 MHz is incident normally on a thick copper sheet. If the incident electric field is 2 V/m, find the electric and magnetic fields inside the sheet. What is the wave velocity inside the sheet? What happens if the signal frequency decreases to 500 Hz? For copper, $\sigma = 5.813 \cdot 10^7$ S/m.

4.18. A wireless communication network is allowed to use a 10 mW/cm^2 power density at 915 MHz. Find the power density in humans working there if the wave is incident normally, and find also the depth over which the fields decrease by $1/e$. Assume that the human body is a semi-infinite plane medium with $\varepsilon_r = 47$ and $\sigma = 2.21$ S/m and that the radiation is in the form of a uniform plane wave. How do these results compare if the radiation frequency decreases to 50 MHz ($\varepsilon_r = 97$ and $\sigma = 0.7$ S/m at this frequency)?

4.19. An LED light source operating at 0.65 μm is generating uniform plane waves that are incident at a semi-infinite glass ($\varepsilon_r = 4.82$). Determine the reflection coefficient of the glass. Suggest a mechanism to reduce this reflection to zero.

4.20. A wireless network is established in an office using a 5.6 GHz signal. The office needs a few cubicals for its staff, and these are to be made from wooden boards of maple ($\varepsilon_r = 2.1$). Find the appropriate thickness of the boards that keeps the partitions from affecting the signal strength. Assume that the network uses the uniform plane waves.

4.21. A 5 mm thick quartz panel ($\varepsilon_r = 2.25$) lies on the x-y plane, as shown in Figure 4.14. An optical uniform plane wave is incident upon the panel's surface at $y = 0$ with an incident angle of 85°. Determine if the panel will guide this wave. What happens if the incident angle reduces to 15°?

4.22. An LED is submerged in a swimming pool. If the light seen on the surface is a circle of radius 3.5 m, find its depth d in the pool. Assume that the water is lossless and its dielectric constant is 1.77 in the optical range.

4.23. A TM-polarized uniform plane wave traveling in water ($\varepsilon_r = 81$) is incident upon its boundary with air. The angle of incidence is 45°. Find the reflection and transmission coefficients and the transmitted electric and magnetic fields into the air.

4.24. The magnetic field intensity of a uniform plane wave propagating through a certain nonmagnetic medium is given by

$$\vec{H}(x, t) = \hat{z}(0.03)\cos(10^9 t - 4x)\ \text{A/m}.$$

Find (a) the direction of wave travel, (b) the velocity of wave, (c) the wavelength, and (d) the associated electric field intensity.

4.25. A uniform plane wave propagating in a lossless nonmagnetic dielectric medium has power density of 5 W/m^2 and an rms electric field intensity of 31.62 V/m. Find (a) the intrinsic impedance, (b) the rms value of the magnetic field intensity, and (c) the velocity of the wave.

4.26. A 100 MHz plane wave is normally incident from air onto a semi-infinite nonmagnetic dielectric slab (region 2). If the standing wave ratio in front of the slab is 1.5 and if $E_{\min}$ is at the boundary, find (a) η_2, (b) ε_{r2}, (c) R, and (d) distance d from the boundary to the nearest $E_{\max}$ of the standing wave pattern.

4.27. Liquid glycerol has the following electrical characteristics at 500 MHz: $\varepsilon_r = 11.4$, $\sigma = 0.235$ S/m, and $\mu_r = 1$. Find (a) the complex propagation constant, (b) the phase velocity, and (c) the penetration depth. (d) Assume that a communication link is set up in this medium and that the transmitting antenna can produce an electric field intensity of 1 mV/m and the receiver can detect signal stronger than 1 μV/m. Under these conditions will it be possible to reliably transmit through 50 cm of glycerol?

4.28. For a uniform plane wave propagating in the $+z$ direction in a nonmagnetic material medium with $\sigma = 10$ S/m and $\varepsilon_r = 9$, the magnetic field intensity in the $z = 0$ plane is given as follows:

$$\vec{\mathscr{H}}(z = 0, t) = \hat{y}0.25\cos(2\pi \cdot 10^8 t - 0.2)\ \text{mA/m}.$$

Find the associated electric field intensity $\vec{\mathscr{E}}(z, t)$ in the medium.

4.29. A uniform plane wave propagating in a lossless nonmagnetic dielectric medium has a power density of 0.25 W/m^2 and an rms electric field intensity of 10 V/m. Find (a) the intrinsic impedance, (b) the rms value of the magnetic field intensity, and (c) the velocity of the wave.

4.30. The magnetic field intensity of a uniform plane wave propagating through a nonmagnetic medium is given as follows:

$$\vec{\mathscr{H}}(x, t) = \hat{z}\, 2.5 \cos(10^9 t - 6x)\ \text{mA/m}.$$

Find (a) the direction of wave propagation, (b) the velocity of wave, (c) its wavelength, and (d) the associated electric field intensity.

4.31. An electromagnetic source radiating in free space sets the instantaneous magnetic field intensity

$$\vec{\mathscr{H}} = \hat{z}5 \cos\left(10^8 t - \beta y + \frac{\pi}{4}\right)\ \mu\text{A/m}.$$

Find (a) β and (b) the associated electric field intensity.

4.32. For a uniform plane wave of 15 GHz propagating in seawater ($\sigma = 4$ S/m, $\varepsilon_r = 80$, and $\mu_r = 1$) compute the propagation constant γ and intrinsic impedance η of the medium. Also find the distance over which the fields attenuate by the factor of e^{-1}.

4.33. The fields of a uniform plane wave of frequency 100 kHz propagating in a material medium are attenuated by a factor e^{-1} over a distance of 1205 m while their phase changes by 2π rad in a distance of 1321 m. Further a ratio of the amplitudes of its electric and magnetic field intensities at a point in the medium is found to be 163.54 Ω. Find the wave's propagation constant γ and complex intrinsic impedance η of the medium.

4.34. A uniform plane wave of 3 GHz propagating through free space is incident normally on a lossless dielectric medium with $\varepsilon_r = 4$ and $\mu_r = 1$. The incident electric field at the interface has a value of 2 mV/m right before it strikes the interface. In free space, find (a) the reflection coefficient, (b) the VSWR, (c) the positions (in meters) of the maxima and minima of the electric field, and (d) the maximum and minimum values of the electric field.

4.35. The electric field intensity of a uniform plane wave propagating through a lossless nonmagnetic medium is given by

$$\vec{\mathscr{E}}(z, t) = \hat{x}0.2 \cos(10^9 t + 2\pi z)\ \text{V/m}.$$

Find (a) the direction of propagation, (b) velocity of the signal, (c) the wavelength, and (d) associated magnetic field intensity.

4.36. A material has an intrinsic impedance of $120\angle 24°$ Ω at 320 MHz. If its relative permeability is 4, find its (a) loss tangent, (b) the dielectric constant, (c) the conductivity, and (d) the complex permittivity.

4.37. Region 1 ($x < 0$) is air whereas the region 2 ($x > 0$) is a nonmagnetic medium characterized by $\sigma = 10^{-4}$ S/m and $\varepsilon_r = 4$. A uniform plane wave, with an electric field intensity as given below, is incident on the interface at $x = 0$ from region 1:

$$\vec{\mathscr{E}}_{\text{in}} = 5\cos\left(2\pi \times 10^8 t - \frac{2\pi}{3}x\right) \text{ V/m}.$$

Find the reflected and transmitted wave electric fields.

4.38. A uniform plane wave propagates in a nondissipative medium in the $+z$ direction. The frequency of the wave is 20 MHz. A probe located at $z = 0$ measures the phase angle of the wave to be 98°. An identical probe located at $z = 2$ m measures the phase angle to be $-15°$. What is the relative permittivity of the medium? (Assume that $\mu = \mu_o$.)

4.39. A uniform plane wave, $\vec{E}(z) = (5\hat{x} + j10\hat{y})e^{-j2z}$ V/m, has a frequency of 50 MHz in a lossless dielectric material for which $\mu_r = 1$. Find v_p, η, and the instantaneous electric field intensity.

4.40. The electric field intensity of a uniform plane wave propagating in a nonmagnetic medium of zero conductivity is given as follows:

$$\vec{\varepsilon}(y, t) = 37.7\cos(9\pi \cdot 10^7 t + 0.3\pi y)\hat{x} \text{ V/m}.$$

Find (a) the frequency, (b) the wavelength, (c) the direction of propagation of the wave, (d) the dielectric constant of the medium, and (e) the associated magnetic field intensity.

4.41. A medium with $\mu = 4\mu_o$ has an intrinsic impedance of $100\angle 22.5°\ \Omega$ at 318.31 MHz. Find its (a) loss tangent, (b) dielectric constant, (c) conductivity, and (d) complex permittivity.

4.42. The instantaneous expression for the magnetic field intensity of a uniform plane wave propagating in the $+y$ direction in free space is given by

$$\vec{\mathscr{H}} = \hat{z}4 \times 10^{-6}\cos\left(10^7 \pi t - \beta y + \frac{\pi}{4}\right) \text{ A/m}.$$

(a) Determine β and the signal wavelength.

(b) Write the instantaneous expression for the electric field intensity.

4.43. A 3 GHz uniform plane wave is propagating in a nonmagnetic medium that has a dielectric constant of 2.5 and a loss tangent of 0.05.

(a) Determine the distance over which the amplitude of the propagating wave will reduce by 50%.

(b) Determine the intrinsic impedance, the wavelength, and the phase velocity of the wave in the medium.

4.44. A uniform plane wave in air with $\vec{E}_{\text{in}}(z) = \hat{x}10e^{-j6z}$ V/m is incident normally on an interface at $z = 0$, with a lossy medium having a dielectric constant 2.25 and a loss tangent 0.3. Find the following:

(a) The phasor expressions for the reflected and transmitted electric and magnetic fields.

(b) The time-average power flow per unit area in the lossy medium.

4.45. The electric field intensity of a uniform plane wave traveling through a material medium ($\varepsilon_r = 4$, $\sigma = 0.1$ S/m, $\mu_r = 1$) is given by $\hat{y}100e^{\gamma x}$ V/m. If the signal frequency is 2.45 GHz, find the associated magnetic field intensity and the propagation constant γ.

4.46. A uniform plane wave traveling through a dielectric medium with $\varepsilon_r = 3$ and $\mu_r = 1$ is incident normally upon a free-space medium. The incident electric field is given by

$$\vec{E}_{\text{in}} = \hat{y}3e^{-j200\pi z} \text{ mV/m}.$$

Find (a) The corresponding magnetic field intensity, (b) the reflected electric field intensity, (c) the reflected magnetic field intensity, (d) the transmitted electric field intensity, and (e) the transmitted magnetic field intensity.

5

TRANSMISSION LINES

Every electrical circuit requires transmission lines to connect the various circuit elements and systems together. Power companies use open-wire lines for distributing electrical power from the generating stations to consumers. The telephone industry commonly employs coaxial lines to connect their subscribers. These typical transmission lines are common at low operating frequencies. The transmission lines can experience significant loss of signal if the frequency is in the RF or microwave range. A specially designed coaxial line, the stripline, the microstrip line, and the waveguide are used at radio and microwave frequencies. There are several other transmission lines that are used at higher frequencies. Among these are the dielectric rods (similar to optical fibers) used in microwave bands. A comprehensive analysis of electromagnetic fields is required for each transmission line in order to learn their signal propagation characteristics. The analysis is significantly simplified for a generalized circuit approach, and the line parameters are found via electromagnetic field analysis of each transmission line.

This chapter begins with a distributed model of a transmission line using line parameters. The characteristic impedance is defined and the transmission line equation is solved to show a signal's behavior over infinitely long lines. The terminated transmission lines are then analyzed, and the concepts of reflection coefficient, return loss, and insertion loss are introduced. After a brief discussion of the quarter-wave impedance transformer technique, the voltage standing wave ratio and the impedance measurement methods are presented. Next the Smith chart is introduced

Practical Electromagnetics: From Biomedical Sciences to Wireless Communication.
By Devendra K. Misra
Copyright © 2007 John Wiley & Sons, Inc.

along with its application in analysis and design of simple circuits. The chapter ends with a brief discussion of the excitation of transients on the transmission line.

5.1 TRANSMISSION LINE EQUATIONS

A transmission line can be modeled as a distributed electrical circuit comprised of inductors and resistors connected in series, and capacitors and resistors connected in parallel, as illustrated in Figure 5.1. These circuit elements are known as the line parameters that are defined per unit length as follows:

$L =$ Inductance per unit length (H/m)
$R =$ Resistance per unit length (Ω/m)
$C =$ Capacitance per unit length (F/m)
$G =$ Conductance per unit length (S/m)

The line parameters are determined theoretically by electromagnetic field analysis of a transmission line. The cross-sectional geometry of a transmission line and the electrical characteristics of the material used in its design determine these parameters. Table 5.1 lists the line parameters of selected transmission lines. The conductivity of the conductors used in the transmission line is σ, and the dielectric constant and loss tangent of nonmagnetic dielectric material used in these lines are ε_r and $\tan\delta$, respectively.

Characteristic Impedance of a Transmission Line

Consider a semi-infinitely long transmission line that is excited by a voltage source. The voltage-to-current ratio at any point on this line is found to be a constant, which is referred to as its characteristic impedance. Another way to look at the characteristic impedance is as follows: An electrical signal continues propagating in the forward direction without hindrance if the transmission line is infinitely long. However, there may be some backreflection when the signal comes across a discontinuity, such as a load that terminates it after a finite length. The reflected signal varies with the terminating load and can go to zero for a certain unique termination. Thus this load absorbs the entire incident signal, and the voltage source

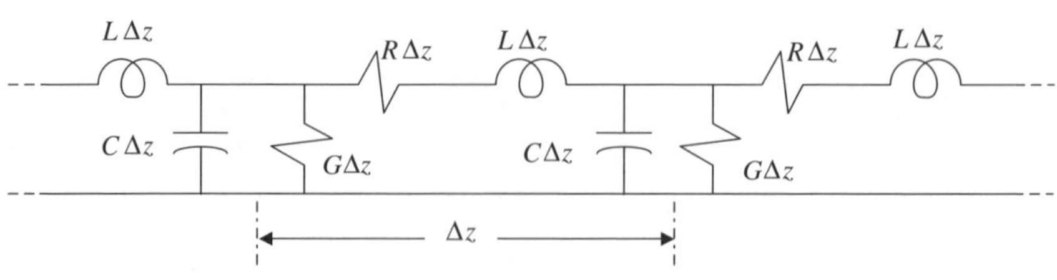

Figure 5.1 Lumped element model of the transmission line.

TABLE 5.1 Selected Line Parameters

	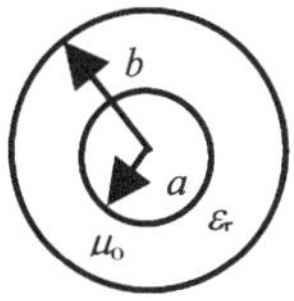 Coaxial Line	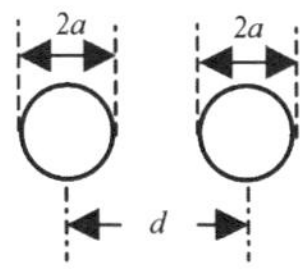 Two-Wire Line	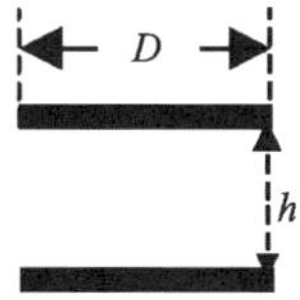 Parallel-Plate Line
L (nH/m)	$200\ \ln(b/a)$	$400\ \cosh^{-1}(d/2a)$	$400\pi h/D$
C (pF/m)	$\dfrac{55.6313\varepsilon_r}{\ln(b/a)}$	$\dfrac{27.8157\varepsilon_r}{\cosh^{-1}(d/2a)}$	$\dfrac{8.854\varepsilon_r D}{h}$
R (mΩ/m)	$\sqrt{\dfrac{0.1f}{\sigma}}\left(\dfrac{1}{a}+\dfrac{1}{b}\right)$	$\dfrac{1}{a}\sqrt{\dfrac{0.4f}{\sigma}}$	$\dfrac{1}{D}\sqrt{\dfrac{1.6f}{\sigma}}$
G (nS/m)	$\dfrac{0.3495\varepsilon_r f\ \tan\delta}{\ln(b/a)}$	$\dfrac{0.1748\varepsilon_r f\ \tan\delta}{\cosh^{-1}(d/2a)}$	$\dfrac{0.0556\varepsilon_r fD\ \tan\delta}{h}$

sees an infinite electrical length. In this situation the voltage-to-current ratio at any point on the line is found to be equal to the terminating impedance. This impedance is called the characteristic impedance of that transmission line.

With line parameters R, L, G, and C of a transmission line, we can define Z and Y as its impedance and admittance per unit length at the signal frequency ω to be

$$Z = R + j\omega L$$

and

$$Y = G + j\omega C.$$

Therefore, using the definition of characteristic impedance and the distributed circuit model shown in Figure 5.1, we can write

$$Z_o = \frac{(Z_o + Z\Delta z)(1/Y\Delta z)}{Z_o + Z\Delta z + (1/Y\Delta z)} = \frac{Z_o + Z\Delta z}{1 + Y\Delta z(Z_o + Z\Delta z)} \Rightarrow Z_oY(Z_o + Z\Delta z) = Z.$$

For $\Delta z \to 0$, it reduces to

$$Z_o = \sqrt{\frac{Z}{Y}} = \sqrt{\frac{R + j\omega L}{G + j\omega C}}. \tag{5.1.1}$$

Note that the characteristic impedance is a complex number for the lossy line. However, it is a real number if the transmission line is lossless, meaning for $R \to 0$ and

$G \rightarrow 0$, $Z_o = \sqrt{L/C}$. Further the characteristic impedance reduces to this value if the operating frequency is such that $\omega L \gg R$ and $\omega C \gg G$.

Example 5.1

The inner and outer conductor diameters of a Teflon-filled ($\varepsilon_r = 3.3$) coaxial line are given as follows: $2a = 0.032$ inch and $2b = 0.176$ inch. Find its characteristic impedance if the operating frequency is such that $\omega L \gg R$ and $\omega C \gg G$.

From the Table 5.1,

$$C = \frac{55.6313\varepsilon_r}{\ln(b/a)}\ \text{pF/m} = \frac{55.6313 \cdot 3.3}{\ln(0.176/0.032)} = 107.7\ \text{pF/m}$$

and

$$L = 200\ \ln\left(\frac{b}{a}\right)\text{nH/m} = 200\ \ln\left(\frac{0.176}{0.032}\right) = 340.95\ \text{nH/m}.$$

Therefore

$$Z_o = \sqrt{\frac{L}{C}} = \sqrt{\frac{340.95 \cdot 10^{-9}}{107.7 \cdot 10^{-12}}} = 56.27\ \Omega.$$

Transmission Line Equations

Figure 5.2 shows an equivalent distributed circuit of the transmission line that is terminated by a load impedance Z_L. A voltage source connected at its other end. Assume that voltages at its nodes A and B are $v(z, t)$ and $v(z + \Delta z, t)$, respectively. Currents leaving these two nodes are assumed to be $i(z, t)$ and $i(z + \Delta z, t)$, respectively.

We can now write the loop equation for a small length Δz of this line as follows:

$$v(z, t) = L\Delta z \frac{\partial i(z, t)}{\partial t} + R\Delta z i(z, t) + v(z + \Delta z, t),$$

or

$$\frac{v(z + \Delta z, t) - v(z, t)}{\Delta z} = -Ri(z, t) - L\frac{\partial i(z, t)}{\partial t}.$$

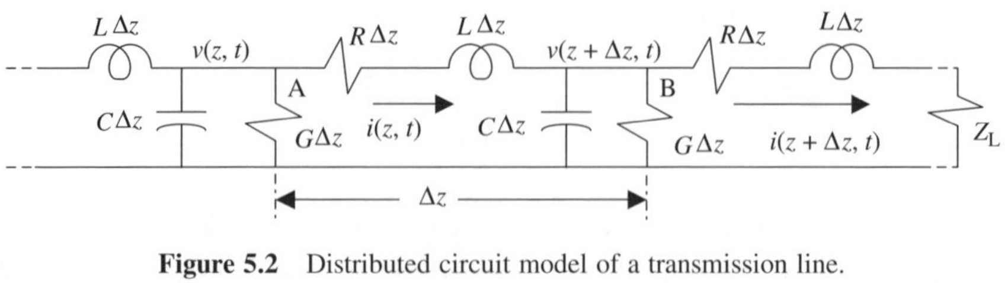

Figure 5.2 Distributed circuit model of a transmission line.

With the limit $\Delta z \to 0$, it reduces to

$$\frac{\partial v(z, t)}{\partial z} = -\left[R \cdot i(z, t) + L\frac{\partial i(z, t)}{\partial t}\right]. \tag{5.1.2}$$

The nodal equation at its node B is found to be

$$i(z, t) = i(z + \Delta z, t) + G\Delta z v(z + \Delta z, t) + C\Delta z\frac{\partial v(z + \Delta z, t)}{\partial t},$$

or

$$\frac{i(z + \Delta z, t) - i(z, t)}{\Delta z} = -\left[G \cdot v(z + \Delta z, t) + C\frac{\partial v(z + \Delta z, t)}{\partial t}\right].$$

Again, under the limit $\Delta z \to 0$, it reduces to

$$\frac{\partial i(z, t)}{\partial z} = -G \cdot v(z, t) - C\frac{\partial v(z, t)}{\partial t}. \tag{5.1.3}$$

Equations (5.1.2) and (5.1.3) can be solved for $v(z, t)$ or $i(z, t)$ as follows:

$$\frac{\partial^2 v(z, t)}{\partial z^2} = RGv(z, t) + (RC + LG)\frac{\partial v(z, t)}{\partial t} + LC\frac{\partial^2 v(z, t)}{\partial t^2} \tag{5.1.4}$$

and,

$$\frac{\partial^2 i(z, t)}{\partial z^2} = RGi(z, t) + (RC + LG)\frac{\partial i(z, t)}{\partial t} + LC\frac{\partial^2 i(z, t)}{\partial t^2}. \tag{5.1.5}$$

For a lossless line, $R = 0$ and $G = 0$. Therefore these two simplify to scalar wave equations as follows:

$$\frac{\partial^2 v(z, t)}{\partial z^2} = LC\frac{\partial^2 v(z, t)}{\partial t^2} \tag{5.1.6}$$

and

$$\frac{\partial^2 i(z, t)}{\partial z^2} = LC\frac{\partial^2 i(z, t)}{\partial t^2}. \tag{5.1.7}$$

Note that the velocity of these waves is $1/\sqrt{LC}$. The partial differential equations (5.1.6) and (5.1.7) are further simplified to ordinary differential equations by using phasor voltages and currents if the source is sinusoidal with time (i.e., time harmonic). Hence

$$\frac{d^2V(z)}{dz^2} = ZYV(z) = \gamma^2 V(z) \tag{5.1.8}$$

and

$$\frac{d^2I(z)}{dz^2} = ZYI(z) = \gamma^2 I(z), \tag{5.1.9}$$

where $V(z)$ and $I(z)$ represent phasor voltage and current, respectively. Z and Y are impedance per unit length and admittance per unit length, respectively, as defined earlier. Therefore

$$\gamma = \sqrt{ZY} = \sqrt{(R + j\omega L)(G + j\omega C)} = \alpha + j\beta. \tag{5.1.10}$$

It is known as the propagation constant of the line. α is the attenuation constant expressed in neper per meter and β is the phase constant in radian per meter. Note the similarities of (5.1.8) and (5.1.9) with (4.1.7). The propagation constant γ is equal to jk, where k is the wave number defined by (4.1.4). Thus (5.1.8) and (5.1.9) are homogeneous Helmholtz equations for voltage and current, respectively, and their solutions are similar to (4.1.10). Once (5.1.8) is solved, the phasor form of (5.1.2) can also be used to find the current on the line.

Solution of the Helmholtz Equation

Assume that $V(z) = e^{Cz}$, and substitute it back into (5.1.8) to find that $C = \pm\gamma$. The complete solution to (5.1.8) is then

$$V(z) = V_{\text{in}} e^{-\gamma z} + V_{\text{ref}} e^{\gamma z}, \tag{5.1.11}$$

where V_{in} and V_{ref} are integration constants that are complex, in general. They are evaluated through the boundary conditions. We can express these two constants in polar form as follows:

$$V_{\text{in}} = v_{\text{in}} e^{j\phi} \qquad \text{and} \qquad V_{\text{ref}} = v_{\text{ref}} e^{j\varphi}.$$

Therefore the voltage $v(z, t)$ is found to be

$$v(z, t) = \text{Re}[V(z)e^{j\omega t}] = \text{Re}[V_{\text{in}} e^{-(\alpha+j\beta)z} e^{j\omega t} + V_{\text{ref}} e^{+(\alpha+j\beta)z} e^{j\omega t}]$$

or

$$v(z, t) = v_{\text{in}} e^{-\alpha z} \cos(\omega t - \beta z + \phi) + v_{\text{ref}} e^{+\alpha z} \cos(\omega t + \beta z + \varphi). \tag{5.1.12}$$

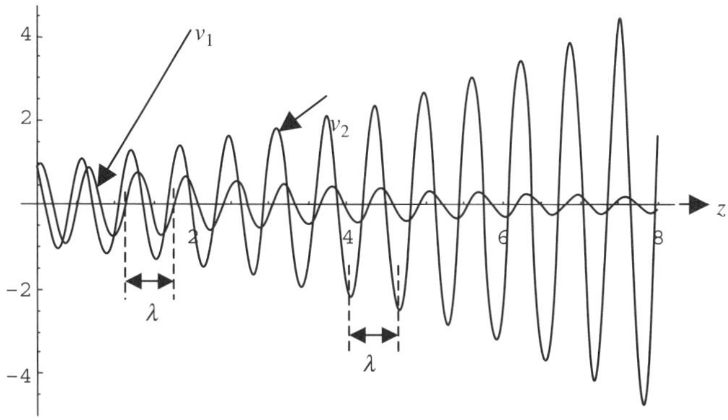

Figure 5.3 Graphical representation of the two solutions of (5.1.12) relative to distance.

Again, note the similarity of this equation to (4.1.15). To understand the behavior of each term on the right-hand side of this equation, we can get help from Figure 4.1, which is redrawn as Figure 5.3 for a ready reference. Note that the first term $v_1(z, t = t_1)$ changes sinusoidally with distance z, while its amplitude decreases exponentially. On the other hand, the amplitude of the second sinusoidal term $v_2(z, t = t_1)$ increases exponentially. Further the argument of the cosine function decreases with distance in the first term while it increases in the second term. When a signal is propagating away from the source along the $+z$-axis, its phase should be delayed. Further, if it is propagating in a lossy medium, then its amplitude should decrease with distance z. Therefore the first term on the right-hand side of equation (5.1.12) represents a wave traveling along the $+z$-axis (an incident or outgoing wave), and the second term represents a wave traveling in the opposite direction (a reflected or incoming wave).

The current $I(z)$ on the line can be found via (5.1.2) and (5.1.11) as follows: Equation (5.1.2) in phasor form is written as

$$\frac{dV(z)}{dz} = -(R + j\omega L)I(z) = -ZI(z). \tag{5.1.13}$$

On substituting (5.1.11), we have

$$-\gamma[V_{\text{in}}e^{-\gamma z} - V_{\text{ref}}e^{\gamma z}] = -ZI(z) \rightarrow I(z) = \frac{\gamma}{Z}[V_{\text{in}}e^{-\gamma z} - V_{\text{ref}}e^{\gamma z}]. \tag{5.1.14}$$

Note that

$$\frac{Z}{\gamma} = \frac{Z}{\sqrt{ZY}} = \sqrt{\frac{Z}{Y}} = Z_o = \frac{V_{\text{in}}}{I_{\text{in}}} = -\frac{V_{\text{ref}}}{I_{\text{ref}}}.$$

Therefore (5.1.14) may be rewritten as follows:

$$I(z) = \frac{V_{\text{in}}}{Z_o} e^{-\gamma z} - \frac{V_{\text{ref}}}{Z_o} e^{\gamma z}. \tag{5.1.15}$$

The incident and reflected waves change sinusoidally with both space and time. The time duration over which the phase angle of a wave goes through a change of 360° (2π radians) is known as its time period. The inverse of the time period in seconds is the signal frequency in Hz. Similarly the distance over which the phase angle of the wave changes by 360° (2π radians) is known as its wavelength (λ). Therefore the phase constant β is equal to 2π divided by the wavelength in meters.

Phase and Group Velocities

As discussed in preceding chapter for uniform plane waves, the velocity with which the phase of a time-harmonic signal moves is called its phase velocity. That is to say, if we tag a phase point of the sinusoidal wave and monitor its velocity, then we obtain the phase velocity, v_p, of this wave. Mathematically

$$v_p = \frac{\omega}{\beta} = f\lambda. \tag{5.1.16}$$

A transmission line has no dispersion if the phase velocity of a propagating signal is independent of frequency. Hence a graphical plot of ω versus β will be a straight line passing through the origin. This kind of plot is called the dispersion diagram of a transmission line. An information-carrying signal is composed of many sinusoidal waves. If the line is dispersive, each of these harmonics will travel at a different velocity. Therefore the information will be distorted at the receiving end. The velocity with which a group of waves travels is called the group velocity, v_g. It is equal to the slope of the dispersion curve of the transmission line.

Consider two sinusoidal signals with angular frequencies $\omega + \delta\omega$ and $\omega - \delta\omega$, respectively. Assume that these waves of equal amplitudes are propagating in the z direction with corresponding phase constants $\beta + \delta\beta$ and $\beta - \delta\beta$. The resultant wave can be found as follows:

$$\begin{aligned} f(z, t) &= \text{Re}\left\{Ae^{j((\omega+\delta\omega)t-(\beta+\delta\beta)z)} + Ae^{j((\omega-\delta\omega)t-(\beta-\delta\beta)z)}\right\} \\ &= 2A\cos(\delta\omega t - \delta\beta z)\cos(\omega t - \beta z). \end{aligned}$$

Hence the resulting wave, $f(z, t)$, is amplitude modulated. The envelope of this signal moves with the group velocity

$$v_g = \frac{\delta\omega}{\delta\beta}. \tag{5.1.17}$$

Example 5.2

A telephone line has the following parameters: $R = 14$ ohm/mile, $G = 0$, $L = 5$ mH/mile, and $C = 0.02$ μF/mile.

For the operating frequency of 1 kHz, calculate its characteristic impedance Z_o and the propagation constant γ. If its characteristic impedance terminates the line and input voltage is 10 V $\angle 0^\circ$, find total voltage at 30 miles from its input:

$$Z_o = \sqrt{\frac{R + j\omega L}{G + j\omega C}} = \sqrt{\frac{14 + j2\pi \cdot 10^3 \cdot 5 \cdot 10^{-3}}{0 + j2\pi \cdot 10^3 \cdot 0.02 \cdot 10^{-6}}}\ \text{ohm} = 523.2\ \Omega \angle -12^\circ$$

$$= 511.8 - j108.9\ \text{ohm}$$

and

$$\gamma = \sqrt{ZY} = 0.0657 \angle 78^\circ\ \text{mile}^{-1} = 0.01368 + j0.0643\ \text{mile}^{-1} = \alpha + j\beta.$$

Therefore $\alpha = 0.01368$ Np/mile and $\beta = 0.0643$ rad/mile.

Since the line terminates in its characteristic impedance, there is only incident wave:

$$V(z = 30\ \text{mile}) = 10e^{-0.01368 \cdot 30} e^{-j0.0643 \cdot 30} = 6.634e^{-j1.929}\ \text{V}.$$

Example 5.3

An infinitely long lossless 50 Ω transmission line is connected to a signal generator with its open-circuit voltage $v(t) = 3\cos(2\pi \cdot 10^8 t)$ V, as shown in Figure 5.4*a*. The generator has an internal resistance of 25 Ω. If the signal propagates with a velocity of $2 \cdot 10^8$ m/s on the line, find the instantaneous voltage and the current at an arbitrary location on the line.

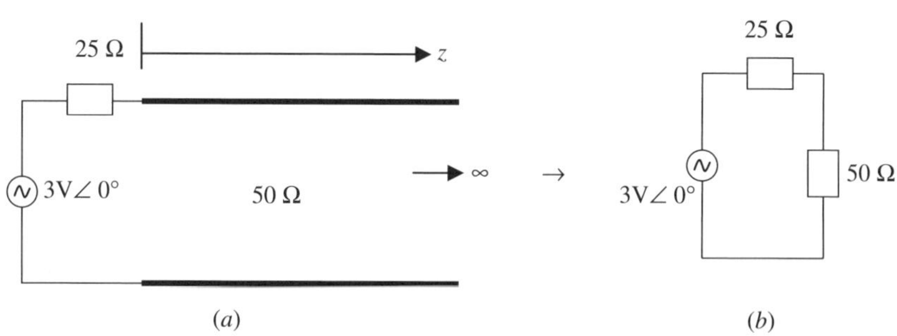

Figure 5.4 (*a*) Transmission line circuit for Example 5.3 and (*b*) its equivalent at the input point.

Since the transmission line extends to infinity, there is only a forward propagating (incident) wave. An equivalent circuit at its input end may be drawn, as shown in Figure 5.4*b*. By the voltage division rule and Ohm's law, incident voltage and current can be determined as follows:

Incident voltage at the input end,

$$V_{\text{in}}(z=0) = \frac{50}{50+25} 3\angle 0^\circ = 2\,\text{V}\angle 0^\circ.$$

Incident current at the input end,

$$I_{\text{in}}(z=0) = \frac{3\angle 0^\circ}{50+25} = 0.04\,\text{A}\angle 0^\circ$$

and

$$\beta = \frac{\omega}{v_{\text{p}}} = \frac{2\pi\cdot 10^8}{2\cdot 10^8} = \pi\ \text{rad/m}.$$

Therefore

$$V(z) = 2e^{-j\pi z}\,\text{V} \qquad \text{and} \qquad I(z) = 0.04e^{-j\pi z}\,\text{A}.$$

Hence

$$v(z,t) = 2\cos(2\pi\cdot 10^8 t - \pi z)\,\text{V}$$

and

$$i(z,t) = 0.04\cos(2\pi\cdot 10^8 t - \pi z)\,\text{A}.$$

5.2 FINITE LENGTH TRANSMISSION LINE

Consider a transmission line of length ℓ and a characteristic impedance Z_{o}. The line terminates in a load impedance Z_{L}, as shown below in Figure 5.5. Assume that the incident and reflected voltages at its input ($z=0$) are V_{in} and V_{ref}, respectively.

If $V(z)$ represents the total phasor voltage at point z on the line and $I(z)$ is the total current at that point, then according to (5.1.11) and (5.1.15) we have

$$V(z) = V_{\text{in}}e^{-\gamma z} + V_{\text{ref}}e^{\gamma z} \tag{5.2.1}$$

and

$$I(z) = \frac{1}{Z_{\text{o}}}(V_{\text{in}}e^{-\gamma z} - V_{\text{ref}}e^{\gamma z}), \tag{5.2.2}$$

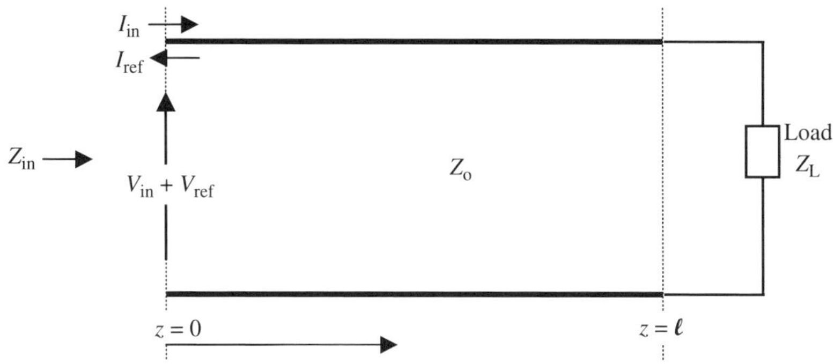

Figure 5.5 Transmission line with termination.

where V_{in} and V_{ref} are the incident and reflected voltages, respectively, at $z = 0$. Therefore the impedance Z_{in} at the input terminals of this transmission line is found by dividing the total voltage by the total current at $z = 0$ as follows:

$$Z_{in} = \frac{V(z=0)}{I(z=0)} = Z_o \frac{V_{in} + V_{ref}}{V_{in} - V_{ref}}$$

or

$$Z_{in} = Z_o \frac{1 + (V_{ref}/V_{in})}{1 - (V_{ref}/V_{in})} = Z_o \frac{1 + \Gamma_o}{1 - \Gamma_o}, \tag{5.2.3}$$

where $\Gamma_o = \rho\, e^{j\varphi}$ is called the input reflection coefficient.

Equation (5.2.3) can be rearranged to define the normalized input impedance $\overline{Z}_{in}$ as follows:

$$\frac{Z_{in}}{Z_o} = \overline{Z}_{in} = \frac{1 + \Gamma_o}{1 - \Gamma_o}. \tag{5.2.4}$$

The voltage and current at $z = \ell$ are related through the load impedance Z_L as follows:

$$Z_L = \frac{V(z=\ell)}{I(z=\ell)} = Z_o \frac{V_{in}e^{-\gamma\ell} + V_{ref}e^{+\gamma\ell}}{V_{in}e^{-\gamma\ell} - V_{ref}e^{+\gamma\ell}} = Z_o \frac{e^{-\gamma\ell} + \Gamma_o e^{+\gamma\ell}}{e^{-\gamma\ell} - \Gamma_o e^{+\gamma\ell}}.$$

Therefore

$$\overline{Z}_L = \frac{e^{-\gamma\ell} + \Gamma_o e^{\gamma\ell}}{e^{-\gamma\ell} - \Gamma_o e^{+\gamma\ell}} \Rightarrow \Gamma_o = \frac{\overline{Z}_L - 1}{\overline{Z}_L + 1} e^{-2\gamma\ell}. \tag{5.2.5}$$

Using (5.2.5), (5.2.4) can be written as follows:

$$\overline{Z}_{\text{in}} = \frac{1 + [(\overline{Z}_{\text{L}} - 1)/(\overline{Z}_{\text{L}} + 1)]e^{-2\gamma\ell}}{1 - [(\overline{Z}_{\text{L}} - 1)/(\overline{Z}_{\text{L}} + 1)]e^{-2\gamma\ell}} = \frac{\overline{Z}_{\text{L}}[1 + e^{-2\gamma\ell}] + [1 - e^{-2\gamma\ell}]}{\overline{Z}_{\text{L}}[1 - e^{-2\gamma\ell}] + [1 + e^{-2\gamma\ell}]}.$$

Since

$$\frac{1 - e^{-2\gamma\ell}}{1 + e^{-2\gamma\ell}} = \frac{e^{+\gamma l} - e^{-\gamma l}}{e^{+\gamma l} + e^{-\gamma l}} = \frac{\sinh(\gamma\ell)}{\cosh(\gamma\ell)} = \tanh(\gamma\ell),$$

we find that

$$\overline{Z}_{\text{in}} = \frac{\overline{Z}_{\text{L}} + \tanh(\gamma\ell)}{1 + \overline{Z}_{\text{L}} \tanh(\gamma\ell)},$$

or

$$Z_{\text{in}} = Z_{\text{o}} \frac{Z_{\text{L}} + Z_{\text{o}} \tanh(\gamma\ell)}{Z_{\text{o}} + Z_{\text{L}} \tanh(\gamma\ell)}. \tag{5.2.6}$$

Further $\gamma = \alpha + j\beta = j\beta$ for a lossless line. Therefore $\tanh(\gamma\ell) = \tanh(j\beta\ell) = j\tan(\beta\ell)$, and (5.2.6) simplifies to

$$Z_{\text{in}} = Z_{\text{o}} \frac{Z_{\text{L}} + jZ_{\text{o}} \tan(\beta\ell)}{Z_{\text{o}} + jZ_{\text{L}} \tan(\beta\ell)}. \tag{5.2.7}$$

Note that Z_{in} repeats periodically over the transmission line's length. In other words, the input impedance on a lossless transmission line is the same for all points at $d \pm n\lambda/2$, where n is an integer. It happens so because of the tangent function. Since

$$\beta\ell = \frac{2\pi}{\lambda}\left(d \pm \frac{n\lambda}{2}\right) = \frac{2\pi d}{\lambda} \pm n\pi,$$

and therefore

$$\tan(\beta\ell) = \tan\left(\frac{2\pi d}{\lambda} \pm n\pi\right) = \tan\left(\frac{2\pi d}{\lambda}\right).$$

There are three cases of special interests here. If $Z_{\text{L}} = 0$ (i.e., a lossless line is short-circuited), then (5.2.7) reduces to

$$Z_{\text{in}} = Z_{\text{sc}} = jZ_{\text{o}} \tan(\beta\ell). \tag{5.2.8}$$

TABLE 5.2 Input Impedance of a 50 Ω Line Terminated by a 10 k Load

Frequency (Hz)	Impedance (Ω)
60	$9999.9994 - j2.5132$
$6 \cdot 10^3$	$9993.6875 - j251.1625$
$60 \cdot 10^3$	$9405.8879 - j2363.8959$
$6 \cdot 10^6$	$15.8901 - j395.1618$
$60 \cdot 10^6$	$0.2764 - j16.2455$

On the other hand, if $Z_L = \infty$ (i.e., a lossless line has an open circuit at the load), then (5.2.7) gives

$$Z_{in} = Z_{oc} = -jZ_o \cot(\beta\ell). \tag{5.2.9}$$

Last, if $\ell = \lambda/4$, then $\beta\ell = \pi/2$, and therefore (5.2.7) simplifies to

$$Z_{in} = \frac{Z_o^2}{Z_L}. \tag{5.2.10}$$

According to (5.2.8) and (5.2.9) a lossless line can be used to synthesize an arbitrary reactance, as is used in microwave circuits. Equation (5.2.10) indicates that a quarter-wavelength-long line of a suitable characteristic impedance can be used to transform a load impedance Z_L to Z_{in}. This kind of transmission line is called the impedance transformer, and it is useful in certain impedance-matching applications. Further this equation can be rearranged as follows:

$$\overline{Z}_{in} = \frac{1}{\overline{Z}_L} = \overline{Y}_L.$$

This says that the value of a normalized impedance at a quarter-wavelength away from the load is same as the normalized load admittance.

Consider a 50 Ω transmission line that is one meter long. A 10 k resistance terminates it. We can find impedance as a function of frequency at its input end using (5.2.7). These results are given in Table 5.2 at selected frequencies. Note that the impedance found at its input terminals differs significantly with the frequencies.

Example 5.4

A lossless transmission line of length d and characteristic impedance Z_o is used as an impedance transformer to match a 1.8 kΩ load to a 200 Ω line, as shown in

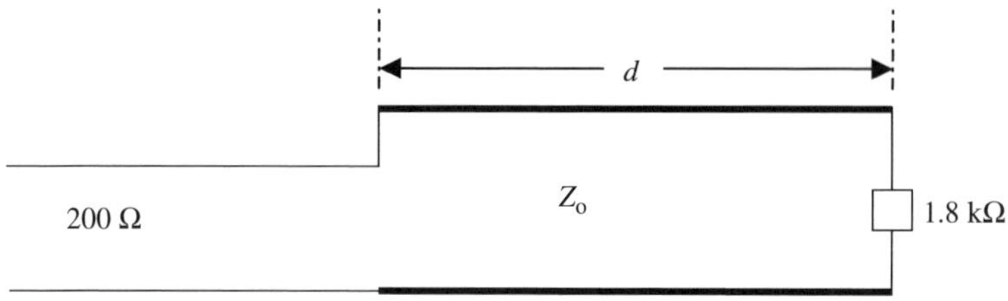

Figure 5.6 Impedance transformer arrangement for Example 5.4.

Figure 5.6. If the signal wavelength is 3 m, find (a) d, (b) Z_o, and (c) the reflection coefficient at the load.

(a) $d = \lambda/4 \rightarrow d = 0.75$ m.
(b) $Z_o = \sqrt{Z_{in}Z_L} \rightarrow Z_o = (1800 \cdot 200)^{1/2} = 600\,\text{ohm}$.
(c)

$$\Gamma_L = \frac{\overline{Z}_L - 1}{\overline{Z}_L + 1} = \frac{Z_L - Z_o}{Z_L + Z_o} = \frac{1800 - 600}{1800 + 600} = 0.5.$$

Example 5.5

A 75 Ω outdoor antenna is being used to receive 850 MHz FM signal. If the receiving circuit uses a 50 Ω line, design a quarter-wavelength transformer to match the antenna with this transmission line. To calculate its length (in cm), assume that the transformer is made from a Teflon-filled ($\varepsilon_r = 2.1$) coaxial line. Also determine the diameter of its inner conductor if the inner diameter of the outer conductor is 0.5 cm.

$$Z_o = \sqrt{Z_{in}Z_L} = \sqrt{75 \cdot 50} = 61.2372\,\Omega$$

and

$$Z_o = \sqrt{\frac{L}{C}} = \sqrt{\frac{200 \times 10^{-9}}{55.63 \times \varepsilon_r \times 10^{-12}}} \ln\left(\frac{b}{a}\right) = 61.2372 \rightarrow \ln\left(\frac{b}{a}\right) = \frac{61.2372}{41.3762}$$

$$= 1.48.$$

Therefore

$$\frac{b}{a} = \frac{2b}{2a} = e^{1.48} = 4.393 \Rightarrow 2a = \frac{2b}{4.393} = \frac{0.5}{4.393} = 0.1138\,\text{cm}.$$

Since $\beta = \omega\sqrt{LC}$ and

$$\frac{1}{\lambda} = \frac{\beta}{2\pi} = \frac{\omega}{2\pi}\sqrt{LC} = f \times \sqrt{200 \times 55.63 \times \varepsilon_r \times 10^{-21}}$$
$$= 850 \times 10^6 \times 10^{-10} \times \sqrt{200 \times 55.63 \times 2.1 \times 0.1}$$
$$= 4.1086\,\text{m}^{-1},$$

then $\lambda = 0.2434$ m and $d = \lambda/4 = 0.0608$ m $= 6.08$ cm.

Example 5.6

A 100 km long telephone line is terminated by a 150–j200 Ω load. The line parameters are as follows: $R = 2$ Ω/km, $G = 0.2$ μS/km, $L = 5$ mH/km, and $C = 8$ nF/km. If this line is being used at 4 kHz, find its (a) characteristic impedance Z_o, (b) propagation constant γ, and (c) input impedance Z_{in}.

(a) $Z = R + j\omega L = 2 + j2\pi \cdot 4 \cdot 10^3 \cdot 5 \cdot 10^{-3} = 2 + j125.6637$ Ω/km and $Y = G + j\omega C = 0.2 \cdot 10^{-6} + j2\pi \cdot 4 \cdot 10^3 \cdot 8 \cdot 10^{-9} = (0.2 + j201.06)$ μS/km. Therefore

$$Z_o = \sqrt{\frac{Z}{Y}} = 790.5973 - j5.8977\,\Omega.$$

(b) $\gamma = \sqrt{ZY} = 1.3439 \cdot 10^{-3} + j0.159\,\text{km}^{-1}$.

(c)

$$Z_{in} = Z_o \frac{Z_L + Z_o \tanh(\gamma\ell)}{Z_o + Z_L \tanh(\gamma\ell)} = 242.5993 - j49.8975\,\Omega.$$

Note that the line parameters are given in kilometers and therefore the propagation constant is also in kilometers. Since the given length is in kilometers as well, there is no need to convert the units to meters.

Reflection Coefficient, Return Loss, and Insertion Loss

The voltage reflection coefficient is defined as the ratio of the reflected to incident phasor voltages at a location in a circuit. In the case of a transmission line terminated by a load Z_L, the voltage reflection coefficient at a distance ℓ from the load is given by (5.2.5). Hence

$$\Gamma = \frac{V_{ref}}{V_{in}} = \frac{Z_L - Z_o}{Z_L + Z_o} e^{-2\gamma\ell} = \rho_L e^{j\theta} e^{-2(\alpha + j\beta)\ell} = \rho_L e^{-2\alpha\ell} e^{-j(2\beta\ell - \theta)}, \qquad (5.2.11)$$

where $\rho_L e^{j\theta} = (Z_L - Z_o)/(Z_L + Z_o)$ is called the load reflection coefficient.

Equation (5.2.11) indicates that the magnitude of the reflection coefficient decreases according to $e^{-2\alpha\ell}$ as the observation point moves away from the load on the transmission line. Further its phase angle changes by $-2\beta\ell$. Therefore a polar (magnitude and phase) plot of this equation will look like a spiral. However, the magnitude of the reflection coefficient remains constant for a lossless line ($\alpha = 0$). As the line length ℓ increases, the reflection coefficient point moves clockwise on a circle of radius equal to its magnitude. It makes one complete revolution for each half-wavelength distance away from the load (because $-2\beta\lambda/2 = -2\pi$).

Similarly the current reflection coefficient Γ_c is defined as the ratio of reflected to incident signal–current phasors. It is related to the voltage reflection coefficient as follows:

$$\Gamma_c = \frac{I_{ref}}{I_{in}} = \frac{-V_{ref}/Z_o}{V_{in}/Z_o} = -\Gamma. \tag{5.2.12}$$

The return loss of a device is defined as the ratio of the reflected power to the incident power at its input. Since power is proportional to the square of a voltage, the return loss may be expressed as a square of the magnitude of the voltage reflection coefficient ρ. Generally, the return loss is expressed in dB as follows:

$$\text{Return loss} = 20\,\log_{10}(\rho)\,\text{dB}. \tag{5.2.13}$$

The insertion loss of a device is defined as the ratio of transmitted power (power available at the output port) to that of the power incident at its input. Since the transmitted power is equal to the difference of incident and reflected powers for a lossless device, the insertion loss can be expressed as follows:

$$\text{Insertion loss of a lossless device} = 10\,\log_{10}(1 - \rho^2)\,\text{dB}. \tag{5.2.14}$$

Example 5.7

A radio-frequency filter exhibits 75 Ω at its input as well as at its output. If the filter is inserted in a 50 Ω system, find its insertion loss.

Since

$$\rho = \left|\frac{75 - 50}{75 + 50}\right| = \frac{25}{125} = 0.2,$$

then

$$\text{Insertion loss} = 10\,\log(1 - 0.2^2) = -0.1773\,\text{dB}.$$

Low-Loss Transmission Lines

Most transmission lines used in practice are not lossless but possess very small losses of the propagating signals. Therefore the propagation constant and the characteristic impedance of such lines can be found as follows:

$$\gamma = \sqrt{ZY} = \sqrt{(R + j\omega L)(G + j\omega C)} = \sqrt{-\omega^2 LC\left(1 + \frac{R}{j\omega L}\right)\left(1 + \frac{G}{j\omega C}\right)}.$$

For $R \ll \omega L$ and $G \ll \omega C$, a first-order approximation is

$$\gamma = \alpha + j\beta \approx j\omega\sqrt{LC}\left(1 + \frac{R}{j2\omega L}\right)\left(1 + \frac{G}{j2\omega C}\right) \approx j\omega\sqrt{LC}\left(1 + \frac{R}{j2\omega L} + \frac{G}{j2\omega C}\right).$$

Therefore

$$\alpha \approx \frac{1}{2}\left(R\sqrt{\frac{C}{L}} + G\sqrt{\frac{L}{C}}\right) \text{ Np/m} \tag{5.2.15}$$

and

$$\beta \approx \omega\sqrt{LC} \text{ rad/m.} \tag{5.2.16}$$

Similarly an approximate characteristic impedance of this line is found to be

$$Z_o = \sqrt{\frac{R + j\omega L}{G + j\omega C}} = \sqrt{\frac{L}{C}}\left(1 + \frac{R}{j\omega L}\right)^{1/2}\left(1 + \frac{G}{j\omega C}\right)^{-1/2},$$

or

$$Z_o \approx \sqrt{\frac{L}{C}}\left(1 + \frac{R}{j2\omega L}\right)\left(1 - \frac{G}{j2\omega C}\right) \approx \sqrt{\frac{L}{C}}\left(1 + \frac{R}{j2\omega L} - \frac{G}{j2\omega C}\right).$$

Hence

$$Z_o = \sqrt{\frac{L}{C}}\left(1 + \frac{1}{j2\omega}\left(\frac{R}{L} - \frac{G}{C}\right)\right). \tag{5.2.17}$$

According to (5.2.15) and (5.2.16), the attenuation constant of a low-loss line is independent of frequency, while its phase constant is close to the case of a lossless line. However, (5.2.17) indicates that the characteristic impedance of a low-loss line is dependent on frequency and therefore it will distort the signal. This frequency dependency of the characteristic impedance can be removed for $RC = GL$. If a low-loss line meets this condition, it is called a distortionless line. Since $G = RC/L$ for such lines, the propagation constant simplifies to

$$\gamma = \sqrt{ZY} = \sqrt{(R + j\omega L)(G + j\omega C)} = (R + j\omega L)\sqrt{\frac{C}{L}} = \alpha + j\beta. \qquad (5.2.18)$$

Example 5.8

A signal attenuates by 0.02 dB for every meter traveled on a 50 Ω distortionless transmission line. If this line has a capacitance of 80 pF per meter, find (a) R, (b) L, (c) G, and (d) v_p.

Since the line is distortionless,

$$Z_o = \sqrt{\frac{L}{C}} = 50.$$

Further 8.6859 dB = 1 Np, and therefore $\alpha = 0.02\ \mathrm{dB/m} \approx 0.02/8.6859\ \mathrm{Np/m} = 2.3026 \cdot 10^{-3}\ \mathrm{Np/m}$. Hence

(a) $R = \alpha\sqrt{L/C} = 2.3026 \cdot 10^{-3} \cdot 50 = 0.1151\ \Omega/\mathrm{m}$.
(b) $L = CZ_o^2 = 80 \cdot 10^{-12} \cdot 50^2\ \mathrm{H/m} = 0.2\ \mu\mathrm{H/m}$.
(c) $G = RC/L = R/Z_o^2 = (0.1151/50^2)\mathrm{S/m} = 46.04\ \mu\mathrm{S/m}$.
(d) $v_p = 1/\sqrt{LC} = 2.5 \cdot 10^8 \mathrm{m/s}$.

Measurement of Characteristic Impedance and Propagation Constant of a Transmission Line

The characteristic impedance and propagation constant of a transmission line can be found after performing open- and short-circuit tests on it as follows: The transmission line under test is kept open at one end, and the impedance at its other end is measured using an impedance bridge. Assume that it is Z_{oc}. The process is repeated after placing a short circuit at its open end, and this impedance is recorded as Z_{sc}. The line's length d is measured. Then from (5.2.6) we find that

$$Z_{oc} = Z_o \coth(\gamma d) \qquad (5.2.19)$$

and

$$Z_{sc} = Z_o \tanh(\gamma d). \qquad (5.2.20)$$

Therefore

$$Z_o = \sqrt{Z_{oc} Z_{sc}} \qquad (5.2.21)$$

and

$$\tanh(\gamma d) = \sqrt{\frac{Z_{sc}}{Z_{oc}}} \Rightarrow \gamma = \frac{1}{d}\tanh^{-1}\left(\sqrt{\frac{Z_{sc}}{Z_{oc}}}\right). \tag{5.2.22}$$

Equations (5.2.21) and (5.2.22) can be solved to determine Z_o and γ. The following identity can be used to facilitate the evaluation of the propagation constant:

$$\tanh^{-1}(Z) = \frac{1}{2}\ln\left(\frac{1+Z}{1-Z}\right).$$

Example 5.9

A load Z_L is connected at one end of the transmission line and its input impedance is measured using an impedance bridge that gives $Z_{in} = 60 - j30\ \Omega$. The experiment is repeated twice with the load replaced first by a short circuit and then by an open circuit. These data are recorded as $j10\ \Omega$ and $-j250\ \Omega$, respectively. Find the characteristic resistance of this line and the load impedance

$$Z_o = \sqrt{Z_{sc} \cdot Z_{oc}} = \sqrt{j10 \cdot (-j250)} = 50\,\Omega.$$

Since the measured input impedance data with both the open and short circuits are pure imaginary and the characteristic impedance is a real number, the line is lossless. Therefore

$$\begin{aligned} Z_{in} &= Z_o \frac{Z_L + jZ_o \tan(\beta\ell)}{Z_o + jZ_L \tan(\beta\ell)} = Z_o \frac{Z_L + Z_{sc}}{Z_o + jZ_L \tan(\beta\ell)}, \\ &= \frac{Z_o}{j\tan(\beta\ell)} \times \frac{Z_L + Z_{sc}}{Z_L + \{Z_o/[j\tan(\beta\ell)]\}} = Z_{oc}\frac{Z_L + Z_{sc}}{Z_L + Z_{oc}}, \end{aligned}$$

and

$$Z_L = Z_{oc}\frac{Z_{sc} - Z_{in}}{Z_{in} - Z_{oc}} = -j250 \times \frac{j10 - (60 - j30)}{60 - j30 - (-j250)} = 75 - j25\,\Omega.$$

Example 5.10

Measurements are made on a 1.5 m long transmission line using an impedance bridge. After short-circuiting at one of the ends, the impedance at the other end is found to be $j103\ \Omega$. Repeating the experiment with the short circuit now replaced by an open circuit gives $-j54.6\ \Omega$. Determine the propagation constant and the characteristic impedance of this line.

$$Z_o = \sqrt{Z_{oc}Z_{sc}} = \sqrt{-j54.6 \times j103} = 74.99 \approx 75\,\Omega.$$

From (5.2.22) we have

$$\tanh(1.5\gamma) = \sqrt{\frac{j103}{-j54.3}} = j1.8969 \Rightarrow 1.5\gamma = \tanh^{-1}(j1.8969)$$

$$= \frac{1}{2}\ln\left(\frac{1+j1.8969}{1-j1.8969}\right),$$

or

$$1.5\gamma = \frac{1}{2}\ln\left(\frac{1+j1.8969}{1-j1.8969}\right) = \frac{1}{2}\ln(1\,\angle 2.1713\,\text{rad}) = \frac{1}{2}\ln(e^{j2.1713}) = j1.08565.$$

Therefore $\gamma = j0.7238\ \text{m}^{-1}$.

Note that the measured data indicate that the line is lossless, so the results obtained are consistent.

Example 5.11

As shown in Figure 5.7, a 5 m long 50 Ω lossless transmission line is terminated by $50 - j100\ \Omega$. The circuit is driven by a 3.1831 MHz signal generator with its open-circuit voltage 100 V $\angle 0°$ and an internal impedance 50 Ω. If the propagating signal has a phase velocity of 200 m/μS, find the impedance at its input end and the phasor voltages at both of its ends.

$$\beta = \frac{\omega}{v_p} = \frac{2\pi \cdot 3.1831 \cdot 10^6}{200 \cdot 10^6} = 0.1\ \text{rad/m}$$

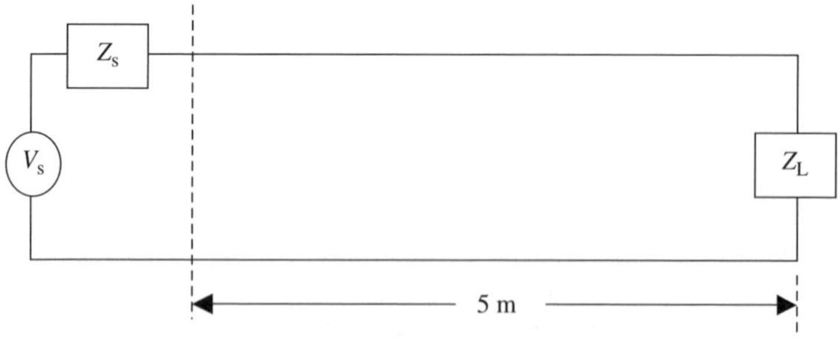

Figure 5.7 Circuit arrangement for Example 5.11.

and

$$Z_{in} = Z_o \frac{Z_L + jZ_o \tan(\beta\ell)}{Z_o + jZ_L \tan(\beta\ell)} = 50 \frac{50 - j100 + j50 \tan(0.5)}{50 + j(50 - j100) \tan(0.5)}$$

$$= 40.7917\,\Omega\angle -1.2236\,\text{rad},$$

or

$$Z_{in} = 13.8799 - j38.3577\,\Omega.$$

The equivalent circuits at its input and at the load can be drawn as shown in Figure 5.8.

Voltage at the input end,

$$V_A = \frac{V_S}{Z_S + Z_{in}} Z_{in} = \frac{100 \cdot 40.7917 \angle -1.2236\,\text{rad}}{50 + 13.8799 - j38.3577},$$

or

$$V_A = 54.7456\,\text{V}\angle -0.6828\,\text{rad}.$$

The equivalent circuit shown in Figure 5.8*b* can be used to determine the load's voltage. Since the source impedance is the same as the characteristic impedance of the line, Z_{th} is found to be

$$Z_{th} = Z_o \frac{Z_s + jZ_o \tan(\beta\ell)}{Z_o + jZ_s \tan(\beta\ell)} = Z_o = 50\,\Omega$$

and

$$V_{th} = (V_{in} + V_{ref})|_{\text{at o.c.}} = (2 \cdot V_{in})|_{\text{at o.c.}} = 100\,\text{V}\angle -0.5\,\text{rad}.$$

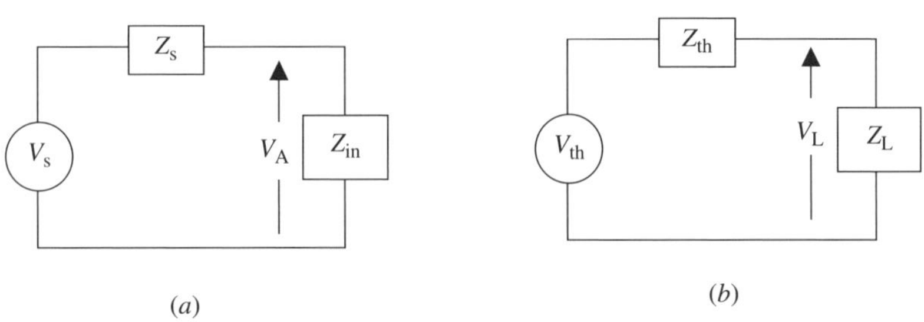

Figure 5.8 Equivalent circuits at the input (*a*) and at the load (*b*).

Therefore

$$V_L = \frac{V_{th}}{Z_{th} + Z_L} Z_L = 79.0569\,\text{V}\angle{-0.8218}\,\text{rad}.$$

Alternately,

$$V(z) = V_{in}e^{-j\beta z} + V_{ref}e^{j\beta z} = V_{in}(e^{-j\beta z} + \Gamma_{in}e^{j\beta z}),$$

where V_{in} is incident voltage at $z = 0$ while Γ_{in} is the input reflection coefficient. Since $V_{in} = 50\ \text{V}\angle 0^\circ$, and

$$\Gamma_{in} = \frac{13.8799 - j38.3577 - 50}{13.8799 - j38.3577 + 50} = 0.7071\ \angle{-1.7854}\,\text{rad},$$

then

$$V_L = V(z = 5\,\text{m}) = 50(e^{-j0.5} + 0.7071e^{j(0.5-1.7854)})$$

$$= 79.0569\,\text{V}\angle{-0.8218}\,\text{rad},$$

Standing Wave and Voltage Standing Wave Ratio

Consider a lossless transmission line that is terminated by a load impedance Z_L, as shown in Figure 5.9. The time-harmonic incident and reflected voltages at its input (i.e., at $z = 0$) are assumed to be V_{in} and V_{ref}, respectively. Therefore the total voltage V(z) can be expressed as follows:

$$V(z) = V_{in}e^{-j\beta z} + V_{ref}e^{+j\beta z}.$$

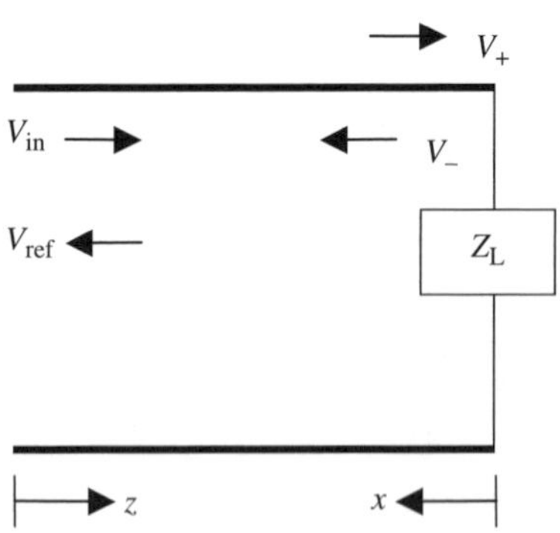

Figure 5.9 Terminated lossless transmission line.

In practice, it is more convenient to use the load as the reference point instead of the source. Therefore we switch to the x variable in place of z and express the total voltage at a point on the line as follows:

$$V(x) = V_+ e^{+j\beta x} + V_- e^{-j\beta x} = V_+\left[e^{+j\beta x} + \Gamma_L e^{-j\beta x}\right],$$

or

$$V(x) = V_+[e^{+j\beta x} + \rho_L e^{-j(\beta x - \phi_L)}], \tag{5.2.23}$$

where

$$\Gamma_L = \rho_L e^{j\phi_L} = \frac{V_-}{V_+}$$

and V_+ and V_- represent incident and reflected wave voltage phasors, respectively, at the load point (i.e., at $x = 0$).

Let us first consider two extreme conditions at the load. In one case, the transmission line has an open-circuit termination while, in the other, it has a short-circuit termination. Therefore the magnitude of the reflection coefficient, ρ_L, is unity in both cases. However, the phase angle, ϕ_L, is zero for the former while it is π for the latter case. Phasor diagrams for these two cases are depicted in Figure 5.10*a* and 5.10*b*, respectively. As distance x from the load increases, phasor $e^{j\beta x}$ rotates counterclockwise because of the increase in its phase angle βx. On the other hand, phasor $e^{-j\beta x}$

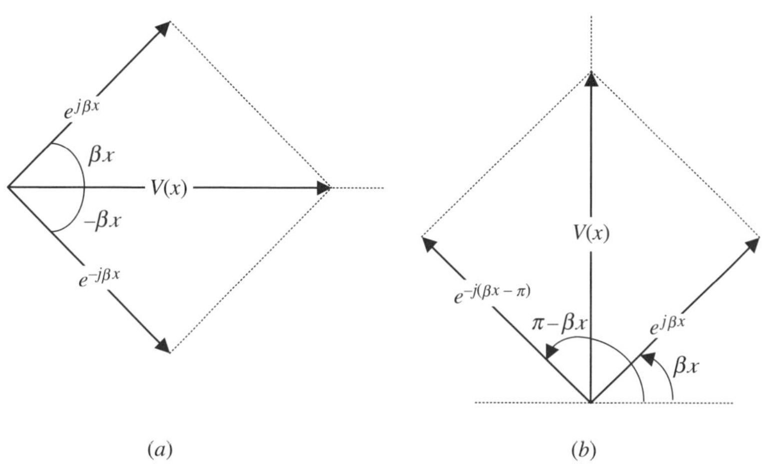

Figure 5.10 Phasor diagrams of line voltage with (*a*) an open-circuit and (*b*) a short-circuit termination.

rotates clockwise by the same amount. Therefore the phase angle of the resultant voltage, $V(x)$, remains constant with space coordinate x while its magnitude varies sinusoidally between $\pm 2V_+$. Since the phase angle of the resultant signal $V(x)$ does not change with distance, it does not represent a propagating wave. Note that there are two waves propagating on this line in opposite directions. However, the resulting signal represents a standing wave.

For an arbitrary termination that is different from characteristic impedance of the line, there are two waves propagating in opposite directions. The interference pattern of these two signals is stationary with time. Assuming that V_+ is unity, a phasor diagram for this case is drawn as shown in Figure 5.11. The magnitude of the resultant signal, $V(x)$, can be determined using the law of parallelogram as follows:

$$|V(x)| = |V_+|\sqrt{1 + \rho_L^2 + 2\rho_L \cos(2\beta x - \phi_L)}. \tag{5.2.24}$$

The reflection coefficient, $\Gamma(x)$, can be found as follows:

$$\Gamma(x) = \frac{V_- e^{-j\beta x}}{V_+ e^{j\beta x}} = \rho_L e^{-j(2\beta x - \phi_L)}. \tag{5.2.25}$$

According to (5.2.24), the magnitude of a standing wave is a function of the location on transmission line. Since x appears only in the argument of the cosine function, the extreme value of the voltage-magnitude occurs whenever this argument is an integer multiple of π. It is a maximum whenever the reflected wave is in phase with the

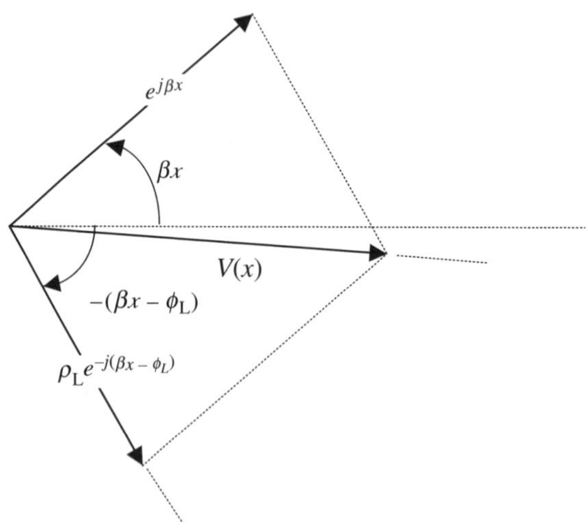

Figure 5.11 Phasor diagram of the line voltage with an arbitrary termination.

incident, and that requires the following condition to be satisfied:

$$(2\beta x - \phi_L) = \pm 2n\pi, \qquad n = 0, 1, 2, \ldots. \tag{5.2.26}$$

On the other hand, $V(x)$ is a minimum wherever the reflected and incident signals are out of phase on the line. Hence x satisfies the following condition at a minimum of the interference pattern:

$$(2\beta x - \phi_L) = \pm(2m+1)\pi, \qquad m = 0, 1, 2, \ldots. \tag{5.2.27}$$

Further these extreme values of the standing wave are

$$|V(x)|_{max} = |V_+|\{1 + \rho_L\} \tag{5.2.28}$$

and

$$|V(x)|_{min} = |V_+|\{1 - \rho_L\}. \tag{5.2.29}$$

The voltage standing wave ratio (VSWR) is defined as follows:

$$\text{VSWR} = S = \frac{|V(x)|_{max}}{|V(x)|_{min}} = \frac{1 + \rho_L}{1 - \rho_L}. \tag{5.2.30}$$

Since $0 \leq \rho_L \leq 1$ for a passive load, minimum value of the VSWR is unity (for a matched load) while its maximum value can be infinity (for total reflection, with a short circuit or an open circuit at the load).

Assume that there is a voltage minimum at x_1 from the load. As one keeps moving toward the source, the next minimum occurs at x_2. In other words, there are two consecutive minimums at x_1 and x_2 with $x_2 > x_1$. Hence

$$2(\beta x_1 - \phi_L) = (2m_1 + 1)\pi$$

and

$$2(\beta x_2 - \phi_L) = [2(m_1 + 1) + 1]\pi.$$

Subtracting the former equation from the latter, we have

$$2\beta(x_2 - x_1) = 2\pi \Rightarrow x_2 - x_1 = \frac{\lambda}{2}, \tag{5.2.31}$$

where $(x_2 - x_1)$ is separation between the two consecutive minimums.

Similarly it can be proved that two consecutive maximums are a half-wavelength apart and also that the separation between the consecutive maximum and minimum is a quarter-wavelength. This information can be used to measure the wavelength of

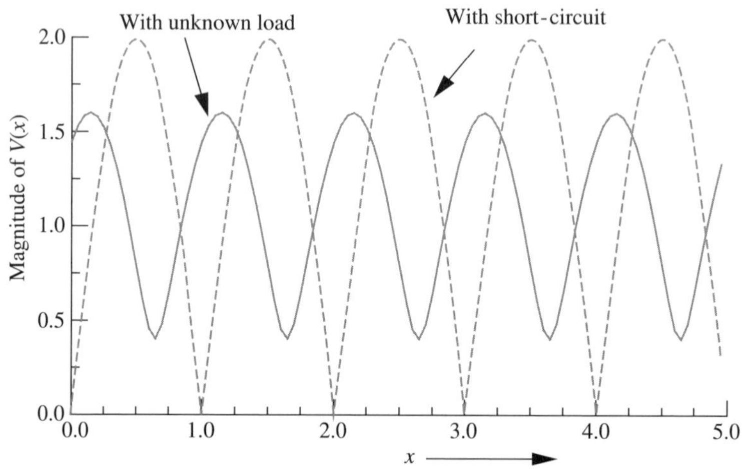

Figure 5.12 Standing wave pattern on a lossless transmission line.

a propagating signal. In practice, the location of a minimum is preferred over that of a maximum. This is mainly because of minimums being sharper in comparison with maximums, as illustrated in Figure 5.12. Further a short (or open) circuit must be used as the load for the best accuracy in the wavelength measurement.

Measurement of Impedance

The impedance of a one-port microwave device can be determined from measurement of the standing wave at its input. The required parameters for that are the VSWR and the location of first minimum (or maximum) from the load. A slotted line that is equipped with a detector probe is connected before the load to facilitate the measurement. Since the output of a detector is proportional to power, the square root of the ratio is taken to find the VSWR. Since it is not possible in most cases to probe up to the input terminals of the load, location of the first minimum is determined as follows: An arbitrary minimum is located on the slotted line with the unknown load. The load is then replaced by a short circuit. As a result there is a shift in minimum, as shown in Figure 5.12. The shift of the original minimum away from the generator is the same as the location of first minimum from the load.

Since the reflected voltage is out of phase with that of the incident signal at the minimum of the standing wave pattern, the relation between the reflection coefficient, $\Gamma_1 = -\rho$, and impedance, Z_1, at this point may be written as follows:

$$\Gamma_1 = -\rho = \frac{\overline{Z}_1 - 1}{\overline{Z}_1 + 1},$$

where $\overline{Z}_1$ is normalized impedance at the location of the first minimum of the voltage's standing wave.

Since the VSWR $S = (1+\rho)/(1-\rho)$,

$$\rho = \frac{S-1}{S+1},$$

and therefore

$$-\frac{S-1}{S+1} = \frac{\overline{Z}_1 - 1}{\overline{Z}_1 + 1} \Rightarrow \overline{Z}_1 = \frac{1}{S}.$$

This normalized impedance is equal to the input impedance of the line that is terminated by load Z_L and has a length d_1. Hence

$$\overline{Z}_1 = \frac{1}{S} = \frac{\overline{Z}_L + j\tan(\beta d_1)}{1 + j\overline{Z}_L \tan(\beta d_1)} \Rightarrow \overline{Z}_L = \frac{1 - jS\tan(\beta d_1)}{S - j\tan(\beta d_1)}. \tag{5.2.32}$$

Similarly the reflected voltage is in phase with the incident signal at the maximum of the standing wave. Assume that the first maximum is located at a distance d_2 from the unknown load and that the impedance at this point is Z_2. Therefore

$$\Gamma_2 = \rho = \frac{\overline{Z}_2 - 1}{\overline{Z}_2 + 1} = \frac{S-1}{S+1} \Rightarrow \overline{Z}_2 = S$$

and

$$\overline{Z}_L = \frac{\overline{Z}_2 - j\tan(\beta d_2)}{1 - j\overline{Z}_2 \tan(\beta d_2)}. \tag{5.2.33}$$

Since the maximum and minimum are measured on a lossless line that feeds the unknown load, the magnitude of the reflection coefficient ρ does not change at different points on the line.

Example 5.12

A 50 Ω transmission line is terminated by a $25 + j100$ Ω load. Find the load reflection coefficient and the voltage standing wave ratio on this line.

$$\Gamma_L = \frac{Z_L - Z_o}{Z_L + Z_o} = \frac{25 + j100 - 50}{25 + j100 + 50} = 0.8246\angle 50.91^\circ$$

and

$$\text{VSWR} = \frac{1+\rho}{1-\rho} = \frac{1 + 0.8246}{1 - 0.8246} = 10.4039.$$

Example 5.13

A 75 Ω transmission line is used to feed a transmitting antenna. In order to find the antenna impedance Z_L, measurements are conducted using a slotted line. It is found that it has a VSWR of 1.8 and the standing wave minimums are 1.5 cm apart. The scale reading at one of these minimums is found to be 2.75 cm. When the antenna is replaced by a short circuit, the minimum moves away from the generator to a point where the scale shows 2.35 cm. Determine the signal wavelength and the load impedance.

Since the minimums are 1.5 cm apart, the signal wavelength $\lambda = 2 \times 1.5 = 3$ cm. Therefore

$$\beta = \frac{2\pi}{\lambda} = \frac{2\pi}{3}\ \text{rad/cm} = 209.4395\ \text{rad/m},$$

$$d_1 = 2.75 - 2.35 = 0.4\ \text{cm},\ \beta d_1 = 0.8378.$$

Hence

$$\overline{Z}_L = \frac{1 - j1.8\tan(0.8378)}{1.8 - j\tan(0.8378)} = 0.8987 - j0.5561$$

or

$$Z_L = 67.4013 - j41.7089\ \Omega.$$

5.3 SMITH CHART

According to (5.2.4), the normalized impedance at a point on the transmission line is related to its reflection coefficient as follows:

$$\overline{Z} = \overline{R} + j\overline{X} = \frac{1+\Gamma}{1-\Gamma} = \frac{1+\Gamma_r + j\Gamma_i}{1-\Gamma_r - j\Gamma_i}, \tag{5.3.1}$$

where Γ_r and Γ_i are real and imaginary parts of the reflection coefficient, respectively, and $\overline{R}$ and $\overline{X}$ are real and imaginary parts of the normalized impedance, respectively. Two real equations are found from (5.3.1) as follows after equating the real and imaginary components on its two sides:

$$\left(\Gamma_r - \frac{\overline{R}}{1+\overline{R}}\right)^2 + \Gamma_i^2 = \left(\frac{1}{1+\overline{R}}\right)^2 \tag{5.3.2}$$

and

$$(\Gamma_r - 1)^2 + \left(\Gamma_i - \frac{1}{\overline{X}}\right)^2 = \left(\frac{1}{\overline{X}}\right)^2. \tag{5.3.3}$$

Equations (5.3.2) and (5.3.3) represent a family of circles on the complex Γ-plane. The circle in (5.3.2) has its center at $\left([\overline{R}/(1+\overline{R})], 0\right)$ and a radius of $1/(1+\overline{R})$. For $\overline{R}=0$, the circle is centered at the origin with unity radius. As $\overline{R}$ increases, the center of the constant resistance circle moves on the positive real axis and its radius decreases. When $\overline{R}=\infty$, the radius reduces to zero and the center of the circle moves to (1, 0). These plots are shown in Figure 5.13. Note that for passive impedance, $0 \le \overline{R} \le \infty$ while $-\infty \le \overline{X} \le +\infty$.

Similarly the circle of (5.3.3) has its center at $\left(1, 1/\overline{X}\right)$ with a radius of $1/\overline{X}$. For $\overline{X}=0$, center of the circle lies at $(1, \infty)$ with an infinite radius. Hence it is a straight line along the Γ_r-axis. As $\overline{X}$ increases on the positive side (i.e., $0 \le \overline{X} \le \infty$), the center of the circle moves toward point (1, 0) along the vertical line defined by $\Gamma_r = 1$, whereas its radius becomes smaller and smaller in size. For $\overline{X}=\infty$, it converges to a point that is located at (1, 0). Similar characteristics are observed for $0 \ge \overline{X} \ge -\infty$. As shown in Figure 5.13, a graphical representation of these two equations for all possible normalized resistance and reactance values is known as the Smith chart. Thus a normalized impedance point on the Smith chart represents the corresponding reflection coefficient in polar coordinates on the complex Γ-plane. According to (5.2.11) the magnitude of the reflection coefficient on a lossless

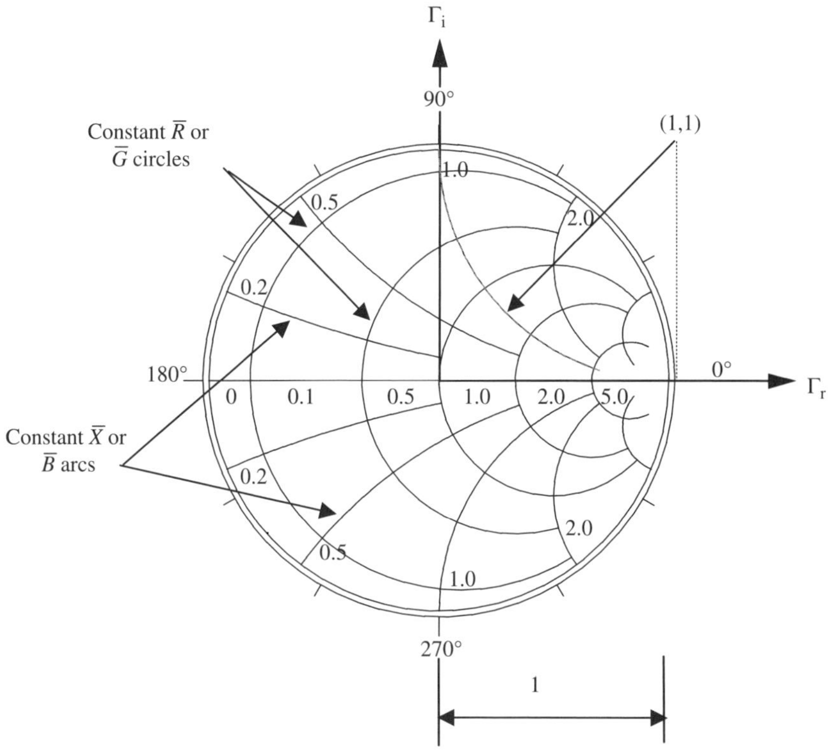

Figure 5.13 The Smith chart.

transmission line remains constant at ρ_L while its phase angle decreases by $-2\beta\ell$. Hence it represents a circle of radius ρ_L. As we move away from the load (i.e., toward the generator), the reflection coefficient point moves clockwise on this circle. Since the reflection coefficient repeats periodically at every half-wavelength, the circumference of the circle is equal to $\lambda/2$. For a given reflection coefficient, the normalized impedance may be found using the impedance scale of the Smith chart. Further $\overline{R}$ in (5.3.1) is equal to the VSWR for $\Gamma_r > 0$ and $\Gamma_i = 0$. On the other hand, it equals the inverse of VSWR for $\Gamma_r < 0$ and $\Gamma_i = 0$. Hence the $\overline{R}$ values on the positive Γ_r-axis also represent the VSWR.

Since inverse of the impedance is the admittance, we find that

$$\bar{Y} = \frac{1}{\bar{Z}} = \frac{1-\Gamma}{1+\Gamma}.$$

Hence a similar analysis can be performed with the normalized admittance as well. This results in a same kind of chart except that the normalized conductance circles replace the normalized resistance circles while the normalized susceptance arcs replace the normalized reactance.

The normalized resistance (or conductance) of each circle is indicated on the Γ_r-axis of the Smith chart. Normalized positive reactance (or susceptance) arcs are shown on the upper half while negative reactance (or susceptance) arcs are seen in the lower half. The Smith chart in conjunction with equation (5.2.5) facilitates the analysis and design of transmission line circuits.

Example 5.14

A load impedance of $100 - j50\ \Omega$ terminates a $50\ \Omega$ lossless quarter-wavelength-long transmission line. Find the impedance at its input end, the load reflection coefficient, and the VSWR on this transmission line.

This problem can be solved using (5.2.4) through (5.2.7) or the Smith chart. The former method proceeds as follows:

$$\beta\ell = \frac{2\pi}{\lambda}\cdot\frac{\lambda}{4} = \frac{\pi}{2} = 90^\circ,$$

$$Z_{in} = Z_o\frac{Z_L + jZ_o\tan(\beta\ell)}{Z_o + jZ_L\tan(\beta\ell)} = 50\frac{(100-j50) + j50\tan(90^\circ)}{50 + j(100-j50)\tan(90^\circ)},$$

or

$$Z_{in} = 50\frac{j50}{j(100-j50)} = \frac{2500}{100-j50} = 20 + j10\,\Omega.$$

Further

$$\Gamma_L = \frac{Z_L - Z_o}{Z_L + Z_o} = \frac{100 - j50 - 50}{100 - j50 + 50} = \frac{50 - j50}{150 - j50} = 0.4472\angle{-26.5651^\circ}$$

and

$$\text{VSWR} = \frac{1 + |\Gamma|}{1 - |\Gamma|} = \frac{1 + 0.4472}{1 - 0.4472} = 2.618.$$

To solve this problem graphically using the Smith chart, the normalized load impedance is determined as follows:

$$\overline{Z}_L = \frac{Z_L}{Z_o} = \frac{100 - j50}{50} = 2 - j1.$$

This point is located on the Smith chart as shown in Figure 5.14. A circle that passes through $2 - j1$ is then drawn with point $1 + j0$ as its center. Since the radius of the Smith chart represents unity magnitude, the radius of this circle is equal to the magnitude of the reflection coefficient ρ_L. In other words, the normalized radius of this circle (i.e., the radius of this circle divided by the radius of the Smith chart) is equal to ρ_L. Note that a clockwise movement on this circle corresponds to a movement away from the load on the transmission line. Hence a point d meters away from the load is located at $-2\beta d$ on the chart. Therefore the input

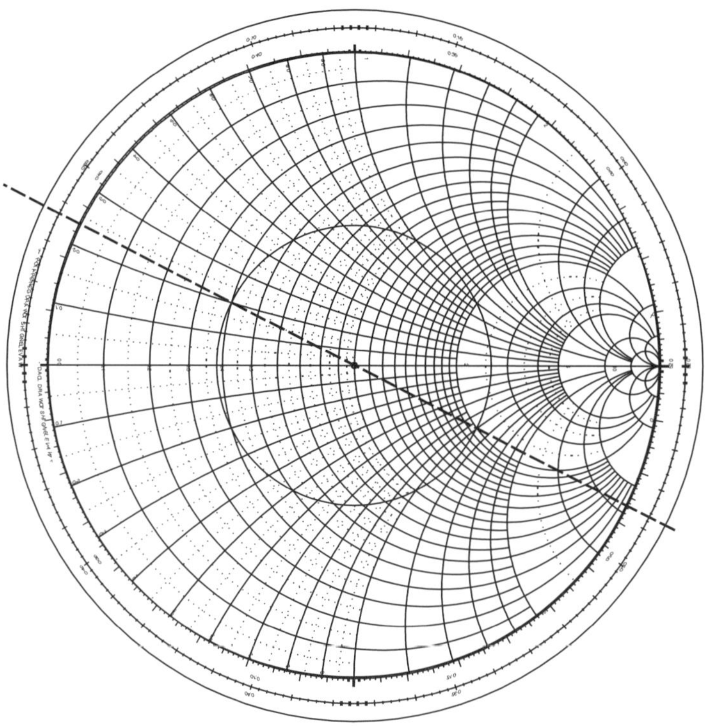

Figure 5.14 Solution to Example 5.14 using a Smith chart.

port of the line that is a quarter-wavelength away from the load (i.e., $d = \lambda/4$) can be located on this circle after moving by $-\pi$ to a point at 0.037λ on "wavelengths toward generator" scale.

From the Smith chart, VSWR = 2.6, $\Gamma_L = 0.45\angle{-26.5^\circ}$, and $\overline{Z}_{in} = 0.4 + j0.2$. Therefore

$$Z_{in} = (0.4 + j0.2) \cdot 50 = 20 + j10\ \Omega.$$

Example 5.15

A 50 Ω lossless transmission line is terminates in $100 - j100\ \Omega$. Using the Smith chart, find (a) Γ_L, (b) VSWR, (c) Z_{in} at a distance of 0.125λ from the load, (d) the shortest length of the line for which impedance is purely resistive, and (e) the value of this resistance.

$$\overline{Z}_L = \frac{100 - j100}{50} = 2 - j2.$$

After locating this normalized impedance point on the Smith chart, the constant VSWR circle is drawn as shown in Figure 5.15.

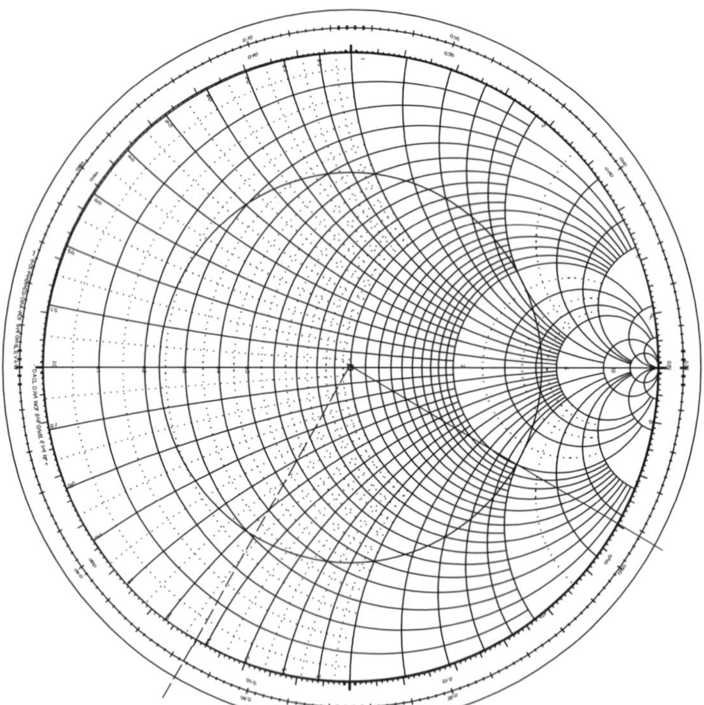

Figure 5.15 Solution to Example 5.15 using a Smith chart.

(a) The magnitude of the reflection coefficient is equal to the radius of the VSWR circle (with the radius of the Smith chart as unity). The angle made by the radial line that connects the load impedance point with the center of the chart is equal to the phase angle of reflection coefficient. Hence

$$\Gamma_L = 0.62 \angle -30^\circ.$$

(b) VSWR is found to be 4.25 from the scale reading for the point where the circle intersects the $+\Gamma_r$-axis.

(c) For $d = 0.125\lambda$, $-2\beta d = -1.5708$ radians $= -90^\circ$ (clockwise from the load). This point is located after moving on the VSWR circle by 0.125λ from the load (at 0.417λ on "wavelengths toward generator" scale). The corresponding normalized impedance is found to be $0.31 - j0.55$. Therefore

$$Z_{\text{in}}(\ell = 0.125\lambda) = (0.31 - j0.55) = 15.5 - j27.5\,\Omega.$$

(d) While moving clockwise from the load point, the VSWR circle crosses the Γ_r-axis for the first time at 0.5λ. The imaginary part of the impedance is zero at this intersection point. Therefore $d = (0.5 - 0.292)\lambda = 0.208\lambda$. The normalized impedance at this point is 0.23. The next point on the transmission line where the impedance is purely real occurs a quarter-wavelength from it (i.e., 0.458λ from load). The normalized impedance at this point is 4.5.

(e) Since normalized resistance at this point is 0.23,

$$R = 0.23 \cdot 50 = 11.5\,\Omega.$$

Example 5.16

A 50 Ω lossless transmission line is terminated by $50 + j50\ \Omega$. Find the location of the first V_{max}, first V_{min}, and the VSWR if the operating wavelength is 10 cm.

$$\overline{Z}_L = \frac{50 + j50}{50} = 1 + j1.$$

As shown in Figure 5.16, this point is located on the Smith chart and the VSWR circle is drawn. From the chart, VSWR $= 2.6$. The scale reading on the "wavelengths toward generator" is 0.162λ at the load point. As one moves away from this point clockwise (toward generator) on this VSWR circle, the voltage maximum is found first at 0.25λ and then a minimum at 0.5λ. If the first voltage maximum is at d_{max} from the load, then $d_{\text{max}} = (0.25 - 0.162)\lambda = 0.088\lambda = 0.88$ cm from the load.

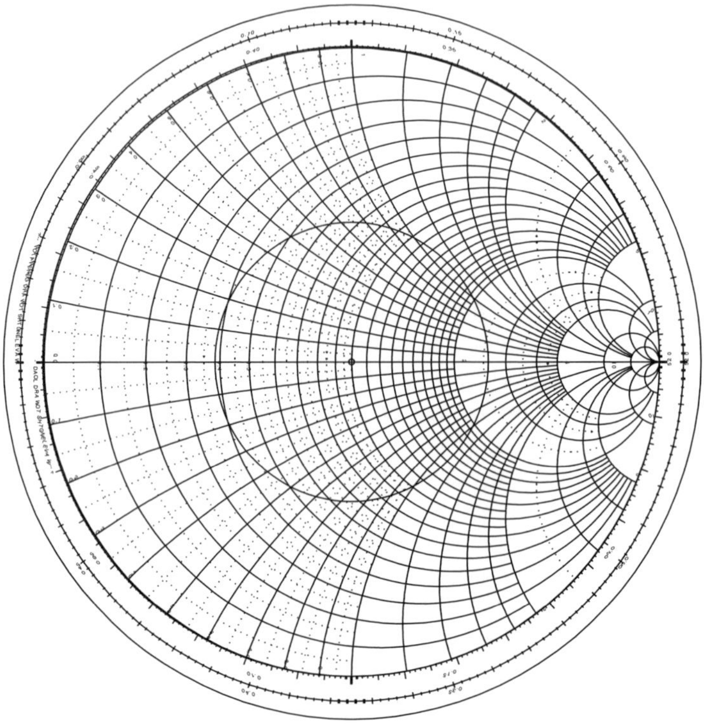

Figure 5.16 Solution to Example 5.16 using a Smith chart.

The first minimum is a quarter-wavelength away from the point of voltage maximum. Hence $d_{\text{min}} = (0.5 - 0.162)\lambda = 0.338\lambda = 3.38$ cm.

Example 5.17

A 100 Ω lossless transmission line is terminated by a 200 Ω load. Find impedance at points 1.65λ and 4.25λ from the termination.

$$\overline{Z}_L = \frac{200}{100} = 2.$$

As illustrated in Figure 5.17, this point is located on the Smith chart and the VSWR circle is drawn. Note that the VSWR on this line is 2 and the load reflection coefficient is about $0.33\angle 0^\circ$. As we move on the transmission line toward the generator, the phase angle of the reflection coefficient changes by $-2\beta d$, where d is the distance away from the load. Hence one revolution around the VSWR circle is completed for every half-wavelength. Therefore the normalized impedance will be 2 at every integer multiple of a half-wavelength from the load. It will be true for a point located at 1.5λ as well as at 4.0λ. For the remaining 0.15λ,

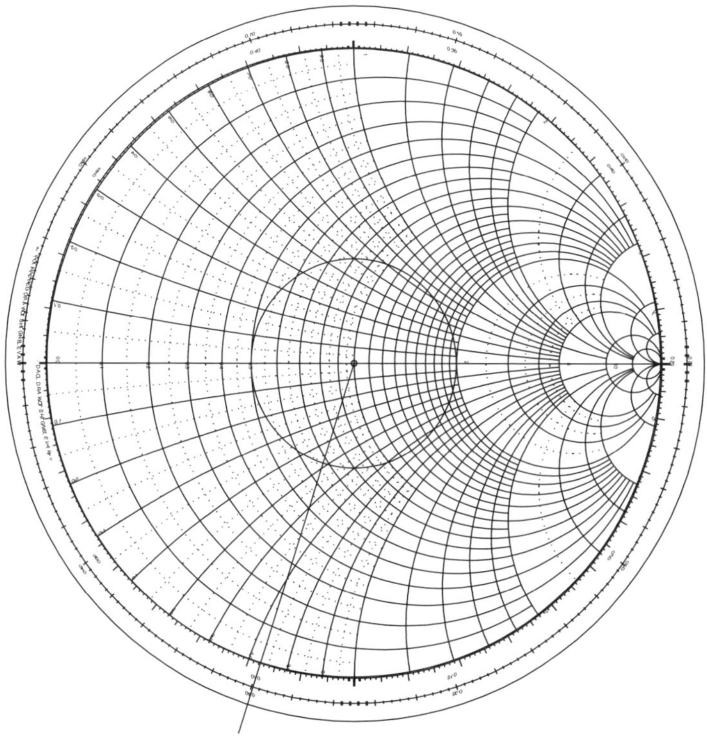

Figure 5.17 Solution to Example 5.17 using a Smith chart.

the impedance point is located on the VSWR circle at 0.40λ (i.e., $0.25\lambda + 0.15\lambda$) on the "wavelengths toward generator" scale. Similarly the point corresponding to 4.25λ from the load is found at 0.5λ.

From the Smith chart, the normalized impedance at 1.65λ is $0.68 - j0.48$, while it is 0.5 at 4.25λ. Therefore, for impedance at 1.65λ,

$$Z_1 = (0.68 - j0.48) \cdot 100 = 68 - j48\,\Omega,$$

and for impedance at 4.25λ,

$$Z_2 = (0.5) \cdot 100 = 50\,\Omega.$$

Example 5.18

A 50 Ω lossless transmission line is terminated by $5 - j5$ mS load. Find the impedance at a point 2.65λ away from load and the VSWR on this line.

$$\overline{Y}_L = \frac{Y_L}{Y_o} = Y_L \cdot Z_o = 0.25 - j0.25.$$

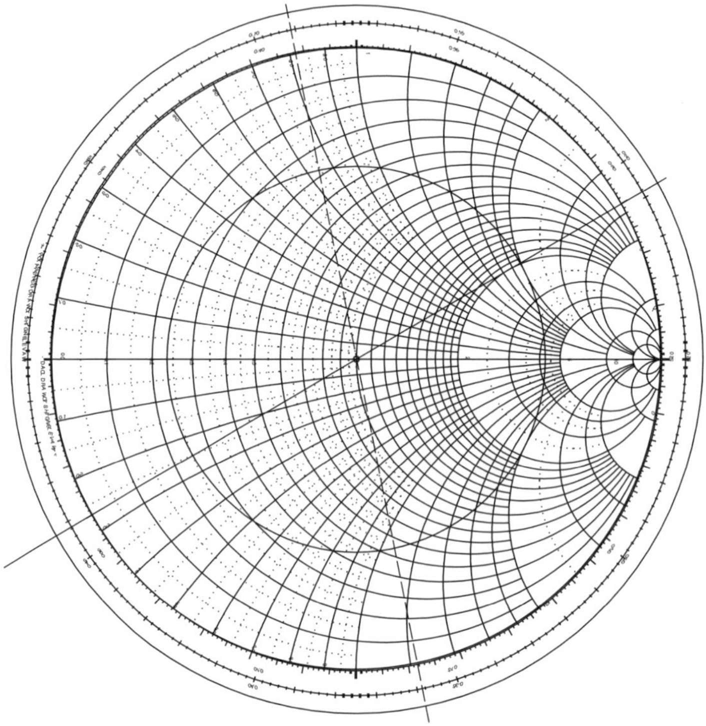

Figure 5.18 Solution to Example 5.18 using a Smith chart.

As before, this normalized admittance point is located on the Smith chart and the VSWR circle is drawn. It is shown in Figure 5.18. There are two choices available at this point. The given normalized load admittance is converted to the corresponding impedance by moving to a point on the diametrically opposite side of the VSWR circle. It shows a normalized load impedance $2 + j2$. Moving from this point by 2.65λ toward the generator, the normalized impedance is found as $0.55 - j1.1$. Alternately, we can first move from the normalized admittance point by 2.65λ toward the generator to a normalized admittance point $0.37 + j0.75$. This is then converted to the normalized impedance by moving to the diametric opposite point on the VSWR circle.

Thus the normalized impedance at a point 2.65λ away from the load is $0.55 - j1.1$. The impedance at this point is $(0.55 - j1.1) \cdot 50 = 27.5 - j55\ \Omega$, and the VSWR is approximately 4.25.

Example 5.19

An experiment is performed using the circuit shown in Figure 5.19. First, a load Z_L is connected at the end of a $50\ \Omega$ transmission line and its VSWR is found to be 1.8. After that, the detector probe is placed at one of the minimums on the line.

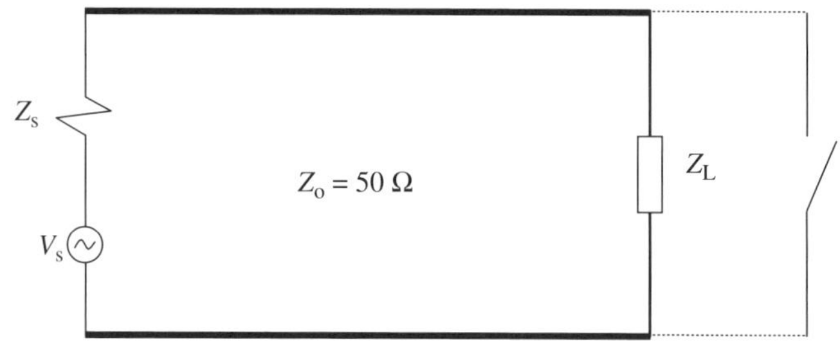

Figure 5.19 Experimental setup used in Example 5.19.

It is found that this minimum shifts toward the load by 9 cm when the load is replaced by a short circuit. Further two consecutive minimums are found to be 30 cm apart. Determine the load impedance.

Since the separation between consecutive minimums is 30 cm, the signal wavelength on the line is 60 cm. Therefore the first minimum of the standing wave pattern occurs at 0.15λ from the load. Further the VSWR on the line is measured to be 1.8. This circle is drawn on the Smith chart as shown in Figure 5.20. As discussed

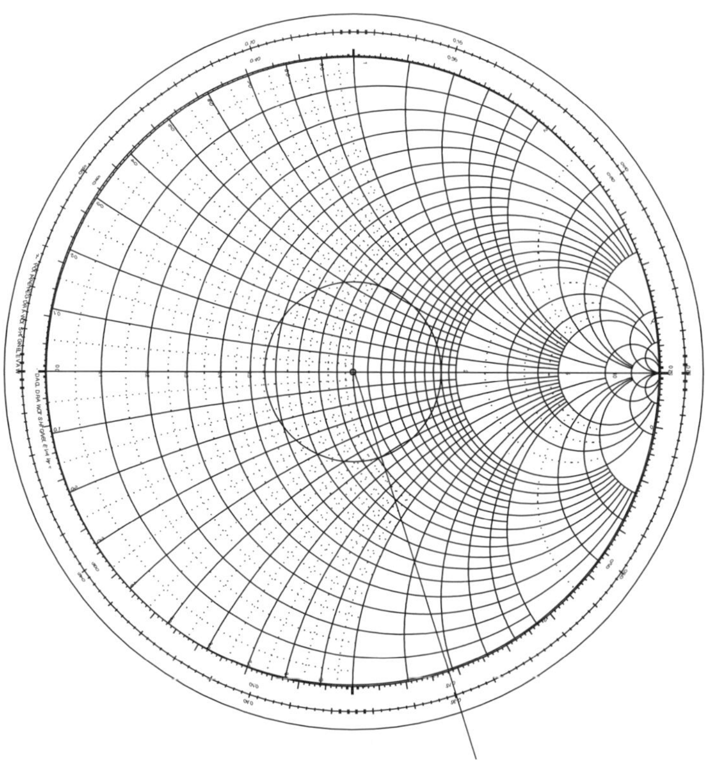

Figure 5.20 Solution to Example 5.19 using a Smith chart.

earlier, a minimum in a voltage standing wave occurs when the phase angle of the reflection coefficient is 180°. It is located on the Smith chart at a point where the VSWR circle intersects the $-\Gamma_r$-axis. From this point we move toward the load by 0.15λ (i.e., counterclockwise) to locate the normalized load impedance point. It is found to be $1.02 - j0.6$.

Therefore,

$$Z_L = (1.02 - j0.6) \cdot 50 = 51 - j30\,\Omega.$$

5.4 TRANSIENTS ON TRANSMISSION LINES

We have analyzed the propagation characteristics of sinusoidal signals on the transmission line. This study was simplified by phasor representations of these signals. As was mentioned earlier, nonsinusoidal signals can be represented by an appropriate sum of sinusoidal signals using a Fourier analysis. Therefore the preceding results can be applied to a variety of cases. We discuss analyzing nonsinusoidal signal characteristics directly in the time domain in this section. However, to simplify the analysis, we assume that the transmission line is ideal. We begin our study with the propagation of narrow pulse signals on a lossless transmission line. The excitation of a finite length transmission line with resistive terminations by step voltages is considered next. The section ends with a step-voltage excitation of a transmission line terminated by a reactive element.

Pulse Excitation of Transmission Line

Consider an ideal transmission line terminated by R_L, as shown in Figure 5.21*a*. A voltage source with internal resistance R_s is connected at its input end. The characteristic impedance of the line is assumed to be Z_o. Recall that the term "impedance" is defined for sinusoidal signals when we use phasor representation. In time-domain analysis all quantities must be real. Since the line is lossless, its characteristic impedance is a real number Z_o, and its use can safely follow the preceding convention. The transmission line's length is ℓ and the wave velocity acting on it is v.

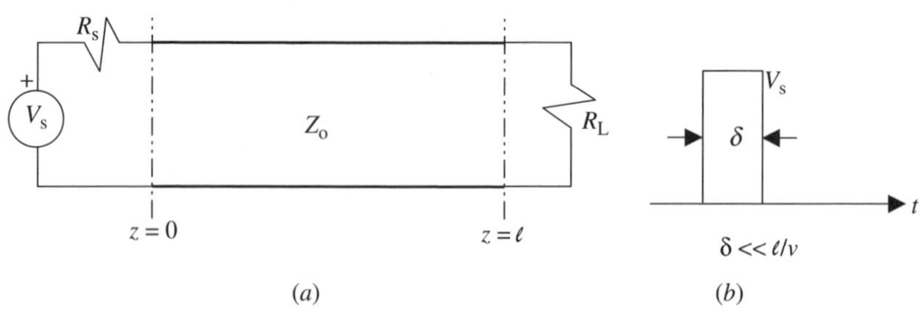

Figure 5.21 (*a*) An ideal transmission line with source and termination and (*b*) a narrow pulse.

Assume that a source excites the transmission line with a narrow pulse, as shown in Figure 5.21*b*. As indicated, the pulse-width is very much narrower than the time it takes to travel from one end of the line to the other. Since the pulse is unaware of the termination, it sees an infinite length to begin with. Therefore voltage V_{in} and current I_{in} at $z = 0$ are found to be

$$V_{in} = \frac{Z_o}{Z_o + R_s} V_s \tag{5.4.1}$$

and

$$I_{in} = \frac{V_s}{Z_o + R_s}. \tag{5.4.2}$$

This signal travels over the transmission line with velocity v and reaches the load R_L after $\tau = \ell/v$ seconds. It sees the discontinuity at this point, and therefore a part of the pulse is reflected back. The reflection coefficient Γ_L can be easily found as follows:

$$\Gamma_L = \frac{R_L - Z_o}{R_L + Z_o}. \tag{5.4.3}$$

The reflected pulse with its voltage and current $V_{ref} = \Gamma_L V_{in}$ and $I_{ref} = -\Gamma_L I_{in}$, respectively, now propagates toward the source. The reflected pulse arrives at $z = 0$ after a time τ where a part of it is again reflected toward the load because of the change in impedance from Z_o to R_s. The reflection coefficient Γ_s for this signal is found to be

$$\Gamma_s = \frac{R_s - Z_o}{R_s + Z_o}. \tag{5.4.4}$$

This process of the pulse bouncing back and forth continues indefinitely. Note that the reflection coefficients are fractional numbers, and therefore the pulse's amplitude goes down with each reflection. A graphical representation helps track this bouncing pulse, as shown in Figure 5.22. This diagram is called a bounce (or reflection) diagram. It can be used to find voltage at any point on a line as a function of time. For example, the bounce diagram shows that three pulses pass through $z = \ell/2$ point with amplitudes V_{in}, $\Gamma_L V_{in}$, and $\Gamma_s \Gamma_L V_{in}$, respectively, during $0 < t < 3\tau$.

Example 5.20

For the circuit shown in Figure 5.21, assume that $R_L = 25\ \Omega$, $R_s = 40\ \Omega$, and $Z_o = 50\ \Omega$. The transmission line is 4 m long and the wave velocity on it is $2 \cdot 10^8$ m/s. If the source excites a narrow pulse of 5 V at $t = 0$, plot the voltage

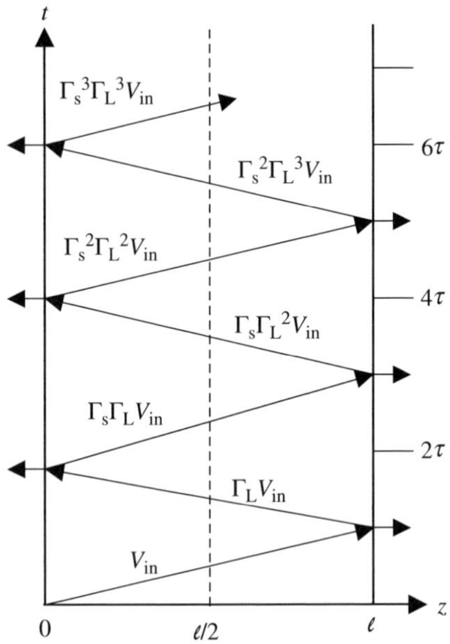

Figure 5.22 Bounce diagram for a transmission line circuit.

and the current at $z = 1$ m as functions of time over first 50 ns.

$$V_{in} = \frac{Z_o}{Z_o + R_s} V_s = \frac{50}{50 + 40} 5 = 2.7778\,\text{V},$$

$$I_{in} = \frac{V_s}{Z_o + R_s} = \frac{5}{50 + 40} = \frac{1}{18}\,\text{A} = 55.6\,\text{mA},$$

$$\Gamma_L = \frac{R_L - Z_o}{R_L + Z_o} = \frac{25 - 50}{25 + 50} = -\frac{1}{3},$$

$$\Gamma_s = \frac{R_s - Z_o}{R_s + Z_o} = \frac{40 - 50}{40 + 50} = -\frac{1}{9},$$

and

$$\tau = \frac{\ell}{v} = \frac{4}{2 \cdot 10^8}\,\text{s} = 20\,\text{ns}.$$

Therefore V_{in} reaches $z = 1$ m in 5 ns, and the pulse after reflecting from the load arrives at this point in another 30 ns. The pulse is reflected back by R_s and arrives in another 10 ns. These three pulses take 45 ns. Next the reflection from the load takes another 30 ns, and therefore it is out of the duration of 50 ns. The various pulses are illustrated in Figure 5.23.

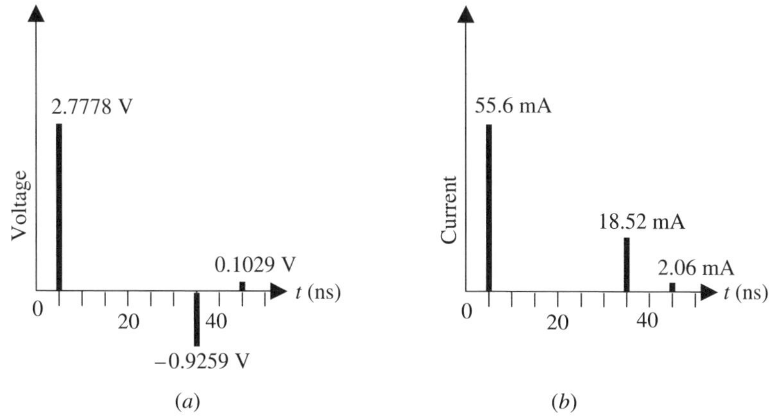

Figure 5.23 Voltage and current pulses on the transmission line of Example 5.20.

Example 5.21

Reconsider the transmission line circuit of Example 5.4. Using time-domain analysis, show that the quarter-wavelength transformer indeed matches the 1.8 kΩ load with a 200 Ω line.

Assume that a sinusoidal signal of unit magnitude is incident at the 600 Ω line at $z = 0$. The reflection and transmission coefficients at this junction are

$$\Gamma_1 = \frac{600 - 200}{600 + 200} = 0.5$$

and

$$T_1 = V_{in} = 1 + \Gamma_1 = 1.5.$$

V_{in} propagates over the $\lambda/4$ section and hits the 1.8 kΩ load where it is partly reflected back with a reflection coefficient Γ_L. The reflected signal bounces back and forth as shown in Figure 5.24*b*. This signal has a reflection coefficient Γ_s at $z = 0$. The reflection coefficients are

$$\Gamma_L = \frac{1800 - 600}{1800 + 600} = 0.5$$

and

$$\Gamma_s = -\Gamma_1 = \frac{200 - 600}{200 + 600} = -0.5.$$

Note that unlike the pulse signal of the previous example, it is a continuous sinusoidal signal. Therefore all components at a point on the line need to be

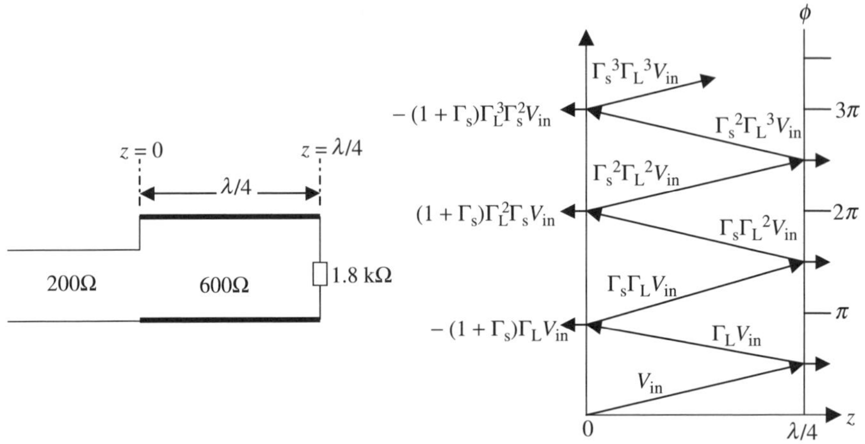

(*a*) (*b*)

Figure 5.24 Circuit arrangement (*a*) and bounce diagram (*b*) for Example 5.21.

summed up in order to find total voltage. Also the signal is delayed by π for each round trip over the $\lambda/4$ section. Thus the total reflected signal into the 200 Ω line is found to be

$$V_{\text{ref}} = \Gamma_1 - V_{\text{in}}\Gamma_{\text{L}}(1+\Gamma_{\text{s}}) + V_{\text{in}}\Gamma_{\text{L}}^2\Gamma_{\text{s}}(1+\Gamma_{\text{s}}) - V_{\text{in}}\Gamma_{\text{L}}^3\Gamma_{\text{s}}^2(1+\Gamma_{\text{s}}) + \cdots,$$

or

$$\begin{aligned} V_{\text{ref}} &= \Gamma_1 - V_{\text{in}}\Gamma_{\text{L}}(1+\Gamma_{\text{s}})[1 - \Gamma_{\text{L}}\Gamma_{\text{s}} + \Gamma_{\text{L}}^2\Gamma_{\text{s}}^2 - \Gamma_{\text{L}}^3\Gamma_{\text{s}}^3 + \cdots] \\ &= \Gamma_1 - V_{\text{in}}\Gamma_{\text{L}}(1+\Gamma_{\text{s}}) \cdot \frac{1}{1+\Gamma_{\text{L}}\Gamma_{\text{s}}}. \end{aligned}$$

Therefore

$$V_{\text{ref}} = \Gamma_1 - \frac{V_{\text{in}}\Gamma_{\text{L}}(1+\Gamma_{\text{s}})}{1+\Gamma_{\text{L}}\Gamma_{\text{s}}} = 0.5 - \frac{1.5 \cdot 0.5 \cdot (1-0.5)}{1+0.5\cdot(-0.5)} = 0.5 - \frac{1.5 \cdot 0.25}{0.75} = 0.$$

As expected, the signal reflected back into 200 Ω line is zero.

Transients on Transmission Line with Resistive Termination

Consider a transmission line of length ℓ terminated by R_{L}, as shown in Figure 5.25. The line is excited by a dc source V_{s} with its internal resistance R_{s} when the switch closes at $t = 0$. Since electrical signal is not aware of the condition at $z = \ell$ at this

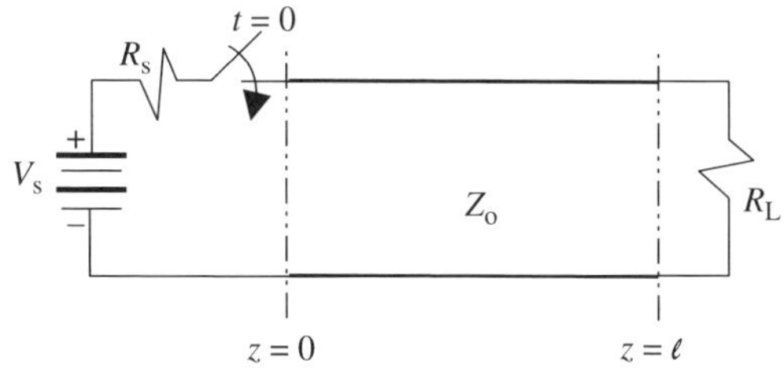

Figure 5.25 Step voltage excitation of a terminated line.

point, a forward traveling step signal V_{in} is introduced at $z = 0$. The voltage V_{in} and the current I_{in} are found to be

$$V_{in} = \frac{Z_o}{R_s + Z_o} V_s \tag{5.4.5}$$

and

$$I_{in} = \frac{V_s}{R_s + Z_o}. \tag{5.4.6}$$

The signal arrives at $z = \ell$ after a time $\tau = \ell / v$, where v is signal velocity on the line. Since the load is different from Z_o, the reflected voltage V_{ref} is sent toward the source. Therefore the following condition must be true at this point:

$$V_{in} + V_{ref} = R_L(I_{in} + I_{ref}) = \frac{R_L}{Z_o}(V_{in} - V_{ref}) \rightarrow V_{ref} = \frac{R_L - Z_o}{R_L + Z_o} V_{in}$$

$$V_{ref} = \Gamma_L V_{in}. \tag{5.4.7}$$

Note that the reflected current due to V_{ref} is $-V_{ref}/Z_o$. At $t = 2\tau$, it is reflected again because the source resistance is different from Z_o. Therefore the total voltage at $z = 0$ is now $V_{in} + V_{ref} + V_{in}^1$, which should meet the following condition:

$$\frac{V_s - (V_{in} + V_{ref} + V_{in}^1)}{R_s} = \frac{V_{in} - V_{ref} + V_{in}^1}{Z_o} \rightarrow V_{ref}^1 = \frac{R_s - Z_o}{R_s + Z_o} V_{in}^1$$

$$V_{ref}^1 = \Gamma_s V_{in}^1. \tag{5.4.8}$$

The signal keeps bouncing back and forth at every τ until steady state is reached. The reflection diagram of Figure 5.22 holds for this situation as well if we sum all the terms up to the desired time. In steady state, voltage V_{line} will be given as follows:

$$V_{line} = \frac{R_L}{R_s + R_L} V_s. \tag{5.4.9}$$

These results can be verified easily after adding all the bouncing terms of Figure 5.22. For $t = \infty$, we have the following infinite series.

$$\begin{aligned} V &= V_{\text{in}}(1+\Gamma_{\text{L}})[1+\Gamma_{\text{s}}\Gamma_{\text{L}}+\Gamma_{\text{s}}^2\Gamma_{\text{L}}^2+\Gamma_{\text{s}}^3\Gamma_{\text{L}}^3+\Gamma_{\text{s}}^4\Gamma_{\text{L}}^4+\Gamma_{\text{s}}^5\Gamma^5+\cdots] \\ &= \frac{V_{\text{in}}(1+\Gamma_{\text{L}})}{1-\Gamma_{\text{s}}\Gamma_{\text{L}}}. \end{aligned} \quad (5.4.10)$$

Substituting for Γ_{s} and Γ_{L} reduces the equation to (5.4.9).

Transients on Transmission Line with Reactive Termination

Consider a transmission line of length ℓ and characteristic impedance Z_{o} that is terminated with a capacitor C, as shown in Figure 5.26*a*. For simplicity we assume that the source is matched with the line (i.e., $R_{\text{s}} = Z_{\text{o}}$). As before, when the switch is closed at $t = 0$, V_{in} and I_{in} at $z = 0$ are found from (5.4.5) and (5.4.6) as follows:

$$V_{\text{in}} = \frac{V_{\text{s}}}{2} \quad (5.4.11)$$

and

$$I_{\text{in}} = \frac{V_{\text{s}}}{2Z_{\text{o}}}. \quad (5.4.12)$$

This forward-traveling step signal reaches at $z = \ell$ in time $\tau = \ell/v$. Since there is a capacitor terminating the line, reflection takes place. Assume that the capacitor has no charge initially, and V_{ref} is the reflected voltage. Therefore the total voltage on the line for $0 < t < \tau$ is V_{in}, and the voltage appearing across the capacitor at $t = \tau$ is $V_{\text{L}} = V_{\text{in}} + V_{\text{ref}}$. A Thevenin equivalent can be found for $t = \tau$, as shown in Figure 5.26*b*. The Thevenin voltage V_{Th} is found to be $2V_{\text{in}} = V_{\text{s}}$ and $Z_{\text{Th}} = Z_{\text{o}}$.

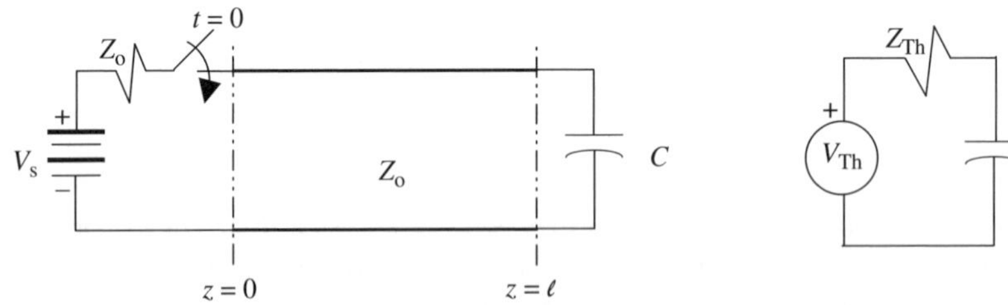

(*a*) (*b*)

Figure 5.26 (*a*) Transient on a transmission line with capacitive termination and (*b*) the Thevenin equivalent at $z = \ell$.

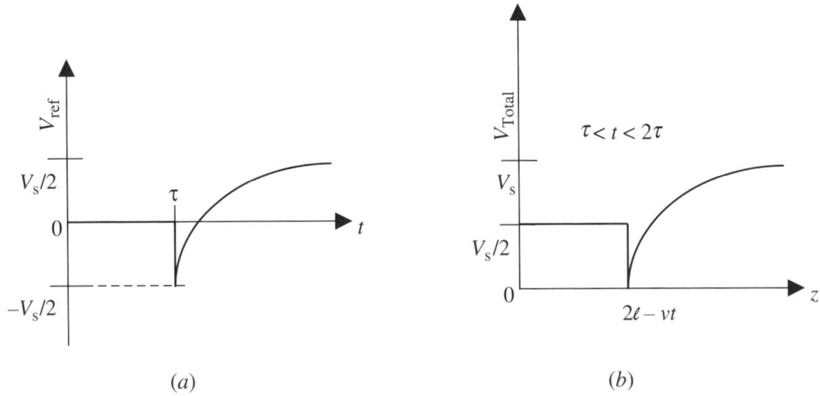

Figure 5.27 (*a*) Reflected voltage $V_{ref}(t)$ at $t = \tau$ and (b) total voltage $V(z)$ for $\tau < t < 2\tau$.

This circuit can now be analyzed by the usual approach as follows:

$$C\frac{dV_L}{dt} = \frac{V_s - V_L}{Z_o} \rightarrow \frac{dV_L}{dt} + \frac{1}{Z_oC}V_L = \frac{1}{Z_oC}V_s. \tag{5.4.13}$$

Equation (5.4.13) can be solved via usual mathematical methods, such as the Laplace transforms, as follows:

$$V_L = (1 - 2e^{-(t-\tau)/z_oC})V_s. \tag{5.4.14}$$

Therefore the reflected voltage V_{ref} at the load is found to be

$$V_{ref} = V_L - V_{in} = (1 - e^{-(t-\tau)/z_oC})V_s - \frac{V_s}{2} = \frac{V_s}{2}(1 - 2e^{-(t-\tau)/Z_oC}). \tag{5.4.15}$$

The V_{ref} propagates toward the source for $\tau < t < 2\tau$ and is absorbed by the source impedance. Remember that the source is matched with the line, so no further reflection takes place. A similar analysis can be carried out for an inductive termination.

PROBLEMS

5.1. The inner and outer conductor diameters of a Teflon-filled ($\varepsilon_r = 3.3$) coaxial line are given as follows: $2a = 0.028$ inch and $2b = 0.182$ inch. Find its characteristic impedance if the operating frequency is such that $\omega L \gg R$ and $\omega C \gg G$.

5.2. A telephone line has the following parameters: $R = 23$ ohm/mile, $G = 0.02$ pS/mile, $L = 3.2$ mH/mile, and $C = 0.01$ μF/mile For the operating frequency of 800 kHz, calculate its characteristic impedance Z_o and the propagation constant γ. If its characteristic impedance terminates the line and the input voltage is 4.8 V $\angle 0^\circ$, find total voltage at 25 miles from its input.

5.3. An infinitely long lossless 75 Ω transmission line is connected to a signal generator with an open-circuit voltage $v(t) = \cos(2\pi \cdot 10^9 t)$ V. The generator has an internal resistance of 50 Ω. If the signal propagates with a velocity of $1.5 \cdot 10^8$ m/s on the line, find instantaneous voltage and current at an arbitrary location on the line.

5.4. A lossless transmission line of length d and characteristic impedance Z_o is used as an impedance transformer to match a 1.2 kΩ load to a 50 Ω line. If the signal wavelength is 5.7 m, find (a) d, (b) Z_o, and (c) the reflection coefficient at the load.

5.5. A 75 Ω outdoor antenna is being used to receive a 95 MHz FM signal. If the receiving circuit uses a 50 Ω line, design a quarter-wavelength transformer to match the antenna with this transmission line. Assume that the transformer is made from a Teflon-filled ($\varepsilon_r = 2.1$) coaxial line to calculate its length (in cm). Also determine the diameter of its inner conductor if the inner diameter of the outer conductor is 6.5 mm.

5.6. A 100 km long telephone line is terminated by $25 + j75$ Ω load. The line parameters are as follows: $R = 1.5$ Ω/km, $G = 150$ nS/km, $L = 6.2$ mH/km, and $C = 9$ nF/km. If this line is being used at 3 kHz, find its (a) characteristic impedance Z_o, (b) propagation constant γ, and (c) input impedance Z_{in}.

5.7. A radio-frequency filter exhibits 50 Ω at its input as well as at its output. If it is inserted in a 75 Ω system, find the filter's insertion loss.

5.8. A signal attenuates by 0.015 dB for each meter traveled on a 50 Ω distortionless transmission line. If this line has a capacitance of 65 pF per meter, find (a) R, (b) L, (c) G, and (d) v_p.

5.9. A load Z_L is connected at one end of the transmission line and its input impedance is measured using an impedance bridge that gives $Z_{in} = 40 + j50$ Ω. The experiment is repeated twice, with the load replaced first by a short circuit and then by an open circuit. The data are recorded as $j20$ Ω and $-j150$ Ω, respectively. Find the characteristic resistance of this line and the load impedance.

5.10. Measurements are made on a 1.5 m long transmission line using an impedance bridge. After short-circuiting at one of its ends, the impedance at the other end is found to be $j130$ Ω. Repeating the experiment with the short circuit now replaced by an open circuit gives $-j61$ Ω. Determine the propagation constant and the characteristic impedance of this line.

5.11. A 2 m long 50 Ω lossless transmission line is terminated by $100 + j50$ Ω. The circuit is driven by a 6.3662 MHz signal generator with its open-circuit voltage 100 V $\angle 0^\circ$ and the internal impedance 50 Ω. If the propagating signal has a phase velocity of 100 m/μS, find the impedance at its input end and the phasor voltages at both its ends.

5.12. A 50 Ω transmission line is terminated by a $75 - j25$ Ω load. Find the load reflection coefficient and the voltage standing wave ratio on this line.

5.13. A 75 Ω transmission line is used to feed a transmitting antenna. In order to find the antenna impedance Z_L, measurements are conducted using a slotted line. It is found that it has a VSWR of 1.4 and the standing wave minimums are 2.3 cm apart. The scale reading at one of these minimums is found to be 3.25 cm. When the antenna is replaced by a short circuit, the minimum moves away from the generator to a point where the scale shows 2.92 cm. Determine the signal wavelength and the load impedance.

5.14. A load impedance of $100 + j50$ Ω terminates a 50 Ω lossless quarter-wavelength-long transmission line. Find the impedance at its input end, the load reflection coefficient, and the VSWR on this transmission line.

5.15. A 50 Ω lossless transmission line is terminates in $100 + j100$ Ω. Using the Smith chart, find (a) Γ_L, (b) VSWR, (c) Z_{in} at a distance of 0.125λ from the load, (d) the shortest length of the line for which impedance is purely resistive, and (e) the value of this resistance.

5.16. A 75 Ω lossless transmission line is terminated by $75 - j75$ Ω. Find the location of the first V_{max}, first V_{min}, and the VSWR, if the operating wavelength is 15 cm.

5.17. A 50 Ω lossless transmission line is terminated by a 150 Ω load. Find the impedance at points 1.65λ and 4.25λ from the termination.

5.18. A 50 Ω lossless transmission line is terminated by $5 + j5$ mS load. Find the impedance at a point 3.15λ away from load and the VSWR on this line.

5.19. An experiment is performed using the circuit shown in Figure 5.19. First, a load Z_L is connected at the end of a 50 Ω transmission line, and its VSWR is found to be 1.45. After that, the detector probe is placed at one of the minimums on the line. It is found that this minimum shifts toward the load by 6.5 cm when the load is replaced by a short circuit. Further two consecutive minimums are found to be 24 cm apart. Determine the load impedance.

5.20. For the circuit shown in Figure 5.21, assume that $R_L = 15$ Ω, $R_s = 50$ Ω, and $Z_o = 75$ Ω. The transmission line is 1 m long and the wave velocity on it is $2 \cdot 10^8$ m/s. If the source excites a narrow pulse of 1 V at $t = 0$, plot the voltage and the current at $z = 50$ cm as functions of time over first 50 ns.

5.21. Consider the transmission line circuit of Problem 5.4. Using time-domain analysis, show that the quarter-wavelength transformer indeed matches the 1.2 kΩ load with a 50 Ω line.

5.22. A given transmission line has the following parameters: $Z_o = 120\angle -8^\circ$ Ω, $\alpha = 2 \cdot 10^{-6}$ dB/m, $v_p = 1.8 \cdot 10^8$ m/s, and $f = 2.0$ MHz. Find the phasor $V(z)$ and $I(z)$, and the corresponding instantaneous values for a wave traveling in the z direction, if maximum value of the current wave at $z = 0$ is 3 mA and it has maximum positive value with respect to time at $t = 0$.

5.23. Determine the characteristic impedance and the phase velocity of a 25 cm long lossless transmission line from the following experimental data: $Z_{sc} = -j78\ \Omega$, $Z_{oc} = j48\ \Omega$, and $f = 900$ MHz. Assume that the line's length is smaller than $\lambda/2$.

5.24. A 17.5 m long lossless transmission line with $Z_o = 50\ \Omega$, is short-circuited at one end, and a voltage source $V_s = 2\cos(50\pi \cdot 10^6 t - \pi/6)$ V is connected at its input terminals. If the source impedance is 50 Ω and the phase velocity on the line is $2 \cdot 10^8$ m/s, find the total currents at its input and through the short circuit.

6

MODIFIED MAXWELL'S EQUATIONS AND POTENTIAL FUNCTIONS

This chapter begins with the concepts of magnetic charge and magnetic current. Vector and scalar potential functions are introduced. Also wave functions in rectangular, cylindrical, and spherical coordinates are presented in order to construct solutions in the various orthogonal coordinates. The discussion includes brief descriptions of Bessel functions, Hankel functions, and Legendre functions. Solutions are constructed for a few simple problems.

6.1 MAGNETIC CHARGE AND CURRENT

Recall equation (3.2.4). It implies that unlike the electric flux there is no possibility of net magnetic flux leaving or entering a closed surface. This is due to the fact that the magnetic charges (north and south poles) do not exist independently. If we break a magnet, each piece will still have both north and south poles. In other words, magnetic monopoles do not exist. The smallest unit of a magnetic source is a magnetic dipole, which is actually an infinitesimal current loop.

Consider a general case where the source consists of two types of currents—a linear current that flows or oscillates in linear direction and a circulatory current that flows or oscillates around an infinitesimal loop. For this case it is rather hard to mathematically describe the electric current density $\vec{J}$ that includes both parts. Mathematically it is sometimes advantageous to recognize the electric current's linear part as $\vec{J}_e$ and imagine its infinitesimal circulatory current as an equivalent

Practical Electromagnetics: From Biomedical Sciences to Wireless Communication.
By Devendra K. Misra
Copyright © 2007 John Wiley & Sons, Inc.

magnetic dipole source. If the circulatory current varies in magnitude and direction or oscillates around the loop with time, the equivalent magnetic dipole strength oscillates with time likewise. When the magnetic dipole oscillates with time, it can be considered an imaginary magnetic current flowing up and down. This is completely analogous to a time-varying electric dipole. Therefore an equivalent magnetic current density $\vec{J}_{\mathrm{m}}$ and a charge density ρ_{m} can represent a time-varying circulatory current.

Imagine now that the magnetic charge ρ_{m} produces a magnetic flux, analogous to the electric charge ρ_{e} producing the electric flux. Then (3.2.4) should be modified to

$$\nabla \cdot \vec{\mathscr{B}} = \rho_{\mathrm{m}}.$$

If (3.2.4) is modified, then (3.2.1) also needs a correction because divergence of the curl of a vector is always zero. However, it is not true here because ρ_{m} is a function of time, which can be found as follows:

$$\nabla \cdot (\nabla \times \vec{\mathscr{E}}) = -\nabla \cdot \left(\frac{\partial \vec{\mathscr{B}}}{\partial t}\right) = -\frac{\partial}{\partial t}(\nabla \cdot \vec{\mathscr{B}}) = -\frac{\partial}{\partial t}\rho_{\mathrm{m}} \neq 0.$$

If it is assumed that the usual continuity equation should hold between ρ_{m} and $\vec{\mathscr{J}}_{\mathrm{m}}$, then

$$\nabla \cdot \vec{\mathscr{J}}_{\mathrm{m}} = -\frac{\partial \rho_{\mathrm{m}}}{\partial t}.$$

Hence equation (3.2.1) should be modified to

$$\nabla \times \vec{\mathscr{E}} = -\vec{\mathscr{J}}_{\mathrm{m}} - \frac{\partial \vec{\mathscr{B}}}{\partial t}.$$

Therefore the modified Maxwell equations for a time-harmonic field and the two continuity equations can be expressed as follows:

$$\nabla \times \vec{E}(\vec{r}) = -\vec{J}_{\mathrm{m}}(\vec{r}) - j\omega\vec{B}(\vec{r}), \tag{6.1.1}$$

$$\nabla \times \vec{H}(\vec{r}) = \vec{J}_{\mathrm{e}}(\vec{r}) + j\omega\vec{D}(\vec{r}), \tag{6.1.2}$$

$$\nabla \cdot \vec{D}(\vec{r}) = \rho_{\mathrm{e}}(\vec{r}), \tag{6.1.3}$$

$$\nabla \cdot \vec{B}(\vec{r}) = \rho_{\mathrm{m}}(\vec{r}), \tag{6.1.4}$$

$$\nabla \cdot \vec{J}_{\mathrm{e}}(\vec{r}) = -j\omega\rho_{\mathrm{e}}(\vec{r}), \tag{6.1.5}$$

and

$$\nabla \cdot \vec{J}_{\mathrm{m}}(\vec{r}) = -j\omega\rho_{\mathrm{m}}(\vec{r}). \tag{6.1.6}$$

6.2 MAGNETIC VECTOR AND ELECTRIC SCALAR POTENTIALS

Potential functions are introduced to transform Maxwell's equations mathematically and facilitate solutions to electromagnetic problems.

Suppose that there are only electric sources (current and charge) present in a volume under consideration (i.e., $\vec{J}_{\mathrm{e}} \neq 0$, $\rho_{\mathrm{e}} \neq 0$, $\vec{J}_{\mathrm{m}} = 0$, and $\rho_{\mathrm{m}} = 0$). Therefore Maxwell's equations for time-harmonic fields and the equation of continuity can be written as follows:

$$\nabla \times \vec{E} = -j\omega\vec{B}, \tag{6.2.1}$$

$$\nabla \times \vec{H} = \vec{J}_{\mathrm{e}} + j\omega\vec{D}, \tag{6.2.2}$$

$$\nabla \cdot \vec{D} = \rho_{\mathrm{e}}, \tag{6.2.3}$$

$$\nabla \cdot \vec{B} = 0, \tag{6.2.4}$$

and

$$\nabla \cdot \vec{J}_{\mathrm{e}} = -j\omega\rho_{\mathrm{e}}. \tag{6.2.5}$$

As defined earlier, the constitutive relations are

$$\vec{B} = \mu\vec{H} \tag{6.2.6}$$

and

$$\vec{D} = \varepsilon\vec{E}. \tag{6.2.7}$$

Since the divergence of the curl of a vector is always zero, the magnetic flux density in (6.2.4) can be assumed to be the curl of vector $\vec{A}$. Hence

$$\nabla \cdot \vec{B} = 0 \Rightarrow \vec{B} = \nabla \times \vec{A}. \tag{6.2.8}$$

Substituting $\vec{B}$ from (6.2.8) into (6.2.1), and recognizing the fact that the curl of a gradient of a scalar is always zero, gives

$$\nabla \times \vec{E} = -j\omega\nabla \times \vec{A} \Rightarrow \nabla \times \{\vec{E} + j\omega\vec{A}\} = 0 \Rightarrow \vec{E} + j\omega\vec{A} = -\nabla\phi_{\mathrm{e}}.$$

Therefore

$$\vec{E} = -\nabla\phi_{\mathrm{e}} - j\omega\vec{A}, \tag{6.2.9}$$

where $\vec{A}$ is called the magnetic vector potential and ϕ_{e} the electric scalar potential.

Combining (6.2.6), (6.2.7), and (6.2.9), we can rewrite (6.2.2) as follows:

$$\nabla \times \left\{\frac{\nabla \times \vec{A}}{\mu}\right\} = \vec{J}_{\mathrm{e}} + j\omega\varepsilon\vec{E} = \vec{J}_{\mathrm{e}} + j\omega\varepsilon\{-\nabla\phi_{\mathrm{e}} - j\omega\vec{A}\}.$$

If μ is not changing with the space coordinates, this relation further simplifies to

$$\nabla \times \nabla \times \vec{A} = \nabla(\nabla \cdot \vec{A}) - \nabla^2 \vec{A} = \mu \vec{J}_e - j\omega\varepsilon\mu\nabla\phi_e + \omega^2 \mu\varepsilon\vec{A}$$

or

$$\nabla^2 \vec{A} + \omega^2 \mu\varepsilon\vec{A} = -\mu\vec{J}_e + j\omega\varepsilon\mu\nabla\phi_e + \nabla(\nabla \cdot \vec{A})$$

or

$$\nabla^2 \vec{A} + \omega^2 \mu\varepsilon\vec{A} = -\mu \vec{J}_e + \nabla\{\nabla \cdot \vec{A} + j\omega\varepsilon\mu\phi_e\}. \tag{6.2.10}$$

Similarly, for ε not changing with space coordinates, (6.2.3), (6.2.7), and (6.2.9) give

$$\nabla \cdot \{-\nabla\phi_e - j\omega\vec{A}\} = \frac{\rho_e}{\varepsilon} \to \nabla^2 \phi_e + j\omega(\nabla \cdot \vec{A}) = -\frac{\rho_e}{\varepsilon}. \tag{6.2.11}$$

Equations (6.2.10) and (6.2.11) represent a pair of coupled partial differential equations for ϕ and $\vec{A}$. These equations can be uncoupled via Helmholtz's theorem, which states that a vector is completely specified by its curl and divergence. Since only the curl of $\vec{A}$ is defined via (6.2.8), we are at liberty to specify its divergence. The Lorentz condition may be used to define it as follows:

$$\nabla \cdot \vec{A} = -j\omega\varepsilon\mu\phi_e. \tag{6.2.12}$$

Thus (6.2.10) and (6.2.11) simplify to

$$\nabla^2 \vec{A} + k^2 \vec{A} = -\mu \vec{J}_e. \tag{6.2.13}$$

and

$$\nabla^2 \phi_e + k^2 \phi_e = -\frac{\rho_e}{\varepsilon}, \tag{6.2.14}$$

where

$$k^2 = \omega^2 \mu\varepsilon \tag{6.2.15}$$

and k is the wave number. Equation (6.2.13) represents a vector Helmholtz equation, whereas (6.2.14) a scalar Helmholtz equation.

Note that (6.2.8) and (6.2.9) include the interdependence of time-varying electric and magnetic fields. As $\omega \to 0$, the two fields become independent of each other. In this case (6.2.8) still holds, whereas (6.2.9) reduces to

$$\vec{E} = -\nabla\phi_e. \tag{6.2.16}$$

Further (6.2.13) and (6.2.14) reduce to

$$\nabla^2 \vec{A} = -\mu \vec{J}_e \tag{6.2.17}$$

and

$$\nabla^2 \phi_e = -\frac{\rho_e}{\varepsilon}. \tag{6.2.18}$$

This is the well-known Poisson equation, which reduces to the Laplace's equation for the source-free case.

6.3 ELECTRIC VECTOR AND MAGNETIC SCALAR POTENTIALS

The electric vector and magnetic scalar potentials are introduced when the presence of only a magnetic current and a magnetic charge is assumed in a region (i.e., $\rho_e = 0$ and $\vec{J}_e = 0$). Maxwell's equations and the equation of continuity for this case are found to be

$$\nabla \times \vec{E} = -\vec{J}_m - j\omega \vec{B}, \tag{6.3.1}$$

$$\nabla \times \vec{H} = j\omega \vec{D}, \tag{6.3.2}$$

$$\nabla \cdot \vec{D} = 0, \tag{6.3.3}$$

$$\nabla \cdot \vec{B} = \rho_m, \tag{6.3.4}$$

and

$$\nabla \cdot \vec{J}_m = -j\omega\rho_m. \tag{6.3.5}$$

As before, $\vec{D} = \varepsilon\vec{E}$, and $\vec{B} = \mu\vec{H}$. Because of (6.3.3) it is assumed that

$$\vec{D} = -\nabla \times \vec{F}, \tag{6.3.6}$$

where $\vec{F}$ is called the electric vector potential.

After substituting (6.3.6) into (6.3.2) and noting that curl of the gradient of a scalar is always zero, we get

$$\nabla \times \vec{H} = -j\omega\nabla \times \vec{F} \rightarrow \nabla \times \{\vec{H} + j\omega\vec{F}\} = 0 \rightarrow \vec{H} + j\omega\vec{F} = -\nabla\phi_m.$$

Therefore

$$\vec{H} = -\nabla\phi_m - j\omega\vec{F}, \tag{6.3.7}$$

where ϕ_m is called the magnetic scalar potential. Now from (6.3.1), (6.3.6), and (6.3.7) we have

$$\nabla \times \left\{-\frac{\nabla \times \vec{F}}{\varepsilon}\right\} = -\vec{J}_m - j\omega\mu\{-\nabla\phi_m - j\omega\vec{F}\},$$

or

$$-\nabla \times \nabla \times \vec{F} = -\varepsilon \vec{J}_{\rm m} + j\omega\mu\varepsilon\nabla\phi_{\rm m} - \omega^2\mu\varepsilon\vec{F},$$

or

$$-\{\nabla(\nabla \cdot \vec{F}) - \nabla^2\vec{F}\} + \omega^2\mu\varepsilon\vec{F} = -\varepsilon\vec{J}_{\rm m} + j\omega\mu\varepsilon\nabla\phi_{\rm m}. \tag{6.3.8}$$

As in the previous case, if the following Lorentz condition

$$\nabla \cdot \vec{F} = -j\omega\mu\varepsilon\phi_{\rm m} \tag{6.3.9}$$

is introduced, then (6.3.8) reduces to

$$\nabla^2\vec{F} + \omega^2\mu\varepsilon\vec{F} = -\varepsilon\vec{J}_{\rm m}. \tag{6.3.10}$$

Further (6.3.4), (6.3.7), and (6.3.9) give

$$\nabla \cdot \{\mu(-\nabla\phi_{\rm m} - j\omega\vec{F})\} = \rho_{\rm m} \Rightarrow -\nabla^2\phi_{\rm m} - j\omega\{-j\omega\mu\varepsilon\phi_{\rm m}\} = \frac{\rho_{\rm m}}{\mu}$$

or

$$\nabla^2\phi_{\rm m} + \omega^2\mu\varepsilon\phi_{\rm m} = -\frac{\rho_{\rm m}}{\mu}. \tag{6.3.11}$$

Note from (6.3.6) and (6.3.7) that interdependency of electric and magnetic fields ceases as $\omega \to 0$. Simplified expressions similar to (6.2.16) through (6.2.18) can be obtained for this case as well.

Any electromagnetic field problem can be divided into these two categories and solved for each case; a complete solution is constructed after superimposing these individual solutions. Thus the electromagnetic fields can be found easily after determining the vector potentials via (6.2.13) and (6.3.10) and the corresponding Lorentz conditions. This procedure is used in the following sections to further formulate the general solution techniques for source-free cases.

6.4 CONSTRUCTION OF A SOLUTION IN RECTANGULAR COORDINATES

The analysis presented in the preceding sections is specialized here for a source-free region with a rectangular coordinate system. It is further divided into two categories. The electric vector potential is assumed to be zero in one, and the magnetic vector potential is zero in the other case, which results in TM^z (transverse magnetic to z) and TE^z (transverse electric to z) mode fields, respectively.

For the transverse magnetic to z case, assume $\vec{A} = \hat{z}A_z$ and $\vec{F} = 0$. Then (6.2.8), (6.2.9), and (6.2.12) give

$$\vec{H} = \frac{1}{\mu}\nabla \times (\hat{z}A_z) = \frac{1}{\mu}\left\{\hat{x}\frac{\partial A_z}{\partial y} + \hat{y}\left(-\frac{\partial A_z}{\partial x}\right) + \hat{z}0\right\} \tag{6.4.1}$$

and

$$\vec{E} = -j\omega\hat{z}A_z + \frac{1}{j\omega\mu\varepsilon}\nabla\left(\frac{\partial A_z}{\partial z}\right). \tag{6.4.2}$$

For this case equation (6.2.13) simplifies to,

$$\nabla^2 A_z + k^2 A_z = 0 \rightarrow \frac{\partial^2 A_z}{\partial x^2} + \frac{\partial^2 A_z}{\partial y^2} + \frac{\partial^2 A_z}{\partial z^2} + k^2 A_z = 0. \tag{6.4.3}$$

After (6.4.3) is solved for a given problem, the field components are found from (6.4.1) and (6.4.2) as follows:

$$H_x = \frac{1}{\mu}\frac{\partial A_z}{\partial_y}, \tag{6.4.4}$$

$$H_y = -\frac{1}{\mu}\frac{\partial A_z}{\partial x}, \tag{6.4.5}$$

$$H_z = 0, \tag{6.4.6}$$

$$E_x = \frac{1}{j\omega\mu\varepsilon}\frac{\partial^2 A_z}{\partial x \partial z}, \tag{6.4.7}$$

$$E_y = \frac{1}{j\omega\mu\varepsilon}\frac{\partial^2 A_z}{\partial y \partial z}, \tag{6.4.8}$$

and

$$E_z = \frac{1}{j\omega\mu\varepsilon}\left(\frac{\partial^2}{\partial z^2} + k^2\right)A_z. \tag{6.4.9}$$

Since the magnetic fields are transverse to the z-axis ($H_z = 0$), these fields are known as TM^z (transverse magnetic to z) mode fields.

For the transverse electric to z case, assume $\vec{F} = \hat{z}F_z$ and $\vec{A} = 0$. Then (6.3.6) and (6.3.7) give

$$\vec{E} = -\frac{1}{\varepsilon}\nabla \times (\hat{z}F_z) = -\frac{1}{\varepsilon}\left\{\hat{x}\frac{\partial F_z}{\partial y} + \hat{y}\left(-\frac{\partial F_z}{\partial x}\right) + \hat{z}0\right\} \tag{6.4.10}$$

and

$$\vec{H} = -j\omega\hat{z}F_z + \frac{1}{j\omega\mu\varepsilon}\nabla\left(\frac{\partial F_z}{\partial z}\right). \tag{6.4.11}$$

For this case equation (6.3.10) simplifies to

$$\nabla^2 F_z + k^2 F_z = 0 \to \frac{\partial^2 F_z}{\partial x^2} + \frac{\partial^2 F_z}{\partial y^2} + \frac{\partial^2 F_z}{\partial z^2} + k^2 F_z = 0. \tag{6.4.12}$$

After (6.4.12) is solved for a given problem, the field components are found from (6.4.10) and (6.4.11) as follows:

$$E_x = -\frac{1}{\varepsilon}\frac{\partial F_z}{\partial y}, \tag{6.4.13}$$

$$E_y = \frac{1}{\varepsilon}\frac{\partial F_z}{\partial x}, \tag{6.4.14}$$

$$E_z = 0, \tag{6.4.15}$$

$$H_x = \frac{1}{j\omega\mu\varepsilon}\frac{\partial^2 F_z}{\partial x \partial z}, \tag{6.4.16}$$

$$H_y = \frac{1}{j\omega\mu\varepsilon}\frac{\partial^2 F_z}{\partial y \partial z}, \tag{6.4.17}$$

and

$$H_z = \frac{1}{j\omega\mu\varepsilon}\left(\frac{\partial^2}{\partial z^2} + k^2\right)F_z. \tag{6.4.18}$$

Since the electric fields are transverse to the z-axis in this case, these fields are known as TE^z (transverse electric to z) mode fields.

Note that similar formulations are possible for the TM and TE mode fields with respect to the x- or y-axis. Alternatively, the rectangular coordinate system can be rotated to match the TM^z and TE^z modes.

The Wave Functions

Since both (6.4.3) and (6.4.12) are similar, we consider the following partial differential equation, which is called a scalar Helmholtz equation:

$$\frac{\partial^2 \varphi(x, y, z)}{\partial x^2} + \frac{\partial^2 \varphi(x, y, z)}{\partial y^2} + \frac{\partial^2 \varphi(x, y, z)}{\partial z^2} + k^2 \varphi(x, y, z) = 0. \tag{6.4.19}$$

Assume

$$\varphi(x, y, z) = h_1(x)h_2(y)h_3(z). \tag{6.4.20}$$

In other words, we assume that solutions to (6.4.19) are the product of three functions of one coordinate each and we use the method of separation of variables. Substituting (6.4.20) into (6.4.19) and rearranging, we have

$$\frac{1}{h_1(x)}\frac{d^2h_1(x)}{dx^2}+\frac{1}{h_2(y)}\frac{d^2h_2(y)}{dy^2}+\frac{1}{h_3(z)}\frac{d^2h_3(z)}{dz^2}=-k^2. \tag{6.4.21}$$

Each term on the left-hand side of this equation depends on only one coordinate. Since each can be changed independently and still the sum remains constant at $-k^2$, each of these terms must remain constant independently. Therefore we conclude that

$$\frac{1}{h_1(x)}\frac{d^2h_1(x)}{dx^2}=-k_x^2, \tag{6.4.22}$$

$$\frac{1}{h_2(y)}\frac{d^2h_2(y)}{dy^2}=-k_y^2, \tag{6.4.23}$$

$$\frac{1}{h_3(z)}\frac{d^2h_3(z)}{dz^2}=-k_z^2, \tag{6.4.24}$$

and

$$k_x^2+k_y^2+k_z^2=k^2, \tag{6.4.25}$$

where k_x, k_y, and k_z are arbitrary constants at this point that will be evaluated for specific boundary conditions of the problem (the boundary-value problem). These specific values are called eigenvalues or characteristic values and the corresponding solutions are known as the eigenfunctions. Equation (6.4.25) is called the separation equation.

Equations (6.4.22) through (6.4.24) have a similar form. We considered this kind of equation earlier in Chapters 4 and 5 where we have found that harmonic functions satisfy this kind of equation. In other words, the solutions to (6.4.22) are

$$h_1(x) \rightarrow e^{jk_xx},\ e^{-jk_xx},\ \sin(k_xx),\ \cos(k_xx). \tag{6.4.26}$$

A linear sum of any two harmonic functions of (6.4.26) is a complete solution to (6.4.22). The selection of some of these functions may simplify the analysis of a particular problem if the characteristics of the function are matched with the physical situations, as listed in Table 6.1. Further the solutions to (6.4.23) and (6.4.24) can be constructed by a similar process. Thus, using (6.4.20) and (6.4.22) through (6.4.24), we can write the solution to (6.4.19) as follows:

$$\varphi_e(x, y, z)=h_1(k_xx)\cdot h_2(k_yy)\cdot h_3(k_zz). \tag{6.4.27}$$

TABLE 6.1 Characteristics of the Harmonic Functions

$h(kx)$	Special Cases of $k = \beta - j\alpha$	Physical Interpretation
$\cos(kx)$	• $\alpha = 0$	• Standing wave
	• $\beta = 0$	• Two evanescent fields
	• $\alpha \neq 0$ and $\beta \neq 0$	• Localized standing waves
$\sin(kx)$	• $\alpha = 0$	• Standing wave
	• $\beta = 0$	• Two evanescent fields
	• $\alpha \neq 0$ and $\beta \neq 0$	• Localized standing waves
e^{-jkx}	• $\alpha = 0$	• Traveling wave along $+x$
	• $\beta = 0$	• Evanescent field
	• $\alpha \neq 0$ and $\beta \neq 0$	• Attenuated traveling wave along $+x$.
e^{jkx}	• $\alpha = 0$	• Traveling wave along $-x$
	• $\beta = 0$	• Evanescent field
	• $\alpha \neq 0$ and $\beta \neq 0$	• Attenuated traveling wave along $-x$.

Together, the preceding solutions are known as the elementary wave function. Note that only two eigenvalues k_i are independent, while the third eigenvalue has to satisfy the separation equation. Since linear combinations of elementary wave functions also satisfy (6.4.19), we can construct more general wave functions by summing over one or two eigenvalues as follows:

$$\varphi(x, y, z) = \sum_{k_x} \sum_{k_y} C_{k_x k_y} h_1(k_x x) \cdot h_2(k_y y) \cdot h_3(k_z z). \qquad (6.4.28)$$

As was mentioned earlier, this formulation facilitates analysis of a large class of problems in the rectangular coordinate system. We consider a few such applications below. A number of other applications will be considered in later chapters.

Metallic Parallel-Plate Waveguide

Consider a parallel-plate waveguide that consists of two perfectly conductive plates extending to infinity on the y-z plane, as illustrated in Figure 6.1. The separation between the two plates is assumed to be a. There is an electromagnetic signal that propagates along the z-axis. Therefore $h_3(k_z z) \rightarrow e^{-jk_z z}$. Further k_y is zero for $\partial/\partial y \rightarrow 0$, meaning the fields remain constant along the y-axis. Therefore the separation equation (6.4.25) reduces to

$$k_x^2 + k_z^2 = k^2. \qquad (6.4.29)$$

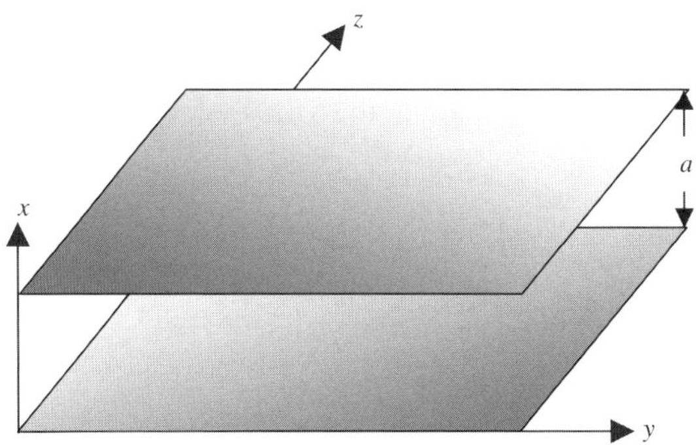

Figure 6.1 Metallic parallel-plate waveguide.

For analyzing the TM modes between the two plates for the conditions specified, the solution to (6.4.3) is determined after selecting an appropriate wave function from (6.4.28) as follows:

$$A_z = \sum_{k_x} C_{k_x} h_1(k_x x) e^{-jk_z z}, \tag{6.4.30}$$

where

$$h_1(k_x x) = C_1 \cos(k_x x) + C_2 \sin(k_x x). \tag{6.4.31}$$

Therefore (6.4.4) through (6.4.9) simplify to

$$H_x = \frac{1}{\mu}\frac{\partial A_z}{\partial y} = 0, \tag{6.4.32}$$

$$H_y = -\frac{1}{\mu}[-C_1 k_x \sin(k_x x) + C_2 k_x \cos(k_x x)]e^{-jk_z z}, \tag{6.4.33}$$

$$H_z = 0, \tag{6.4.34}$$

$$E_x = \frac{-jk_z k_x}{j\omega\mu\varepsilon}[-C_1 \sin(k_x x) + C_2 \cos(k_x x)]e^{-jk_z z}, \tag{6.4.35}$$

$$E_y = \frac{1}{j\omega\mu\varepsilon}\frac{\partial^2 A_z}{\partial y \partial z} = 0, \tag{6.4.36}$$

and

$$E_z = \frac{1}{j\omega\mu\varepsilon}(-k_z^2 + k^2)[C_1 \cos(k_x x) + C_2 \sin(k_x x)]e^{-jk_z z}. \tag{6.4.37}$$

Since the boundary conditions require that the tangential electric fields be zero on the conducting surfaces at $x = 0$ and at $x = a$, E_z must be zero on these surfaces. Note that E_y is already zero according to (6.4.36), and E_x is normal to conducting surfaces. Hence

$$E_z\Big|_{\substack{x=0\\x=a}} = 0 \rightarrow \begin{cases} C_1 = 0, \\ k_x = \dfrac{m\pi}{a}, \quad m = 0, 1, 2, \ldots . \end{cases} \tag{6.4.38}$$

Also (6.4.29) gives

$$k_z = \sqrt{k^2 - \left(\frac{m\pi}{a}\right)^2}. \tag{6.4.39}$$

Therefore the signal will propagate without attenuation if $k > m\pi/a$, and it will attenuate for $k < m\pi/a$. The cutoff occurs at $k = m\pi/a$.

Next the field components (6.4.32) through (6.4.37) can be expressed as follows:

$$H_x = 0, \tag{6.4.40}$$

$$H_y = -\frac{C_2}{\mu}\frac{m\pi}{a}\cos\left(\frac{m\pi}{a}x\right)e^{-jk_z z}, \tag{6.4.41}$$

$$H_z = 0, \tag{6.4.42}$$

$$E_x = C_2\frac{k_z}{\omega\mu\varepsilon}\frac{m\pi}{a}\cos\left(\frac{m\pi}{a}x\right)e^{-jk_z z}, \tag{6.4.43}$$

$$E_y = 0, \tag{6.4.44}$$

TABLE 6.2 Signal Propagation in Parallel-Plate Waveguides

	TE_{m0} Modes	TM_{m0} Modes
k_c	$\dfrac{m\pi}{a}$	$\dfrac{m\pi}{a}$
k_z	$\sqrt{k_c^2 - k_o^2}$	$\sqrt{k_c^2 - k_o^2}$
$H_z(x, z)$	$-\dfrac{k_c}{j\omega\mu}H_o\cos(k_c x)e^{-jk_z z}$	0
$E_z(x, z)$	0	$jE_o\dfrac{k_c}{\omega\varepsilon}\sin(k_c a)e^{-jk_z z}$
$H_x(x, z)$	$-\dfrac{k_z}{\omega\mu}H_o\sin(k_c x)e^{-jk_z z}$	0
$H_y(x, z)$	0	$E_o\cos(k_c x)e^{-jk_z z}$
$E_x(x, z)$	0	$\dfrac{k_z}{\omega\varepsilon}E_o\cos(k_c x)e^{-jk_z z}$
$E_y(x, z)$	$H_o\sin(k_c x)e^{-jk_z z}$	0

and

$$E_z = \frac{C_2}{j\omega\mu\varepsilon}\left(\frac{m\pi}{a}\right)^2 \sin\left(\frac{m\pi}{a}x\right)e^{-jk_z z}. \tag{6.4.45}$$

These results are summarized in Table 6.2, assuming that $-C_2(k_c/\mu) = E_o$. A similar procedure can be used to determine the characteristics of the TE mode fields propagating through this waveguide. The final results are summarized in Table 6.2.

Example 6.1

A metallic parallel-plate waveguide is air-filled, and the separation between the plates is 4.5 cm. Investigate the characteristics of a 12 GHz signal propagating in the TM_{m0} modes.

$$k_c = \frac{m\pi}{a} \rightarrow \lambda_c = \frac{2\pi}{k_c} = \frac{2a}{m} = \frac{2 \times 4.5 \times 10^{-2}}{m} \text{ meters}$$

and

$$f_c = \frac{3 \times 10^8}{\lambda_c} = \frac{3 \times 10^8 \times m}{9 \times 10^{-2}} \text{ Hz} = 3.3333m \text{ GHz}$$

Hence the cutoff frequencies for first four TM modes are found as follows:

$TM_{10} \rightarrow f_c = 3.3333$ GHz
$TM_{20} \rightarrow f_c = 6.6667$ GHz
$TM_{30} \rightarrow f_c = 10$ GHz
$TM_{40} \rightarrow f_c = 13.3333$ GHz

Since the cutoff frequency for TM_{40} mode is higher than the signal frequency of 12 GHz, only the TM_{10}, TM_{20}, and TM_{30} modes will exist for this signal. The corresponding cutoff wavelengths are 9 cm, 4.5 cm, and 3 cm, respectively. The signal wavelength inside the guide for each mode can be found as follows:

$$TM_{10} \rightarrow \lambda_g = \frac{\lambda_o}{\sqrt{1-(\lambda_o/\lambda_c)^2}} = \frac{2.5}{\sqrt{1-(2.5/9)^2}} \text{ cm} = 2.6024 \text{ cm},$$

$$TM_{20} \rightarrow \lambda_g = \frac{\lambda_o}{\sqrt{1-(\lambda_o/\lambda_c)^2}} = \frac{2.5}{\sqrt{1-(2.5/4.5)^2}} \text{ cm} = 3.0067 \text{ cm},$$

and

$$TM_{30} \rightarrow \lambda_g = \frac{\lambda_o}{\sqrt{1-(\lambda_o/\lambda_c)^2}} = \frac{2.5}{\sqrt{1-(2.5/3)^2}} \text{ cm} = 4.5227 \text{ cm}.$$

The Dielectric Slab Guide

As indicated in Chapter 4, nonconducting structures such as dielectric slabs and rods can guide electromagnetic waves as well. Dielectric rods in the form of fibers are commonly used at optical frequencies. Here we consider a slab as shown in Figure 6.2 and analyze its waveguiding characteristics.

As in the preceding case we consider TM^z and TE^z modes of propagation in this slab. Further we subdivide the wave functions in each case into even and odd along the x-axis. Therefore $h_1(k_x x)$ will be $\cos(k_x x)$ for even modes and $\sin(k_x x)$ for the odd modes inside the slab. However, the fields must attenuate with x in the air regions. Therefore k_{xo} must be equal to $-j\alpha_{xo}$. Assume that the signal is guided along the z-axis and therefore $h_3(k_z z) \to e^{-jk_z z}$. Since the fields must satisfy the boundary conditions for all z, $h_3(k_z z)$ is the same in the dielectric as well as in air. Further k_y is zero for $\partial/\partial y \to 0$, meaning the fields remain constant along the y-axis. Therefore the separation equations (6.4.25) for the dielectric and air regions are found to be

$$k_{xd}^2 + k_z^2 = k_d^2 = \omega^2 \mu_d \varepsilon_d \tag{6.4.46}$$

and

$$-\alpha_{xo}^2 + k_z^2 = k_o^2 = \omega^2 \mu_o \varepsilon_o. \tag{6.4.47}$$

For odd TM^z modes, A_z in the dielectric and air is found to be

$$A_{zd}^{o} = C_d \sin(k_{xd} x) e^{-jk_z z}, \qquad |x| < \frac{a}{2} \tag{6.4.48}$$

and

$$A_{zo}^{o} = \begin{cases} C_o e^{-\alpha_{xo} x} e^{-jk_z z}, & x > \dfrac{a}{2}, \\[2ex] -C_o e^{\alpha_{xo} x} e^{-jk_z z}, & x < \dfrac{-a}{2}. \end{cases} \tag{6.4.49}$$

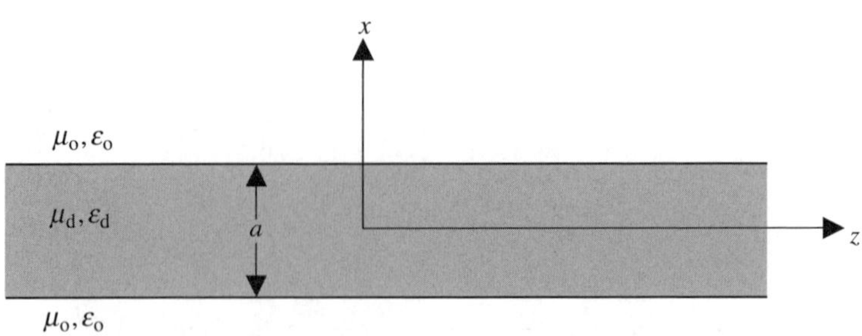

Figure 6.2 Dielectric slab in air.

As we saw in (6.4.32) through (6.4.37), in (6.4.4) through (6.4.9) there are only H_y, E_x, and E_z in the two media, and E_x is normal to the air–dielectric interfaces. These field expressions simplify to

$$H_y = -\frac{C}{\mu}\frac{d}{dx} h_1(k_x x) e^{-jk_z z}, \tag{6.4.50}$$

$$E_x = -\frac{k_z}{\omega\mu\varepsilon} C \frac{d}{dx} h_1(k_x x) e^{-jk_z z}, \tag{6.4.51}$$

and

$$E_z = \frac{C}{j\omega\mu\varepsilon}(k^2 - k_z^2) h_1(k_x x) e^{-jk_z z}. \tag{6.4.52}$$

Similarly the tangential fields for odd modes are found to be

$$E_z = \begin{cases} \dfrac{k_{xd}^2}{j\omega\mu_d\varepsilon_d} C_d \sin(k_{xd} x) e^{-jk_z z}, & |x| < \dfrac{a}{2}, \\[2ex] -\dfrac{\alpha_{xo}^2}{j\omega\mu_o\varepsilon_o} C_o e^{-\alpha_{xo} x} e^{-jk_z z}, & x > \dfrac{a}{2}, \\[2ex] \dfrac{\alpha_{xo}^2}{j\omega\mu_o\varepsilon_o} C_o e^{\alpha_{xo} x} e^{-jk_z z}, & x < \dfrac{-a}{2}, \end{cases} \tag{6.4.53}$$

and

$$H_y = \begin{cases} -\dfrac{k_{xd}}{\mu_d} C_d \cos(k_{xd}\, x) e^{-jk_z z}, & |x| < \dfrac{a}{2}, \\[2ex] \dfrac{\alpha_{xo}}{\mu_o} C_o e^{-\alpha_{xo}|x|} e^{-jk_z z}, & |x| > \dfrac{a}{2}. \end{cases} \tag{6.4.54}$$

The boundary conditions require continuity of tangential fields at $x = \pm a/2$. By the continuity of H_y, we have

$$-\frac{k_{xd}}{\mu_d} C_d \cos\left(\frac{k_{xd} a}{2}\right) = \frac{\alpha_{xo}}{\mu_o} C_o e^{-\alpha_{xo} a/2} \tag{6.4.55}$$

Similarly, by continuity of E_z,

$$\frac{k_{xd}^2}{j\omega\mu_d\varepsilon_d} C_d \sin\left(\frac{k_{xd} a}{2}\right) = -\frac{\alpha_{xo}^2}{j\omega\mu_o\varepsilon_o} C_o e^{-\alpha_{xo} a/2}. \tag{6.4.56}$$

On dividing (6.4.56) by (6.4.55), we have

$$\frac{\varepsilon_o}{\varepsilon_d}\frac{k_{xd}a}{2}\tan\left(\frac{k_{xd}a}{2}\right) = \frac{\alpha_{xo}a}{2}. \tag{6.4.57}$$

In case of even TM^z modes, A_z in the dielectric and the air can be found as follows:

$$A^e_{zd} = C_{de}\cos(k_{xd}x)e^{-jk_z z}, \qquad |x| < \frac{a}{2} \tag{6.4.58}$$

and

$$A^o_{zo} = C_{oe}e^{-\alpha_{xo}|x|}e^{-jk_z z}, \qquad |x| > \frac{a}{2}. \tag{6.4.59}$$

The separation of parameter equations (6.4.46) and (6.4.47) still hold, and the fields are given by equations (6.4.50) through (6.4.52). Following a procedure similar to the preceding case, the continuity of the tangential fields is enforced at the boundaries, yielding

$$-\frac{\varepsilon_o}{\varepsilon_d}\frac{k_{xd}a}{2}\cot\left(\frac{k_{xd}a}{2}\right) = \frac{\alpha_{xo}a}{2}. \tag{6.4.60}$$

Similarly the nonzero TE^z fields can be found from (6.4.12) through (6.4.18) as follows:

$$F_z = Ch_1(k_x x)e^{-jk_z z}. \tag{6.4.61}$$

Therefore

$$E_y = \frac{C}{\varepsilon}\frac{d}{dx}h_1(k_x x)e^{-jk_z z}, \tag{6.4.62}$$

$$H_x = -\frac{C}{\omega\mu\varepsilon}\frac{d}{dx}h_1(k_x x)e^{-jk_z z}, \tag{6.4.63}$$

and

$$H_z = \frac{C}{j\omega\mu\varepsilon}(k^2 - k_z^2)h_1(k_x x)e^{-jk_z z}. \tag{6.4.64}$$

After the boundary conditions are matched, the characteristic equations for the odd and the even TE^z modes, respectively, can be found:

$$\frac{\mu_o}{\mu_d}\frac{k_{xd}a}{2}\tan\left(\frac{k_{xd}a}{2}\right) = \frac{\alpha_{xo}a}{2} \tag{6.4.65}$$

and

$$\frac{\mu_o}{\mu_d}\frac{k_{xd}a}{2}\cot\left(\frac{k_{xd}a}{2}\right) = \frac{\alpha_{xo}a}{2}. \tag{6.4.66}$$

Note that the propagation characteristics (and hence the field distributions) can be found after evaluating k_{xd}, k_z, and α_{xo} via (6.4.46) and (6.4.47) along with (6.4.57) for odd TM^z, or (6.4.60) for even TM^z, or (6.4.65) for odd TE^z, or (6.4.66) for even TE^z modes. Thus (6.4.46) and (6.4.47) are common in all four cases, whereas the third equation is specific to the mode under consideration. Eliminating k_z from (6.4.46) and (6.4.47), we have

$$k_{xd}^2 + \alpha_{xo}^2 = k_d^2 - k_o^2. \tag{6.4.67}$$

Now k_{xd} and α_{xo} can be evaluated by (6.4.67), and by (6.4.57) for odd TM^z, by (6.4.60) for even TM^z, by (6.4.65) for odd TE^z, or by (6.4.66) for even TE^z fields. Numerical techniques can be used for this purpose, or alternatively, a graphical method can be used to find the roots of these equations. For the graphical approach, (6.4.67) can be re-arranged as follows:

$$\left(\frac{k_{xd}a}{2}\right)^2 + \left(\frac{\alpha_{xo}a}{2}\right)^2 = (k_d^2 - k_o^2)\left(\frac{a}{2}\right)^2 = \left(\frac{k_o a}{2}\right)^2 \left(\frac{\mu_d \varepsilon_d}{\mu_o \varepsilon_o} - 1\right). \tag{6.4.68}$$

This is the equation of a circle on the $k_{xd}a/2 - \alpha_{xo}a/2$ plane. Any of the other four equations are plotted on the same plane, and the intersections of two curves are found as roots of the equations. Examples are included below to demonstrate the procedure.

In general, $k_z = \beta_z - j\alpha_z$. However, α_z must be zero for a propagating mode. Therefore α_{xo} in (6.4.47) can be only a real or an imaginary number (i.e., it cannot be a complex number). If the fields are guided along the z-axis, then α_{xo} cannot be an imaginary number and therefore k_z must be larger than k_o. Further k_{xd} has to be a real number in (6.4.46) for real k_z and k_d. Hence $k_o \le k_z \le k_d$ if the fields are propagating in the z direction. On the other hand, α_{xo} becomes imaginary for k_z smaller than k_o, and therefore these modes start propagating continuously along the x-axis as well. This phenomenon is present in dielectric antennas. Thus the guided fields are cut off for $k_z \to k_o$. Therefore (6.4.47) and (6.4.46) reduce to $\alpha_{xo} = 0$ and $k_{xd} = \sqrt{k_d^2 - k_o^2}$, respectively. Consequently (6.4.57) and (6.4.65) require that

$$\tan\left(\sqrt{k_d^2 - k_o^2}\,\frac{a}{2}\right) = 0. \tag{6.4.69}$$

Similarly (6.4.60) and (6.4.66) at the cutoff give

$$\cot\left(\sqrt{k_d^2 - k_o^2}\,\frac{a}{2}\right) = 0. \tag{6.4.70}$$

Combining these two requirements, we find that

$$\frac{a}{2}\sqrt{k_d^2 - k_o^2} = \frac{n\pi}{2} \to \frac{k_o a}{2}\sqrt{\left(\frac{k_d}{k_o}\right)^2 - 1} = \frac{n\pi}{2} \to$$
$$\frac{2\pi a}{2\lambda_c}\sqrt{\frac{\mu_d \varepsilon_d}{\mu_o \varepsilon_o} - 1} = \frac{n\pi}{2}, \qquad n = 0, 1, 2, \ldots.$$

Here we have used $k_o = 2\pi/\lambda_c$ because it is associated with the cutoff condition. Therefore the cutoff wavelength λ_c is found to be

$$\lambda_c = \frac{2a}{n}\sqrt{\frac{\mu_d\varepsilon_d}{\mu_o\varepsilon_o} - 1}. \tag{6.4.71}$$

The cutoff frequency f_c for nth mode is found to be

$$f_c = \frac{1}{\lambda_c\sqrt{\mu_o\varepsilon_o}} = \frac{n}{2a\sqrt{\mu_d\varepsilon_d - \mu_o\varepsilon_o}}. \tag{6.4.72}$$

Note that the lowest order modes have zero cutoff frequency. Hence the TE_0 and TM_0 modes exist in all dielectric guides, along with possibly other higher orders that the slab thickness can support.

Example 6.2

A 0.76 cm thick polystyrene slab is being used at 30 GHz. If its dielectric constant is 2.56, calculate the cutoff frequencies of the TM modes that this structure supports. Determine the propagation parameters of these modes. Repeat your calculations for a slab thickness of 0.5 cm, and compare the two sets of results.

Note that polystyrene is nonmagnetic, and therefore $\mu_d = \mu_o$. We know that the cutoff frequency for the TM_0 mode is zero. For TM_1, we find from (6.4.72) that it is 15.8021 GHz for 0.76 cm, and 24.0192 GHz for 0.5 cm. For $n \geq 2$, the cutoff frequencies are found to be higher than 30 GHz, and therefore those modes are not supported at either thickness.

Equations (6.4.57) and (6.4.60) are graphed on the $k_{xd}a/2 - \alpha_{xo}a/2$ plane, as shown in Figure 6.3. In order to plot (6.4.68) on it, radius of the circle is found to be

$$r_1 = \left(\frac{k_o a}{2}\right)\sqrt{\frac{\mu_d\varepsilon_d}{\mu_o\varepsilon_o} - 1} = \frac{2\pi c a}{f2}\sqrt{\frac{\mu_d\varepsilon_d}{\mu_o\varepsilon_o} - 1} = \frac{2\pi \cdot 3 \cdot 10^{10} \cdot 0.76}{30 \cdot 10^9 \cdot 2}\sqrt{2.56 - 1}$$
$$= 2.9821 \text{ rad.}$$

In graphing this circle, the solution points 1 and 2 are identified. The approximate coordinate values of these points may be further improved using an iterative numerical method. Thus the coordinates of these two points are found to be (1.37122, 2.6481) and (2.56151, 1.5269), respectively. The corresponding k_{xd} and α_{xo} for the TM_0 mode are found to be 3.6085 rad/cm and 6.9687 Np/cm, respectively. Similarly the second point gives k_{xd} and α_{xo} for the TM_1 mode as 6.7408 rad/cm and 4.0181 Np/cm, respectively. Equation (6.4.46) or (6.4.47) can now be used to find k_z for the TM_0 and TM_1 modes as 9.383 rad/cm and 7.458 rad/cm, respectively.

When the slab thickness is 0.5 cm, a new circle can be drawn on the same graph. Its radius r_2 is found to be

$$r_2 = \left(\frac{k_o a}{2}\right)\sqrt{\frac{\mu_d \varepsilon_d}{\mu_o \varepsilon_o} - 1} = \frac{2\pi c a}{f2}\sqrt{\frac{\mu_d \varepsilon_d}{\mu_o \varepsilon_o} - 1} = \frac{2\pi \cdot 3 \cdot 10^{10} \cdot 0.5}{30 \cdot 10^9 \cdot 2}\sqrt{2.56 - 1}$$

$$= 1.9619\,\text{rad}.$$

The precise coordinates for solution points 3 and 4 are found to be (1.25606, 1.5071) and (1.94, 0.2931), respectively. The corresponding k_{xd} and α_{xo} for the TM_0 mode are found to be 5.02424 rad/cm and 6.0284 Np/cm, respectively. Similarly point 4 gives k_{xd} and α_{xo} for the TM_1 mode as 7.76 rad/cm and 1.1724 Np/cm, respectively. Hence k_z for the TM_0 and TM_1 modes are found to be 8.7075 rad/cm and 6.3916 rad/cm, respectively.

A comparison of the two set of results indicates that k_{xd} for a given mode increased with decreasing thickness whereas the corresponding α_{xo} decreased. Therefore the fields will extend farther along the x-axis in air for the thin slab.

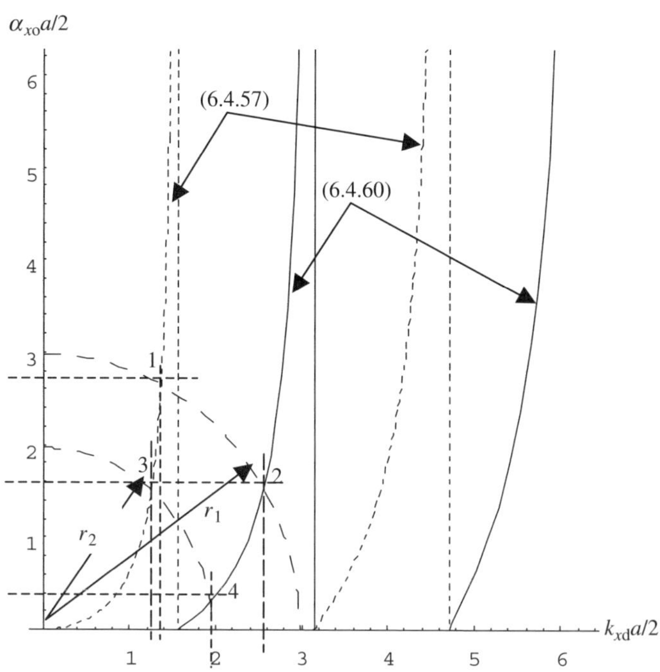

Figure 6.3 Graphical solutions for Example 6.2 to determine the propagation parameters.

Example 6.3

Repeat the preceding example for TE^z modes.

Note that the cutoff frequencies for TE^z modes are same as for the corresponding TM^z fields. Hence the TE_0 mode has no cutoff frequency at either thickness of the slab. For the TE_1 mode, it is 15.8021 GHz for 0.76 cm and 24.0192 GHz for a 0.5 cm thick slab. As shown in Figure 6.4, (6.4.65) and (6.4.66) are graphed on the $k_{xd}a/2-\alpha_{xo}a/2$ plane. Since (6.4.68) still holds, this circle of radius 2.9821 is traced on this graph and the solution points 1 and 2 are identified.

The coordinates of points 1 and 2 are found to be (1.16826, 2.74374) and (2.27423, 1.92894), respectively. The corresponding k_{xd} and α_{xo} for the TE_0 mode are found to be 3.0744 rad/cm and 7.2204 Np/cm, respectively. Similarly the second point gives k_{xd} and α_{xo} for the TE_1 mode as 5.9848 rad/cm and 5.0712 Np/cm, respectively. As before, (6.4.46) or (6.4.47) can be used now to find k_z for the TE_0 and TE_1 modes as 9.5714 rad/cm and 8.0743 rad/cm, respectively.

When the slab thickness is 0.5 cm, a new circle of radius 1.9619 is drawn on the same graph, and solution points 3 and 4 are identified. The precise coordinates of these points are found to be (1.02254, 1.67436) and (1.873, 0.58389), respectively. The corresponding k_{xd} and α_{xo} for the TE_0 mode are found to be 4.0902 rad/cm

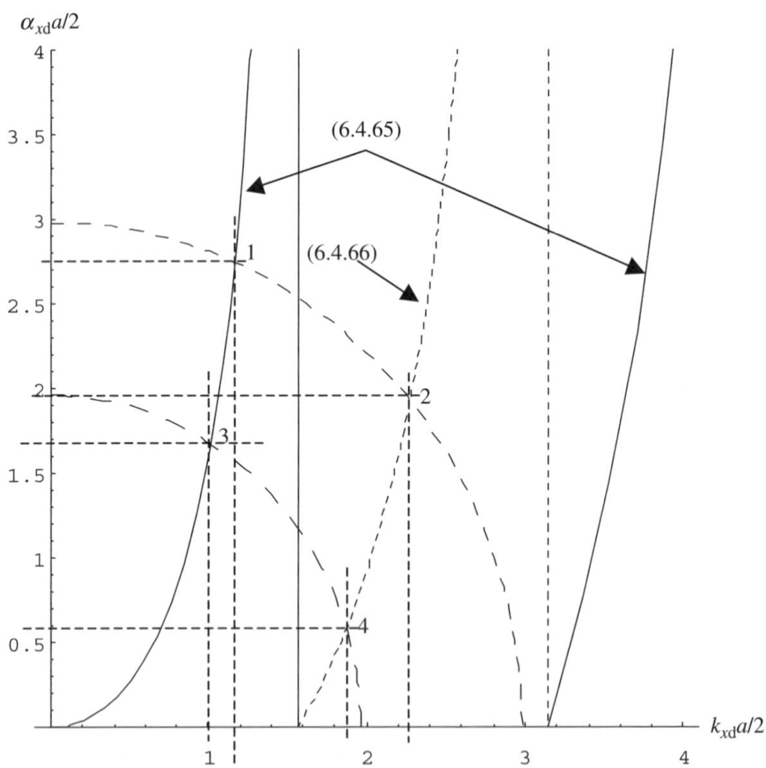

Figure 6.4 Graphical solutions for Example 6.3 to determine the propagation parameters.

and 6.6974 Np/cm, respectively. Similarly point 4 gives k_{xd} and α_{xo} for the TE_1 mode as 7.492 rad/cm and 2.3355 Np/cm, respectively. Hence k_z for the TE_0 and TE_1 modes are found to be 9.1833 rad/cm and 6.7032 rad/cm, respectively.

As in the case of TM^z, a comparison of the two set of results indicates that k_{xd} for a given mode increases with decreasing thickness whereas the corresponding α_{xo} decreases. Therefore in air the fields will extend farther along the x-axis for the thin slab.

6.5 CONSTRUCTION OF A SOLUTION IN CYLINDRICAL COORDINATES

As in the preceding section, the solutions in cylindrical coordinates are categorized as TM^z and TE^z. These two modes are analyzed as follows.

For TM^z, assume $\vec{A} = \hat{z}A_z$ and $\vec{F} = 0$. Then from (6.2.8), (6.2.9), and (6.2.12), get

$$\vec{H} = \frac{1}{\mu}\nabla \times (\hat{z}A_z) = \frac{1}{\mu}\left\{\hat{\rho}\frac{1}{\rho}\frac{\partial A_z}{\partial \phi} + \hat{\phi}\left(-\frac{\partial A_z}{\partial \rho}\right) + \hat{z}0\right\} \tag{6.5.1}$$

and

$$\vec{E} = -j\omega\hat{z}A_z + \frac{1}{j\omega\mu\varepsilon}\nabla\left(\frac{\partial A_z}{\partial z}\right). \tag{6.5.2}$$

Equation (6.2.13) is rewritten in cylindrical coordinates as follows:

$$\nabla^2 A_z + k^2 A_z = 0 \rightarrow \frac{1}{\rho}\frac{\partial}{\partial \rho}\left(\rho\frac{\partial A_z}{\partial \rho}\right) + \frac{1}{\rho^2}\frac{\partial^2 A_z}{\partial \phi^2} + \frac{\partial^2 A_z}{\partial z^2} + k^2 A_z = 0. \tag{6.5.3}$$

Once the solutions to (6.5.3) are found, the electromagnetic fields are determined from (6.5.1) and (6.5.2) as follows:

$$H_\rho = \frac{1}{\mu\rho}\frac{\partial A_z}{\partial \phi}, \tag{6.5.4}$$

$$H_\phi = -\frac{1}{\mu}\frac{\partial A_z}{\partial \rho}, \tag{6.5.5}$$

$$H_z = 0, \tag{6.5.6}$$

$$E_\rho = \frac{1}{j\omega\mu\varepsilon}\frac{\partial^2 A_z}{\partial \rho \partial z}, \tag{6.5.7}$$

$$E_\phi = \frac{1}{j\omega\mu\varepsilon\rho}\frac{\partial^2 A_z}{\partial \phi \partial z}, \tag{6.5.8}$$

and

$$E_z = \frac{1}{j\omega\mu\varepsilon}\left\{\frac{\partial^2}{\partial z^2} + k^2\right\}A_z. \tag{6.5.9}$$

Because there is no magnetic field along the z-axis, this is the case of the TM^z mode (magnetic field transverse to the z-axis) in cylindrical coordinate system.

For TE^z, assume $\vec{F} = \hat{z}F_z$ and $\vec{A} = 0$. From (6.3.6), (6.3.7), and (6.3.9) obtain

$$\vec{E} = -\frac{1}{\varepsilon}\nabla \times (\hat{z}F_z) = -\frac{1}{\varepsilon}\left\{\hat{\rho}\frac{1}{\rho}\frac{\partial F_z}{\partial \phi} + \hat{\phi}\left(-\frac{\partial F_z}{\partial \rho}\right) + \hat{z}0\right\} \tag{6.5.10}$$

and

$$\vec{H} = -j\omega\hat{z}F_z + \frac{1}{j\omega\mu\varepsilon}\nabla\left(\frac{\partial F_z}{\partial z}\right). \tag{6.5.11}$$

Then (6.3.10) reduces to

$$\nabla^2 F_z + k^2 F_z = 0 \rightarrow \frac{1}{\rho}\frac{\partial}{\partial \rho}\left(\rho\frac{\partial F_z}{\partial \rho}\right) + \frac{1}{\rho^2}\frac{\partial^2 F_z}{\partial \phi^2} + \frac{\partial^2 F_z}{\partial z^2} + k^2 F_z = 0. \tag{6.5.12}$$

After (6.5.12) is solved, the field components are found from (6.5.10) and (6.5.11) as follows:

$$E_\rho = -\frac{1}{\varepsilon\rho}\frac{\partial F_z}{\partial \phi}, \tag{6.5.13}$$

$$E_\phi = \frac{1}{\varepsilon}\frac{\partial F_z}{\partial \rho}, \tag{6.5.14}$$

$$E_z = 0, \tag{6.5.15}$$

$$H_\rho = \frac{1}{j\omega\mu\varepsilon}\frac{\partial^2 F_z}{\partial \rho \partial z}, \tag{6.5.16}$$

$$H_\phi = \frac{1}{j\omega\mu\varepsilon\rho}\frac{\partial^2 F_z}{\partial \phi \partial z}, \tag{6.5.17}$$

and

$$H_z = \frac{1}{j\omega\mu\varepsilon}\left(\frac{\partial^2}{\partial z^2} + k^2\right)F_z. \tag{6.5.18}$$

This is the case of the TE^z mode (electric field transverse to the z-axis) in the cylindrical coordinate system because there is no electric field component in the z direction.

Note that unlike z (and all three in rectangular coordinates), the unit vectors along the ρ and ϕ axes change their directions from point to point. In other words, these two unit vectors do not stay parallel from one point to the other. This adds some complication to the analysis. However, the unit vector along the z-axis stays in the same direction at every point. Therefore only the TM^z and TE^z modes are considered. We defer to the next section the case where all three unit vectors of the spherical coordinates change directions from point to point, and where we have no choice but to deal with the situation directly.

The Wave Functions

Since both (6.5.3) and (6.5.12) are similar, we consider the following partial differential equation, which is a scalar Helmholtz equation in cylindrical coordinates:

$$\frac{1}{\rho}\frac{\partial}{\partial\rho}\left\{\rho\frac{\partial}{\partial\rho}\varphi(\rho,\phi,z)\right\}+\frac{1}{\rho^2}\frac{\partial^2\varphi(\rho,\phi,z)}{\partial\phi^2}+\frac{\partial^2\varphi(\rho,\phi,z)}{\partial z^2}+k^2\varphi(\rho,\phi,z)=0. \quad (6.5.19)$$

Assume

$$\varphi(\rho,\phi,z)=f_1(\rho)f_2(\phi)f_3(z). \quad (6.5.20)$$

In other words, assume that the solutions to (6.5.19) are the product of three functions of one coordinate each and use the method of separation of variables. Substituting (6.5.20) into (6.5.19) and rearranging gives

$$\frac{1}{f_1(\rho)}\frac{d^2f_1(\rho)}{d\rho^2}+\frac{1}{\rho f_1(\rho)}\frac{df_1(\rho)}{d\rho}+\frac{1}{\rho^2f_2(\phi)}\frac{d^2f_2(\phi)}{d\phi^2}+\frac{1}{f_3(z)}\frac{d^2f_3(z)}{dz^2}=-k^2. \quad (6.5.21)$$

Note that first three terms on the left-hand side of (6.5.21) are dependent on both ρ and ϕ. However, the last term is only dependent on z. Hence the last term can be equated to a constant k_z as follows:

$$\frac{1}{f_3(z)}\frac{d^2f_3(z)}{dz^2}=-k_z^2\rightarrow\frac{d^2f_3(z)}{dz^2}+k_z^2f_3(z)=0. \quad (6.5.22)$$

Next (6.5.21) can be rearranged to get

$$\frac{\rho^2}{f_1(\rho)}\frac{d^2f_1(\rho)}{d\rho^2}+\frac{\rho}{f_1(\rho)}\frac{df_1(\rho)}{d\rho}+\frac{1}{f_2(\phi)}\frac{d^2f_2(\phi)}{d\phi^2}=-\left(k^2-k_z^2\right)\rho^2 \quad (6.5.23)$$

Since the third term of (6.5.23) is only dependent on ϕ, write

$$\frac{1}{f_2(\phi)}\frac{d^2f_2(\phi)}{d\phi^2}=-n^2\rightarrow\frac{d^2f_2(\phi)}{d\phi^2}+n^2f_2(\phi)=0. \quad (6.5.24)$$

Therefore (6.5.23) reduces to

$$\rho^2\frac{d^2f_1(\rho)}{d\rho^2}+\rho\frac{df_1(\rho)}{d\rho}+\left\{(k^2-k_z^2)\rho^2-n^2\right\}f_1(\rho)=0,$$

or

$$\rho^2\frac{d^2f_1(\rho)}{d\rho^2}+\rho\frac{df_1(\rho)}{d\rho}+\left\{(k_\rho\rho)^2-n^2\right\}f_1(\rho)=0, \quad (6.5.25)$$

where

$$k_\rho^2=k^2-k_z^2\rightarrow k_\rho^2+k_z^2=k^2. \quad (6.5.26)$$

As was considered earlier, harmonic functions satisfy (6.5.22) and (6.5.24). Therefore we can denote solutions to these equations by $h_3(k_z z)$ and $h_2(n\phi)$, respectively. Appropriate harmonic functions can be selected following the characteristics given in Table 6.1. Equation (6.5.25) is known as Bessel's equation of order n. Satisfying this equation are Bessel's functions of the first kind $J_n(k_\rho \rho)$, Bessel's functions of the second kind $Y_n(k_\rho \rho)$ that are known also as Neumann functions, Hankel's functions of first kind $H_n^{(1)}(k_\rho \rho)$, and Hankel's functions of second kind $H_n^{(2)}(k_\rho \rho)$, as can be seen in Appendix C. Any two of these functions are linearly independent solutions to the Bessel equation. We denote these by $Z_n(k_\rho \rho)$. Hence

$$Z_n(k_\rho \rho) \rightarrow J_n(k_\rho \rho),\ Y_n(k_\rho \rho),\ H_n^{(1)}(k_\rho \rho),\ H_n^{(2)}(k_\rho \rho). \tag{6.5.27}$$

The characteristics of these functions are summarized in Table 6.3. Appendix C includes a number of relevant relations, graphical characteristics of a few Bessel and Neumann functions, tabulated zeros of $J_n(x)$ and $Y_n(x)$, and zeros of their derivatives.

Therefore the elementary wave functions satisfying (6.5.19) are

$$\varphi_e(\rho, \phi, z) = Z_n(k_\rho \rho) \cdot h_2(n\phi) \cdot h_3(k_z z). \tag{6.5.28}$$

Note that (6.5.26) relates the eigenvalues k_z and k_ρ, so only one of these two is independent. We can construct more general wave functions by summing the elementary wave functions over one of these eigenvalues and n as follows:

$$\varphi(\rho, \phi, z) = \sum_n \sum_{k_z} C_{nk_z} Z_n(k_\rho \rho) \cdot h_2(n\phi) \cdot h_3(k_z z) \tag{6.5.29}$$

or

$$\varphi(\rho, \phi, z) = \sum_n \sum_{k_\rho} C_{nk_\rho} Z_n(k_\rho \rho) \cdot h_2(n\phi) \cdot h_3(k_z z). \tag{6.5.30}$$

If the region under consideration includes all ϕ from 0 to 2π, then n in $h_2(n\phi)$ has to be an integer so that the wave function is single-valued. As the graphs in Appendix C show, only Bessel functions of the first kind have finite values at the origin. This is the only possible $Z_n(k_\rho \rho)$ if the fields are to be finite at $\rho = 0$. Similarly $H_n^{(1)}(k_\rho \rho)$ is selected for $Z_n(k_\rho \rho)$ if the wave is incoming, whereas it is $H_n^{(2)}(k_\rho \rho)$ if the wave is traveling outward. Note that this selection of the Hankel functions is based on our assumption of time-harmonic variation to be $e^{j\omega t}$. In case of $e^{-j\omega t}$ time variation, $H_n^{(1)}(k_\rho \rho)$ will represent the outgoing wave and $H_n^{(2)}(k_\rho \rho)$, the incoming wave. This behavior can be easily understood by the waves' asymptotic formulas given in Table 6.3. Further the modified Bessel functions $I_n(x)$ and $K_n(x)$ are used if $k_\rho = jx$. The characteristics of these functions are included in Appendix C. As shown in Figure C.3, magnitude of $I_n(x)$ increases (similar to e^x) with x but deceases (similar to e^{-x}) in case of $K_n(x)$.

TABLE 6.3 Properties of Bessel and Hankel Functions

$Z_n(k\rho)$		$\lvert k\rho\rvert \to 0$ Approximation	$\lvert k\rho\rvert \to \infty$ Approximation	Physical Interpretation
$J_n(k\rho)$	Bessel function of first kind and order n	1 for $n = 0$; $\frac{(k\rho)^n}{2^n n!}$ for $n > 0$	$\sqrt{\frac{2}{\pi k\rho}} \cos\left(k\rho - \frac{n\pi}{2} - \frac{\pi}{4}\right)$	Standing wave if k is real Two evanescent fields if k is imaginary
$Y_n(k\rho)$	Bessel function of second kind and order n	$-\frac{2}{\pi} \ln\left(\frac{1.1229}{k\rho}\right)$, $n = 0$ $-\frac{2^n(n-1)!}{\pi(k\rho)^n}$, $n > 0$	$\sqrt{\frac{2}{\pi k\rho}} \sin\left(k\rho - \frac{n\pi}{2} - \frac{\pi}{4}\right)$	Standing wave if k is real Two evanescent fields if k is imaginary
$H_n^{(1)}(k\rho)$	Hankel function of first kind and order n	$1 - j\frac{2}{\pi} \ln\left(\frac{1.1229}{k\rho}\right)$, $n = 0$ $\frac{(k\rho)^n}{2^n n!} - j\frac{2^n(n-1)!}{\pi(k\rho)^n}$, $n > 0$	$\sqrt{\frac{2}{\pi k\rho}} e^{j[k\rho-(n\pi/2)-(\pi/4)]}$	Inward traveling wave if k is real Evanescent field if k is imaginary Attenuated traveling wave if k is complex
$H_n^{(2)}(k\rho)$	Hankel function of second kind and order n	$1 + j\frac{2}{\pi} \ln\left(\frac{1.1229}{k\rho}\right)$, $n = 0$ $\frac{(k\rho)^n}{2^n n!} + j\frac{2^n(n-1)!}{\pi(k\rho)^n}$, $n > 0$	$\sqrt{\frac{2}{\pi k\rho}} e^{-j[k\rho-(n\pi/2)-(\pi/4)]}$	Outward traveling wave if k is real Evanescent field if k is imaginary Attenuated traveling wave if k is complex

The Parallel-Plate Radial Waveguide

The two conducting parallel plates considered in the preceding section can also support radial waves. In the present analysis we consider the parallel-plate geometry shown in Figure 6.5. We assume that the distance a separates the two plates that extend to infinity.

Note in the figure that ϕ ranges from 0 to 2π. Therefore only integer values of n are possible in this case. Further only $H_n^{(2)}(k_\rho \rho)$ are possible solutions to Bessel's equation for the outgoing waves. Therefore the suitable solution to (6.5.3) for A_z is

$$A_z = \{C_1 \cos(k_z z) + C_2 \sin(k_z z)\}\{D_1 \cos(n\phi) + D_2 \sin(n\phi)\}H_n^{(2)}(k_\rho \rho). \quad (6.5.31)$$

Since the boundary conditions require that the tangential electric field components must be zero on conducting plates, E_ρ and E_ϕ must be zero at $z = 0$ and $z = a$. Therefore (6.5.7) and (6.5.8) require that

$$\left.\frac{\partial A_z}{\partial z}\right|_{\substack{z=0\\z=a}} = 0 \rightarrow \begin{cases} C_2 = 0 \\ k_z = \dfrac{m\pi}{a}, \qquad m = 0, 1, 2, \ldots \end{cases} \quad (6.5.32)$$

Then (6.5.31) and (6.5.26) give

$$A_z = \left\{C_1 \cos\left(\frac{m\pi}{a} z\right)\right\}\{D_1 \cos(n\phi) + D_2 \sin(n\phi)\}H_n^{(2)}(k_\rho \rho) \quad (6.5.33)$$

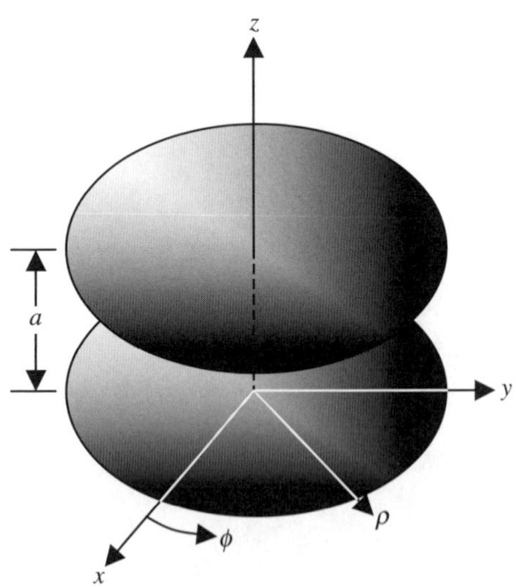

Figure 6.5 Geometry of a parallel-plate radial waveguide.

and

$$k_\rho = \sqrt{k^2 - \left(\frac{m\pi}{a}\right)^2}. \tag{6.5.34}$$

Note that there are no boundary conditions along ϕ to satisfy, so D_1 and D_2 cannot be evaluated. Since the sine functions represent 90° phase-shifted cosine functions (and vice-versa), the field distributions obtained with the sine function simply change to the other as we rotate the coordinate system by 90°. The complete set of field components can be found now using (6.5.33) and (6.5.4) through (6.5.9).

For $m = 0$ and $n = 0$, (6.5.33) reduces to

$$A_{z00} = C_{00} H_0^{(2)}(k\rho). \tag{6.5.35}$$

Therefore it is found from (6.5.4) through (6.5.9) and (6.5.35) that only the following field components exist for the TM_{00}^z mode:

$$H_\phi = -\frac{C_{00}}{\mu}\frac{d}{d\rho} H_0^{(2)}(k\rho) \tag{6.5.36}$$

and

$$E_z = \frac{k^2}{j\omega\mu\varepsilon} C_{00} H_0^{(2)}(k\rho). \tag{6.5.37}$$

Note that these fields represent an electromagnetic wave that is TEM to ρ. It is a transmission line mode because it is close to the plane transmission line mode. By the corresponding formula in Appendix C, (6.5.36) can be simplified to

$$Z_n'(x) = -Z_{n+1}(x) + \frac{n}{x} Z_n(x) \rightarrow H_\phi = \frac{kC_{00}}{\mu} H_1^{(2)}(k\rho). \tag{6.5.38}$$

Similarly F_z can be found via the suitable wave functions, and the TE^z mode fields can be determined subsequently from (6.5.13) through (6.5.18), as follows:

$$F_z = \{C_1 \cos(k_z z) + C_2 \sin(k_z z)\}\{D_1 \cos(n\phi) + D_2 \sin(n\phi)\} H_n^{(2)}(k_\rho \rho). \tag{6.5.39}$$

Since the boundary conditions require that the tangential electric field components must vanish on the conducting plates, E_ρ and E_ϕ must be zero at $z = 0$ and $z = a$. Therefore (6.5.13) and (6.5.14) require that

$$F_z|_{\substack{z=0\\z=a}} = 0 \rightarrow \begin{cases} C_1 = 0, \\ k_z = \dfrac{m\pi}{a}, \quad m = 0, 1, 2, \ldots. \end{cases} \tag{6.5.40}$$

Therefore

$$F_z = C_2 \sin\left(\frac{m\pi}{a} z\right)\{D_1 \cos(n\phi) + D_2 \sin(n\phi)\}H_n^{(2)}(k_\rho \rho). \tag{6.5.41}$$

Equation (6.5.34) holds for TE^z modes as well. A mode is propagating if k_ρ is real, and it is evanescent if k_ρ is imaginary. The cutoff occurs for $k = k_c$ where

$$k_c = \frac{m\pi}{a} \rightarrow \lambda_C = \frac{2a}{m}. \tag{6.5.42}$$

For an air-filled waveguide, the cutoff frequency f_c is found to be

$$f_c = \frac{3 \cdot 10^8 \cdot m}{2a} \text{ Hz}. \tag{6.5.43}$$

Note that F_z in (6.5.41) goes to zero for $m = 0$, so TE$^z_{00}$ mode does not exist. Therefore the lowest order mode (known as the dominant mode) in the parallel-plate waveguide is TM$^z_{00}$.

Example 6.4

Find the separation between the two plates of a radial waveguide that guarantees lowest order single TM^z mode up to 500 MHz. Repeat your calculations for the lowest order TE^z mode. Assume that the medium between the plates is air.

Since the lowest order TM^z mode is for $m = 0$ and has no cutoff, we need to make sure that cutoff frequency for $m = 1$ mode is higher than 500 MHz. Hence (6.5.43) gives

$$a \le \frac{3 \cdot 10^8}{2 \cdot 5 \cdot 10^8} = 0.3 \text{ m}.$$

In the case of TE^z modes the lowest order mode that exists is for $m = 1$. Therefore we need to stop $m = 2$ mode. Hence (6.5.43) gives

$$a \le \frac{3 \cdot 10^8 \cdot 2}{2 \cdot 5 \cdot 10^8} = 0.6 \text{ m}.$$

The Wedge Radial Waveguide

Consider two inclined conducting sheets that also support propagation of radial waves. As shown in Figure 6.6, that angle between the two planes is ϕ_o and the conducting sheets extend to infinity along the ρ- and z-axes. Therefore consider it to be a two-dimensional problem with no variation of the field along the z-axis and the wave propagating along ρ. Equation (6.5.26) reduces to $k_\rho = k$. By the appropriate wave functions, A_z for the TM^z modes is found to be

$$A_z = \{D_1 \cos(n\phi) + D_2 \sin(n\phi)\}H_n^{(2)}(k\rho). \tag{6.5.44}$$

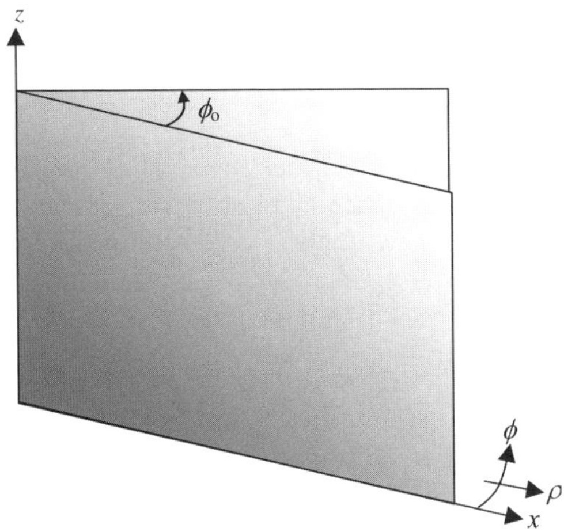

Figure 6.6 Geometry of a wedge radial waveguide.

The boundary conditions require that the tangential electric field components must be zero on the conducting plates. Hence E_ρ and E_z must be zero at $\phi = 0$ and $\phi = \phi_o$. Therefore (6.5.7) and (6.5.8) require that

$$A_z|_{\substack{\phi=0\\ \phi=\phi_o}} = 0 \longrightarrow \begin{cases} D_1 = 0, \\ n = \dfrac{m\pi}{\phi_o}, \qquad m = 1, 2, \ldots. \end{cases} \tag{6.5.45}$$

Then (6.5.44) reduces to

$$A_z = D_2 \sin\left(\frac{m\pi}{\phi_o}\phi\right) H^{(2)}_{m\pi/\phi_o}(k\rho), \qquad m = 1, 2, \ldots. \tag{6.5.46}$$

TM^z mode fields can now be found after substituting (6.5.46) in (6.5.4) through (6.5.9). In this case TM_1^z is the lowest order mode, and the corresponding field components are found to be

$$H_{\rho 1} = \frac{1}{\mu\rho}\frac{\partial}{\partial\phi}\left[D_2 \sin\left(\frac{\pi}{\phi_o}\phi\right) H^{(2)}_{\pi/\phi_o}(k\rho)\right] = \frac{D_2}{\mu\rho}\frac{\pi}{\phi_o}\cos\left(\frac{\pi}{\phi_o}\phi\right) H^{(2)}_{\pi/\phi_o}(k\rho), \tag{6.5.47}$$

$$H_{\phi 1} = -\frac{1}{\mu}\frac{\partial}{\partial\rho}\left[D_2 \sin\left(\frac{\pi}{\phi_o}\phi\right) H^{(2)}_{\pi/\phi_o}(k\rho)\right] = -\frac{D_2}{\mu}\sin\left(\frac{\pi}{\phi_o}\phi\right)\frac{d}{d\rho} H^{(2)}_{\pi/\phi_o}(k\rho), \tag{6.5.48}$$

and

$$E_{z1} = \frac{k^2}{j\omega\mu\varepsilon} D_2 \sin\left(\frac{\pi}{\phi_o}\phi\right) H^{(2)}_{\pi/\phi_o}(k\rho). \tag{6.5.49}$$

Note that the remaining TM_1^z field components are zero.

Similarly the appropriate F_z for the TE^z modes is written as follows:

$$F_z = \{D_1 \cos(n\phi) + D_2 \sin(n\phi)\}H_n^{(2)}(k\rho). \tag{6.5.50}$$

The boundary conditions require that the tangential electric field components be zero on the conducting plates. Hence E_ρ must be zero at $\phi = 0$ and $\phi = \phi_o$. Note from (6.5.15) that E_z is already zero in this case. Therefore we find from (6.5.13) that

$$\left.\frac{\partial F_z}{\partial \phi}\right|_{\substack{\phi=0\\ \phi=\phi_o}} = 0 \rightarrow \begin{cases} D_2 = 0, \\ n = \dfrac{m\pi}{\phi_o}, \qquad m = 1, 2, \ldots. \end{cases} \tag{6.5.51}$$

Then (6.5.50) reduces to

$$F_z = D_1 \cos\left(\frac{m\pi}{\phi_o}\phi\right)H_{m\pi/\phi_o}^{(2)}(k\rho), \qquad m = 0, 1, 2, \ldots. \tag{6.5.52}$$

The TE^z mode fields can now be found after substituting (6.5.52) in (6.5.13) through (6.5.18). In this case TE$_0^z$ is the lowest order mode, and the corresponding field components are found to be

$$E_{\phi 0} = \frac{1}{\varepsilon}\frac{\partial}{\partial \rho}\left[D_1 H_0^{(2)}(k\rho)\right] = \frac{D_1}{\varepsilon}\frac{d}{d\rho}\left[H_0^{(2)}(k\rho)\right] = \frac{D_1 k}{\varepsilon}H_1^{(2)}(k\rho) \tag{6.5.53}$$

and

$$H_{z0} = \frac{k^2}{j\omega\mu}D_1 H_0^{(2)}(k\rho). \tag{6.5.54}$$

The remaining field components of this mode are zero. Note that (6.5.53) and (6.5.54) are analogous to (6.5.37) and (6.5.38), and therefore this mode is the dual of TM$_0^z$ mode of the parallel-plate waveguide. These fields are TEM to ρ representing the transmission line mode.

6.6 CONSTRUCTION OF A SOLUTION IN SPHERICAL COORDINATES

As was mentioned earlier, the procedure used in the preceding sections does not work when we work in spherical coordinates. This is because all three unit vectors change directions from point to point, and (6.2.13) and (6.3.10) cannot be separated into scalar wave equations. Nevertheless, there is a way to formulate the wave equations in the radial components A_r and F_r. The electric and magnetic field intensities can be expressed in terms of these components of the vector potentials. If there is only the radial component A_r of the magnetic vector potential nonzero, then (6.2.8) gives

$$\vec{B} = \nabla \times (\hat{r}A_r) = \hat{\theta}\frac{1}{r\sin\theta}\frac{\partial A_r}{\partial \phi} - \hat{\phi}\frac{1}{r}\frac{\partial A_r}{\partial \theta} \tag{6.6.1}$$

and

$$\nabla \times \vec{B} = \frac{1}{\mu}\nabla \times \vec{H} = \nabla \times \nabla \times \vec{A} = \nabla \times \left\{\hat{\theta}\frac{1}{r\sin\theta}\frac{\partial A_r}{\partial \phi} - \hat{\phi}\frac{1}{r}\frac{\partial A_r}{\partial \theta}\right\},$$

or

$$\begin{aligned}\nabla \times \nabla \times \vec{A} = \hat{r}\frac{1}{r\sin\theta}&\left\{\frac{\partial}{\partial\theta}\left(-\frac{\sin\theta}{r}\frac{\partial A_r}{\partial\theta}\right) - \frac{\partial}{\partial\phi}\left(\frac{1}{r\sin\theta}\frac{\partial A_r}{\partial\phi}\right)\right\}\\ &+ \hat{\theta}\left\{-\frac{1}{r}\frac{\partial}{\partial r}\left(-\frac{\partial A_r}{\partial\theta}\right)\right\} + \hat{\phi}\frac{1}{r}\left\{\frac{\partial}{\partial r}\left(\frac{1}{\sin\theta}\frac{\partial A_r}{\partial\phi}\right)\right\}.\end{aligned} \tag{6.6.2}$$

Similarly, if only radial component F_r is nonzero in the electric vector potential, then (6.3.6) reduces to

$$\vec{D} = -\nabla \times (\hat{r}F_r) = -\hat{\theta}\frac{1}{r\sin\theta}\frac{\partial F_r}{\partial \phi} + \hat{\phi}\,\frac{1}{r}\frac{\partial F_r}{\partial \theta} \tag{6.6.3}$$

and

$$\nabla \times \vec{D} = \frac{1}{\varepsilon}\nabla \times \vec{E} = -\nabla \times \nabla \times \vec{F} = \nabla \times \left\{\hat{\theta}\frac{1}{r\sin\theta}\frac{\partial F_r}{\partial \phi} - \hat{\phi}\frac{1}{r}\frac{\partial F_r}{\partial \theta}\right\},$$

or

$$\begin{aligned}-\nabla \times \nabla \times \vec{F} = \hat{r}\frac{1}{r\sin\theta}&\left\{\frac{\partial}{\partial\theta}\left(-\frac{\sin\theta}{r}\frac{\partial F_r}{\partial\theta}\right) - \frac{\partial}{\partial\phi}\left(\frac{1}{r\sin\theta}\frac{\partial F_r}{\partial\phi}\right)\right\}\\ &+ \hat{\theta}\left\{-\frac{1}{r}\frac{\partial}{\partial r}\left(-\frac{\partial F_r}{\partial\theta}\right)\right\} + \hat{\phi}\frac{1}{r}\left\{\frac{\partial}{\partial r}\left(\frac{1}{\sin\theta}\frac{\partial F_r}{\partial\phi}\right)\right\}.\end{aligned} \tag{6.6.4}$$

Note that the medium is assumed to be linear, isotropic, and homogeneous. For a source-free region, $\vec{J}_e = 0$ and $\vec{J}_m = 0$, and (6.6.1) through (6.6.4), along with (6.1.1) and (6.1.2), give

$$\vec{E} = \frac{1}{j\omega\mu\varepsilon}\nabla \times \nabla \times \vec{A} - \frac{1}{\varepsilon}\nabla \times \vec{F} \tag{6.6.5}$$

and

$$\vec{H} = \frac{1}{j\omega\mu\varepsilon}\nabla \times \nabla \times \vec{F} + \frac{1}{\mu}\nabla \times \vec{A}. \tag{6.6.6}$$

Therefore (6.6.5), along with (6.6.1) through (6.6.4), gives

$$E_r = \frac{1}{j\omega\mu\varepsilon r \sin\theta}\left\{\frac{\partial}{\partial\theta}\left(-\frac{\sin\theta}{r}\frac{\partial A_r}{\partial\theta}\right) - \frac{\partial}{\partial\phi}\left(\frac{1}{r\sin\theta}\frac{\partial A_r}{\partial\phi}\right)\right\}, \tag{6.6.7}$$

$$E_\theta = \frac{1}{j\omega\mu\varepsilon r}\frac{\partial^2 A_r}{\partial r\partial\theta} - \frac{1}{\varepsilon r\sin\theta}\frac{\partial F_r}{\partial\phi}, \tag{6.6.8}$$

$$E_\phi = \frac{1}{j\omega\mu\varepsilon r\sin\theta}\frac{\partial^2 A_r}{\partial r\partial\phi} + \frac{1}{\varepsilon r}\frac{\partial F_r}{\partial\theta}. \tag{6.6.9}$$

Similarly (6.6.6), along with (6.6.1) through (6.6.4), gives

$$H_r = \frac{1}{j\omega\mu\varepsilon r \sin\theta}\left\{\frac{\partial}{\partial\theta}\left(-\frac{\sin\theta}{r}\frac{\partial F_r}{\partial\theta}\right) - \frac{\partial}{\partial\phi}\left(\frac{1}{r\sin\theta}\frac{\partial F_r}{\partial\phi}\right)\right\}, \tag{6.6.10}$$

$$H_\theta = \frac{1}{j\omega\mu\varepsilon r}\frac{\partial^2 F_r}{\partial r\partial\theta} + \frac{1}{\mu r\sin\theta}\frac{\partial A_r}{\partial\phi}, \tag{6.6.11}$$

$$H_\phi = \frac{1}{j\omega\mu\varepsilon r\sin\theta}\frac{\partial^2 F_r}{\partial r\partial\phi} - \frac{1}{\mu r}\frac{\partial A_r}{\partial\theta}. \tag{6.6.12}$$

Note that (6.6.7) can be simplified further using the formulation included below for (6.6.18) as

$$\frac{1}{r\sin\theta}\left\{\frac{\partial}{\partial\theta}\left(-\frac{\sin\theta}{r}\frac{\partial A_r}{\partial\theta}\right) - \frac{\partial}{\partial\phi}\left(\frac{1}{r\sin\theta}\frac{\partial A_r}{\partial\phi}\right)\right\} = \frac{\partial^2 A_r}{\partial r^2} + k^2 A_r$$

A similar expression for F_r can be used to simplify (6.6.10) as well.

Hence the electric and magnetic field components are found in (6.6.7) through (6.6.12) in terms of radial components of two vector potentials. The wave equations for A_r and F_r are formulated as follows: For $F_r = 0$, (6.6.5) and (6.2.9) give

$$\nabla \times \nabla \times (\hat{r}A_r) = j\omega\mu\varepsilon\{-\nabla\phi_e - \hat{r}j\omega A_r\}. \tag{6.6.13}$$

Similarly, for $A_r = 0$, (6.6.6) and (6.3.7) give

$$\nabla \times \nabla \times (\hat{r}F_r) = j\omega\mu\varepsilon\{-\nabla\phi_m - \hat{r}j\omega F_r\}. \tag{6.6.14}$$

These equations can be expanded using (6.6.2) and (6.6.4), respectively. Since both equations have a similar form, we consider below only (6.6.13). A similar process can be used for (6.6.14).

Equating the θ components on the two sides of (6.6.13), we have

$$\frac{1}{r}\frac{\partial}{\partial r}\left(\frac{\partial A_r}{\partial \theta}\right) = -j\omega\mu\varepsilon\frac{\partial \phi_e}{\partial \theta} \rightarrow \frac{\partial}{\partial \theta}\left\{\frac{\partial A_r}{\partial r} + j\omega\mu\varepsilon\phi_e\right\} = 0. \tag{6.6.15}$$

Similarly, equating the ϕ components on the two sides of (6.6.13) gives

$$\frac{1}{r\sin\theta}\frac{\partial}{\partial r}\left(\frac{\partial A_r}{\partial \phi}\right) = -j\omega\mu\varepsilon\frac{1}{r\sin\theta}\frac{\partial \phi_e}{\partial \phi} \rightarrow \frac{\partial}{\partial \phi}\left\{\frac{\partial A_r}{\partial r} + j\omega\mu\varepsilon\phi_e\right\} = 0. \tag{6.6.16}$$

Both conditions (6.6.15) and (6.6.16) are satisfied if

$$\frac{\partial A_r}{\partial r} = -j\omega\mu\varepsilon\phi_e. \tag{6.6.17}$$

Now equating the r components on the two sides of (6.6.13), we have

$$\frac{1}{r\sin\theta}\left\{\frac{\partial}{\partial \theta}\left(-\frac{\sin\theta}{r}\frac{\partial A_r}{\partial \theta}\right) - \frac{\partial}{\partial \phi}\left(\frac{1}{r\sin\theta}\frac{\partial A_r}{\partial \phi}\right)\right\} = -j\omega\mu\varepsilon\frac{\partial \phi_e}{\partial r} + \omega^2\mu\varepsilon A_r,$$

or

$$\frac{1}{r\sin\theta}\left\{\frac{\partial}{\partial \theta}\left(-\frac{\sin\theta}{r}\frac{\partial A_r}{\partial \theta}\right) - \frac{\partial}{\partial \phi}\left(\frac{1}{r\sin\theta}\frac{\partial A_r}{\partial \phi}\right)\right\} = \frac{\partial^2 A_r}{\partial r^2} + k^2 A_r,$$

or

$$\frac{\partial^2 A_r}{\partial r^2} + \frac{1}{r^2\sin\theta}\frac{\partial}{\partial \theta}\left(\sin\theta\frac{\partial A_r}{\partial \theta}\right) + \frac{1}{r^2\sin^2\theta}\frac{\partial^2 A_r}{\partial \phi^2} + k^2 A_r = 0,$$

or

$$\frac{\partial^2}{\partial r^2}\left(\frac{A_r}{r}\right) + \frac{2}{r}\frac{\partial}{\partial r}\left(\frac{A_r}{r}\right) + \frac{1}{r^2\sin\theta}\frac{\partial}{\partial \theta}\left\{\sin\theta\frac{\partial}{\partial \theta}\left(\frac{A_r}{r}\right)\right\}$$
$$+ \frac{1}{r^2\sin^2\theta}\frac{\partial^2}{\partial \phi^2}\left(\frac{A_r}{r}\right) + k^2\left(\frac{A_r}{r}\right) = 0. \tag{6.6.18}$$

The equation can also be expressed in compact form using the Laplacian in spherical coordinates (2.9.3) as follows:

$$(\nabla^2 + k^2)\left(\frac{A_r}{r}\right) = 0. \tag{6.6.19}$$

By a similar procedure with (6.6.14), we can find another equation like (6.6.18) except that F_r replaces A_r. Therefore we attempt to construct the solutions to the following wave equation:

$$\frac{\partial^2}{\partial r^2}\varphi + \frac{2}{r}\frac{\partial}{\partial r}\varphi + \frac{1}{r^2 \sin\theta}\frac{\partial}{\partial\theta}\left\{\sin\theta\frac{\partial}{\partial\theta}\varphi\right\} + \frac{1}{r^2 \sin^2\theta}\frac{\partial^2}{\partial\phi^2}\varphi + k^2\varphi = 0. \tag{6.6.20}$$

As before, we use the separation of variables technique to solve this equation. We assume that φ is a product of three functions $R(r)$, $\Theta(\theta)$, and $\Phi(\phi)$. After substituting $\varphi(r, \theta, \phi) = R(r)\Theta(\theta)\Phi(\phi)$ into (6.6.20), and following the same procedure we used earlier to separate the equations, we get

$$\frac{d^2}{d\phi^2}\Phi(\phi) + m^2\Phi(\phi) = 0, \tag{6.6.21}$$

$$\frac{1}{\sin\theta}\frac{d}{d\theta}\left\{\sin\theta\frac{d}{d\theta}\Theta(\theta)\right\} + \left\{n(n+1) - \frac{m^2}{\sin^2\theta}\right\}\Theta(\theta) = 0, \tag{6.6.22}$$

and

$$r^2\frac{d^2}{dr^2}R(r) + 2r\frac{d}{dr}R(r) + [(kr)^2 - n(n+1)]R(r) = 0. \tag{6.6.23}$$

The first two terms of the last equation can be rearranged to get

$$\frac{d}{dr}\left\{r^2\frac{d}{dr}R(r)\right\} + [(kr)^2 - n(n+1)]R(r) = 0. \tag{6.6.24}$$

The odd choice of $n(n+1)$ as the separation constant in (6.6.22) is made because this leads to a standard differential equation known as an associated Legendre equation. This time, unlike the preceding cases, the separation constants m and n are not related.

Note that (6.6.21) is the familiar harmonic equation, so $\sin(m\phi)$, $\cos(m\phi)$, $e^{jm\phi}$, and $e^{jm\phi}$ are its solutions. We can select any two of these harmonic functions as a complete solution to (6.6.21). As in the case of cylindrical coordinates, m has to be an integer if φ is single-valued for $0 \leq \phi \leq 2\pi$.

A comparison of (6.6.22) with (D.1) of Appendix D confirms that it is an associated Legendre equation; its solutions are called associated Legendre functions. There are two kinds of associated Legendre functions included in Appendix D. $P_n^m(\cos\theta)$ are the associated Legendre functions of the first kind and $Q_n^m(\cos\theta)$ are the associated Legendre functions of the second kind. In general, we can represent both types of solutions as $L_n^m(\cos\theta)$. Further, it is found that out of all solutions of this equation only $P_n^m(\cos\theta)$ is finite at $\theta = 0$ and $\theta = \pi$, provided that n is an integer.

Finally, a comparison of (6.6.23) with (C.39) of Appendix C indicates that its solutions must be a spherical Bessel type of function, which is related to ordinary Bessel functions as follows:

$$z_n(kr) = \sqrt{\frac{\pi}{2kr}} Z_{n+1/2}(kr). \tag{6.6.25}$$

As summarized in Appendix C, the characteristics of spherical Bessel functions are similar to the corresponding cylindrical Bessel functions. Hence $j_n(kr)$ and $y_n(kr)$ represent standing waves, $h_n^{(2)}(kr)$ represents outward traveling waves, and $h_n^{(1)}(kr)$ represents an inward traveling wave. The elementary wave functions that satisfy (6.6.20) are found to be

$$\varphi_{mn} = h(m\phi)L_n^m(\cos\theta)z_n(kr), \tag{6.6.26}$$

where the complete solution is

$$\varphi = \sum_m \sum_n C_{mn} h(m\phi)L_n^m(\cos\theta)z_n(kr). \tag{6.6.27}$$

Since φ represents A_r/r or F_r/r, a new kind of spherical Bessel functions is commonly used for this purpose. As indicated in Appendix C, the components of these functions are generally identified by a caret, and they are related with cylindrical Bessel functions as follows:

$$\hat{Z}_n(kr) = \sqrt{\frac{\pi kr}{2}} Z_{n+1/2}(kr). \tag{6.6.28}$$

The qualitative characteristics of spherical Bessel functions are also similar to the corresponding cylindrical Bessel functions. Only the spherical Bessel function of the first kind is finite at the origin (i.e., at $r = 0$). The spherical Hankel functions of the second kind likewise represent outgoing waves. The zeros of the spherical Bessel functions of the first kind as well as their derivatives are included in Tables C.5 and C.6, respectively.

The Conducting Spherical Cavity

Figure 6.7 shows the geometry of a conducting spherical shell that forms a resonant circuit (known as the cavity resonator) at microwave frequencies. The radius of this cavity is assumed to be a. When properly excited, an electromagnetic field can arise inside the shell. We divide this problem into TE^r and TM^r modes. For TE^r modes, A_r is zero and only F_r exists.

Since $0 \le \phi \le 2\pi$, m has to be an integer. Further there are no boundary conditions to satisfy in this direction, and therefore the integration constants used with solutions cannot be determined. We can use $e^{\pm jm\phi}$ or sine and cosine functions as solutions to this harmonic equation. Sometimes only one of the sine and cosine functions is used, and the modes are subdivided into odd (with sine functions)

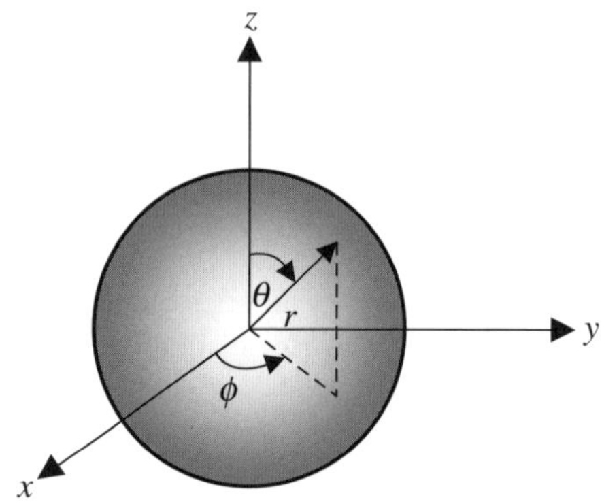

Figure 6.7 Geometry of a spherical cavity.

and even (with cosine functions) types. We select $P_n^m(\cos\theta)$ as the appropriate associate Legendre function because the fields have to be finite at $\theta = 0$ and $\theta = \pi$. Similarly we select spherical Bessel functions of first kind to meet the requirement that fields have to be finite at $r = 0$. Hence

$$F_r = C_{mn}\hat{J}_n(kr)P_n^m(\cos\theta)\begin{cases}\cos(m\phi) & \text{for even modes,}\\ \sin(m\phi) & \text{for odd modes.}\end{cases} \tag{6.6.29}$$

In the equation C_{mn} is a constant. The electric and magnetic fields can be found now using (6.6.7) through (6.6.12) with A_r zero. Since E_r is zero in this case, these fields can be designated as TE^r modes. The boundary conditions require that the tangential electric field components be zero on the conducting surface at $r = a$; therefore

$$E_\theta|_{r=a} = 0 \rightarrow -\frac{1}{\varepsilon r\sin\theta}\frac{\partial F_r}{\partial\phi}\bigg|_{r=a} = 0 \tag{6.6.30}$$

and

$$E_\phi|_{r=a} = 0 \rightarrow \frac{1}{\varepsilon r}\frac{\partial F_r}{\partial\theta}\bigg|_{r=a} = 0. \tag{6.6.31}$$

Both conditions will be satisfied if

$$\hat{J}_n(ka) = 0 = \hat{J}_n(v_{np}) \tag{6.6.32}$$

where v_{np} is the pth zero of the nth order spherical Bessel function of the first kind. There is an infinite set of these zeros. Lower order zeros are listed in Table C.5 of Appendix C. Hence

$$k = \frac{v_{np}}{a} \rightarrow \frac{2\pi}{\lambda_r} = \frac{v_{np}}{a} \rightarrow f_{mnp}^{\text{TE}^r} = \frac{v_{np}}{2\pi a\sqrt{\mu\varepsilon}}, \qquad m \le n, \tag{6.6.33}$$

where λ_r is wavelength at the resonance and $f_{mnp}^{\mathrm{TE^r}}$ is the resonant frequency for $\mathrm{TE^r}$ modes; μ is the permeability and ε is the permittivity of medium filling the cavity.

Similarly we select $F_r = 0$ and A_r nonzero to find the $\mathrm{TM^r}$ mode fields. In this case

$$A_r = D_{mn}\hat{J}_n(kr)P_n^m(\cos\theta)\begin{cases}\cos(m\phi) & \text{for even modes,}\\ \sin(m\phi) & \text{for odd modes.}\end{cases} \tag{6.6.34}$$

D_{mn} is a constant. The electric and magnetic fields can now be found using (6.6.7) through (6.6.12) with F_r zero. Since the tangential electric field components must be zero on the conducting surface at $r = a$, we find that

$$E_\theta|_{r=a} = 0 \rightarrow \left.\frac{1}{j\omega\mu\varepsilon r}\frac{\partial^2 A_r}{\partial r\partial\theta}\right|_{r=a} = 0 \tag{6.6.35}$$

and

$$E_\phi|_{r=a} = 0 \rightarrow \left.\frac{1}{j\omega\mu\varepsilon r\sin\theta}\frac{\partial^2 A_r}{\partial r\partial\phi}\right|_{r=a} = 0. \tag{6.6.36}$$

Both conditions will be satisfied if

$$\hat{J}'_n(ka) = 0 = \hat{J}'_n(v'_{np}), \tag{6.6.37}$$

where the prime over the Bessel function is used to indicate the derivative, and v'_{np} is pth zero of the derivative of the nth-order spherical Bessel function of first kind. There is an infinite set of these zeros. Lower order zeros are listed in Table C.6 of Appendix C. Hence

$$k = \frac{v'_{np}}{a} \rightarrow \frac{2\pi}{\lambda_r} = \frac{v'_{np}}{a} \rightarrow f_{mnp}^{\mathrm{TM^r}} = \frac{v'_{np}}{2\pi a\sqrt{\mu\varepsilon}}, \qquad m \le n, \tag{6.6.38}$$

where λ_r is wavelength at the resonance and $f_{mnp}^{\mathrm{TM^r}}$ is the resonant frequency for the $\mathrm{TM^r}$ modes; μ is the permeability and ε is the permittivity of the medium filling the cavity.

Example 6.5

The radius of an air-filled spherical cavity is 3 cm. Find the resonant frequencies of first five modes that it supports.

From Tables C.5 and C.6 we find that first zero of the spherical Bessel function (for $\mathrm{TE^r}$ modes) occurs at $v_{np} = 4.4934$ for $n = p = 1$. However, there are two zeros of derivatives (for the $\mathrm{TM^r}$ modes) that occur at $v'_{np} = 2.7437$ and 3.8702 for $n = 1$, and 2 with $p = 1$. Thus we can write these modes in ascending order as TM^r_{011}, TM^r_{111} (even and odd), TM^r_{021}, TM^r_{121} (even and odd), and TM^r_{221} (even and odd). Note that we have five modes. If even and odd modes are counted separately, then we will have eight modes.

$$f_{011}^{TM^r} = f_{111}^{TM^r(\text{even+odd})} = \frac{2.7437 \cdot 3 \cdot 10^8}{2\pi \cdot \cdot 0.03} = 4.3667 \cdot 10^9 \text{ Hz} = 4.3667 \text{ GHz}$$

and

$$f_{021}^{TM^r} = f_{121}^{TM^r(\text{even+odd})} = f_{221}^{TM^r(\text{even+odd})} = \frac{3.8702 \cdot 3 \cdot 10^8}{2\pi \cdot \cdot 0.03} = 6.1596 \cdot 10^9 \text{ Hz}$$

$$= 6.1596 \text{ GHz}.$$

PROBLEMS

6.1. A metallic parallel-plate waveguide is air filled, and the separation between its plates is 3.5 cm. Investigate the characteristics of a 9 GHz signal propagating in TM_{m0} modes.

6.2. A 0.6 cm thick polystyrene slab is being used at 25 GHz. If its dielectric constant is 2.56, calculate the cutoff frequencies of the TM modes this structure supports. Determine the propagation parameters of these modes. Repeat your calculations for the slab thickness of 0.35 cm, and compare the two sets of results.

6.3. Repeat Problem 6.2 for TE^z modes.

6.4. Find the separation between the two plates of a radial waveguide that guarantees lowest order single TM^z mode up to 950 MHz. Repeat your calculations for the lowest order TE^z mode. Assume that the medium between the plates is air.

6.5. Radius of an air-filled spherical cavity is 4.5 cm. Find the resonant frequencies of first five modes that is supports.

7

SOURCE IN INFINITE SPACE

This chapter begins with an analysis of electromagnetic fields of an infinitesimal current element. The magnetic vector potential set up by the element is evaluated to determine the various fields. An expression for the complex power density is formulated to find the real power flow. After introducing the basic parameters of antennas, linear antennas are analyzed. A section on linear antenna arrays follows it. The chapter ends with a section on the Friis transmission formula and the radar range equation.

7.1 FIELDS OF AN INFINITESIMAL SOURCE

Consider an infinitesimal current element of total length $d\ell$ and cross section da, as shown in Figure 7.1. At the origin the source element is aligned symmetrically on the z-axis and carries a current $i(t) = I\cos(\omega t)$A. The current density phasor may be expressed as follows:

$$\vec{J} = \hat{z}J_z = \hat{z}\frac{I}{da} \tag{7.1.1}$$

Our goal is to solve (6.2.13) and find the magnetic and electric fields using (6.2.8), (6.2.9), and (6.2.12) as given by

$$\vec{H} = \frac{1}{\mu}\nabla \times \vec{A} \tag{7.1.2}$$

Practical Electromagnetics: From Biomedical Sciences to Wireless Communication.
By Devendra K. Misra
Copyright © 2007 John Wiley & Sons, Inc.

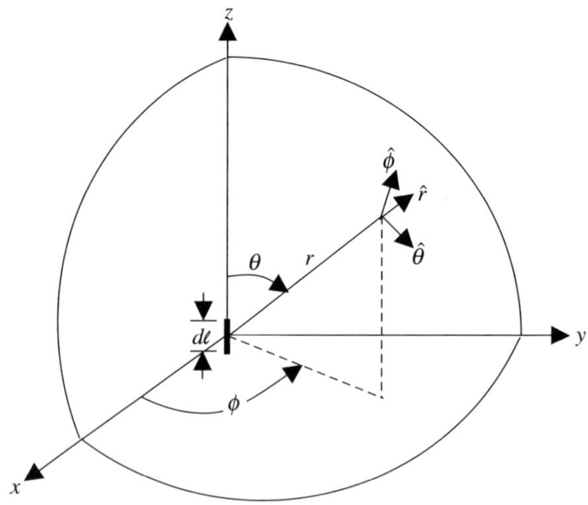

Figure 7.1 Infinitesimal current element in the spherical coordinate system.

and

$$\vec{E} = -j\omega\vec{A} - \nabla\phi_e = -j\omega\vec{A} + \frac{1}{j\omega\mu\varepsilon}\nabla(\nabla\cdot\vec{A}) = \frac{1}{j\omega\varepsilon}\nabla\times\vec{H}. \tag{7.1.3}$$

A z-directed current will produce the vector potential in the same direction. Hence (6.2.13) simplifies to

$$\nabla^2 A_z + k^2 A_z = -\mu J_z. \tag{7.1.4}$$

Further the source element occupies an infinitesimal volume $dv = da \cdot d\ell$, so, there is spherical symmetry in the source's distribution. In other words, $\partial/\partial\theta = 0$ and $\partial/\partial\phi = 0$. For the time being, we exclude a small spherical volume of radius r_o that contains the source. Therefore (7.1.4) reduces to

$$\frac{1}{r^2}\frac{d}{dr}\left(r^2\frac{dA_z}{dr}\right) + k^2 A_z = 0,$$

or

$$\frac{d^2A_z}{dr^2} + \frac{2}{r}\frac{dA_z}{dr} + k^2 A_z = 0. \tag{7.1.5}$$

Assume $A_z = \varphi/r$. Then

$$\frac{dA_z}{dr} = \frac{1}{r}\frac{d\varphi}{dr} - \frac{\varphi}{r^2} \tag{7.1.6}$$

and

$$\frac{d^2 A_z}{dz^2} = \frac{1}{r}\frac{d^2 \varphi}{dr^2} - \frac{2}{r^2}\frac{d\varphi}{dr} + \frac{2}{r^3}\varphi. \tag{7.1.7}$$

Substitution of (7.1.6) and (7.1.7) into (7.1.5) gives

$$\frac{d^2 \varphi}{dr^2} + k^2 \varphi = 0. \tag{7.1.8}$$

This is the harmonic equation that we have analyzed earlier. A general solution may be written as follows:

$$\varphi = C_1 e^{-jkr} + C_2 e^{jkr}. \tag{7.1.9}$$

Since the fields due to this source must be bounded (zero at the infinity) and represent only an outward traveling wave, the constant C_2 must be zero. Therefore

$$A_z = \frac{\varphi}{r} = C_1 \frac{e^{-jkr}}{r}. \tag{7.1.10}$$

The constant C_1 must be related to source strength. To do so, we integrate both sides of (7.1.4) over the small volume of radius r_o that we excluded earlier and force the radius to zero. Then with $dv = r^2 \sin\theta dr d\theta d\phi$, we have

$$\lim_{r_o \to 0} \int_{\phi=0}^{\phi=2\pi} \int_{\theta=0}^{\theta=\pi} \int_{r=0}^{r=r_o} (\nabla^2 A_z + k^2 A_z) r^2 \sin\theta \, dr d\theta d\phi$$

$$= -\mu \lim_{r_o \to 0} \int_{\phi=0}^{\phi=2\pi} \int_{\theta=0}^{\theta=\pi} \int_{r=0}^{r=r_o} J_z r^2 \sin\theta \, dr d\theta d\phi. \tag{7.1.11}$$

By the divergence theorem, the first term on the left-hand side of this equation reduces to

$$\lim_{r_o \to 0} \int_{\phi=0}^{\phi=2\pi} \int_{\theta=0}^{\theta=\pi} \int_{r=0}^{r=r_o} \nabla^2 A_z r^2 \sin\theta \, dr d\theta d\phi = \lim_{r_o \to 0} \int_{\phi=0}^{2\pi} \int_{\theta=0}^{\pi} \frac{d}{dr}\left(C_1 \frac{e^{-jkr}}{r}\right)$$

$$\times r^2 \sin\theta \, d\theta d\phi \Big|_{r=r_o} = \lim_{r_o \to 0} \int_{\phi=0}^{2\pi} \int_{\theta=0}^{\pi} C_1 \left(\frac{-jkr - 1}{r^2}\right) e^{-jkr} r^2 \sin\theta \, d\theta d\phi \Big|_{r=r_o}$$

$$= -C_1 \int_{\phi=0}^{2\pi} \int_{\theta=0}^{\pi} \sin\theta \, d\theta d\phi = -4\pi C_1. \tag{7.1.12}$$

The second term on the left-hand side of (7.1.11) reduces to

$$\lim_{r_o\to 0}\int_{\phi=0}^{2\pi}\int_{\theta=0}^{\pi}\int_{r=0}^{r_o} k^2 C_1 \frac{e^{-jkr}}{4\pi r} r^2 \sin\theta\, dr d\theta d\phi = 0, \tag{7.1.13}$$

and the right-hand side of (7.1.11) gives

$$\begin{aligned} &-\mu \lim_{r_o\to 0}\int_{\phi=0}^{\phi=2\pi}\int_{\theta=0}^{\theta=\pi}\int_{r=0}^{r=r_o} J_z r^2 \sin\theta\, dr d\theta d\phi \\ &= -\mu \lim_{r_o\to 0}\int_V \frac{I}{da} d\ell da = -\mu I d\ell. \end{aligned} \tag{7.1.14}$$

Substituting (7.1.12) through (7.1.14) into (7.1.11), we find that

$$C_1 = \frac{\mu I\, d\ell}{4\pi}. \tag{7.1.15}$$

Therefore the magnetic vector potential associated with the current element may be expressed in spherical coordinate system as follows:

$$\vec{A} = \hat{z}A_z = \hat{z}\frac{\mu I d\ell}{4\pi r} e^{-jkr} = \left(\hat{r}\cos\theta - \hat{\theta}\sin\theta\right)\frac{\mu I d\ell}{4\pi r} e^{-jkr}. \tag{7.1.16}$$

Thus the magnetic vector potential represents an outward traveling wave whose amplitude decreases inversely with distance. This kind of wave is called a spherical wave (the amplitude and phase remain constant on an spherical surface). In (7.1.16), for convenience, we already transformed the unit vector from rectangular to spherical coordinates.

We can now evaluate the magnetic and electric fields using (7.1.2) and (7.1.3) as follows:

$$\vec{H} = \frac{1}{\mu}\nabla\times\left[\left(\hat{r}\cos\theta - \hat{\theta}\sin\theta\right)\frac{\mu I d\ell}{4\pi r} e^{-jkr}\right] = \hat{\phi}\frac{jkI d\ell \sin\theta}{4\pi r} e^{-jkr}\left(1 - j\frac{1}{kr}\right),$$

or

$$\vec{H} = \hat{\phi}H_\phi = \hat{\phi}\,\frac{jkI\, d\ell \sin\theta}{4\pi r} e^{-jkr}\left[1 - j\frac{1}{kr}\right], \tag{7.1.17}$$

and

$$\begin{aligned} \vec{E} = \frac{1}{j\omega\varepsilon}\nabla\times\left(\hat{\phi}H_\phi\right) &= \frac{1}{j\omega\varepsilon}\left[\hat{r}\frac{1}{r\sin\theta}\frac{\partial}{\partial\theta}\left(\hat{\theta}H_\phi\right) - \hat{\theta}\frac{1}{r}\frac{\partial}{\partial r}(rH_\phi)\right] \\ &= \hat{r}E_r + \hat{\theta}\,E_\theta + \hat{\phi}\,E_\phi, \end{aligned}$$

where

$$E_r = \frac{\eta I d\ell \cos\theta}{2\pi r^2}\left(1 - j\frac{1}{kr}\right)e^{-jkr}, \tag{7.1.18}$$

$$E_\theta = j\frac{\eta k I d\ell \sin\theta}{4\pi r}\left(1 - j\frac{1}{kr} - \frac{1}{k^2 r^2}\right)e^{-jkr} \tag{7.1.19}$$

and

$$E_\phi = 0. \tag{7.1.20}$$

The complex Poynting vector (or the complex power density) associated with these fields can be evaluated as follows:

$$\vec{S} = \tfrac{1}{2}\left(\vec{E} \times \vec{H}^*\right) = \tfrac{1}{2}\left(\hat{r}E_r + \hat{\theta}E_\theta\right) \times \left(\hat{\phi}H_\phi^*\right) = \tfrac{1}{2}\left(\hat{r}E_\theta H_\phi^* - \hat{\theta}E_r H_\phi^*\right)$$

$$= \hat{r}S_r + \hat{\theta}S_\theta, \tag{7.1.21}$$

where

$$S_r = \frac{1}{2}\eta\left(\frac{k|I|d\ell \sin\theta}{4\pi r}\right)^2\left(1 - j\frac{1}{k^3 r^3}\right) = \frac{1}{8}\eta\left(\frac{|I|d\ell \sin\theta}{\lambda r}\right)^2\left[1 - j\frac{1}{(kr)^3}\right] \tag{7.1.22}$$

and

$$S_\theta = j\eta\frac{k|Id\ell|^2 \sin\theta \cos\theta}{16\pi^2 r^3}\left[1 + \frac{1}{(kr)^2}\right]. \tag{7.1.23}$$

Here $\eta = k^2/\omega\varepsilon = \sqrt{\mu/\varepsilon}$ is the intrinsic impedance of the medium, as defined earlier in Chapter 4. Note that S_r is complex and S_θ is purely imaginary. Further the reactive power densities (imaginary parts of the two) decrease at much faster rate with distance r in comparison to the real part of S_r. There is a real power flow only in the radial direction. This represents the power radiated by this current element. The field components responsible for the radiated power are identified in (7.1.17) through (7.1.20) to be

$$\vec{H}_{\text{rad}} = \hat{\phi}\frac{jkI\,d\ell \sin\theta}{4\pi r}e^{-jkr} = \hat{\phi}H_{\phi\,\text{rad}} \tag{7.1.24}$$

and

$$\vec{E}_{\text{rad}} = \hat{\theta}j\frac{\eta k I d\ell \sin\theta}{4\pi r}e^{-jkr} = \hat{\theta}E_{\theta\,\text{rad}}. \tag{7.1.25}$$

Dividing (7.1.24) by (7.1.25), we have

$$\frac{E_{\theta\,\mathrm{rad}}}{H_{\phi\,\mathrm{rad}}} = \eta. \tag{7.1.26}$$

The total radiated power P_{rad} can be found after integrating the power density over a spherical surface as follows:

$$P_{\mathrm{rad}} = \int_{\phi=0}^{2\pi}\int_{\theta=0}^{\pi} \frac{1}{8}\eta\frac{|Id\ell|^2\sin^2\theta}{\lambda^2 r^2} r^2 \sin\theta d\theta d\phi = \frac{2\pi}{8}\eta\frac{|I\,d\ell|^2}{\lambda^2}\int_{\theta=0}^{\pi}\sin^3\theta d\theta,$$

or

$$P_{\mathrm{rad}} = \frac{\pi\eta|I\,d\ell|^2}{3\lambda^2}. \tag{7.1.27}$$

The infinitesimal current element considered in this section is a building block for the linear antenna analyzed in Section 7.3. It is also called an ideal electric dipole or a Hertzian dipole.

Example 7.1

A 1 cm long Hertzian dipole antenna is located at the origin of the spherical coordinates. It carries a current of 1 A at 100 MHz. Find the power density per unit area that leaves a spherical surface of 1 km radius. Plot its normalized

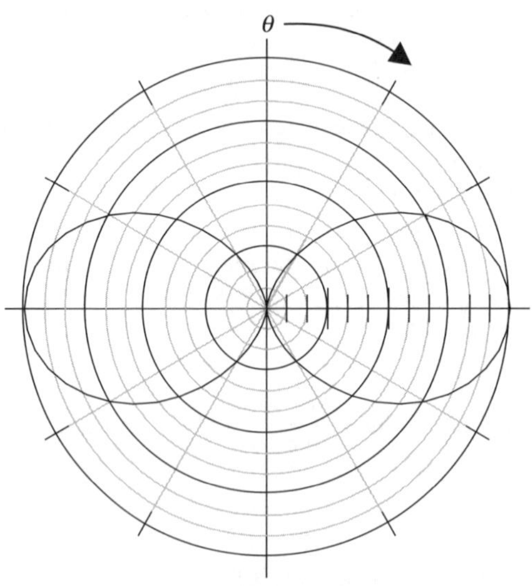

Figure 7.2 Power density of a current element as a function of θ.

characteristics (divide by the maximum power per unit area) on the $\phi = 90^\circ$ and $\theta = 90^\circ$ planes.

Since $\lambda = (3 \cdot 10^8)/(100 \cdot 10^6) = 3\,\text{m}$ and $\eta = 377\ \Omega$ for the free space, from (7.1.22) we have

$$\text{Re}(S_r) = \frac{1}{8}\eta\left(\frac{|I|d\ell \sin\theta}{\lambda r}\right)^2 = \frac{377}{8}\left(\frac{1 \cdot 10^{-2}}{3 \cdot 10^3}\sin\theta\right)^2 = 5.24 \cdot 10^{-10}\sin^2\theta\ \text{W/m}^2.$$

This expression is independent of ϕ. Therefore the normalized power characteristics on the $\theta = 90^\circ$ plane will be a circle of unity radius; only the $\phi = 90^\circ$ plane is illustrated in Figure 7.2.

7.2 ANTENNA PARAMETERS

Figure 7.3 illustrates some of the antennas that are commonly used in wireless communication systems. These can be grouped as wire-type antennas and aperture-type antennas. Electric dipole and the loop (also known as the magnetic dipole) antennas belong to the former group, whereas horn, patch, reflector, and lens belong to the latter group. The aperture antennas can be further subdivided into primary and secondary (or passive) antennas. Primary antennas are directly excited by a source and can be used independently for the transmission or reception of signals. A secondary antenna requires another antenna as its feeder. Horn and patch antennas fall in the first category and the reflector and lens in the second. Various kinds of horn antennas are commonly used as feeders in reflector and lens antennas.

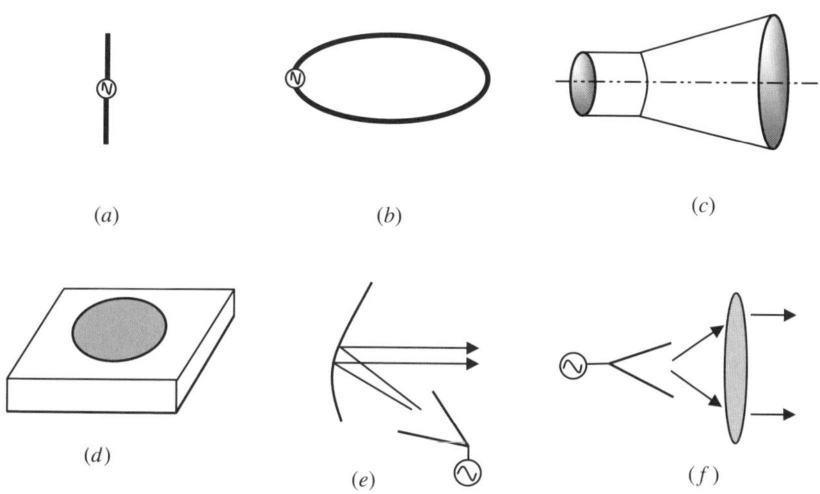

Figure 7.3 Some of the commonly used antennas: (*a*) electric dipole, (*b*) loop, (*c*) conical horn, (*d*) patch (*e*) reflector, and (*f*) lens.

When an antenna is energized, it generates two types of electromagnetic fields. As (7.1.22) and (7.1.23) indicate, the reactive power stays near the antenna whereas the real power propagates outward as radiation. The propagating power represents radiation fields that are inversely related to distance r. Nonpropagating power represents the reactive (capacitive or inductive) components of electromagnetic fields that decrease in magnitude at the rate of r^{-2} and r^{-3}. Thus the space surrounding an antenna can be divided into three regions. The reactive fields dominate in nearby regions but reduce in strength at a faster rate in comparison with fields associated with the propagating signal. If the largest dimension of an antenna is D and the signal wavelength is λ, then the reactive fields dominate up to about $0.62\sqrt{(D^3/\lambda)}$ and diminish after $2D^2/\lambda$. The region beyond $2D^2/\lambda$ is called the far-field (or radiation field) region.

Power radiated by an antenna per unit solid angle is known as the radiation intensity U. It is a far-field parameter that is related to power density (power per unit area) W_{rad} and distance r as follows:

$$U = r^2 W_{\text{rad}}. \tag{7.2.1}$$

Directive Gain and Directivity

If an antenna radiates uniformly in all directions, then it is called an *isotropic antenna*. This is a hypothetical antenna that helps in defining the characteristics of a real one. The directive gain D_G is defined as the ratio of radiation intensity due to the test antenna to that of an isotropic antenna. It is assumed that total radiated power remains the same in the two cases. Hence

$$D_G = \frac{U}{U_o} = \frac{4\pi U}{P_{\text{rad}}}, \tag{7.2.2}$$

where U is radiation intensity due to the test antenna in watts per unit solid angle, U_o is radiation intensity due to the isotropic antenna in watts per unit solid angle, and P_{rad} is the total radiated power in watts. Since U is a directional dependent quantity, the directive gain of an antenna depends on the angles θ and ϕ. If the radiation intensity assumes its maximum value, then the directive gain is called the directivity D_o. That is,

$$D_o = \frac{U_{\max}}{U_o} = \frac{4\pi U_{\max}}{P_{\text{rad}}}. \tag{7.2.3}$$

Example 7.2

Find the directive gain and the directivity of an infinitesimal current element.

Since we have already analyzed the fields generated by the current element, we can use those expressions to find its directive gain. Note that real part of S_r in (7.1.22) is W_{rad} used in (7.2.1). Hence

$$U = r^2 \frac{\eta}{8}\left(\frac{|I|d\ell \sin\theta}{\lambda r}\right)^2 = \frac{\eta}{8\lambda^2}(|I|d\ell \sin\theta)^2.$$

Using (7.1.27) and (7.2.2), we have

$$D_G = \frac{U}{U_o} = 4\pi\frac{U}{P_{rad}} = 4\pi\frac{\eta(|I|d\ell \sin\theta)^2/8\lambda^2}{\pi\eta|Id\ell|^2/3\lambda^2} = \frac{3}{2}\sin^2\theta.$$

Therefore, the directivity of the current element is

$$D_o = D_G|_{max} = 1.5.$$

Gain of an Antenna

The power gain G of an antenna is defined as the ratio of its radiation intensity $U(\theta, \phi)$ at a point to the radiation intensity that results from a uniform radiation of the same input power P_{in}. We can express it as follows:

$$G = 4\pi\frac{\text{Radiation intensity}}{\text{Total input power}} = 4\pi\frac{U(\theta, \phi)}{P_{in}}. \tag{7.2.4}$$

Most of the time we deal with relative gain. Relative gain is defined as a ratio of the power gain of the test antenna in a given direction to the power gain of a reference antenna. Both antennas must have the same input power. The reference antenna may be a dipole, horn, or any other antenna of known gain or it can be calculated. However, it is a lossless isotropic radiator in most cases. Hence

$$G = 4\pi\frac{U(\theta, \phi)}{P_{in}(\text{Lossless isotropic antenna})}. \tag{7.2.5}$$

When the direction is not stated, the power gain is usually taken in the direction of maximum radiation.

Radiation Patterns and Half-power Beam Width (HPBW)

Far-field power distribution at a distance r from the antenna depends on the spatial coordinates θ and ϕ. The graphical representations of these distributions on the orthogonal plane (θ plane or ϕ plane) at a constant distance r from the antenna are called its radiation patterns. Figure 7.2 shows the radiation pattern of an infinitesimal current element on the θ plane. Its ϕ-plane pattern is a circle that can be found easily after rotating this pattern about the $\theta = 0$ axis. The three-dimensional picture of this radiation pattern is doughnut shaped. Similarly the power distributions of other antennas generally show peaks and valleys in the radiation zone. The highest peak between the two valleys is termed the *main lobe* and the others peaks the *side lobes*. The total angle about the main peak over which power falls by 50% of its maximum value is called the *half-power beam width* on that plane.

The following relations are used to estimate the power gain G and the half-power beam width HPBW (or BW) of an aperture antenna:

$$G = \frac{4\pi}{\lambda^2} A_e = \frac{4\pi}{\lambda^2} A\kappa \tag{7.2.6}$$

and

$$BW(\text{in degrees}) = \frac{65 \times \lambda}{d}, \tag{7.2.7}$$

where A_e is the effective area of radiating aperture in square meters, A is its physical area ($\pi d^2/4$, for a reflector antenna dish with its diameter d), κ is the efficiency of antenna (ranges from 0.6 to 0.65), and λ is the signal wavelength in meters.

Example 7.3

Calculate the power gain (in dBi) and the half-power beam width of a parabolic dish antenna of 25 m in diameter that is radiating at 3 GHz.

The signal wavelength and the area of the aperture are

$$\lambda = \frac{3 \cdot 10^8}{3 \cdot 10^9} = 0.1 \text{ m}$$

and

$$A = \frac{\pi d^2}{4} = \pi \frac{25^2}{4} = 490.8738 \, \text{m}^2.$$

If we assume that the aperture efficiency is 0.6, the antenna gain and the half-power beam width are found to be

$$G = \frac{4\pi}{0.1^2} 490.8738 \cdot 0.6 = 370110.1257 = 10 \log(370110.1257) = 55.6833 \, \text{dBi}$$

and

$$BW = \frac{65 \times 0.1}{25} = 0.26 \text{ deg}.$$

Radiation Resistance

Radiation resistance is an equivalent resistance dissipating the same amount of power that an antenna radiates when the current flowing through the resistance is equal to the current input to the antenna.

Example 7.4

Find the radiation resistance R_{rad} of an infinitesimal current element as a function of its length $d\ell$ and the signal wavelength λ.

Since we have already formulated in (7.1.27) an expression for the power that this current element radiates, we can find R_{rad} as follows:

$$\frac{1}{2}|I|^2 R_{rad} = \frac{\pi\eta}{3}|I|^2\left(\frac{d\ell}{\lambda}\right)^2 \rightarrow R_{rad} = \frac{2\pi\eta}{3}\left(\frac{d\ell}{\lambda}\right)^2.$$

Since $\eta \approx 120\pi\ \Omega$ ($\approx$377 Ω) for free space, this expression simplifies to

$$R_{rad} = 80\pi^2\left(\frac{d\ell}{\lambda}\right)^2.$$

Antenna Efficiency

If an antenna is not matched with its feeder, then a part of the signal available from the source is reflected back. This is considered reflection (or mismatch) loss. Reflection (or mismatch) efficiency is defined as a ratio of the power input to the antenna to the power available from the source. Since the ratio of reflected power to the power available from the source is equal to the square of the magnitude of the voltage reflection coefficient, the reflection efficiency e_r is given by

$$e_r = 1 - |\Gamma|^2,$$

where

$$\Gamma = \text{Voltage reflection coefficient} = \frac{Z_A - Z_o}{Z_A + Z_o},$$

Z_A is the antenna impedance, and Z_o is the characteristic impedance of the feeding line.

Besides the power loss due to mismatch, a signal energy may dissipate in an antenna because of an imperfect conductor or dielectric material. These efficiencies are hard to compute. However, the combined conductor and dielectric efficiency e_{cd} can be experimentally determined after measuring the input power P_{in} and the radiated power P_{rad}. This is given as:

$$e_{cd} = \frac{P_{rad}}{P_{in}}.$$

The overall efficiency e_o is a product of the efficiencies above. Therefore

$$e_o = e_r\, e_{cd}. \tag{7.2.8}$$

Example 7.5

A 75 Ω transmission line feeds a lossless one-half wavelength long dipole antenna. If the antenna impedance is 73 Ω and its radiation intensity $U(\theta, \phi)$ is given as follows, find the maximum overall gain:

$$U = B_o \sin^3(\theta).$$

The maximum radiation intensity, U_{max}, is B_o, which occurs at $\theta = \pi/2$. The total radiated power is found to be

$$P_{rad} = \int_0^{2\pi} \int_0^{\pi} B_o \sin^3\theta \, \sin\theta d\theta d\phi = \tfrac{3}{4}\pi^2 B_o.$$

Hence

$$D_o = 4\pi \frac{U_{max}}{P_{rad}} = \frac{4\pi B_o}{3\pi^2 B_o/4} = \frac{16}{3\pi} = 1.6977,$$

or

$$D_o(\text{dB}) = 10 \log_{10}(1.6977)\,\text{dB} = 2.2985\,\text{dB}.$$

Since the antenna is lossless, there is no conductor or dielectric loss. Therefore its radiation efficiency e_{cd} is unity (0 dB). However, there is power loss due to its mismatch with the feeding line. The mismatch efficiency is computed as follows:

The voltage reflection coefficient at the input of antenna is found to be

$$\Gamma = \frac{Z_A - Z_o}{Z_A + Z_o} = \frac{73 - 75}{73 + 75} = -\frac{2}{148} = -0.0135.$$

The mismatch efficiency of this antenna then is

$$e_r = 1 - (0.0135)^2 = 0.9998 = 10 \log_{10}(0.9998)\,\text{dB} = -0.0008\,\text{dB}.$$

The overall gain G_o (in dB) is obtained as:

$$G_o(\text{dB}) = 2.2985 - 0 - 0.0008 = 2.2977\,\text{dB}.$$

Bandwidth

Antenna characteristics, such as gain, radiation pattern, and impedance are frequency dependent. The bandwidth of an antenna is defined as the frequency band over which its performance with respect to some characteristic (HPBW, directivity, etc.) conforms to a specified standard.

Polarization

The polarization of an antenna is same as the polarization of its radiating wave. As considered in Chapter 4, it is the property of an electromagnetic wave describing the time-varying direction and relative magnitude of an electric field vector. The curve traced by an instantaneous electric field vector over time is the polarization of that wave. The polarization characteristics are summarized below for the ready reference:

- *Linear polarization.* If over time the tip of the electric field intensity traces a straight line in some direction, then the wave is linearly polarized.
- *Circular polarization.* If over time the end of the electric field traces a circle in space then the electromagnetic wave is circularly polarized. Further the wave may be right-handed circularly polarized (RHCP) or left-handed circularly polarized (LHCP), depending on whether the electric field vector rotates clockwise or counterclockwise.
- *Elliptical polarization.* If over time the tip of the electric field intensity traces an ellipse in space, then the wave is elliptically polarized. As in the preceding case, the wave may experience either right-handed or left-handed elliptical polarization (RHEP and LHEP).

In a receiving system the polarization of the antenna and the incoming wave need to be matched for maximum response. When this is not the case, there will be some signal loss, which is known as *polarization loss*. For example, if there is a vertically polarized wave incident on a horizontally polarized antenna, then the induced voltage available across the terminals will be zero. In this case the antenna is characterized as *cross-polarized with an incident wave*. The square of the cosine of the angle between the wave-polarization and antenna-polarization is a measure of the polarization loss. It can be determined by squaring the scalar product of the unit vectors representing the two polarizations.

Example 7.6

A uniform plane electromagnetic wave propagating in the z direction is incident upon an antenna that is polarized as follows:

$$\vec{E}_a(\vec{r}) = (\hat{x} + \hat{y})E(\vec{r}).$$

The electric field intensity of the incoming electromagnetic wave is

$$\vec{E}(\vec{r}, t) = \hat{y}E_o \cos(\omega t - kz) \text{ V/m}.$$

Find the polarization loss factor.

In this case the incident wave is linearly polarized along y-axis whereas the receiving antenna is linearly polarized at 45° from it. Therefore one-half of the incident field is cross-polarized with the antenna, and its polarization loss can be found as follows:

The unit vector along the polarization of incident wave is

$$\hat{u}_i = \hat{y}.$$

The unit vector along the antenna polarization is

$$\hat{u}_{\mathrm{a}} = \frac{1}{\sqrt{2}}(\hat{x} + \hat{y}).$$

Hence the polarization loss factor is

$$|\hat{u}_{\mathrm{i}} \cdot \hat{u}_{\mathrm{a}}|^2 = 0.5 = -3.01\,\mathrm{dB}.$$

Effective Isotropic Radiated Power

Effective isotropic radiated power (EIRP) is a measure of the power gain of an antenna. It is equal to the power that an isotropic antenna needs to provide the same radiation intensity at a given point as the directional antenna. If the power input to the feeding line is P_{t} and the antenna gain is G_{t}, then EIRP is defined as follows:

$$\mathrm{EIRP} = \frac{P_{\mathrm{t}} G_{\mathrm{t}}}{L}, \tag{7.2.9}$$

where L is the input-to-output power ratio of a transmission line that connects the output of the transmitter and the antenna. It is given by

$$L = \frac{P_{\mathrm{t}}}{P_{\mathrm{ant}}} \tag{7.2.10}$$

Alternatively, the EIRP can be expressed in dBW as follows:

$$\mathrm{EIRP(dBW)} = P_{\mathrm{t}}(\mathrm{dBW}) - L(\mathrm{dB}) + G_{\mathrm{t}}(\mathrm{dB}). \tag{7.2.11}$$

Example 7.7

A transmitter has 1 kW power at the output of its final high-power amplifier. If the gain of the transmitting antenna is 60 dB and its feeding line has an attenuation of 20%, find the EIRP in dBW.

$$P_{\mathrm{t}} = 1000\,\mathrm{W} = 30\,\mathrm{dBW},$$

$$P_{\mathrm{ant}} = 0.8 \times 1000 = 800\,\mathrm{W},$$

$$G_{\mathrm{t}} = 60\,\mathrm{dB} = 10^6,$$

and

$$L = \frac{1000}{800} = 1.25 = 10\log(1.25) = 0.9691\,\mathrm{dB}.$$

Hence

$$\mathrm{EIRP(dBW)} = 30 - 0.9691 + 60 = 89.0309\,\mathrm{dBW},$$

or

$$\text{EIRP} = \frac{1000 \cdot 10^6}{1.25} = 800 \cdot 10^6\,\text{W} = 89.0309\,\text{dBW}.$$

Space Loss

A transmitting antenna radiates in all directions depending on its radiation characteristics. However, a receiving antenna receives only the power that is incident on it. Hence the rest of the power is not used and is lost in space. It is regarded as space loss.

Space loss can be determined as follows: The power density w_t of a signal transmitted by an isotropic antenna is given by

$$w_t = \frac{P_t}{4\pi r^2}\,\text{W/m}^2, \tag{7.2.12}$$

where P_t is the transmitted power in watts and r is the distance from the antenna in meters. The power received by a unity gain antenna located at r is found to be

$$P_r = w_t A_{eu}, \tag{7.2.13}$$

where A_{eu} is the effective area of an isotropic antenna.

From (7.2.6), for an isotropic antenna

$$G = \frac{4\pi}{\lambda^2} A_{eu} = 1,$$

or

$$A_{eu} = \frac{\lambda^2}{4\pi}.$$

Hence (7.2.13) can be rewritten as

$$P_r = \frac{P_t}{4\pi R^2} \cdot \frac{\lambda^2}{4\pi}, \tag{7.2.14}$$

and the space loss ratio is found to be

$$\frac{P_r}{P_t} = \left(\frac{\lambda}{4\pi R}\right)^2. \tag{7.2.15}$$

This is usually expressed in dB as follows:

$$\text{Space loss ratio} = 20\,\log_{10}\left(\frac{\lambda}{4\pi R}\right)\,\text{dB}. \tag{7.2.16}$$

Example 7.8

A geostationary satellite is 35860 km away from the earth's surface. Find the space loss ratio if it is operating at 6 GHz.

$$R = 35860000\,\text{m}$$

and

$$\lambda = \frac{3 \cdot 10^8}{6 \cdot 10^9} = 0.05\,\text{m}.$$

Hence

$$\text{Space loss ratio} = \left(\frac{0.05}{4\pi \cdot 35860000}\right)^2 = 1.2311 \times 10^{-20} = -199.1\,\text{dB}.$$

7.3 LINEAR ANTENNAS

Equation (7.1.16) gives the magnetic vector potential due to a z-directed infinitesimal current element located at the origin. This can be generalized easily for the z-directed current source located at $\vec{r}\,'$ as follows:

$$A_z = \int_{V'} \mu J_z \frac{e^{-jkR}}{4\pi R} dv', \tag{7.3.1}$$

where $R = \left|\vec{r} - \vec{r}\,'\right|$ is the distance from source point to the field point (observation point) as shown in Figure 7.4. In the figure the primed coordinates represent the source point, the unprimed coordinates are the field points.

Similar expressions can be used for the x- and y-directed current sources. A general formulation based on the impulse response (Green's function) and the superposition integral are summarized in Appendix B. A general expression for the magnetic vector potential is

$$\vec{A}(\vec{r}) = \int_{V'} \mu\vec{J} \frac{e^{-jkR}}{4\pi R} dv'. \tag{7.3.2}$$

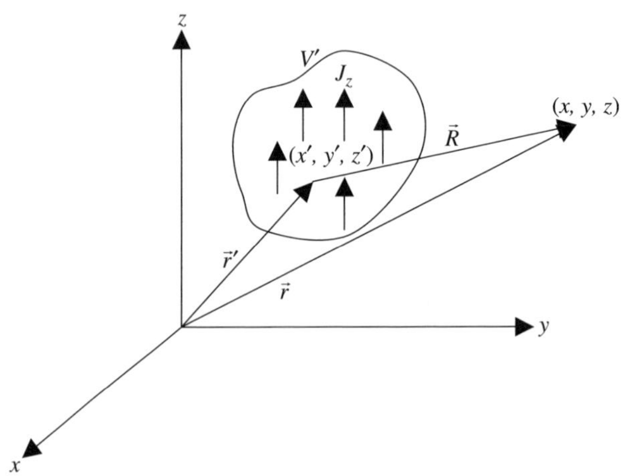

Figure 7.4 Coordinate system used for fields due to a z-directed current source.

In this section we consider the radiation characteristics of only thin linear antennas. We assume that the antenna is a thin wire lying along the z-axis and carrying a current I, with $x' = y' \approx 0$. Hence (7.3.1) can be simplified by the coordinate geometry illustrated in Figure 7.5.

Consider an incremental length dz' of thin linear wire antenna centered on point $(0, 0, z')$. The coordinates of the field point are (x, y, z). Therefore the separation R between the source and the field points is found to be

$$R = |\vec{r} - \vec{r}'| = \sqrt{x^2 + y^2 + (z - z')^2}$$
$$= \sqrt{x^2 + y^2 + z^2 - 2z'z + z'^2} = \sqrt{r^2 - 2z'r\cos\theta + z'^2}$$

Here we have switched the field point (unprimed) coordinates from rectangular to spherical coordinates. Using binomial expansion (formula 47 in Appendix A) and simplifying it we obtain

$$R = r - z'\cos\theta + \frac{z'^2}{2r}\sin^2\theta + \frac{z'^3}{2r^3}\cos\theta\sin^2\theta + \cdots. \tag{7.3.3}$$

Since $r \gg z'$, the magnitude of A_z in (7.3.1) is not significantly altered if we keep just the first term of this expansion as the approximate value of R. However, we also keep the second term in its phase term $e^{jkz'\cos\theta}$ because it contributes significantly. For example, if $z'\cos\theta$ changes from 0 to $\lambda/2$ (where λ is the signal wavelength), then $e^{jkz'\cos\theta}$ changes from 1 to -1. As illustrated in Figure 7.5b, this approximation leads to $\vartheta \approx \theta$. Hence (7.3.1) is approximated as follows for determining the field radiated by a linear antenna:

$$A_z = \frac{\mu}{4\pi r}e^{-jkr}\int_{-\ell/2}^{\ell/2} I(z')e^{jkz'\cos\theta}\,dz'. \tag{7.3.4}$$

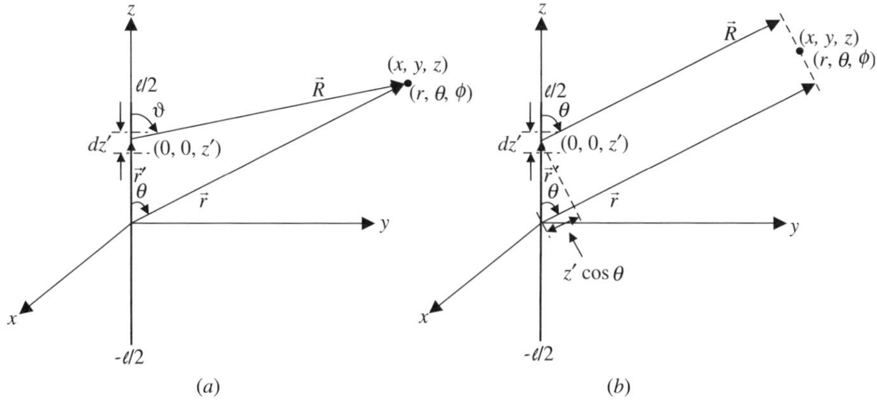

Figure 7.5 Linear antenna geometry (a) and an approximation of its radiation analysis (b).

After (7.3.4) is evaluated for a given current distribution on a linear antenna, its fields are found from (7.1.2) and (7.1.3). Alternatively, (7.1.24) and (7.1.25) can be used as the radiated fields from an infinitesimal element that is integrated over the length of the linear antenna. To use this approach, we need to define the incremental electric field in the radiation zone via (7.1.25). The approximation proceeds as follows:

$$\Delta\vec{E}_{\rm rad} \approx \hat{\theta} j \frac{\eta k I dz' \sin\theta}{4\pi r} e^{-jkr} e^{jkz'\cos\theta}. \tag{7.3.5}$$

Therefore

$$\vec{E}_{\rm rad} \approx \hat{\theta}\,\frac{j\eta k \sin\theta}{4\pi r} e^{-jkr} \int_{-\ell/2}^{\ell/2} I(z') e^{jkz'\cos\theta} dz'. \tag{7.3.6}$$

Since the current must be zero at the ends of the thin wire, the following distribution of current is usually assumed for a center-driven linear antenna. Assume that the peak value of the current is I_o. Then

$$I(z') = I_o \sin\left[k\left(\frac{\ell}{2} - |z'|\right)\right], \qquad -\frac{\ell}{2} \le z' \le \frac{\ell}{2}. \tag{7.3.7}$$

Hence

$$\vec{E}_{\rm rad} = \hat{\theta}\,\frac{j\eta k I_o \sin\theta}{4\pi r} e^{-jkr} \int_{-\ell/2}^{\ell/2} \sin\left[k\left(\frac{\ell}{2} - |z'|\right)\right] e^{jkz'\cos\theta} dz'. \tag{7.3.8}$$

This integral can be evaluated with the help of integration formula 28 in Appendix A. Alternatively, because the sine function in the integrand of (7.3.8) is an even function (because of $|z'|$) and exponential term can be split into even and odd terms via Euler's formula (equation 24 in Appendix A), this integral can be simplified to

$$\vec{E}_{\rm rad} = \hat{\theta}\,\frac{j\eta k I_o \sin\theta}{4\pi r} e^{-jkr} 2 \int_{0}^{\ell/2} \sin\left[k\left(\frac{\ell}{2} - z'\right)\right] \cos(kz'\cos\theta) dz'. \tag{7.3.9}$$

Now this integral can be easily evaluated using software with symbolic mathematics capability (e.g., Mathcad®). This gives

$$\vec{E}_{\rm rad} = \hat{\theta}\,\frac{j\eta I_o}{2\pi r} e^{-jkr} \frac{\cos[(k\ell/2)\cos\theta] - \cos(k\ell/2)}{\sin\theta}. \tag{7.3.10}$$

The corresponding magnetic field can be found from (7.1.26) as follows:

$$\vec{H}_{\rm rad} = \hat{\phi}\,\frac{jI_o}{2\pi r} e^{-jkr} \frac{\cos[(k\ell/2)\cos\theta] - \cos(k\ell/2)}{\sin\theta}. \tag{7.3.11}$$

Therefore the radiation power density (power flow per unit area) W_{rad} is found to be

$$W_{rad} = \frac{\eta I_o^2}{8\pi^2 r^2}\left\{\frac{\cos[(k\ell/2)\cos\theta] - \cos(k\ell/2)}{\sin\theta}\right\}^2. \tag{7.3.12}$$

The radiated power P_{rad} can be obtained now:

$$\begin{aligned} P_{rad} &= \int_{\theta=0}^{\pi}\int_{\phi=0}^{2\pi} W_{rad} r^2 \sin\theta d\theta d\phi \\ &= \frac{\eta I_o^2}{4\pi}\int_{\theta=0}^{\pi}\frac{[\cos[(k\ell/2)\cos\theta] - \cos(k\ell/2)]^2}{\sin\theta} d\theta. \end{aligned} \tag{7.3.13}$$

At this point we consider a special case where the total length of the antenna is equal to one-half of the signal wavelength:

$$\ell = \frac{\lambda}{2} \rightarrow \frac{k\ell}{2} = \frac{1}{2}\cdot\frac{2\pi}{\lambda}\cdot\frac{\lambda}{2} = \frac{\pi}{2}.$$

Therefore (7.3.10) through (7.3.13) reduce to

$$\vec{E}_{rad} = \hat{\theta}\,\frac{j\eta I_o}{2\pi r} e^{-jkr}\frac{\cos[(\pi/2)\cos\ \theta]}{\sin\theta}, \tag{7.3.14}$$

$$\vec{H}_{rad} = \hat{\phi}\,\frac{jI_o}{2\pi r} e^{-jkr}\frac{\cos[(\pi/2)\cos\ \theta]}{\sin\theta}, \tag{7.3.15}$$

$$W_{rad} = \frac{\eta I_o^2}{8\pi^2 r^2}\left\{\frac{\cos[(\pi/2)\cos\ \theta]}{\sin\theta}\right\}^2, \tag{7.3.16}$$

and

$$P_{rad} = \frac{\eta I_o^2}{4\pi}\int_{\theta=0}^{\pi}\frac{\{\cos[(\pi/2)\cos\theta]\}^2}{\sin\theta} d\theta = \frac{\eta I_o^2}{4\pi}\cdot 1.2186. \tag{7.3.17}$$

Note that this definite integral requires numerical integration. It can be easily carried out on a programmable calculator. The radiation resistance R_{rad} can then be found as follows:

$$\frac{1}{2} I_o^2 R_{rad} = P_{rad} = \frac{\eta I_o^2}{4\pi}\cdot 1.2186 \rightarrow R_{rad} = \frac{0.6093}{\pi}\eta\Omega = 73.1\ \Omega. \tag{7.3.18}$$

Radiation patterns of selected linear antennas are shown in Figure 7.6.

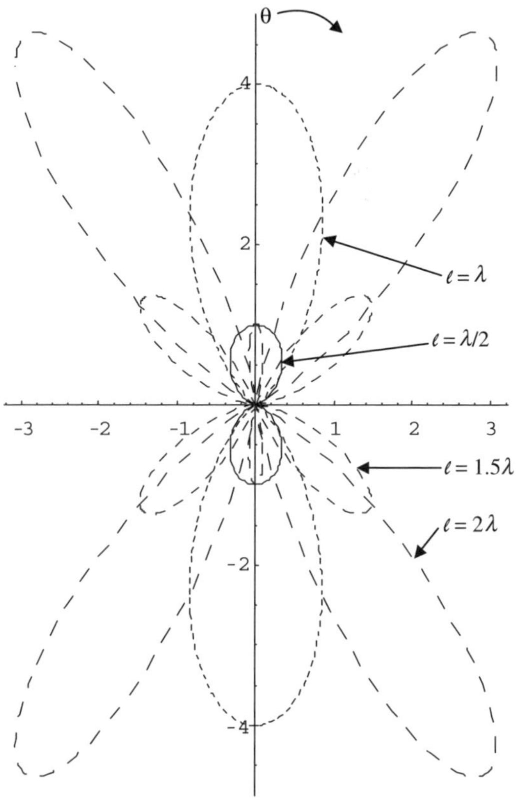

Figure 7.6 Radiation patterns of linear antennas on the $\phi = 0$ plane.

Next, the directivity D_o of a half-wavelength dipole can be found from (7.2.1), (7.2.2), (7.3.16), and (7.3.17) as follows:

$$U_{max} = \frac{\eta I_o^2}{8\pi^2} \left\{ \frac{\cos[(\pi/2)\cos\theta]}{\sin\theta} \right\}^2 \Bigg|_{max} = \frac{\eta I_o^2}{8\pi^2}$$

and

$$D_o = 4\pi \frac{\eta I_o^2/8\pi^2}{(\eta I_o^2/4\pi)\cdot 1.2186} = \frac{2}{1.2186} = 1.6412. \tag{7.3.19}$$

7.4 ANTENNA ARRAYS

Consider an array of two infinitesimal current elements along the x-axis, as shown in Figure 7.7. The separation between the two elements is d. Using (7.1.25), we can

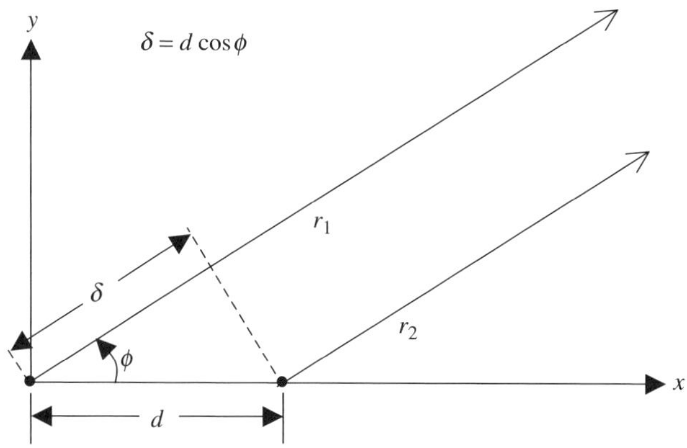

Figure 7.7 Linear array of two antennas along the x-axis.

express the total radiated electric field on the x-y plane (i.e., $\theta = 90^\circ$) as follows:

$$E_\theta = \frac{j\eta k I d\ell}{4\pi}\left[e^{j\alpha_1}\frac{e^{-jkr_1}}{r_1} + e^{j\alpha_2}\frac{e^{-jkr_2}}{r_2}\right]. \tag{7.4.1}$$

As indicated in Figure 7.7, the field point is far away from the two sources such that the two signal paths can be assumed to be almost parallel. Therefore the distances r_1 and r_2 are assumed to be equal in the two magnitude terms. However, r_2 is approximated by $r_1 - d \cos \phi$ in the phase term (similar to the approximation used for linear antennas in the preceding section) for better accuracy. As a result (7.4.1) reduces to

$$\begin{aligned} E_\theta &= \frac{j\eta k I d\ell}{4\pi r_1} e^{-j(kr_1 - \alpha_1)}[1 + e^{j(\alpha_2 - \alpha_1)} \cdot e^{jkd\cos\phi}] \\ &= \frac{j\eta k I d\ell}{4\pi r_1} e^{-j(kr_1 - \alpha_1)}[1 + e^{j(kd\cos\phi + \alpha)}], \end{aligned} \tag{7.4.2}$$

where $\alpha = \alpha_2 - \alpha_1$.

A comparison of the multiplying coefficient on the right-hand side of (7.4.2) with that of radiation field given in (7.1.25) shows that the two are identical. This term is called the element factor, and the remaining term (inside the square brackets) is called the array factor (AF). Hence

$$\begin{aligned} AF &= 1 + e^{j(kd\cos\phi + \alpha)} = e^{j(kd\cos\phi+\alpha)/2}\left[e^{-j(kd\cos\phi+\alpha)/2} + e^{j(kd\cos\phi+\alpha)/2}\right] \\ &= 2e^{j(kd\cos\phi+\alpha)/2}\cos\left(\frac{kd\cos\phi + \alpha}{2}\right). \end{aligned}$$

For $kd\cos\phi + \alpha = \psi$ this relation can be expressed as follows:

$$AF = 2e^{j\psi/2}\cos\left(\frac{\psi}{2}\right). \tag{7.4.3}$$

Note that the $e^{j\psi/2}$ term in AF reduces to 1 if the reference is moved to the center of the array. Since the maximum possible magnitude of AF is 2, it can be normalized as follows:

$$AF_{\mathrm{n}} = \frac{|AF|}{2} = \cos\left(\frac{\psi}{2}\right). \tag{7.4.4}$$

The array factor depends only on the separation between the two radiators and the relative phase of their excitation. If we use different antennas in place of infinitesimal current elements, only the element factor will change. The total radiated field is found after multiplying the new element factor by the array factor of (7.4.3). This is known as the principle of pattern multiplication, which states that the radiation pattern of an array is the product of the element factor of an individual antenna with the array factor.

Example 7.9

Find the radiation field pattern on the x-y plane of a two-element short dipole array with in-phase excitation (i.e., $\alpha = 0$) and $d = 0.5\lambda$.

From (7.4.4) we have

$$AF_{\mathrm{n}} = \cos\left(\frac{\psi}{2}\right) = \cos\left(\frac{kd\cos\phi + \alpha}{2}\right) = \cos\left(\frac{2\pi d}{2\lambda}\cos\phi + \frac{\alpha}{2}\right) = \cos\left(\frac{\pi}{2}\cos\phi\right).$$

This AF_{n} is displayed in Figure 7.8. Since the element factor in this case is simply a circle, the radiation pattern of this array is simply the array factor shown. Note that the maximums are at $\phi = \pm 90^\circ$ and nulls along the $\pm x$-axis (i.e., $\phi = 0^\circ$ and 180°). This characteristic of the pattern can be explained easily via wave propagation as follows: Since the two antennas are excited in-phase (i.e., $\alpha = 0$), waves reaching in the direction of $\phi = \pm 90^\circ$ from the two antennas are added. The two cancel out along the $\pm x$-axis because of $d = 0.5\lambda$ (hence the phase delay of 180° of the wave arriving at one antenna from the other). This kind of arrangement is known as a broadside array.

Example 7.10

Find the radiation field pattern on the x-y plane of a two-element short dipole array with $\alpha = \pi/2$ and $d = 0.25\lambda$.

In this case (7.4.4) gives

$$\begin{aligned} AF_{\mathrm{n}} &= \cos\left(\frac{\psi}{2}\right) = \cos\left(\frac{2\pi d}{2\lambda}\cos\phi + \frac{\alpha}{2}\right) = \cos\left(\frac{\pi\cdot\lambda}{\lambda\cdot 4}\cos\phi + \frac{\pi}{4}\right) \\ &= \cos\left(\frac{\pi}{4}\cos\phi + \frac{\pi}{4}\right). \end{aligned}$$

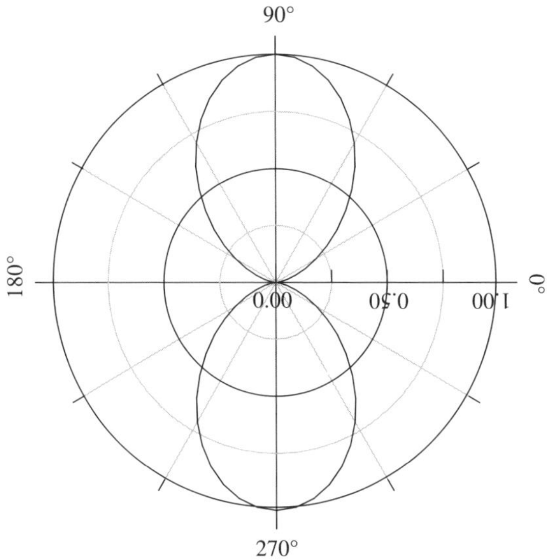

Figure 7.8 Pattern of a two-element linear array of Example 7.9 on the *x*-*y* plane.

This AF_n is displayed in Figure 7.9. As before, AF_n represents the radiation pattern of this array because the element factor is simply a circle. Note that there is only one maximum in this case occurring at $\phi = 180^\circ$ and a null along the *x*-axis (i.e., $\phi = 0^\circ$). This characteristic of the pattern can be explained easily via wave propagation as follows: Since the second antenna is excited by the $\pi/2$ phase advanced with respect to the first antenna and the separation d

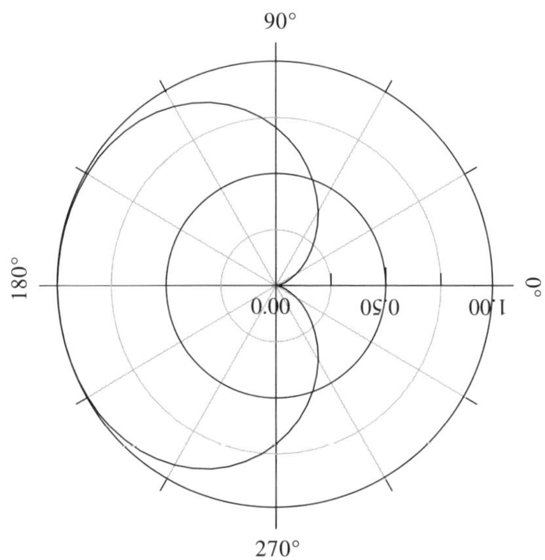

Figure 7.9 Pattern of a two-element linear array of Example 7.10 on the *x*-*y* plane.

introduces a phase delay of $\pi/2$, the waves from the two antennas cancel out each other along $\phi = 0^\circ$. However, the two waves are added up along $\phi = 180^\circ$. This kind of arrangement is known as an end-fire array.

Example 7.11

Find the radiation field pattern on the *x-y* plane of a two-element short dipole array with $\alpha = -\pi/2$ and $d = 0.25\lambda$.

In this case (7.4.4) gives

$$AF_{\mathrm{n}} = \cos\left(\frac{\psi}{2}\right) = \cos\left(\frac{kd\cos\phi + \alpha}{2}\right) = \cos\left(\frac{\pi\cdot\lambda}{\lambda\cdot 4}\cos\phi - \frac{\pi}{4}\right)$$

$$= \cos\left(\frac{\pi}{4}\cos\phi - \frac{\pi}{4}\right).$$

This AF_{n} is displayed in Figure 7.10. As before, it represents the radiation pattern of this array because the element factor is simply a circle. Note that this time there is only one maximum occurring at $\phi = 0^\circ$ and a null along the $-x$-axis (i.e., $\phi = 180^\circ$). Since the second antenna is excited by a $\pi/2$ phase delay with respect to the first antenna and the separation d introduces a phase delay of $\pi/2$, waves from the two antennas cancel out each other along $\phi = 180^\circ$. However, the two waves are added up along $\phi = 0^\circ$. This is another end-fire array arrangement.

These examples demonstrate a few significant characteristics that have a number of practical applications. The radiation beam width (or *HPBW*) of the array is much smaller than that of the individual antenna. This plays important role in many

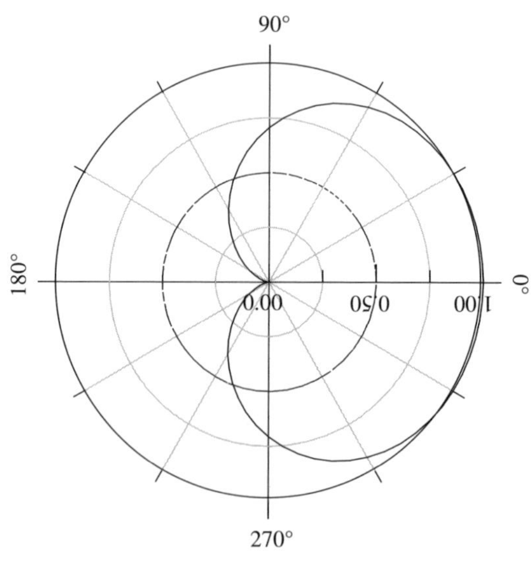

Figure 7.10 Pattern of a two-element linear array of Example 7.11 on *x-y* plane.

applications including the radar and point-to-point communication systems. Note that separation d between the elements is the same in Examples 7.10 and 7.11. However, the radiation beam peaks along $+x$ in one case and along $-x$ in the other. This occurs only because the excitation phase α changes from $-\pi/2$ to $\pi/2$. Because of this phenomenon the main beam can be moved electronically from one location to other continuously as we vary α. It is by this principle that the phased array radar works to track various targets without mechanically moving its antenna system.

Example 7.12

Find the radiation field pattern on the x-y plane of a two-element short dipole array with in-phase excitation (i.e., $\alpha = 0$) for $d = 0.5\lambda$, 0.9λ, 1λ, and 2λ.

In this case (7.4.4) gives

$$AF_{\mathrm{n}} = \cos\left(\frac{\psi}{2}\right) = \cos\left(\frac{kd\cos\phi + \alpha}{2}\right) = \cos\left(\frac{\pi \cdot d}{\lambda}\cos\phi\right).$$

This relation is graphed in Figure 7.11 for four different cases with $-\pi \leq \phi \leq \pi$. As the figure shows, there are two broadside main lobes (peaks) for $d = \lambda/2$. As the separation increases to 0.9λ, the broadside lobes become narrower. However, there is another lobe located at $\phi = 0$ with a peak strength of over 0.9. The main lobes on the broadside further narrow down as the separation between the two elements increases. However, the strength of the lobe at $\phi = 0$ not only becomes equal to that of the main lobe, but also there are two more lobes of equal strength for $d = 2\lambda$. The extra lobes with main beam strengths are referred to as grating lobes. In most applications only one main lobe with narrow beam width is desired. This example indicates that the element separation must be kept smaller than a wavelength to avoid grating lobes.

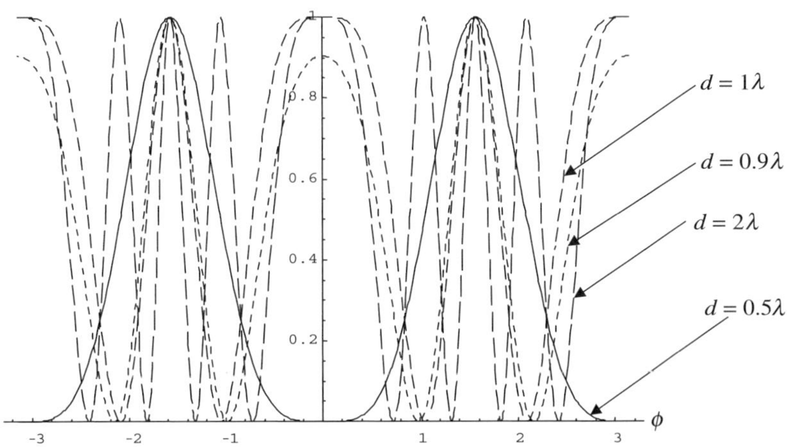

Figure 7.11 Pattern of a two-element linear array of Example 7.12 on x-y plane.

Linear Array of N Elements

Consider an array of N uniformly spaced antennas, as shown in Figure 7.12. The excitation magnitude is assumed to be same, but there is a progressive phase increase. As in the preceding analysis of a two-element array, we can find the array factor as follows:

$$AF = 1 + e^{j(kd\cos\phi+\alpha)} + e^{j2(kd\cos\phi+\alpha)} + \cdots + e^{j(N-1)(kd\cos\phi+\alpha)} = \sum_{n=1}^{N} e^{j(n-1)(kd\cos\phi+\alpha)},$$

or

$$AF = \sum_{n=1}^{N} e^{j(n-1)\psi}, \tag{7.4.5}$$

where

$$\psi = kd\cos\phi + \alpha. \tag{7.4.6}$$

Equation (7.4.5) represents a geometric series that can be added (see Appendix A, formula 46) to get

$$AF = \frac{1-e^{jN\psi}}{1-e^{j\psi}} = \frac{e^{jN\psi/2}\left(e^{-jN\psi/2}-e^{jN\psi/2}\right)}{e^{j\psi/2}(e^{-j\psi/2}-e^{j\psi/2})} = e^{j(N-1)\psi/2}\frac{\sin(N\psi/2)}{\sin(\psi/2)}. \tag{7.4.7}$$

If the reference point is moved to the midpoint, the phase shift term (first coefficient) disappears. Therefore (7.4.7) reduces to

$$AF = \frac{\sin(N\psi/2)}{\sin(\psi/2)}. \tag{7.4.8}$$

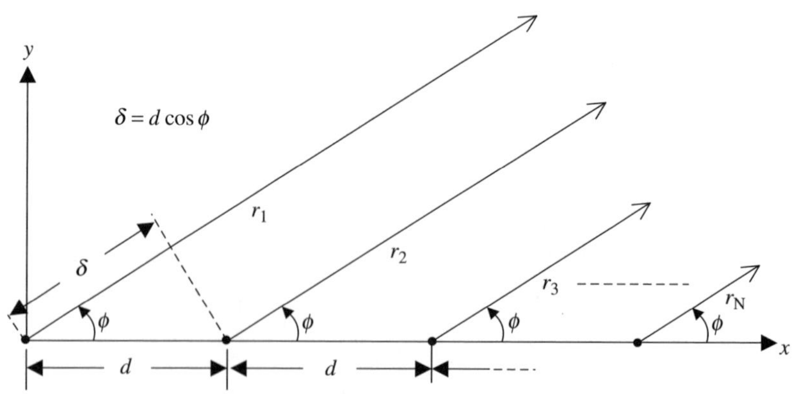

Figure 7.12 Linear array of N antennas.

The maximum value of this array factor is found to be N, which occurs when $\psi = 0$. Hence the normalized array factor is found to be

$$AF_n = \frac{\sin(N\psi/2)}{N\sin(\psi/2)}. \tag{7.4.9}$$

When the number of antenna elements N is large (typically 10 or so), the sine term in the denominator varies much more slowly around $\psi = 0$ in comparison with the numerator. Therefore (7.4.9) can be approximated around the maximum at $\psi = 0$ as follows:

$$AF_n \approx \frac{\sin(N\psi/2)}{N\psi/2}. \tag{7.4.10}$$

Note that $-1 \leq \cos\phi \leq 1$, and therefore $-kd + \alpha \leq \psi \leq kd + \alpha$. This is called the visible range of the array. As we saw in Example 7.12, grating lobes are possible if the element separation d is larger than a wavelength.

The array factor peaks whenever the signals radiated from various antennas arrive in phase (constructive interference of waves). However, it falls when the signals arriving are out of phase (destructive interference of waves). The maximums occur at

$$\sin\left(\frac{\psi}{2}\right) = 0 = \sin(\pm m\pi), \qquad m = 0, 1, 2, \ldots.$$

Therefore

$$kd\cos\phi_p + \alpha = \pm 2m\pi \rightarrow \phi_p = \cos^{-1}\left[\frac{\lambda}{2\pi d}(\pm 2m\pi - \alpha)\right]. \tag{7.4.11}$$

The peaks occur at $\phi = \phi_p$. The major maximum occurs at $m = 0$. There are smaller peaks that are called side lobes occurring whenever the numerator of AF is maximized. This happens when

$$\sin\left(\frac{N\psi}{2}\right) = \pm 1 \rightarrow \frac{N}{2}(kd\cos\phi_p + \alpha) = \pm(2p+1)\frac{\pi}{2}, \qquad p = 1, 2, 3 \ldots$$

Hence

$$\cos\phi_p = \frac{\lambda}{2\pi d}\left[\pm(2p+1)\frac{\pi}{2} - \alpha\right], \tag{7.4.12}$$

or

$$\phi_p = \cos^{-1}\left\{\frac{\lambda}{2\pi d}\left[\pm(2p+1)\frac{\pi}{2} - \alpha\right]\right\}. \tag{7.4.13}$$

Alternatively, (7.4.12) can be expressed as follows:

$$\sin\left(\frac{\pi}{2} - \phi_{\rm p}\right) = \frac{\lambda}{2\pi d}\left[\pm(2p+1)\frac{\pi}{2} - \alpha\right]. \tag{7.4.14}$$

Therefore

$$\phi_{\rm p} = \frac{\pi}{2} - \sin^{-1}\left\{\frac{\lambda}{2\pi d}\left[\pm(2p+1)\frac{\pi}{2} - \alpha\right]\right\}. \tag{7.4.15}$$

Note that the maximum of the first minor lobe occurs at $p = 1$. Therefore

$$\frac{N}{2}(kd\cos\phi_{\rm p1} + \alpha) \approx \pm\frac{3\pi}{2} \rightarrow kd\cos\phi_{\rm p1} + \alpha = \pm\frac{3\pi}{N}. \tag{7.4.16}$$

Further the peak magnitude of the first side lobe is found to be

$$AF_{\rm n}\Big|_{\rm p1} \approx \frac{\sin(N\psi/2)}{N\psi/2}\Bigg|_{(N\psi/2)=\pm 3\pi/2} = \frac{2}{3\pi} = 0.2122. \tag{7.4.17}$$

Since the maximum value of the normalized AF is unity, the first side lobe is fairly small. It can be expressed in dB as follows:

$$AF_{\rm n}\Big|_{\rm P1(dB)} = 20\log\left(\frac{2}{3\pi}\right) = -13.4648\,\text{dB}. \tag{7.4.18}$$

The array factor is zero at

$$\sin\left(\frac{N\psi}{2}\right) = 0 = \sin(\pm n\pi), \qquad n = 1, 2, 3, \ldots, \text{but } n \neq N, 2N, 3N, \ldots.$$

Therefore

$$kd\cos\phi + \alpha|_{\phi=\phi_z} = \pm\frac{2n\pi}{N} \rightarrow \phi_z = \cos^{-1}\left[\frac{\lambda}{2\pi d}\left(\pm\frac{2n\pi}{N} - \alpha\right)\right]. \tag{7.4.19}$$

The principal lobe maximum occurs at $m = 0$. Hence

$$\phi_{\max} = \cos^{-1}\left(\frac{\lambda\alpha}{2\pi d}\right). \tag{7.4.20}$$

The *HPBW* (3 dB beam width) of this array can be found if we can determine $\phi = \phi_{\rm h}$ when peak power reduces by one-half. Therefore

$$\left(\frac{\sin(N\psi/2)}{N\psi/2}\right)^2\Bigg|_{\phi=\phi_{\rm h}} = \frac{1}{2} \rightarrow \sin\left(\frac{N}{2}kd\cos\phi_{\rm h} + \alpha\right) = \frac{1}{\sqrt{2}}\frac{N}{2}kd\cos\phi_{\rm h} + \alpha. \tag{7.4.21}$$

Equation (7.4.21) can be solved numerically using software (or a graphing calculator) to find

$$\frac{N}{2}(kd\cos\phi_{\rm h}+\alpha)=\pm 1.3916 \rightarrow \cos\phi_{\rm h}=\frac{\lambda}{2\pi d}\left(\pm\frac{2.7832}{N}-\alpha\right)=\sin\left(\frac{\pi}{2}-\phi_{\rm h}\right).$$

Therefore

$$\phi_{\rm h}=\cos^{-1}\left[\frac{\lambda}{2\pi d}\left(\pm\frac{2.7832}{N}-\alpha\right)\right]=\frac{\pi}{2}-\sin^{-1}\left[\frac{\lambda}{2\pi d}\left(\pm\frac{2.7832}{N}-\alpha\right)\right] \tag{7.4.22}$$

and

$$HPBW=2\left|\phi_{\max}-\phi_{\rm h}\right|. \tag{7.4.23}$$

As we saw earlier with the two-element array, there are two special cases of particular interest: broadside and end-fire arrays.

Broadside Arrays

For $\alpha = 0$, (7.4.11) gives

$$\psi = kd\cos\phi_{\rm p} = 0 \rightarrow \cos\phi_{\rm p} = 0 \rightarrow \phi_{\rm p} = \pm\frac{\pi}{2}. \tag{7.4.24}$$

Therefore, if the antennas are excited in phase, the main beam is on the broadside for all N and d. However, there will be grating lobes in the visible range if the separation between the elements is larger than a wavelength.

End-Fire Arrays

For $\alpha = -kd$, (7.4.11) gives

$$\psi = kd\cos\phi_{\rm p} - kd = 0 \rightarrow kd(\cos\phi_{\rm p} - 1) = 0 \rightarrow \cos\phi_{\rm p} = 1 \rightarrow \phi_{\rm p} = 0. \tag{7.4.25}$$

Similarly, if $\alpha = kd$, then from (7.4.11) we have

$$\psi = kd\cos\phi_{\rm p} + kd = 0 \rightarrow kd(\cos\phi_{\rm p} + 1) = 0 \rightarrow \cos\phi_{\rm p} = -1 \rightarrow \phi_{\rm p} = \pi. \tag{7.4.26}$$

These two cases give the end-fire arrays with two elements considered in Examples 7.10 and 7.11.

Example 7.13

Find the phase of excitation for an 8-element array that has its main beam at 60° from the array's axis. Let the separation between the elements be 0.3183λ.

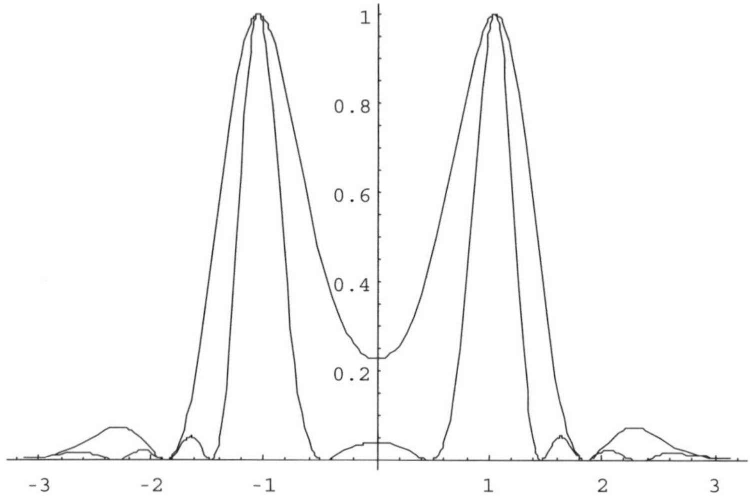

Figure 7.13 Radiation patterns of the array in Example 7.13.

How will this change if only four elements are used instead? Plot the array patterns for both cases.

Assume that the array is along the x-axis. Therefore, for the main beam at $\phi_{\text{p}} = 60^\circ$,

$$\psi = kd \cos \phi_{\text{p}} + \alpha = 0 \rightarrow \alpha = -\frac{2\pi \cdot 0.3183\lambda}{\lambda} \cos(60^\circ) = -1.$$

The normalized array factor for 8 elements is found to be

$$AF_{\text{n}} = \frac{\sin(N\psi/2)}{N \sin(\psi/2)} = \frac{\sin(8 \cos \phi - 4)}{8 \sin(\cos \phi - 0.5)}.$$

When there are only 4 elements in the array, the array factor changes to

$$AF_{\text{n}} = \frac{\sin(N\psi/2)}{N \sin(\psi/2)} = \frac{\sin(4 \cos \phi - 2)}{4 \sin(\cos \phi - 0.5)}.$$

Figure 7.13 shows the array patterns for both cases.

7.5 FRIIS TRANSMISSION FORMULA AND THE RADAR RANGE EQUATION

The analysis and design of communication and monitoring systems often require an estimation of transmitted and received powers. The Friis transmission formula and the radar range equation provide the means for such calculations. The former is applicable to a one-way communication system where the signal transmitted at one end is received at the other end of the link. In the case of the radar range equation the transmitted signal hits a target, and the reflected signal is generally received at the location of transmission. We consider these two formulations in this section.

Friis Transmission Equation

Consider the simplified communication link illustrated in Figure 7.14. A distance R separates the transmitter and the receiver. The effective apertures of the transmitting and receiving antennas are A_{et} and A_{er}, respectively. Further the two antennas are assumed to be polarization matched.

If the power input to the transmitting antenna is P_t, then isotropic power density w_o at a distance R from the antenna is given as follows:

$$w_o = \frac{P_t e_t}{4\pi R^2}, \tag{7.5.1}$$

where e_t is the radiation efficiency of the transmitting antenna. For a directional transmitting antenna, the power density w_t is found to be

$$w_t = \frac{P_t G_t}{4\pi R^2} = \frac{P_t e_t D_t}{4\pi R^2}, \tag{7.5.2}$$

where G_t is the gain and D_t is the directivity of transmitting antenna. The power collected by the receiving antenna is

$$P_r = A_{er} w_t. \tag{7.5.3}$$

From (7.2.6),

$$A_{er} = \frac{\lambda^2}{4\pi} G_r, \tag{7.5.4}$$

where the receiving antenna gain is G_r. Therefore from (7.5.2) through (7.5.4) we have

$$P_r = \frac{\lambda^2}{4\pi} G_r w_t = \frac{\lambda^2}{4\pi} G_r \frac{P_t G_t}{4\pi R^2}$$

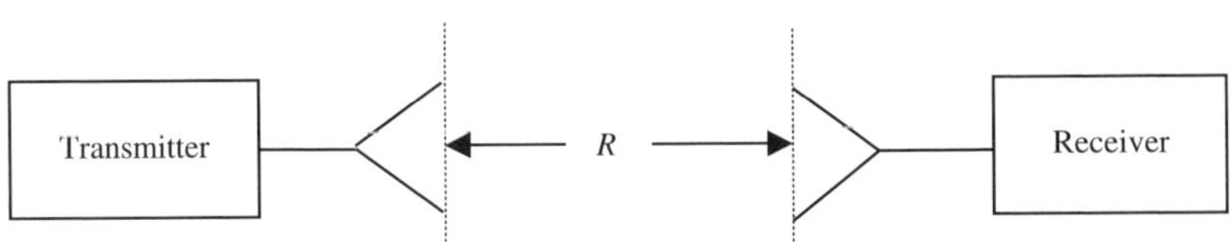

Figure 7.14 Simplified block diagram of a communication link.

or

$$\frac{P_r}{P_t} = \left(\frac{\lambda}{4\pi R}\right)^2 G_r G_t = e_r e_t \left(\frac{\lambda}{4\pi R}\right)^2 D_r D_t. \qquad (7.5.5)$$

If the signal frequency is f, then for a link in free space we find that

$$\frac{\lambda}{4\pi R} = \frac{3 \cdot 10^8}{4\pi f R},$$

where f is in Hz and R is in meters.

Generally, the link distance is long and the signal frequency is high such that kilometer and megahertz are more convenient units than the usual meter and hertz, respectively. For R in km and f in MHz, we find that

$$\frac{\lambda}{4\pi R} = \frac{3 \cdot 10^8}{4\pi \cdot 10^6 \cdot f_{\text{MHz}} \cdot 10^3 \cdot R_{\text{km}}} = \frac{0.3}{4\pi} \cdot \frac{1}{f_{\text{MHz}} \cdot R_{\text{km}}}.$$

Hence (7.5.5) may be expressed as follows:

$$P_r(\text{dBm}) = P_t(\text{dBm}) + 20\log_{10}\left(\frac{0.3}{4\pi}\right) - 20\log_{10}(f_{\text{MHz}} R_{\text{km}}) + G_t(\text{dB}) + G_r(\text{dB}),$$

or

$$P_r(\text{dBm}) = P_t(\text{dBm}) + G_t(\text{dB}) + G_r(\text{dB}) - 20\log_{10}(f_{\text{MHZ}} R_{\text{km}}) - 32.4418, \quad (7.5.6)$$

where the transmitted and received powers are in dBm while the antenna gains are in dB.

Example 7.14

A 20 GHz transmitter on board a satellite uses a parabolic antenna that is 45.7 cm in diameter. The antenna gain is 37 dB and its radiated power is 2 W. The ground station, which is 36941.031 km away from the transmitter, has an antenna gain of 45.8 dB. Find the power collected by the ground station. How much power will be collected at the ground-station if there are isotropic antennas on both sides?

The transmitted power P_t (dBm) = 10 $\log_{10}$ (2000) = 33.0103 dBm, and

$$20\log_{10}(f_{\text{MHZ}} R_{\text{km}}) = 20\log_{10}(20 \cdot 10^3 \cdot 36941.031) = 177.3708\,\text{dB}.$$

Hence the power received at the earth station is found as follows:

$$P_r(\text{dBm}) = 33.0103 + 37 + 45.8 - 177.3708 - 32.4418 = -94.0023\,\text{dBm},$$

or

$$P_r = 3.979 \cdot 10^{-10}\,\text{mW}.$$

If the two antennas are isotropic, $G_t = G_r = 1$ (or 0 dB), and therefore

$$P_r(\text{dBm}) = 33.0103 + 0 + 0 - 177.3708 - 32.4418 = -176.8023\,\text{dBm},$$

or

$$P_r = 2.0882 \cdot 10^{-18}\,\text{mW}.$$

Radar Equation

In a radar system the transmitted signal is scattered by a target in all possible directions. The receiving antenna collects part of the energy that is scattered back toward it. Generally, a single antenna is employed for both the transmitter and the receiver, as shown in Figure 7.15.

If the power input to a transmitting antenna is P_t and its gain is G_t, then the power density w_{inc} incident on the target is

$$w_{inc} = \frac{P_t G_t}{4\pi R^2} = \frac{P_t A_{et}}{\lambda^2 R^2}, \tag{7.5.7}$$

where A_{et} is the effective aperture of the transmitting antenna.

The radar cross section σ of an object is defined as the area intercepting that amount of power that, when scattered isotropically, produces at the receiver a power density that is equal to that scattered by the actual target. Hence

$$\text{Radar cross section} = \frac{\text{Scattered power}}{\text{Incident power density}}\,\text{m}^2,$$

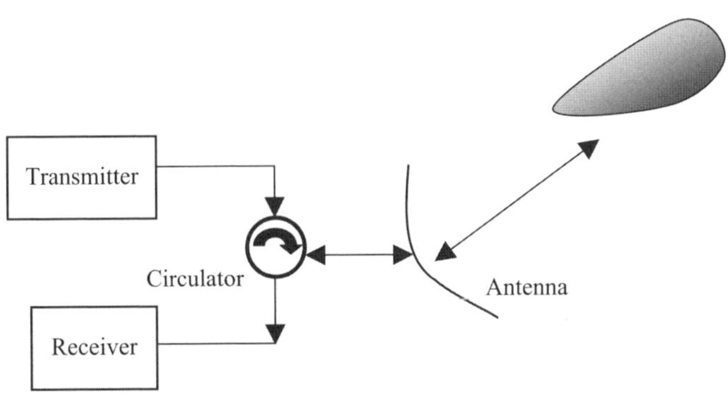

Figure 7.15 Simplified block diagram of a radar system.

TABLE 7.1 Radar Cross Sections of Selected Objects

Object	Radar Cross Section (m^2)
Pickup truck	200
Automobile	100
Jumbo-jet airliner	100
Large bomber	40
Large fighter aircraft	6
Small fighter aircraft	2
Adult male	1
Conventional winged missile	0.5
Bird	0.01
Insect	0.00001
Advanced tactical fighter	0.000001

or

$$\sigma = \frac{4\pi R^2 w_r}{w_{inc}}, \tag{7.5.8}$$

where w_r is the isotropically backscattered power density at distance R and w_{inc} is the power density incident on the object. Thus the radar cross section of an object is its effective area, which intercepts an incident power density w_{inc} and gives an isotropically scattered power of $4\pi R^2 w_r$ for a backscattered power density. Radar cross sections of selected objects are listed in Table 7.1.

From the radar cross section of a target the power intercepted by the object can be found as follows:

$$P_{inc} = \sigma w_{inc} = \frac{\sigma P_t G_t}{4\pi R^2}. \tag{7.5.9}$$

The power density arriving back at the receiver is

$$w_{scatter} = \frac{P_{inc}}{4\pi R^2}, \tag{7.5.10}$$

and the power available at the receiver input is

$$P_r = A_{er} w_{scatter} = \frac{G_r \lambda^2 \sigma P_t G_t}{4\pi (4\pi R^2)^2} = \frac{\sigma A_{er} A_{et} P_t}{4\pi \lambda^2 R^4}. \tag{7.5.11}$$

Example 7.15

As shown in Figure 7.16 a distance of 100 λ separates two lossless X-band horn antennas. The reflection coefficients at the terminals of the transmitting and receiving antennas are 0.1 and 0.2, respectively. The maximum directivities of the transmitting and receiving antennas are 16 and 20 dB, respectively. Assume that the input power in the lossless transmission line connected to the

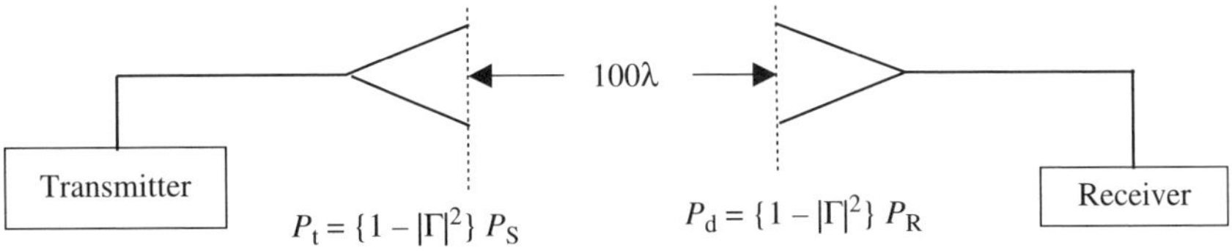

Figure 7.16 X-band communication link for Example 7.15.

transmitting antenna is 2 W, and that the two antennas are aligned for maximum radiation between them and are polarization matched. Find the power input to the receiver.

As discussed in Chapter 5, an impedance discontinuity generates an echo signal very similar to that of an acoustical echo. Hence the signal power available beyond the discontinuity is reduced. The ratio of the reflected signal voltage to that of the incident is called the *reflection coefficient*. Since power is proportional to the square of voltage, the power reflected from the discontinuity is equal to the square of the reflection coefficient times the incident power. Therefore the power transmitted in the forward direction will be given by

$$P_t = [1 - |\Gamma|^2]P_{in}.$$

The power radiated by the transmitting antenna is then found to be

$$P_t = \{1 - 0.1^2\}2 = 1.98\,\text{W}.$$

Since the Friis transmission equation requires antenna gain as a ratio instead of in dB, G_t and G_r are calculated as follows:

$$G_t = 16\,\text{dB} = 10^{1.6} = 39.8107,$$

$$G_r = 20\,\text{dB} = 10^{2.0} = 100.$$

Hence from (7.5.5) we get

$$P_r = \left(\frac{\lambda}{4\pi \cdot 100\lambda}\right)^2 \cdot 100 \cdot 39.8107 \cdot 1.98,$$

or

$$P_r = 5\,\text{mW},$$

and the power delivered to the receiver, P_d, is

$$P_d = (1 - 0.2^2)5 = 4.8\,\text{mW}.$$

Example 7.16

Radar operating at 12 GHz transmits 25 kW through an antenna of a 25 dB gain. A target with its radar cross section at 8 m^2 is located at 10 km from the radar. If the same antenna is used for the receiver, determine the received power.

$$
\begin{aligned}
P_t &= 25\,\text{kW},\\
f &= 12\text{ GHz} \rightarrow \lambda = \frac{3\cdot 10^8}{12\cdot 10^9} = 0.025\text{ m},\\
G_r &= G_t = 25\,\text{dB} \rightarrow 10^{2.5} = 316.2278,\\
R &= 10\,\text{km},\\
\sigma &= 8\,\text{m}^2.
\end{aligned}
$$

Hence

$$P_r = \frac{G_r G_t P_t \sigma \lambda^2}{4\pi(4\pi R^2)^2} = \frac{316.2278^2 \cdot 25000 \cdot 8 \cdot 0.025^2}{(4\pi)^3 \cdot (10^4)^4} = 6.3 \cdot 10^{-13}\text{ W},$$

or

$$P_r = 0.63\text{ pW}.$$

Doppler Radar

An electrical signal propagating in free space can be represented by the simple expression

$$v(z, t) = A\cos(\omega t - kz). \tag{7.5.12}$$

The signal frequency is ω radians per second, and k is its wavenumber (equal to ω/c, where c is speed of light in free space) in radians per meter. Assume that there is a receiver located at $z = R$, as shown in Figure 7.14, and R is changing with time (the receiver may be moving toward or away from the transmitter). In this situation the receiver response v_o(t) is given as follows:

$$v_o(t) = V\cos(\omega t - kR). \tag{7.5.13}$$

The angular frequency ω_o, of v_o(t) can be easily determined after differentiating the argument of the cosine function with respect to time. Hence

$$\omega_o = \frac{d}{dt}(\omega t - kR) = \omega - k\frac{dR}{dt}. \tag{7.5.14}$$

Note that k is time independent, and the time derivative of R represents the velocity, v_r, of the receiver with respect to the transmitter. Hence (7.5.14) can be written as follows:

$$\omega_o = \omega - \frac{\omega v_r}{c} = \omega\left(1 - \frac{v_r}{c}\right). \tag{7.5.15}$$

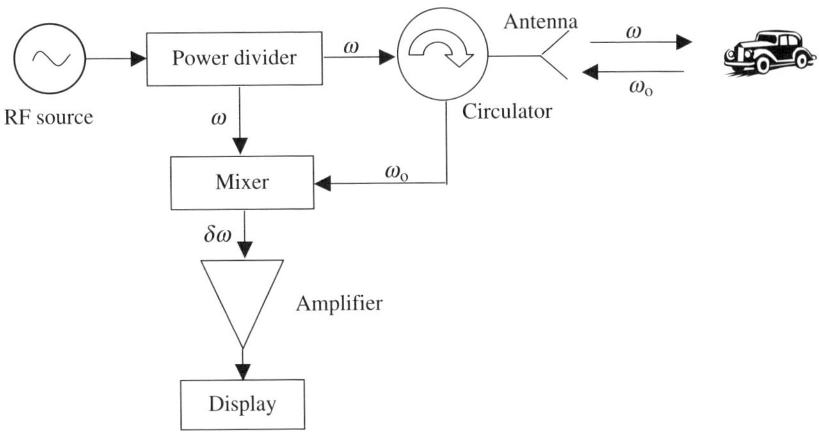

Figure 7.17 Simplified block diagram of a Doppler radar.

If the receiver is closing in, then v_r will be negative (negative slope of R), and therefore the received signal will indicate a signal frequency higher than ω. On the other hand, it will show a lower frequency if R is increasing with time. It is the Doppler frequency shift that is employed to design the Doppler radar.

Consider the simplified block diagram of the radar illustrated in Figure 7.17. The microwave signal generated by the oscillator is split into two parts via the power divider. The circulator feeds one part of this power to the antenna that illuminates the target while the mixer uses the remaining fraction as its reference signal. Further the antenna intercepts a part of the signal that is scattered by the object. It is then directed to the mixer through the circulator. The output of the mixer includes a difference frequency signal that can be filtered out for further processing. The two inputs to the mixer will have the same frequency if the target is stationary, and therefore the Doppler shift $\delta\omega$ will be zero. The mixer's output will have Doppler frequency if the target is moving. Note that the signal travels twice over the same distance, and therefore the Doppler frequency shift in this case will be twice that found via (7.5.15). Mathematically

$$\omega_o = \omega\left(1 - \frac{2v_r}{c}\right) \tag{7.5.16}$$

and

$$\delta\omega = \frac{2\omega v_r}{c}. \tag{7.5.17}$$

PROBLEMS

7.1. A 2 cm long Hertzian dipole antenna is located at the origin of the spherical coordinates. It carries a current of 2 A at 10 MHz. Find the power density per unit area that leaves a spherical surface of 1 km radius. Plot its normalized characteristics (divide by the maximum power per unit area) on $\phi = 90^\circ$ and $\theta = 90^\circ$ planes.

7.2. Calculate the power gain (in dBi) and the half-power beam width of a parabolic dish antenna of 35 m in diameter that is radiating at 5.9 GHz.

7.3. Find the radiation resistance R_{rad} of a 1 cm long current element operating at 10 MHz in free space.

7.4. A 50 Ω transmission line feeds a lossless dipole antenna. If the antenna impedance is 76 Ω and its radiation intensity, $U(\theta, \phi)$, is given as follows, find the maximum overall gain:

$$U(\theta,\phi) = B_o \sin^2\theta \cos\theta.$$

7.5. A uniform plane electromagnetic wave propagating in the z direction is incident upon an antenna that is polarized as follows:

$$\vec{E}_a(\vec{r}) = \hat{y}E(\vec{r}).$$

The electric field intensity of the incoming electromagnetic wave is

$$\vec{E}(\vec{r},t) = (\hat{x} + \hat{y})E_o \cos(\omega t - kz) \text{ V/m}.$$

Find the polarization loss factor.

7.6. A transmitter has 100 kW power at the output of its final high-power amplifier. If the gain of the transmitting antenna is 15 dB and its feeding line has an attenuation of 5%, find the EIRP in dBW.

7.7. A geostationary satellite is 35,860 km away from the earth's surface. Find the space-loss ratio if the satellite is operating at 26 GHz.

7.8. Find the radiation field pattern on the x-y plane of a two-element short dipole array with in-phase excitation (i.e., $\alpha = 0$) and $d = 0.35\lambda$.

7.9. Find the radiation field pattern on the x-y plane of a two-element short dipole array with $\alpha = \pi/2$ and $d = 0.5\lambda$.

7.10. Find the radiation field pattern on the x-y plane of a two-element short dipole array with $\alpha = -\pi/2$ and $d = 0.5\lambda$.

7.11. Find the radiation field pattern on the x-y plane of a two-element short dipole array with $\alpha = \pi/2$ for $d = 0.9\lambda$, 1λ, and 2λ.

7.12. Find the phase of excitation for a 10 element array that has its main beam at 45° from the array's axis. Assume that the separation between the elements is 0.3183λ. How will this change if only four elements are used instead? Plot the array patterns of both cases.

7.13. A 30 GHz transmitter on board the satellite uses a parabolic antenna. Its gain is 45 dB and the radiated power is 2 W. The ground station that is 36941.031 km away from it has an antenna gain of 50 dB. Find the power collected by the ground station. How much power will be collected at the ground-station if there are isotropic antennas on both sides?

7.14. A distance of 150 λ separates two lossless X-band horn antennas. The reflection coefficients at the terminals of the transmitting and receiving antennas are 0.05 and 0.15, respectively. The maximum directivities of the transmitting and receiving antennas are 18 dB and 22 dB, respectively. Assuming that the input power in a lossless transmission line connected to the transmitting antenna is 2.5 W, and that the two antennas are aligned for maximum radiation between them and are polarization matched, find the power input to the receiver.

7.15. A radar operating at 26 GHz transmits 5 kW through an antenna of a 45 dB gain. A target with its radar cross section at 4 m^2 is located 10 km from the radar. If the same antenna is used for the receiver, determine the received power.

7.16. An AM broadcast station is to be located east of the area it is to serve. Design an antenna for this station that gives a broad coverage to the west (from SW through W to NW) with reduced field intensity in the other directions. However, to obtain FCC approval, the pattern must have a null NE at 120° from W. The antenna is to consist of an in line array of $\lambda/4$ vertical elements oriented along an east–west line. A minimum number of elements must be used.

8

ELECTROSTATIC FIELDS

When the electrical charge and current sources remain constant over time, the associated fields also do not change over time. As (3.2.1) and (3.2.2) indicate, the electric and magnetic fields become independent in this situation. In other words, the electrical charge that is constant over time is responsible for the electric fields, and the electrical current that is constant over time is responsible for the magnetic fields. These are known respectively, as electrostatic and magnetostatic fields. This chapter begins with the various laws used to analyze electrostatic fields. Poisson's and Laplace's equations are solved for a few boundary value problems. Electrical capacitors are considered, and the energy storage in an electrical field is formulated. A number of practical applications are included throughout the chapter.

8.1 LAWS OF ELECTROSTATIC FIELDS

The electric scalar potential ϕ_e due to a volume charge density ρ_v in unbounded space is found in Appendix B (B.35) as follows:

$$\phi_e = \int_V \frac{\rho_v e^{-jkR}}{4\pi\varepsilon R} dv', \tag{8.1.1}$$

Practical Electromagnetics: From Biomedical Sciences to Wireless Communication.
By Devendra K. Misra
Copyright © 2007 John Wiley & Sons, Inc.

where

$$R = \left|\vec{r} - \vec{r}'\right|. \tag{8.1.2}$$

For an electrical charge that is constant over time, $\omega \to 0$ and $k \to 0$. Therefore (8.1.1) reduces to

$$\phi_{\mathrm{e}} = \int_V \frac{\rho_{\mathrm{v}}}{4\pi\varepsilon R} dv'. \tag{8.1.3}$$

If there is a point charge of Q C located at the origin $R = \left|\vec{r}\right| = r$, then (8.1.3) gives

$$\phi_{\mathrm{e}} = \frac{Q}{4\pi\varepsilon r}. \tag{8.1.4}$$

The corresponding electric field intensity can be found from (6.2.16) as follows:

$$\vec{E} = -\nabla\phi_{\mathrm{e}} = -\nabla\int_V \frac{\rho_{\mathrm{v}}}{4\pi\varepsilon R} dv' = -\frac{1}{4\pi\varepsilon}\int_V \rho_{\mathrm{v}}\nabla\left(\frac{1}{R}\right) dv'. \tag{8.1.5}$$

From Example 2.20 we know that

$$\nabla\left(\frac{1}{R}\right) = -\frac{\vec{R}}{R^3} = -\frac{\hat{R}}{R^2}.$$

Therefore (8.1.5) simplifies to

$$\vec{E} = \frac{1}{4\pi\varepsilon}\int_V \rho_{\mathrm{v}}\frac{\hat{R}}{R^2} dv'. \tag{8.1.6}$$

The Lorentz force on a test point charge Q_{t} at $\vec{r}$ can found via (3.5.1) as follows:

$$\vec{F} = Q_{\mathrm{t}}\vec{E} = \frac{Q_{\mathrm{t}}}{4\pi\varepsilon}\int_V \rho_{\mathrm{v}}\frac{\hat{R}}{R^2} dv'. \tag{8.1.7}$$

If the source of electric field is a point charge Q located at $\vec{r}'$, the force of the test point charge Q_{t} at $\vec{r}$ reduces to

$$\vec{F} = \hat{R}\frac{QQ_{\mathrm{t}}}{4\pi\varepsilon R^2}. \tag{8.1.8}$$

This relation is known as Coulomb's law. Here it is assumed that the test charge Q_{t} is small enough such that it does not disturb the electric field produced by Q.

Example 8.1

A thin circular ring of radius a carries an electrical charge with density ρ_ℓ C/m. Find the electrical potential and the electric field intensity at a point (0, 0, z) on its axis.

The field and source points in this case are

$$\vec{r} = z\hat{z}$$

and

$$\vec{r}' = a\hat{\rho}.$$

Therefore

$$|\vec{R}| = |\vec{r} - \vec{r}'| = \sqrt{a^2 + z^2}.$$

As shown in Figure 8.1, incremental length along the ring is given by $d\ell = a d\phi'$. Hence the potential at a point on the axis can be found from (8.1.3) as follows:

$$\phi_e(0, 0, z) = \int_c \frac{\rho_\ell d\ell}{4\pi\varepsilon|\vec{R}|} = \int_0^{2\pi} \frac{\rho_\ell a d\phi'}{4\pi\varepsilon\sqrt{a^2 + z^2}} = \frac{\rho_\ell a}{2\varepsilon\sqrt{a^2 + z^2}} \text{ V}.$$

The electric field intensity is found to be

$$\vec{E}(0, 0, z) = -\nabla\phi_e = -\hat{z}\frac{d}{dz}\left[\frac{\rho_\ell a}{2\varepsilon\sqrt{a^2 + z^2}}\right] = \hat{z}\frac{\rho_\ell a}{2\varepsilon}\left[\frac{z}{(a^2 + z^2)^{3/2}}\right] \text{ V/m}.$$

Example 8.2

A point charge Q of 10 μC is located at point (2, 3, 4). Find the electric field intensity at (−2, −3, −4) and the force on a charge q of −10 nC at this point.

Position vectors for the field and the source points are

$$\vec{r} = -2\hat{x} - 3\hat{y} - 4\hat{z}$$

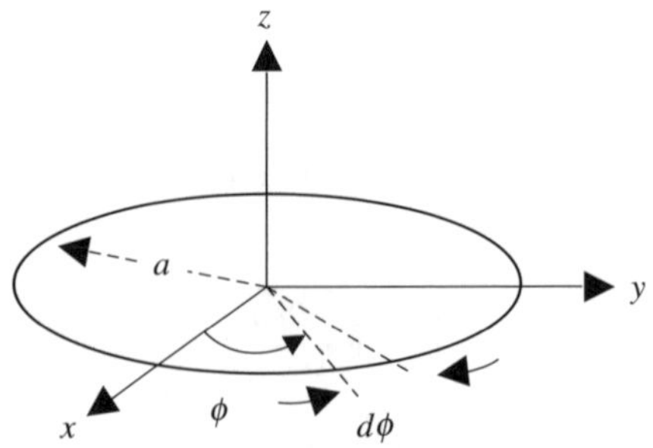

Figure 8.1 Thin circular ring with uniform charge.

and

$$\vec{r}' = 2\hat{x} + 3\hat{y} + 4\hat{z}.$$

Therefore

$$\vec{R} = \vec{r} - \vec{r}' = -4\hat{x} - 6\hat{y} - 8\hat{z}$$

and

$$|\vec{R}|^2 = 4^2 + 6^2 + 8^2 = 116 \rightarrow |\vec{R}| = \sqrt{116} = 10.7703.$$

The unit vector $\hat{R}$ is found to be

$$\hat{R} = \frac{\vec{R}}{|\vec{R}|} = \frac{-4\hat{x} - 6\hat{y} - 8\hat{z}}{10.7703} = -0.3714\hat{x} - 0.5571\hat{y} - 0.7428\hat{z}.$$

Therefore the electric field intensity due to Q is

$$\vec{E} = \frac{Q}{4\pi\varepsilon|\vec{R}|^2}\hat{R} = \frac{10^{-5}}{4\pi\varepsilon \cdot 116}(-0.3714\hat{x} - 0.5571\hat{y} - 0.7128\hat{z})\,\text{V/m}.$$

The force on the test charge q located at $\vec{R}$ is found to be

$$\vec{F} = q\vec{E} = \frac{qQ}{4\pi\varepsilon|\vec{R}|^2}\hat{R} = \frac{(-10^{-8}) \cdot 10^{-5}}{4\pi\varepsilon \cdot 116}(-0.3714\hat{x} - 0.5571\hat{y} - 0.7128\hat{z})\,\text{N}.$$

Example 8.3

A uniformly charged thin conductor is lying along the z-axis, as shown in Figure 8.2. If the conductor is infinitely long and carries a straight-line charge density of ρ_ℓ C/m, find its electric field intensity.

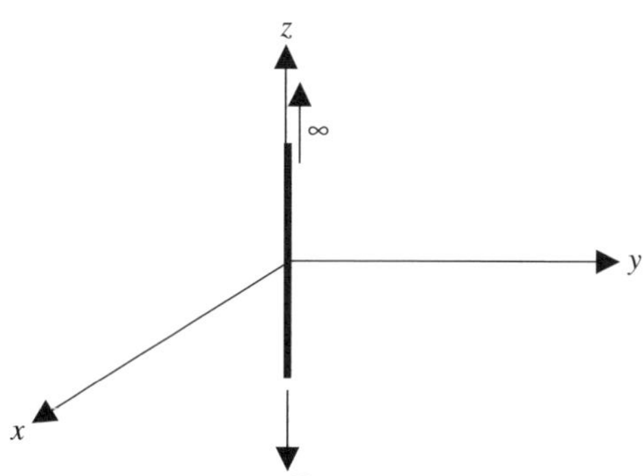

Figure 8.2 Geometry of the infinitely long straight-line charge.

For this geometry the position vectors in the field and the source points are found to be

$$\vec{r} = \hat{\rho}\rho$$

and

$$\vec{r}' = \hat{z}z'.$$

Therefore

$$\vec{R} = \vec{r} - \vec{r}' = \hat{\rho}\rho - \hat{z}z'.$$

Using (8.1.7) we get

$$\vec{E} = \frac{1}{4\pi\varepsilon}\int_V \rho_v \frac{\hat{R}}{|\vec{R}|^2}\,dv' = \frac{1}{4\pi\varepsilon}\int_{-\infty}^{\infty} \frac{\rho_\ell[\hat{\rho}\rho - \hat{z}z']}{[\rho^2 + z'^2]^{3/2}}\,dz' = \hat{\rho}E_\rho + \hat{z}E_z.$$

Since E_z is an odd function of z', it goes to zero on integration. However, the integral associated with E_ρ is an even function of z' that can be evaluated using the integration formula 4 of Appendix A to get

$$E_\rho = \frac{\rho_\ell \rho}{4\pi\varepsilon}\, 2\int_0^{\infty} \frac{1}{[\rho^2 + z'^2]^{3/2}}\,dz' = \frac{\rho_\ell}{2\pi\varepsilon\rho}\ \text{V/m}.$$

Example 8.4

A uniformly charged disk of radius a lies on the x-y plane, as shown in Figure 8.3. If the surface charge density on the disk is ρ_s C/m^2, find scalar electrical potential and the electric field intensity on its axis.

Since $\vec{r} = z\hat{z}$ and $\vec{r}' = \hat{\rho}\rho'$,

$$|\vec{R}| = |\vec{r} - \vec{r}'| = \sqrt{\rho'^2 + z^2},$$

and $ds' = \rho' d\rho' d\phi'$.

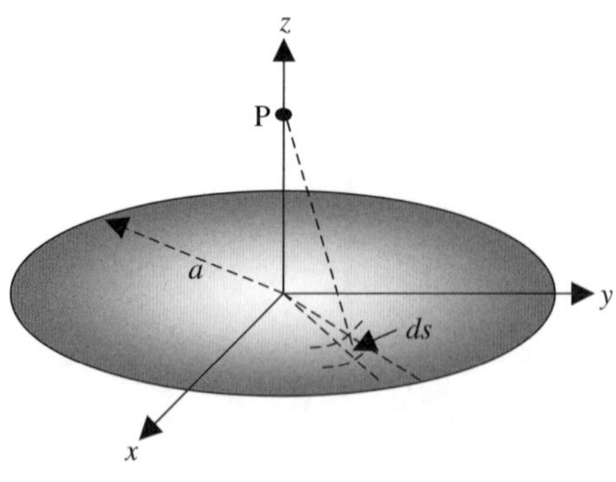

Figure 8.3 Geometry of the uniformly charged disk.

Therefore

$$\phi_e(0, 0, z) = \frac{\rho_s}{4\pi\varepsilon_o}\int_0^{2\pi}\int_{\rho'=0}^{a}\frac{\rho' d\rho' d\phi'}{(\rho'^2 + z^2)^{1/2}} = \frac{\rho_s}{2\varepsilon_o}\int_0^{a}\frac{\rho' d\rho'}{(\rho'^2 + z^2)^{1/2}}.$$

The last integral can be evaluated easily via the integration formula 12 from Appendix A to get

$$\phi_e(0, 0, z) = \frac{\rho_s}{2\varepsilon_o}\int_0^{a}\frac{\rho' d\rho'}{(\rho'^2 + z^2)^{1/2}} = \frac{\rho_s}{2\varepsilon_o}\left[(\rho'^2 + z^2)^{1/2}\right]_{\rho'=0}^{a}$$

$$= \frac{\rho_s}{2\varepsilon_o}\begin{cases}\sqrt{a^2+z^2} - z, & z \geq 0,\\ \sqrt{a^2+z^2} + z, & z \leq 0.\end{cases}$$

Note that the sign in front of z is selected such that the potential is zero at $\pm\infty$. Now the electric field intensity can be found as follows:

$$\vec{E}(0, 0, z) = -\nabla\phi_e(0, 0, z) = -\hat{z}\frac{d}{dz}\phi_e = -\frac{\rho_s}{2\varepsilon_o}\begin{cases}\dfrac{z}{\sqrt{a^2+z^2}} - 1, & z \geq 0,\\[2ex] \dfrac{z}{\sqrt{a^2+z^2}} + 1, & z \leq 0.\end{cases}$$

Therefore

$$\vec{E}(0, 0, z) = \frac{\rho_s}{2\varepsilon_o}\begin{cases}1 - \dfrac{z}{\sqrt{a^2+z^2}}, & z \geq 0,\\[2ex] 1 + \dfrac{z}{\sqrt{a^2+z^2}}, & z \leq 0.\end{cases}$$

Example 8.5

Two equal and opposite point charges $+q$ and $-q$ are separated by a small distance d, as shown in Figure 8.4. This kind of arrangement is known as the

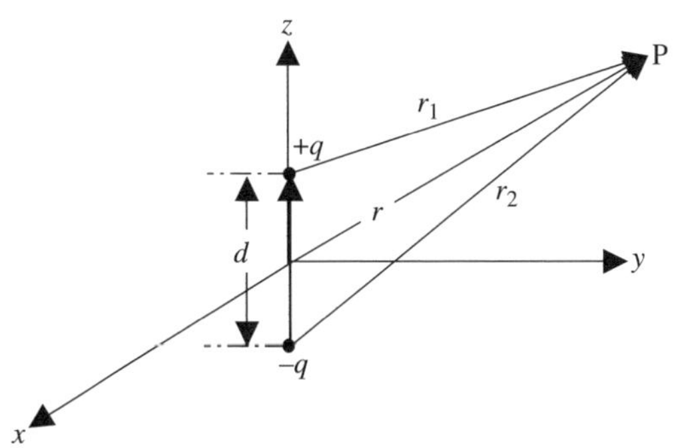

Figure 8.4 Geometry of the electric dipole charge.

electric dipole. Find scalar potential and electric field intensity at an arbitrary far point P ($r \gg d$).

$$\phi_e = \frac{q}{4\pi\varepsilon r_1} - \frac{q}{4\pi\varepsilon r_2} = \frac{q}{4\pi\varepsilon}\frac{(r_2 - r_1)}{r_1 r_2}.$$

For $\theta_1 \approx \theta_2 \approx \theta$, $r_1 r_2 \approx r^2$, and $(r_2 - r_1) \approx d\cos\theta$,

$$\phi_e \approx \frac{q}{4\pi\varepsilon}\frac{d\cos\theta}{r^2}$$

and

$$\vec{E} = -\nabla\phi_e = -\hat{r}\frac{\partial\phi_e}{\partial r} - \hat{\theta}\frac{1}{r}\frac{\partial\phi_e}{\partial\theta} = \frac{qd}{4\pi\varepsilon r^3}(\hat{r}2\cos\theta + \hat{\theta}\sin\theta).$$

Here qd is called the electric dipole moment.

The electric dipole source is commonly employed as a simplest representation in bioelectric sources where a current source and a current sink are quite close. For example, the current may be flowing out of the membrane of a cell and sinking in another one nearby. Assume that the source strength at one point is I_o and at the other it is $-I_o$. The two points are separated by a small distance d. The biological medium is highly conducting (a general characteristic) with its conductivity σ. Following the procedure of this example, the electric potential ϕ_e at a distance R is found to be

$$\phi_e = \frac{I_o d\cos\theta}{4\pi\sigma R^2} = \frac{p\cos\theta}{4\pi\sigma R^2}, \tag{8.1.9}$$

where $p = I_o d$ is the dipole moment.

8.2 GAUSS'S LAW

It is one of the four Maxwell's equations (Gauss's law for the electric field) that we have already used earlier in Section 3.1. It is restated below both in integral and in differential forms:

$$\oint_S \vec{D} \cdot d\vec{s} = \int_V \rho_v \, dv \tag{8.2.1}$$

and

$$\nabla \cdot \vec{D} = \rho_v. \tag{8.2.2}$$

This simply states that the electric flux emanating from a closed surface is equal to the net electrical charge enclosed by that surface.

Example 8.6

Reconsider Example 8.3 in which an infinitely long charge filament has a charge density of ρ_ℓ C/m. Use Gauss's law to determine its electric field intensity. Using (8.2.1) over a unit length of the charge filament, we get

$$\oint_S \vec{D} \cdot d\vec{s} = \int_V \rho_v \, dv \rightarrow \int_0^{2\pi} D_\rho \rho \, d\phi = \rho_\ell \rightarrow D_\rho = \frac{\rho_\ell}{2\pi\rho}$$

and

$$E_\rho = \frac{D_\rho}{\varepsilon} = \frac{\rho_\ell}{2\pi\rho\varepsilon}.$$

Thus we got the same result as in Example 8.3 with relatively less effort.

8.3 POISSON'S AND LAPLACE'S EQUATIONS

It was noted in Chapter 6 that (6.2.14) reduces to (6.2.18) when the source is constant over time. This is known as Poisson's equation, as given again below:

$$\nabla^2 \phi_e = -\frac{\rho_v}{\varepsilon}. \tag{8.3.1}$$

In the case of a source-free region, (8.3.1) reduces to

$$\nabla^2 \phi_e = 0. \tag{8.3.2}$$

This is known as Laplace's equation. Both equations provide significant understanding of many scientific phenomena including the analyses of various electronic devices, biomedical systems, sensors, and actuators. The basic principles of a few of these applications are included in the examples presented in this chapter.

Example 8.7

A perfectly conducting sphere of radius a, as shown in Figure 8.5, has an electrical potential V_o. Use Laplace's equation to find the electrical charge density on its surface and the electric field intensity around it.

If we rotate the sphere along θ or ϕ, nothing seems to be changing and we can conclude that it is a case of symmetry. Therefore $\partial/\partial\phi = 0$ and $\partial/\partial\theta = 0$, and Laplace's simplifies to

$$\nabla^2 \phi_e = 0 \rightarrow \frac{1}{r}\frac{\partial}{\partial r}\left(r^2 \frac{\partial \phi_e}{\partial r}\right) = \frac{1}{r}\frac{d}{dr}\left(r^2 \frac{d\phi_e}{dr}\right) = 0.$$

Integrating once, we get

$$r^2 \frac{d\phi_e}{dr} = K_1,$$

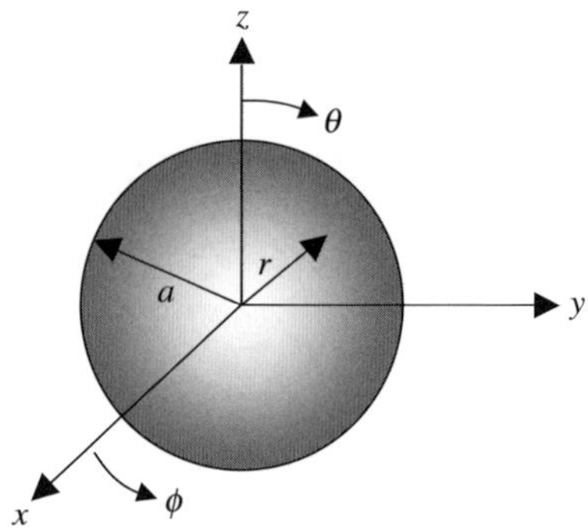

Figure 8.5 Geometry of the conducting sphere.

where K_1 is the integration constant. Integrating one more time, we find that

$$\phi_e = \int \frac{K_1}{r^2}\, dr = -\frac{K_1}{r} + K_2,$$

where K_2 is another integration constant.

Both integration constants need to be evaluated such that the solution satisfies the conditions at its boundaries. The boundary conditions are that the potential must be zero at infinity (because there is no source at infinity), and the potential is V_o on the surface of sphere. Hence

$$\phi_e\big|_{r\to\infty} = 0 \;\rightarrow\; K_2 = 0$$

and

$$\phi_e\big|_{r=a} = V_o \;\rightarrow\; K_1 = -aV_o.$$

Therefore

$$\phi_o = \frac{aV_o}{r}, \qquad r \geq a.$$

Now we can find the electric field intensity as follows:

$$\vec{E} = -\nabla\phi_e = -\hat{r}\frac{d\phi_e}{dr} = -\hat{r}\frac{d}{dr}\left(\frac{aV_o}{r}\right) = \hat{r}\frac{aV_o}{r^2}, \qquad r \geq a.$$

The electric flux density is found to be

$$\vec{D} = \varepsilon\vec{E} = \hat{r}\frac{\varepsilon aV_o}{r^2}, \qquad r \geq a.$$

Since electric field originates from the positive charge and terminates on the negative, (3.4.7) can be used to find the surface charge on the conducting sphere as follows:

$$\rho_{sa} = D_r|_{r=a} = \frac{\varepsilon a V_o}{a^2} = \frac{\varepsilon V_o}{a}.$$

8.4 CAPACITORS AND ENERGY STORAGE

An electrical voltage source connected between two conductors sets up the electrical lines of force that originate from the conductor with positive charge and terminate on the other with negative charge. Further these lines of force are always perpendicular to the conducting surface because the tangential component of electric field has to be zero on a conducting surface. A change in the source voltage changes the charge on the conductors accordingly. This voltage-to-charge relation is expressed as follows:

$$C = \frac{Q}{V}, \tag{8.4.1}$$

where Q is the charge in coulombs on each conductor, V in volts is the applied voltage, and the constant C is known as the capacitance in farads. This device is known as the capacitor.

Example 8.8

Find the capacitance of the conducting sphere considered in Example 8.7.

Recall from Example 8.7 that the surface charge density on the sphere is given as follows:

$$\rho_{sa} = D_r|_{r=a} = \frac{\varepsilon a V_o}{a^2} = \frac{\varepsilon V_o}{a}.$$

Therefore the total charge Q on the sphere is

$$Q = \int_S \rho_{sa} ds = \int_{\theta=0}^{\pi} \int_{\phi=0}^{2\pi} \frac{\varepsilon V_o}{a} a^2 \sin\theta \, d\theta d\phi = 4\pi a \varepsilon V_o.$$

The capacitance C is now found to be

$$C = \frac{Q}{V_o} = 4\pi a \varepsilon.$$

Energy Storage in Electrical Systems

The electric field intensity due to a point charge q_1 is found to be

$$\vec{E} = \hat{R}\frac{q_1}{4\pi\varepsilon R^2}. \tag{8.4.2}$$

If there is another point charge q_2 located at R, then it experiences a repulsive force that is found to be

$$\vec{F} = q_2\vec{E} = \hat{R}\frac{q_1 q_2}{4\pi\varepsilon R^2}. \tag{8.4.3}$$

The work required against this force in moving q_2 to R from infinity is

$$W_{12} = -\int_{\infty}^{R_{12}} \vec{F}\cdot\hat{R}dR = -\int_{\infty}^{R_{12}} \frac{q_1 q_2}{4\pi\varepsilon R^2}dR = \frac{q_1 q_2}{4\pi\varepsilon R_{12}}. \tag{8.4.4}$$

This represents a gain in energy that is now stored in the system. Since there are two charges now, their total potential field is found to be

$$\phi_{e12} = \frac{q_1}{4\pi\varepsilon R_1} + \frac{q_2}{4\pi\varepsilon R_2}. \tag{8.4.5}$$

Next consider the case where another charge q_3 is brought in from infinity. It requires more work, and therefore the total energy stored increases to

$$W_{1-3} = \frac{q_1 q_3}{4\pi\varepsilon R_{13}} + \frac{q_2 q_3}{4\pi\varepsilon R_{23}} + \frac{q_1 q_2}{4\pi\varepsilon R_{12}}. \tag{8.4.6}$$

If generalized for N charges, the total electrical energy stored is found to be

$$U_E = \frac{1}{2}\sum_{\substack{m=1\\ m\neq n}}^{N}\sum_{n=1}^{N}\frac{q_m q_n}{4\pi\varepsilon R_{mn}}. \tag{8.4.7}$$

First summation in (8.4.7) is identified as the electric potential ϕ_e due to $N-1$ charges (because it excludes nth charge). Therefore (8.4.7) reduces to

$$U_E = \frac{1}{2}\sum_{n=1}^{N}\phi_e q_n. \tag{8.4.8}$$

Now consider the situation where the charge is distributed over a volume V with a density ρ_v C/m^3. The total energy stored may be found after summing the contributions of each infinitesimal volume as follows:

$$U_E = \frac{1}{2}\int_V \phi_e\rho_v\, dv. \tag{8.4.9}$$

Since $\nabla\cdot\vec{D} = \rho_v$ (Gauss's law in the point form), we can write

$$U_E = \frac{1}{2}\int_V \phi_e(\nabla\cdot\vec{D})dv = \frac{1}{2}\left[\int_V \nabla\cdot(\phi_e\vec{D})dv - \int_V \nabla\phi_e\cdot\vec{D}\,dv\right]. \tag{8.4.10}$$

Here we used vector identity 6 from Appendix A. This can be further simplified by the divergence theorem (Appendix A) to get

$$U_E = \frac{1}{2}\left[\oint_S \phi_e \vec{D} \cdot d\vec{s} - \int_V \nabla\phi_e \cdot \vec{D}\, dv\right]. \tag{8.4.11}$$

Since the integration is carried over the entire volume V that extends to infinity, the first integral goes to zero when evaluated over the surface S that bounds this volume. It is so because the potential ϕ_e is zero at the infinity. Using the relations among the electric potential, the electric field intensity, and the electric flux density, the second integral may be expressed as follows:

$$U_E = \frac{1}{2}\int_V \vec{E} \cdot \vec{D}\, dv = \frac{1}{2}\int_V \vec{E} \cdot (\varepsilon\vec{E})\, dv = \frac{1}{2}\int_V \varepsilon\left|\vec{E}\right|^2 dv. \tag{8.4.12}$$

Note that this relation was part of (3.6.5) when we considered the Poynting vector.

For example, take the energy stored in the capacitor formed by the sphere of Example 8.7. From the results obtained in preceding two examples, we find that

$$U_E = \frac{1}{2}\int_V \varepsilon\left|\vec{E}\right|^2 dv = \frac{1}{2}\int_{r=a}^{\infty}\int_{\theta=0}^{\pi}\int_{\phi=0}^{2\pi} \varepsilon\left(\frac{aV_o}{r^2}\right)^2 r^2 \sin\theta\, dr d\theta d\phi = \frac{1}{2}\varepsilon V_o^2 4\pi a$$

$$= \frac{1}{2}CV_o^2.$$

This is a general result in the sense that it holds good for all kinds of capacitors.

Example 8.9

A distance d separates two conducting sheets, each with a surface area of A as shown in Figure 8.6. A dielectric material of permittivity ε fills the space between the two plates. Neglect the fringing fields at the edges to find the capacitance of this arrangement. Also find the energy stored in this capacitor.

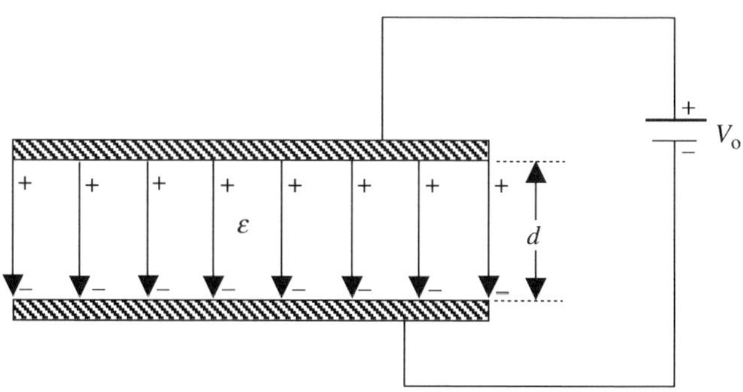

Figure 8.6 Parallel-plate capacitor.

Note that the electric field distribution shown in Figure 8.6 is uniform. This is still true in the region away from the edges. However, this distribution changes at its edges in order to meet the continuity of tangential electric fields across the boundary. Evaluation of the fields at the edges is rather complex, and therefore these fringing fields are ignored here. If the surface area of each plate is large, the error introduced by this assumption is negligible.

In order to determine the capacitance, we need to find the charge on one plate that is set up by V_o. Laplace's equation can be solved for the space in between the two plates to find the potential. This leads to the electric field intensity and the electric flux density in the space. The surface charge on the conductor can be found via (3.4.7). Alternatively, we enclose one conductor by a surface s and apply Gauss's law as follows:

$$\oint_S \vec{D} \cdot d\vec{s} = \int_V \rho_v \, dv \rightarrow DA = \rho_s A \rightarrow \varepsilon EA = Q.$$

Since the electric field is uniform, $E = V_o/d$, and therefore this relation gives

$$\varepsilon \frac{V_o}{d} A = Q \rightarrow \frac{Q}{V_o} = C = \varepsilon \frac{A}{d}.$$

We can find the energy stored in this capacitor as follows:

$$U_E = \frac{1}{2}\int_V \varepsilon |\vec{E}|^2 dv = \frac{1}{2}\int_V \varepsilon \left(\frac{V_o}{d}\right)^2 dv = \frac{1}{2}\varepsilon \frac{V_o^2}{d^2}\int_V dv = \frac{1}{2}\varepsilon \frac{V_o^2}{d^2} Ad = \frac{1}{2} C V_o^2.$$

Example 8.10

Two conducting parallel plates are used to form a capacitor. Each plate has an area of 1 m^2 and a distance of 1 mm separates the plates. The space between the two plates is filled with a dielectric material. However, its dielectric constant is not known. Based on the capacitance bridge, its capacitance is found to be 22.135 nF. Find the dielectric constant of the material.

Since $C = \varepsilon(A/d)$,

$$\varepsilon = \frac{Cd}{A} = \frac{22.135 \cdot 10^{-9} \cdot 10^{-3}}{1} = 22.135 \cdot 10^{-12} \text{ F/m}.$$

The dielectric constant ε_r is found to be

$$\varepsilon_r = \frac{\varepsilon}{\varepsilon_o} = \frac{22.135 \cdot 10^{-12}}{8.854 \cdot 10^{-12}} = 2.5.$$

This kind of technique can be used to find the dielectric constant of a material.

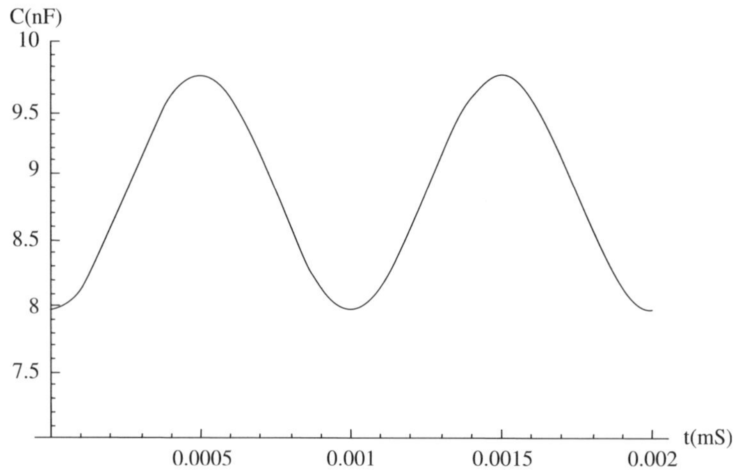

Figure 8.7 Variation of capacitance over time.

Example 8.11

The parallel-plate capacitor of Example 8.10 uses air as its dielectric. If separation d of the parallel-plate capacitor is changing over time, as given below, find the capacitance as function of time and plot the characteristics.

$$d = 1 + 0.1\cos(2000\pi t)\,\text{mm},$$
$$C = \varepsilon\frac{A}{d} = \varepsilon_o\frac{1}{[1 + 0.1\cos(2000\pi t)]\cdot 10^{-3}}$$
$$= 8.854\cdot 10^{-9}[1 + 0.1\cos(2000\pi t)]^{-1}\,\text{F},$$

or

$$C \approx 8.854[1 - 0.1\cos(2000\pi t)]\,\text{nF}.$$

This shows that the capacitance variation follows the change in separation closely. This is the principle on which the capacitive microphones work. The capacitance variation with time is displayed in Figure 8.7.

Example 8.12

Reconsider the capacitor of Example 8.10. However, the space in between the plates has two different dielectric materials, as shown in Figure 8.8. If a 0.9 mm thick dielectric sheet of $\varepsilon_r = 2.5$ is used as one medium and the rest is filled with air, find its capacitance.

Assume that the area of each plate is A, and electric field intensities in two dielectric media are E_1 and E_2, respectively. As before, we ignore the fringing

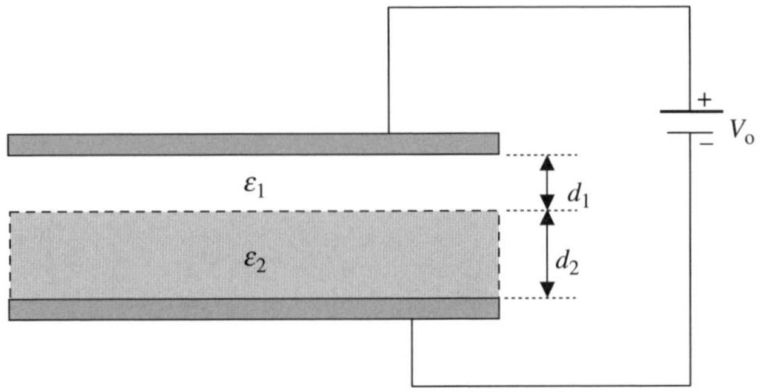

Figure 8.8 Parallel-plate capacitor with two dielectrics.

field at its edges, and the uniform electric field is directed from the top plate to the bottom plate. Since the normal component of electric flux density D is continuous across an interface, we find that

$$D_1 = D_2 \rightarrow \varepsilon_1 E_1 = \varepsilon_2 E_2 \rightarrow E_2 = \frac{\varepsilon_1}{\varepsilon_2} E_1,$$

$$V_o = E_1 d_1 + E_2 d_2 = E_1 d_1 + \frac{\varepsilon_1}{\varepsilon_2} E_1 d_2 = E_1 \left(d_1 + \frac{\varepsilon_1}{\varepsilon_2} d_2 \right).$$

Total charge Q on the top plate is found to be

$$Q = A\rho_s = AD_1 = A\varepsilon_1 E_1.$$

Therefore

$$C = \frac{Q}{V_o} = \frac{A\varepsilon_1 E_1}{E_1[d_1 + (\varepsilon_1/\varepsilon_2)d_2]} = \frac{A\varepsilon_1\varepsilon_2}{d_1\varepsilon_2 + \varepsilon_1 d_2} = \frac{1}{(d_1/A\varepsilon_1) + (d_2/A\varepsilon_2)}$$

$$= \frac{1}{(1/C_1) + (1/C_2)} = \frac{C_1 C_2}{C_1 + C_2}.$$

Thus the two dielectric layers work as two capacitors connected in series, even though there is no conducting sheet at the interface.

Using the given data, we find that

$$C_1 = \varepsilon_o \frac{A}{d} = 8.854 \cdot 10^{-12} \frac{1}{0.1 \cdot 10^{-3}} \text{ F} = 88.54 \text{ nF}$$

and

$$C_2 = \varepsilon_2 \frac{A}{d} = 2.5 \cdot 8.854 \cdot 10^{-12} \frac{1}{0.9 \cdot 10^{-3}} \text{ F} = 24.5944 \text{ nF}$$

Therefore

$$C = \frac{C_1 C_2}{C_1 + C_2} = \frac{88.54 \cdot 24.5944}{88.54 + 24.5944}\,\text{nF} = 19.25\,\text{nF}.$$

Example 8.13

Actuators are employed in nanotechnology area to perform various tasks. One such arrangement is shown in Figure 8.9 in which a conducting plate makes an angle of ϕ_o with the anode. This kind of system has been used in the MEMS-based electrostatic projection displays. Assume that the plate width is W and the fringing fields negligible. Find the capacitance of this arrangement.

For $\partial/\partial\rho \to 0$ and $\partial/\partial z \to 0$, Laplace's equation simplifies as follows:

$$\nabla^2 \phi_e = 0 \to \frac{1}{\rho}\frac{\partial^2 \phi_e}{\partial \phi^2} = 0 \to \frac{d^2 \phi_e}{d\phi^2} = 0.$$

On integrating it once, we get

$$\frac{d\phi_e}{d\phi} = K_1,$$

where K_1 is an integration constant.

Integrating it one more time, we find that

$$\phi_e = K_1 \phi + K_2.$$

K_2 is another integration constant. These two integration constants are evaluated via the boundary conditions of the problem that require the potential ϕ_e to be V_o for $\phi = 0$, and it is zero at $\phi = \phi_o$. Hence

$$\phi_e\big|_{\phi=0} = V_o \to K_2 = V_o$$

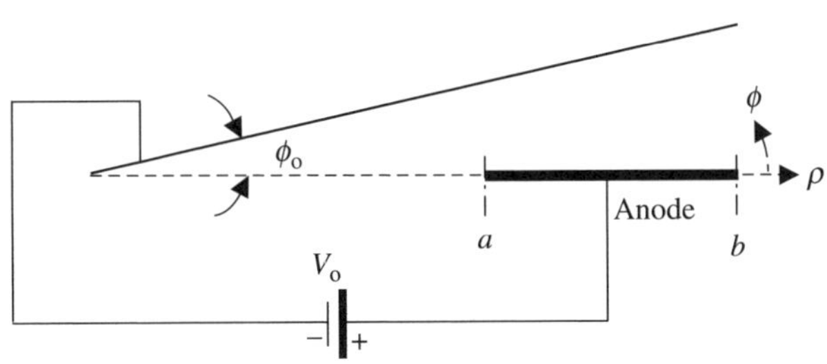

Figure 8.9 Actuator arrangement.

and

$$\phi_e\big|_{\phi=\phi_o} = 0 \rightarrow K_1 = -\frac{K_2}{\phi_o} = -\frac{V_o}{\phi_o}.$$

Therefore

$$\phi_e = V_o\left(1 - \frac{\phi}{\phi_o}\right).$$

The electric field intensity in the region is found to be

$$\vec{E} = -\nabla\phi_e = -\hat{\phi}\frac{1}{\rho}\frac{\partial\phi_e}{\partial\phi} = -\hat{\phi}\frac{1}{\rho}\frac{d}{d\phi}\left[V_o\left(1 - \frac{\phi}{\phi_o}\right)\right] = \hat{\phi}\frac{V_o}{\rho\phi_o}\ \text{V/m}.$$

Assuming that ε is the permittivity of the medium, the electric flux density is found to be

$$\vec{D} = \varepsilon\vec{E} = \hat{\phi}\frac{\varepsilon V_o}{\rho\phi_o}\ \text{C/m}^2.$$

Therefore the surface charge ρ_s on the anode is

$$\rho_s = D_\phi = \frac{\varepsilon V_o}{\rho\phi_o}\ \text{C/m}^2.$$

Total charge Q is found to be

$$Q = \frac{\varepsilon V_o}{\phi_o}\int_a^b \frac{W d\rho}{\rho} = \frac{\varepsilon V_o W}{\phi_o}\ln\left(\frac{b}{a}\right).$$

Therefore its capacitance is

$$C = \frac{Q}{V_o} = \frac{\varepsilon W}{\phi_o}\ln\left(\frac{b}{a}\right).$$

Example 8.14

A coaxial line of inner and outer conductor radii a and b, respectively, has a voltage V_o as shown in Figure 8.10. The uniform surface charge density on its inner conductor is ρ_{sa}, whereas on inside surface of the outer conductor it is ρ_{sb}. Using Gauss's law, find the potential distribution in the space between its conductors, the electric field intensity, and its capacitance per meter of its length.

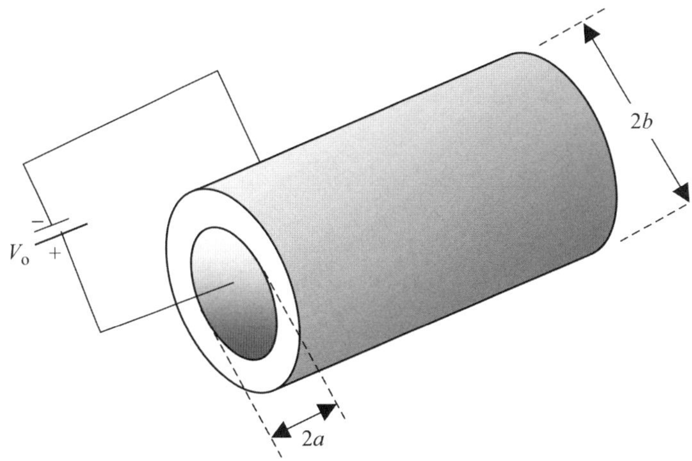

Figure 8.10 Geometry of a coaxial line.

From Gauss's law, we find that

$$\oint_S \vec{D} \cdot d\vec{s} = \int_V \rho_v \, dv \rightarrow \int_{\phi=0}^{2\pi} D_\rho \rho \, d\phi$$

$$= \int_{\phi'=0}^{2\pi} \rho_{sa} a \, d\phi' \rightarrow D_\rho = \frac{\rho_{sa} a}{\rho}, \qquad a \leq \rho \leq b,$$

and

$$D_\rho = 0, \quad \rho > b, \quad \text{and} \quad \rho < a.$$

Therefore

$$E_\rho = \frac{D_\rho}{\varepsilon} = \frac{a\rho_{sa}}{\varepsilon\rho}, \qquad a \leq \rho \leq b,$$

and

$$\phi_e = -\int E_\rho \, d\rho = -\int \frac{a\rho_{sa}}{\varepsilon\rho} \, d\rho = -\frac{a\rho_{sa}}{\varepsilon} \ln \rho + K_1,$$

where K_1 is the integration constant that can be found from the condition that $\phi_e = 0$ at $\rho = b$. Hence

$$\phi_e = 0 = -\frac{a\rho_{sa}}{\varepsilon} \ln b + K_1 \rightarrow K_1 = \frac{a\rho_{sa}}{\varepsilon} \ln b.$$

Therefore

$$\phi_{eb} = \frac{a\rho_{sa}}{\varepsilon} [\ln b - \ln \rho]$$

and

$$\phi_{\text{eab}} = V_{\text{o}} = \frac{a\rho_{\text{sa}}}{\varepsilon}[\ln b - \ln a] = \frac{a\rho_{\text{sa}}}{\varepsilon}\ln\left(\frac{b}{a}\right).$$

The capacitance per unit length C is found to be

$$C = \frac{Q}{V_{\text{o}}} = \frac{2\pi a\rho_{\text{sa}}}{(a\rho_{\text{sa}}/\varepsilon)\ln(b/a)} = \frac{2\pi\varepsilon}{\ln(b/a)}\ \text{F/m}.$$

Example 8.15

The inner and outer sphere radii of two conducting concentric spheres are a and b, respectively, as shown in Figure 8.11. The inner sphere is at a potential V_{o} with respect to outer one. Use Gauss's law to find the potential distribution, the electric field intensity, and the capacitance of this system.

Since the problem has symmetry in the θ and ϕ directions (rotating the spheres makes no difference in these directions), Gauss's law $\oint_S \vec{D} \cdot d\vec{s} = \int_V \rho_{\text{v}} dv$ for the space between the two spheres gives

$$\int_{\phi=0}^{2\pi}\int_{\theta=0}^{\pi} D_{\text{r}} r^2 \sin\theta\, d\theta d\phi = \int_{\phi=0}^{2\pi}\int_{\theta=0}^{\pi} \rho_{\text{sa}} a^2 \sin\theta\, d\theta d\phi,$$

or

$$4\pi r^2 D_{\text{r}} = 4\pi a^2 \rho_{\text{sa}} = Q \rightarrow D_{\text{r}} = \frac{Q}{4\pi r^2},$$

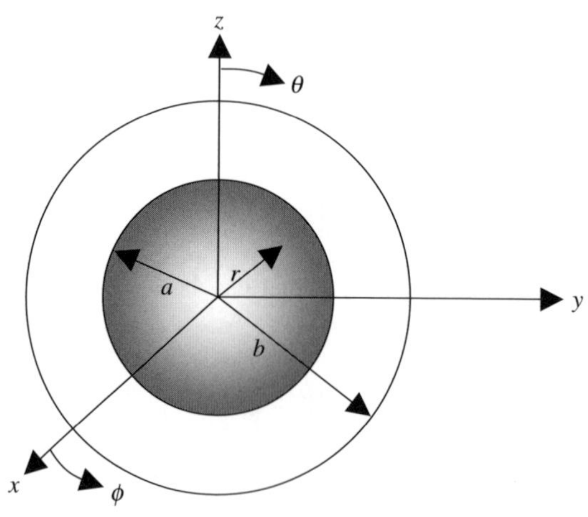

Figure 8.11 Geometry of the concentric spheres.

where ρ_{sa} is the charge density on the inner sphere and Q is the total charge on it. Note that the electric flux density is zero for $r > b$ as well as for $r < a$ because there is no net charge enclosed in these cases.

Therefore the electric field intensity in the space between the two spheres is

$$\vec{E} = \hat{r}\frac{Q}{4\pi\varepsilon r^2}.$$

The corresponding electric potential is found to be

$$\phi_e = -\int \vec{E}\cdot d\vec{\ell} = -\int \frac{Q}{4\pi\varepsilon r^2}dr = \frac{Q}{4\pi\varepsilon r} + K_1.$$

The integration constant K_1 is evaluated from the boundary condition that the potential is zero at $r = b$, meaning $\phi_e\big|_{r=b} = 0$. Hence

$$\frac{Q}{4\pi\varepsilon b} + K_1 = 0 \rightarrow K_1 = -\frac{Q}{4\pi\varepsilon b}.$$

Therefore

$$\phi_e = \frac{Q}{4\pi\varepsilon}\left[\frac{1}{r} - \frac{1}{b}\right], \qquad a \leq r \leq b.$$

Using this relation, we find that

$$V_o = \frac{Q}{4\pi\varepsilon}\left[\frac{1}{a} - \frac{1}{b}\right]$$

and

$$C = \frac{Q}{V_o} = \frac{4\pi\varepsilon}{[(1/a) - (1/b)]}.$$

Example 8.16

Reconsider Example 8.14. Use Laplace's equation to find the potential's distribution inside the coaxial line, the electric field intensity, and its capacitance per meter of its length.

Since $\partial/\partial\phi = 0$ and $\partial/\partial z = 0$ in this case, Laplace's equation simplifies to

$$\nabla^2\phi_e = 0 \rightarrow \frac{1}{\rho}\frac{d}{d\rho}\left(\rho\frac{d\phi_e}{d\rho}\right) = 0.$$

On integrating once, we get

$$\rho\frac{d\phi_e}{d\rho} = K_1.$$

K_1 is the integration constant. Integrating it one more time, we find that

$$\phi_e = \int \frac{K_1}{\rho} d\rho = K_1 \ln(\rho) + K_2.$$

K_2 is another integration constant. Both integration constants are evaluated from the boundary conditions $\phi_e = 0$ at $\rho = b$, and $\phi_e = V_o$ at $\rho = a$. Hence

$$0 = K_1 \ln(b) + K_2$$

and

$$V_o = K_1 \ln(a) + K_2.$$

These two equations are solved to find the integration constants as follows:

$$K_1 = \frac{V_o}{\ln(a/b)}$$

and

$$K_2 = -\frac{V_o}{\ln(a/b)} \ln(b).$$

Therefore

$$\phi_e = \frac{V_o}{\ln(a/b)} [\ln(\rho) - \ln(b)] = \frac{V_o}{\ln(a/b)} \ln(\rho/b), \qquad a \leq \rho \leq b,$$

and

$$\vec{E} = -\nabla \phi_e = -\hat{\rho} \frac{\partial \phi_e}{\partial \rho} = -\hat{\rho} \frac{V_o}{\rho \ln(a/b)}.$$

The corresponding electric flux density is found to be

$$\vec{D} = \varepsilon \vec{E} = -\hat{\rho} \frac{\varepsilon V_o}{\rho \ln(a/b)}.$$

Since

$$\rho_{sa} = D_\rho \big|_{\rho=a} = -\frac{\varepsilon V_o}{a \ln(a/b)} = \frac{\varepsilon V_o}{a \ln(b/a)} \text{ C/m}^2,$$

$$Q = 2\pi a \rho_{sa} = \frac{2\pi \varepsilon V_o}{\ln(b/a)} \text{ C/m}.$$

Therefore

$$C = \frac{Q}{V_o} = \frac{2\pi\varepsilon}{\ln(b/a)} \text{ F/m}.$$

As expected, the results are same as we found earlier in Example 8.14.

Example 8.17

Reconsider Example 8.15 and use Laplace's equation to find the potential's distribution in between the concentric spheres, the electric field intensity, and its capacitance.

If we rotate the sphere along θ or ϕ, then nothing seems to be changing and we can conclude that it is a case of symmetry. Therefore $\partial/\partial\phi = 0$ and $\partial/\partial\theta = 0$, and Laplace's equation simplifies to

$$\nabla^2\phi_e = 0 \rightarrow \frac{1}{r}\frac{\partial}{\partial r}\left(r^2\frac{\partial\phi_e}{\partial r}\right) = \frac{1}{r}\frac{d}{dr}\left(r^2\frac{d\phi_e}{dr}\right) = 0.$$

Integrating once, we get

$$r^2\frac{d\phi_e}{dr} = K_1,$$

where K_1 is the integration constant. Integrating it one more time, we find that

$$\phi_e = \int\frac{K_1}{r^2}\,dr = -\frac{K_1}{r} + K_2,$$

where K_2 is another integration constant.

Both integration constants need to be evaluated such that the solution satisfies the conditions at its boundaries. The boundary conditions are that the potential is zero on the outer spherical surface and it is V_o on the surface of inner sphere. Hence

$$\phi_e\big|_{r\rightarrow b} = 0 \rightarrow -\frac{K_1}{b} + K_2 = 0$$

and

$$\phi_e\big|_{r=a} = V_o \rightarrow -\frac{K_1}{a} + K_2 = V_o.$$

Thcsc two cquations are solved to find K_1 and K_2, as follows:

$$K_1 = -\frac{V_o}{[(1/a) - (1/b)]}$$

and

$$K_2 = -\frac{V_o}{b[(1/a) - (1/b)]}.$$

Therefore

$$\phi_e = \frac{V_o}{[(1/a) - (1/b)]}\left[\frac{1}{r} - \frac{1}{b}\right], \qquad a \le r \le b.$$

The corresponding electric field intensity and the electric flux density are found to be

$$\vec{E} = -\nabla\phi_e = -\hat{r}\frac{\partial\phi_e}{\partial r} = \hat{r}\frac{V_o}{[(1/a) - (1/b)]}\frac{1}{r^2}$$

and

$$\vec{D} = \varepsilon\vec{E} = \hat{r}\frac{V_o\varepsilon}{[(1/a) - (1/b)]}\frac{1}{r^2}.$$

Hence the surface charge density on the inner sphere is

$$\rho_{sa} = D_r|_{r=a} = \frac{V_o\varepsilon}{[(1/a) - (1/b)]}\frac{1}{a^2}.$$

The total charge Q on this sphere is found to be

$$Q = 4\pi a^2\rho_{sa} = 4\pi a^2\frac{V_o\varepsilon}{[(1/a) - (1/b)]}\frac{1}{a^2} = \frac{4\pi V_o\varepsilon}{[(1/a) - (1/b)]}.$$

Hence its capacitance can be found now as follows:

$$C = \frac{Q}{V_o} = \frac{4\pi\varepsilon}{[(1/a) - (1/b)]}\ \text{F}.$$

8.5 FURTHER APPLICATIONS OF POISSON'S AND LAPLACE'S EQUATIONS

As was mentioned earlier, Poisson's and Laplace's equations are used to solve a number of scientific and engineering problems. Here we consider two such cases. One of them is a pn-junction diode used in electronic circuits and the other is a simple model for understanding electroencephalography (EEG).

The pn-Junction Diode

Poisson's equation is employed to analyze semiconductor diodes and transistors. Here we consider a simple case of the pn-junction diode with an abrupt junction. A p-type semiconductor has a deficiency of electrons (hence it is said to have holes) whereas the n-type has excess electrons. These semiconductors are known to have acceptor and donor impurities, respectively. When a junction of these two is formed, the excess electrons start leaving the donor atoms to combine with the holes of the acceptors. Similarly the holes try to leave from the p-side and combine with electrons on the n-side. The electrons moving from the n-side leave behind positively charged donor atoms and the holes leave behind negatively charged acceptors, as illustrated in Figure 8.12*a*. The donor and acceptor atoms are bound charges that cannot move, and therefore a space charge layer is formed across the junction with positive charges on the n-side and negative charges on the p-side. Since there are no free charges left in this region for conduction, this layer is called the depletion layer. The bound charges in this depletion region give rise to an electric field that is directed from the positive charges left on the n-side to the negative charges on the p-side of the junction. The electric field opposes the movement of electrons and holes across the junction, thus creating equilibrium.

Assume that the donor and acceptor impurity concentrations are N_{d} and N_{a} per cubic meters, respectively. The depletion layer extends up to x_{n} meters on the n-side and up to $-x_{\mathrm{p}}$ on the p-side. With the electronic charge denoted as q_{e}, the charge distribution in depletion region is shown in Figure 8.12*b*. Since semiconductor is neutral,

$$q_{\mathrm{e}} N_{\mathrm{a}} x_{\mathrm{p}} A = q_{\mathrm{e}} N_{\mathrm{d}} x_{\mathrm{n}} A \;\rightarrow\; N_{\mathrm{a}} x_{\mathrm{p}} = N_{\mathrm{d}} x_{\mathrm{n}}. \tag{8.5.1}$$

Poisson's equation for the space charge region may be written as follows:

$$\frac{d^2\psi_{\mathrm{e}}}{dx^2} = \begin{cases} -\dfrac{q_{\mathrm{e}} N_{\mathrm{d}}}{\varepsilon}, & 0 < x < x_{\mathrm{n}}, \\ \dfrac{q_{\mathrm{e}} N_{\mathrm{a}}}{\varepsilon}, & -x_{\mathrm{p}} < x < 0. \end{cases} \tag{8.5.2}$$

Integrating it once, we find that

$$\frac{d\psi_{\mathrm{e}}}{dx} = \begin{cases} -\dfrac{q_{\mathrm{e}} N_{\mathrm{d}}}{\varepsilon} x + K_1, & 0 < x < x_{\mathrm{n}}, \\ \dfrac{q_{\mathrm{e}} N_{\mathrm{a}}}{\varepsilon} x + K_2, & -x_{\mathrm{p}} < x < 0. \end{cases} \tag{8.5.3}$$

Since $d\psi_{\mathrm{e}}/dx = -E_{\mathrm{x}}$ is the electric field intensity that starts from positive charge and terminates on the negative,

$$E_{\mathrm{x}}\Big|_{\substack{x=x_{\mathrm{n}} \\ x=-x_{\mathrm{p}}}} = -\left.\frac{d\psi_{\mathrm{e}}}{dx}\right|_{\substack{x=x_{\mathrm{n}} \\ x=-x_{\mathrm{p}}}} = 0. \tag{8.5.4}$$

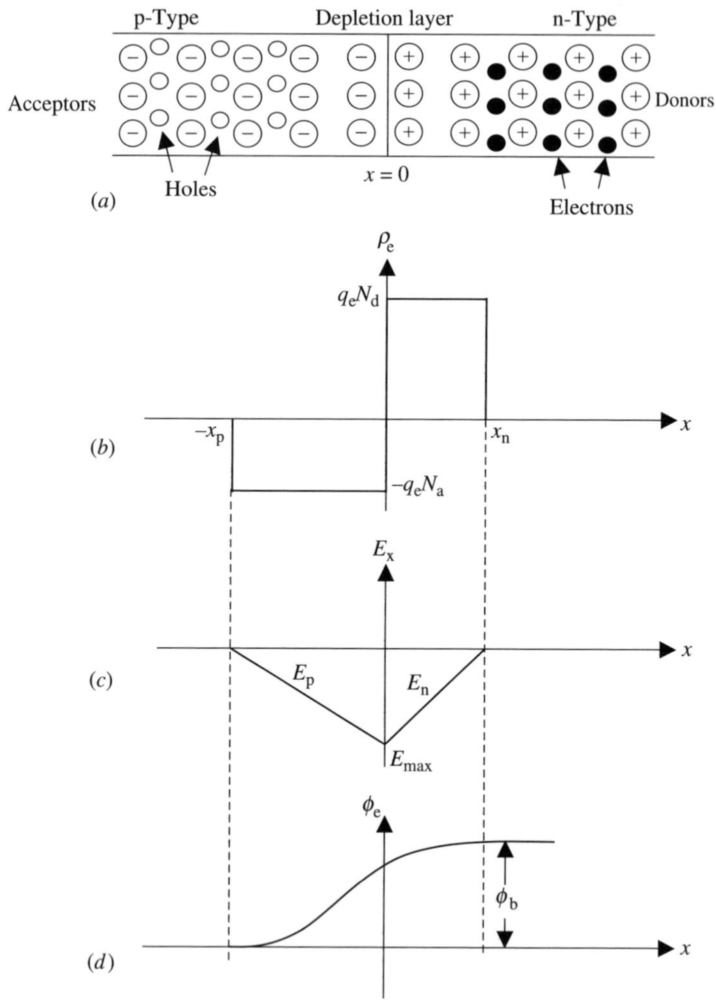

Figure 8.12 A pn-junction diode (*a*), the charge density (*b*), the electric field intensity (*c*), and the potential's distribution in depletion region (*d*).

From these two conditions the integration constants K_1 and K_2 are found to be

$$K_1 = \frac{q_e N_d x_n}{\varepsilon} \tag{8.5.5}$$

and

$$K_2 = \frac{q_e N_a x_p}{\varepsilon}. \tag{8.5.6}$$

Therefore (8.5.3) can be expressed as

$$\frac{d\psi_e}{dx} = \begin{cases} \dfrac{q_e N_d}{\varepsilon}(x_n - x), & 0 < x < x_n, \\ \dfrac{q_e N_a}{\varepsilon}(x_p + x), & -x_p < x < 0. \end{cases} \tag{8.5.7}$$

Integrating (8.5.7), we get

$$\psi_e = \begin{cases} -\dfrac{q_e N_d}{2\varepsilon}(x_n - x)^2 + K_3, & 0 < x < x_n, \\ \dfrac{q_e N_a}{2\varepsilon}(x_p + x)^2 + K_4, & -x_p < x < 0. \end{cases} \tag{8.5.8}$$

For an arbitrary reference point at $x = -x_p$, the integration constant K_4 is found to be

$$\psi_e|_{x=x_p} = 0 \rightarrow K_4 = 0. \tag{8.5.9}$$

Further the potential ψ_e must be continuous at $x = 0$ because $E_x = d\psi_e/dx$ is finite at this point. Hence the integration constant K_3 is found to be

$$-\frac{q_e N_d}{2\varepsilon}x_n^2 + K_3 = \frac{q_e N_a}{2\varepsilon}x_p^2 \rightarrow K_3 = \frac{q_e}{2\varepsilon}\left[N_d x_n^2 + N_a x_p^2\right]. \tag{8.5.10}$$

Therefore (8.5.8) gives

$$\psi_e = \begin{cases} -\dfrac{q_e N_d}{2\varepsilon}(x_n - x)^2 + \dfrac{q_e}{2\varepsilon}\left[N_d x_n^2 + N_a x_p^2\right], & 0 < x < x_n, \\ \dfrac{q_e N_a}{2\varepsilon}(x_p + x)^2, & -x_p < x < 0. \end{cases} \tag{8.5.11}$$

This equation represents the potential in the depletion layer with respect to $x = -x_p$. Therefore the potential difference ψ_{eo} across the junction is found to be

$$\psi_{eo} = \frac{q_e}{2\varepsilon}\left[N_d x_n^2 + N_a x_p^2\right] = \frac{q_e}{2\varepsilon}\left[N_d \frac{N_a + N_d}{N_a + N_d}x_n^2 + N_a \frac{N_a + N_d}{N_a + N_d}x_p^2\right],$$

or

$$\psi_{eo} = \frac{q_e}{2\varepsilon(N_a + N_d)}\left[N_d N_a x_n^2 + N_d N_a x_p^2 + N_d^2 x_n^2 + N_a^2 x_p^2\right]. \tag{8.5.12}$$

Since $N_a x_p = N_d x_n$ from (8.5.1), we find that

$$N_d^2 x_n^2 + N_a^2 x_p^2 = N_d \frac{N_a x_p}{x_n}x_n^2 + N_a \frac{N_d x_n}{x_p}x_p^2 = 2N_d N_a x_p x_n.$$

Hence (8.5.12) can be expressed as

$$\psi_{eo} = \frac{q_e}{2\varepsilon} \cdot \frac{N_d N_a}{N_d + N_a} [x_n^2 + x_p^2 + 2x_n x_p] = \frac{q_e}{2\varepsilon} \cdot \frac{N_d N_a}{N_d + N_a} [x_n + x_p]^2. \qquad (8.5.13)$$

The total length $d_o = x_n + x_p$ of the depletion region is found to be

$$d_o^2 = \frac{2\varepsilon\psi_{eo}}{q_e} \left[\frac{N_a + N_d}{N_a N_d} \right] \to d_o = \sqrt{\frac{2\varepsilon\psi_{eo}}{q_e} \left[\frac{1}{N_d} + \frac{1}{N_a} \right]}. \qquad (8.5.14)$$

Example 8.18

A silicon pn-junction diode has $N_d = 10^{22}\,\mathrm{m}^{-3}$ and $N_a = 4 \cdot 10^{24}\,\mathrm{m}^{-3}$ on its n- and p-sides, respectively. If the built-in potential is 0.8531 V, find the length d_o of the depletion region and the lengths of the space-charge regions on each side of the junction. Assume that the dielectric constant of silicon is 11.8.

$$d_o = \sqrt{\frac{2\varepsilon\psi_{eo}}{q_e} \left[\frac{1}{N_d} + \frac{1}{N_a} \right]} = \sqrt{\frac{2 \cdot 11.8 \cdot 8.854 \cdot 10^{-12} \cdot 0.8531}{1.602 \cdot 10^{-19}} \left[\frac{1}{10^{22}} + \frac{1}{4 \cdot 10^{24}} \right]},$$

or

$$d_o = 333.99 \cdot 10^{-9}\,\mathrm{m} = 333.99\,\mathrm{nm},$$

$$\frac{x_p}{x_n} = \frac{N_d}{N_a} = \frac{10^{22}}{4 \cdot 10^{24}} = 0.0025 \to x_p = 0.0025\,x_n,$$

$$x_n + x_p = d_o = 333.99\,\mathrm{nm},$$

$$x_n = 333.16\,\mathrm{nm},$$

and

$$x_p = 0.83\,\mathrm{nm}.$$

The Electroencephalography (EEG)

The electrical activities occurring within the human body produce potentials on the skin. These electrical potentials are monitored for diagnostic and other purposes. For example, electrical potentials measured on the chest provide the information about activities of the heart. Over time these recordings became known as electrocardiograms (ECG or EKG). Similarly recordings of the electrical potentials on the scalp provide information about the electrical activities of the brain. These are called electroencephalograms (EEG). The amplitudes of the EEG signals range from

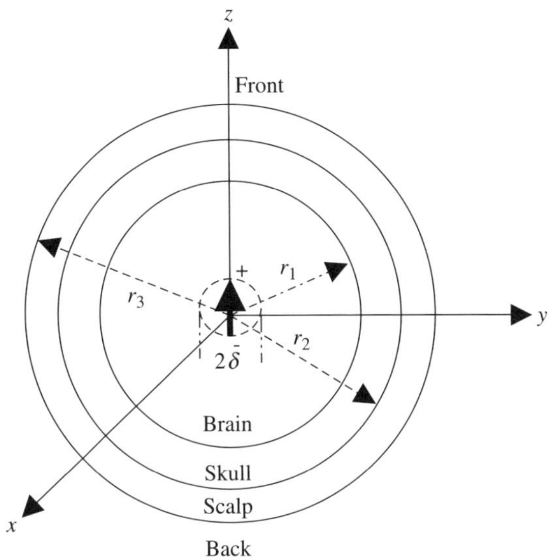

Figure 8.13 Simplified model of the human head.

about 1 μV to 100 μV peak to peak at the cranial surface and can increase by a factor of 10 on the surface of the cerebrum. The frequencies of these signals range from 0.5 to 100 Hz. On the other hand, EKG chest surface signals are about 0.5 to 100 mV peak to peak. Since EEG signals are much smaller, special care is needed in placing the electrodes and in designing the associated electronics. Theoretical modeling, computer simulations, and experiments help in this respect. Figure 8.13 shows a simplified model of the human head that consists of three concentric spheres with an electric dipole source at the center. This model provides the most simple way to understand the potential field setup on the surface of the scalp.

As illustrated in Figure 8.13, the dipole source of the three concentric spheres is confined to a small sphere of radius δ. The brain is considered to be a sphere of radius r_1 and conductivity σ_1. The skull is a spherical shell with a thickness $r_2 - r_1$ and conductivity σ_2. The scalp is a spherical shell with a thickness of $r_3 - r_2$ and a conductivity of σ_3. Usually σ_1 is nearly equal to σ_3, and σ_2 is much smaller than σ_1. Excluding the small spherical volume that encloses the dipole source, Laplace's equation can be used to find the potential at any point in the brain, skull, and scalp.

Assume that the potential does not change azimuthally, (i.e., $\partial/\partial\phi \rightarrow 0$). Then Laplace's equation simplifics to

$$\frac{1}{r^2}\frac{\partial}{\partial r}\left(r^2\frac{\partial\phi_e}{\partial r}\right) + \frac{1}{r^2 \sin\theta}\frac{\partial}{\partial\theta}\left(\sin\theta\frac{\partial\phi_e}{\partial\theta}\right) = 0. \tag{8.5.15}$$

This assumption holds if the polar axis of the spherical coordinates is chosen to be parallel with the axis of dipole source. If we assume that $u = \cos(\theta)$, then

$$\frac{\partial}{\partial\theta} = \left(\frac{\partial u}{\partial\theta}\right)\frac{\partial}{\partial u} = -\sin\theta\frac{\partial}{\partial u} = -\sqrt{1-u^2}\frac{\partial}{\partial u} \tag{8.5.16}$$

and

$$\frac{1}{\sin\theta}\frac{\partial}{\partial\theta} = -\frac{\partial}{\partial u}. \tag{8.5.17}$$

With this change of variable, (8.5.15) becomes

$$\frac{\partial}{\partial r}\left(r^2\frac{\partial\phi_e}{\partial r}\right) + \frac{\partial}{\partial u}\left[(1-u^2)\frac{\partial\phi_e}{\partial u}\right] = 0. \tag{8.5.18}$$

Since the potential $\phi_e(r, \theta)$ is a function of r and θ only, (8.5.18) can be solved by the method of separation of variables as follows: Assume

$$\phi_e = f_1(r)f_2(u), \tag{8.5.19}$$

where $f_1(r)$ is a function of r only and $f_2(u)$ is a function of only u (or $\cos\theta$). Substitution of (8.5.19) into (8.5.18) results in

$$f_2(u)\frac{d}{dr}\left[r^2\frac{df_1(r)}{dr}\right] + f_1(r)\frac{d}{du}\left[(1-u^2)\frac{df_2(u)}{du}\right] = 0,$$

or

$$\frac{1}{f_1(r)}\frac{d}{dr}\left[r^2\frac{df_1(r)}{dr}\right] + \frac{1}{f_2(u)}\frac{d}{du}\left[(1-u^2)\frac{df_2(u)}{du}\right] = 0. \tag{8.5.20}$$

Since (8.5.20) holds for all r and θ, each term must be equal to a constant. Therefore

$$\frac{1}{f_1(r)}\frac{d}{dr}\left[r^2\frac{df_1(r)}{dr}\right] = c = n(n+1) \tag{8.5.21}$$

and

$$\frac{1}{f_2(u)}\frac{d}{du}\left[(1-u^2)\frac{df_2(u)}{du}\right] = -c = -n(n+1). \tag{8.5.22}$$

Mathematically c can be any constant. However, n should be an integer, including zero for $f_2(u)$ to be well-behaved (see Appendix D). The negative value of n is degenerate and therefore is not considered. Hence (8.5.22) can be expressed as

$$(1-u^2)\frac{d^2f_2(u)}{du^2} - 2u\frac{df_2(u)}{du} + n(n+1)f_2(u) = 0. \tag{8.5.23}$$

This is Legendre's equation, and solutions for $f_2(u)$ are the Legendre polynomials considered in Appendix D. Hence

$$f_2(u) = P_n(u) = P_n(\cos\theta). \tag{8.5.24}$$

Equation (8.5.21) can be rearranged as follows:

$$r^2\frac{d^2f_1(r)}{dr^2} + 2r\frac{df_1(r)}{dr} - n(n+1)f_1(r) = 0. \tag{8.5.25}$$

Assume that

$$f_1(r) = r^\ell. \tag{8.5.26}$$

On substituting (8.5.26) into (8.5.25), we have

$$r^2\ell(\ell-1)r^{\ell-2} + 2r\ell r^{\ell-1} - n(n+1)r^\ell = r^\ell\lfloor \ell^2 + \ell - n(n+1)\rfloor = 0.$$

The quadratic equation in ℓ gives two solutions as follows:

$$[\ell^2 + \ell - n(n+1)] = 0 \rightarrow \ell = \begin{cases} n, \\ -(n+1). \end{cases}$$

Hence two possible solutions to (8.5.25) are found:

$$f_1(r) = \begin{cases} r^n, \\ r^{-(n+1)}. \end{cases} \tag{8.5.27}$$

Then, on substituting (8.5.24) and (8.5.27) into (8.5.19), also two possible solutions of (8.5.15) are found:

$$\phi_e = \begin{cases} r^n P_n(\cos\theta), \\ r^{-(n+1)}P_n(\cos\theta). \end{cases} \tag{8.5.28}$$

Since n can be 0 or any positive integer, the general solution of (8.5.15) is

$$\phi_e = \sum_{n=0}^{\infty}[a_n r^n P_n(\cos\theta) + b_n r^{-(n+1)}P_n(\cos\theta)], \tag{8.5.29}$$

where a_n and b_n are arbitrary constants to be evaluated by the boundary conditions.

Since the source is a dipole, only a discrete mode of solution is allowed in (8.5.29). This is because the finite boundaries have little effects at points close to the dipole source (i.e., $r \rightarrow \delta$) and therefore the potential setup by the dipole must behave as if it is in an infinite medium. From Example 8.5, we know that the potential of a dipole in an infinite region behaves as follows:

$$\phi_e|_{r\to\delta} \sim \frac{1}{r^2}.$$

That means only $n = 1$ mode is the possible solution in (8.5.29) for this dipole source. Therefore (8.5.29) reduces to

$$\phi_e = (a_1 r + b_1 r^{-2}) P_1(\cos\theta) = (a_1 r + b_1 r^{-2}) \cos\theta. \tag{8.5.30}$$

Using this solution, we can express the potential field in each region as follows:

In brain: $\delta < r < r_1$,

$$\phi_{e1} = (A_1 r + B_1 r^{-2}) \cos\theta. \tag{8.5.31}$$

In skull, $r_1 < r < r_2$,

$$\phi_{e2} = (A_2 r + B_2 r^{-2}) \cos\theta. \tag{8.5.32}$$

In scalp, $r_2 < r < r_3$,

$$\phi_{e3} = (A_3 r + B_3 r^{-2}) \cos\theta. \tag{8.5.33}$$

In space, $r > r_3$,

$$\phi_{e4} = B_4 r^{-2} \cos\theta. \tag{8.5.34}$$

The $A_4 r$ term is not allowed in (8.5.34) because the potential ϕ_{e4} must be zero as $r \to \infty$.

There are seven unknown integration constants in (8.5.31) through (8.5.34). The coefficient B_1 can be related to the intensity of the dipole and the remaining six coefficients can be determined from six independent boundary conditions at $r = r_1$, $r = r_2$, and $r = r_3$. As mathematically expressed below, these boundary conditions require that the potential and the normal component of current density be continuous across the boundaries.

At $r = r_1$,

$$\phi_{e1}(r_1, \theta) = \phi_{e2}(r_1, \theta), \tag{8.5.35}$$

$$\sigma_1 \frac{\partial \phi_{e1}}{\partial r}\bigg|_{r=r_1} = \sigma_2 \frac{\partial \phi_{e2}}{\partial r}\bigg|_{r=r_1}. \tag{8.5.36}$$

At $r = r_2$,

$$\phi_{e2}(r_2, \theta) = \phi_{e3}(r_2, \theta), \tag{8.5.37}$$

$$\sigma_2 \frac{\partial \phi_{e2}}{\partial r}\bigg|_{r=r_2} = \sigma_3 \frac{\partial \phi_{e3}}{\partial r}\bigg|_{r=r_2}. \tag{8.5.38}$$

At $r = r_3$,

$$\phi_{e3}(r_3, \theta) = \phi_{e4}(r_3, \theta), \tag{8.5.39}$$

$$\frac{\partial \phi_{e3}}{\partial r}\bigg|_{r=r_3} = 0. \tag{8.5.40}$$

Equation (8.5.40) is based on the fact that no normal component of the electric field (or normal component of the current) exists at the boundary between the biological body and air. This condition is also evident if the conductivity of the air is assumed to be zero.

Substitution of (8.5.31) through (8.5.34) into (8.5.35) through (8.5.40) leads to the following six equations for the unknown coefficients:

$$r_1 A_1 - r_1 A_2 - \frac{1}{r_1^2} B_2 = -\frac{1}{r_1^2} B_1, \tag{8.5.41}$$

$$\sigma_1 A_1 - \sigma_2 A_2 + \frac{2\sigma_2}{r_1^3} B_2 = \frac{2\sigma_1}{r_1^3} B_1, \tag{8.5.42}$$

$$r_2 A_2 - r_2 A_3 + \frac{1}{r_2^2} B_2 - \frac{1}{r_2^2} B_3 = 0, \tag{8.5.43}$$

$$\sigma_2 A_2 - \sigma_3 A_3 - \frac{2\sigma_2}{r_2^3} B_2 + \frac{2\sigma_3}{r_2^3} B_3 = 0, \tag{8.5.44}$$

$$r_3 A_2 + \frac{1}{r_3^2} B_3 - \frac{1}{r_3^2} B_4 = 0, \tag{8.5.45}$$

$$A_3 - \frac{2}{r_3^3} B_3 = 0. \tag{8.5.46}$$

From (8.5.41) through (8.5.46), A_1, A_2, A_3, B_2, B_3, and B_4 can be expressed in terms of B_1, which, in turn, can be related to the magnitude of the dipole source. We are mainly interested in A_3 and B_3 because these describe the scalp potential distribution that is the source of the EEG signal. Therefore we determine only A_3 and B_3 here as follows.

From (8.5.46),

$$B_3 = \frac{r_3^3}{2} A_3 \tag{8.5.47}$$

Substituting (8.5.47) into (8.5.43) and (8.5.44) and then eliminating A_3 from (8.5.43) and (8.5.44), we find that

$$B_2 = MA_2, \tag{8.5.48}$$

where

$$M = \frac{r_2^3\big[(1-\sigma_3/\sigma_2) + (r_3^3/r_2^3)[(\sigma_3/\sigma_2) + (1/2)]\big]}{[(\sigma_3/\sigma_2) + 2] - (r_3^3/r_2^3)[(\sigma_3/\sigma_2) - 1]}. \tag{8.5.49}$$

At the same time, A_3 is found to be

$$A_3 = \frac{1 + (M/r_2^3)}{1 + (r_3^3/2r_2^3)} A_2. \tag{8.5.50}$$

Next, substituting (8.5.48) into (8.5.41) and (8.5.42), and then eliminating A_1, we have

$$A_2 = \frac{3}{[1 - (\sigma_2/\sigma_1)]r_1^3 + [1 + (2\sigma_2/\sigma_1)]M} B_1. \tag{8.5.51}$$

Combining (8.5.50) and (8.5.51), we find that

$$A_3 = \frac{9B_1}{M_1 + M_2}. \tag{8.5.52}$$

where

$$M_1 = \left(1 - \frac{\sigma_2}{\sigma_1}\right)\left[\left(\frac{\sigma_3}{\sigma_2} + 2\right) - \frac{r_3^3}{r_2^3}\left(\frac{\sigma_3}{\sigma_2} - 1\right) r_1^3\right] \tag{8.5.53}$$

and

$$M_2 = \left(1 + \frac{2\sigma_2}{\sigma_1}\right)\left[\left(1 - \frac{\sigma_3}{\sigma_2}\right) + \frac{r_3^3}{r_2^3}\left(\frac{\sigma_3}{\sigma_2} + \frac{1}{2}\right) r_2^3\right]. \tag{8.5.54}$$

As was mentioned earlier, B_1 is related to the magnitude of the dipole source. Recall from (8.1.9) that the potential produced at a point (r, θ) by the dipole immersed in an infinite medium of conductivity σ_1 is given as

$$\phi_e(r, \theta) = \frac{I_o d}{4\pi\sigma_1 r^2} \cos\theta. \tag{8.1.9}$$

The potential near the dipole source located at the center of the brain is expected to behave similarly. Hence

$$\phi_{e1}(r, \theta)|_{r\to 0} = \frac{B_1}{r^2}\cos\theta \approx \frac{I_o d}{4\pi\sigma_1} \cdot \frac{1}{r^2} \cos\theta. \tag{8.5.55}$$

Then the determination of B_1 proceeds as follows:

$$B_1 = \frac{I_o d}{4\pi\sigma_1}. \tag{8.5.56}$$

Therefore the potential's distribution on the scalp is found to be

$$\begin{aligned}\phi_{e3}(r, \theta) &= \left(A_3 r + \frac{B_3}{r^2}\right)\cos\theta = A_3\left(r + \frac{r_3^3}{2r^2}\right)\cos\theta\\ &= \frac{r_3}{2}A_3\left[2\left(\frac{r}{r_3}\right) + \left(\frac{r_3}{r}\right)^2\right]\cos\theta,\end{aligned}$$

or

$$\phi_{e3}(r, \theta) = \frac{I_o d}{4\pi\sigma_1}N\left[2\left(\frac{r}{r_3}\right) + \left(\frac{r_3}{r}\right)^2\right]\cos\theta, \tag{8.5.57}$$

where

$$N = \frac{9r_3}{2(M_1 + M_2)}. \tag{8.5.58}$$

Potential difference between a lead pair placed at the front and the back of scalp (or aligned in the direction of dipole moment) is found to be

$$V = \phi_{e3}(r_3, 0) - \phi_{e3}(r_3, \pi) = \frac{3I_o d}{2\pi\sigma_1}N. \tag{8.5.59}$$

The potential difference between any two other points can be found directly from (8.5.57).

PROBLEMS

8.1. The radius of a thin circular ring is 30 cm. It carries an electrical charge with density 5 C/m. Find the electrical potential and the electric field intensity at a point (0, 0, 10 m) on its axis.

8.2. A point charge Q of 25 μC is located at point (3, 1, 5). Find electric field intensity at (7, 9, 1) and the force on a charge q of 2 nC at this point.

8.3. A uniformly charged thin conductor is lying along the z-axis. If the conductor is infinitely long and carries a straight-line charge density of 15 C/m, find its electric field intensity.

8.4. The radius of a uniformly charged disk is 1 m. If the surface charge density on the disk is 3.5 C/m^2, find the scalar electrical potential and the electric field intensity on its axis.

8.5. Two charges of 5 C and -5 C are separated by 1 cm. Find the scalar potential and electric field intensity at an arbitrary point 2 m away from it.

8.6. An infinitely long charge filament with a charge density of 5 C/m lies along the z-axis. Use Gauss's law to determine the electric field intensity.

8.7. The radius of a perfectly conducting sphere is 50 cm. It has an electrical potential of 50 V. Use Laplace's equation to find the electrical charge density on the sphere's surface and the electric field intensity around it.

8.8. Find the capacitance of the conducting sphere of Problem 8.7.

8.9. A distance of 1 cm separates two conducting sheets, each with a surface area of 4 m^2. A perfect dielectric material of $\varepsilon_r = 2.3$ fills the space between the two plates. Neglect the fringing fields at the edges to find the capacitance of this arrangement. Also find the energy stored in this capacitor.

8.10. Two conducting parallel plates are used to form a capacitor. Each plate has an area of 0.5 m^2 and a distance of 1 mm separates the plates. The space between the two plates is filled with a dielectric material. However, its dielectric constant is not known. By way of a capacitance bridge, its capacitance is found to be 120 nF. Find the dielectric constant of the material.

8.11. The parallel-plate capacitor of Problem 8.10 uses air as its dielectric. If the separation d of the parallel-plate capacitor is changing over time as given below, find the capacitance as a function of time and plot the characteristics:

$$d = 0.9 - 0.1\sin^2(60\pi t)\ \text{mm}.$$

8.12. Consider the capacitor of Example 8.10. However, the space between the plates has two different dielectric materials. A 0.6 mm thick dielectric sheet of $\varepsilon_r = 4.3$ is used as one medium and another dielectric fills the remaining space. If second medium has a dielectric constant of 2.1, find the total capacitance.

8.13. An air-filled coaxial line of inner and outer conductor radii 1 mm and 3 mm, respectively, has a voltage of 5 V. Using Gauss's law, find potential distribution in the space between its conductors, the electric field intensity, and the capacitance per meter of its length.

8.14. The inner and outer radii of two conducting concentric spheres are 1 mm and 3 mm, respectively. The inner sphere is at a potential of 5 V with respect to the outer one. Use Gauss's law to find the potential's distribution, the electric field intensity, and the capacitance of this system.

8.15. Consider Problem 8.13 and use Laplace's equation to find the potential's distribution inside the coaxial line, the electric field intensity, and the capacitance per meter of its length.

8.16. Reconsider Example 8.14 and use Laplace's equation to find the potential's distribution between the concentric spheres, the electric field intensity, and its capacitance.

8.17. A silicon pn-junction diode has $N_d = 10^{21}\ m^{-3}$ and $N_a = 10^{25}\ m^{-3}$ on its n- and p-sides, respectively. If the built-in potential is 0.62 V, find length d_o of its depletion region and the lengths of the space-charge regions on each side of the junction. Assume that the dielectric constant of silicon is 11.8.

9

MAGNETOSTATIC FIELDS

As was mentioned earlier, electric and magnetic fields become independent of each other when the electrical charge and current sources producing them are constant over time. In other words, the electrical charge that is constant over time is responsible for the electric field and the electrical current that is constant over time is responsible for the magnetic field. These fields of constant charge and current are known as the electrostatic and magnetostatic fields, respectively. Electrostatic fields were considered in Chapter 8. Here we cover magnetostatic fields. The chapter begins with a section on the laws that govern magnetostatic fields. Concepts of inductance and energy storage in magnetic fields are presented. A brief discussion on magnetic materials is included in a separate section. The chapter ends with an introduction to magnetic circuits.

9.1 LAWS OF MAGNETOSTATIC FIELDS

For an electrical source constant over time Ampere's law (3.1.2) reduces to

$$\oint_C \vec{H} \cdot d\vec{\ell} = \int_S \vec{J} \cdot d\vec{s} = I. \tag{9.1.1}$$

Practical Electromagnetics: From Biomedical Sciences to Wireless Communication.
By Devendra K. Misra
Copyright © 2007 John Wiley & Sons, Inc.

Similarly its differential form (3.2.9) is

$$\nabla \times \vec{H} = \vec{J}. \tag{9.1.2}$$

Gauss's law for the magnetic field (3.1.4) in a homogeneous medium gives

$$\oint_S \vec{\mathscr{B}} \cdot d\vec{s} = 0 \to \int_V \nabla \cdot (\mu \vec{H}) dv = 0 \to \oint_S \vec{H} \cdot d\vec{s} = 0. \tag{9.1.3}$$

Its differential form (3.2.11) gives

$$\nabla \cdot \vec{B} = 0 \to \nabla \cdot (\mu \vec{H}) = 0 \to \nabla \cdot \vec{H} = 0. \tag{9.1.4}$$

Equation (9.1.2) is used to find the so-called vector Poisson's equation:

$$\nabla \times \nabla \times \vec{A} = -\mu \vec{J}_{\mathrm{e}}. \tag{9.1.5}$$

The magnetic vector potential given by (B.33) simplifies for magnetostatic fields as follows:

$$\vec{A}(\vec{r}) = \frac{\mu}{4\pi} \int_V \frac{\vec{J}_{\mathrm{e}}}{R} dv'. \tag{9.1.6}$$

The magnetic flux φ_{m} passing through a given area S bound by C is found to be

$$\varphi_{\mathrm{m}} = \int_S \vec{B} \cdot d\vec{s}' = \int_S (\nabla \times \vec{A}) \cdot d\vec{s}' = \oint_c \vec{A} \cdot d\vec{\ell}. \tag{9.1.7}$$

Example 9.1

An infinitely long solenoid of radius a has n turns per unit length of a filamentary wire, as shown in Figure 9.1. It carries a current I. Find its magnetic field using Ampere's law.

Two different views of the solenoid are included in Figure 9.1. As the problem statement says, the wire is assumed to be thin and the turns are close to each other. Since its geometry is circularly symmetric and the solenoid is infinitely long, the magnetic field that it produces is independent of ϕ and z. Hence its magnetic field intensity can be expressed as follows:

$$\vec{H} = \hat{\rho} H_\rho(\rho) + \hat{\phi} H_\phi(\rho) + \hat{z} H_z(\rho).$$

In order to determine the radial component of magnetic field, we imagine a cylindrical volume of radius ρ and length ℓ inside the solenoid, as shown in

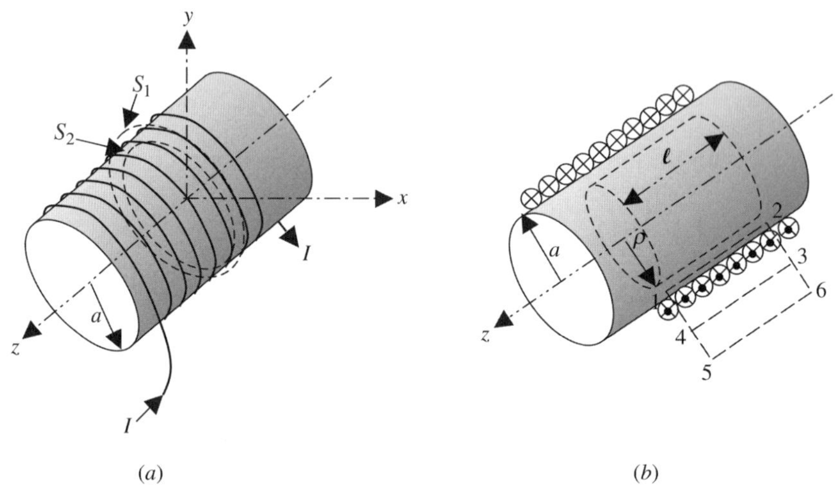

Figure 9.1 Geometry of a solenoid.

Figure 9.1*b*. Using (9.1.4), we find that the surface integrals on its two ends cancel out each other because H_z enters one end surface and leaves from the other. On the lateral surface we have

$$\oint_S \vec{H} \cdot d\vec{s} = \int_{z=0}^{\ell} \int_{\phi=0}^{2\pi} H_\rho(\rho)\hat{\rho} \cdot \hat{\rho}\rho d\phi dz = 2\pi\rho\ell H_\rho(\rho) = 0.$$

We conclude that

$$H_\rho(\rho) = 0.$$

Next, we consider the two circular surfaces S_1 and S_2 that are bounded by C_1 and C_2, respectively, shown in Figure 9.1*a*. As indicated, surface S_2 is inside the solenoid whereas S_1 extends outside. The current flowing through the winding is along the boundary, and therefore it does not leave (or cut) these surfaces. Hence applying (9.1.1) on either surface gives

$$\oint_C \vec{H} \cdot d\vec{\ell} = \oint_{C_1 \text{ or } C_2} H_\phi(\rho)\rho d\phi = 0 \rightarrow H_\phi(\rho) = 0.$$

Now using (9.1.1) on the area bound by 3–4–5–6, as shown is Figure 9.1*b*, we find that

$$\oint_C \vec{H} \cdot d\vec{\ell} = \int_3^4 H_z(\rho)dz + \int_4^5 H_\rho(\rho)d\rho + \int_5^6 H_z(\rho)dz + \int_6^3 H_\rho(\rho)d\rho = 0.$$

Since we have already found that the radial magnetic field is zero, this expression reduces to

$$\int_3^4 H_z(\rho)dz + \int_5^6 H_z(\rho)dz = \ell\Big[H_z(\rho)\big|_3^4 - H_z(\rho)\big|_5^6\Big] = 0.$$

This condition is satisfied even when H_z is constant in this region. However, the field is bounded, so it has to be zero at infinity. Thus we can conclude that it is zero everywhere outside the solenoid.

Finally, using (9.1.1) over the area bound by 1–2–3–4, as shown in Figure 9.1*b*, we have

$$\oint_C \vec{H} \cdot d\vec{\ell} = \int_2^1 H_z(\rho)dz + \int_1^4 H_\rho(\rho)d\rho + \int_4^3 H_z(\rho)dz + \int_3^2 H_\rho(\rho)d\rho = n\ell I.$$

Note that the radial magnetic field is zero everywhere and also H_z is zero outside the solenoid. Therefore this expression reduces to

$$\int_2^1 H_z(\rho)dz = H_z(\rho)\ell = n\ell I \rightarrow H_z(\rho) = nI, \qquad \rho < a.$$

We can conclude that

$$\vec{H} = \begin{cases} \hat{z}nI, & \rho < a, \\ 0, & \rho > a. \end{cases}$$

The corresponding magnetic flux density is give by

$$\vec{B} = \begin{cases} \hat{z}\mu nI, & \rho < a, \\ 0, & \rho < a. \end{cases}$$

Example 9.2

The mean radius of the toroid shown in Figure 9.2 is *a*, and it has *N* turns of a filamentary wire covering it. The cross section of the core is circular with a radius *b*. There is a current *I* flowing through the winding. Find the magnetic flux density using Ampere's law.

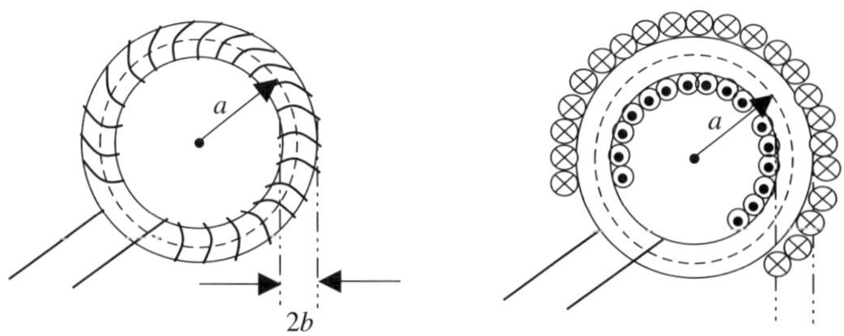

Figure 9.2 Geometry of a toroid.

The toroid may be considered a solenoid with two ends brought together as shown in Figure 9.2. Therefore Ampere's law gives

$$\oint_C \vec{H} \cdot d\vec{\ell} = \begin{cases} H_\phi(\rho) \int_{\phi=0}^{2\pi} \rho d\phi = 2\pi\rho H_\phi(\rho) = NI, & a - b < \rho < a + b, \\ 0, & \rho < a - b, \ \rho > a + b. \end{cases}$$

The magnetic flux density is found to be

$$\vec{B} = \begin{cases} \hat{\phi} \dfrac{\mu NI}{2\pi\rho}, & a - b < \rho < a + b, \\ 0, & \rho < a - b, \ \rho > a + b. \end{cases}$$

For $a \gg b$ this relation simplifies to

$$\vec{B} \approx \begin{cases} \hat{\phi} \dfrac{\mu NI}{2\pi a}, & a - b < \rho < a + b, \\ 0, & \rho < a - b, \ \rho > a + b. \end{cases}$$

Biot-Savart Law

Ampere's law is easy to use when the structure under consideration possesses certain symmetry. It is not that useful if the symmetry does not exist. All kinds of problems can be solved using the potential functions discussed in Chapter 6. In magnetostatic fields we work with (9.1.6) as follows: If an incremental length of the conductor with its cross section S' carries a current I, then we find that

$$\vec{J} dv' = JS' d\vec{\ell}' = I d\vec{\ell}'.$$

Therefore (9.1.6) may be written as follows:

$$\vec{A}(\vec{r}) = \frac{\mu_o I}{4\pi} \int_c \frac{1}{R} d\vec{\ell}'. \tag{9.1.8}$$

The corresponding magnetic flux density is found to be

$$\vec{B} = \nabla \times \vec{A}(\vec{r}) = \frac{\mu_o I}{4\pi} \int_c \nabla \times \left(\frac{d\vec{\ell}'}{R} \right). \tag{9.1.9}$$

Since

$$\nabla \times \left(\frac{d\vec{\ell}'}{R} \right) = \frac{1}{R} \nabla \times d\vec{\ell}' + \nabla\left(\frac{1}{R}\right) \times d\vec{\ell}' = \nabla\left(\frac{1}{R}\right) \times d\vec{\ell}' = -\frac{\hat{R}}{R^2} \times d\vec{\ell}' = \frac{d\vec{\ell}' \times \hat{R}}{R^2},$$

(9.1.9) simplifies to

$$\vec{B} = \frac{\mu_o I}{4\pi} \int_c \frac{d\vec{\ell}' \times \hat{R}}{R^2} = \frac{\mu_o I}{4\pi} \int_c \frac{d\vec{\ell}' \times \vec{R}}{R^3}. \tag{9.1.10}$$

This relation is known as the Biot-Savart law because it was originally proposed by these two great scientists. The Biot-Savart relation can be used to find the magnetic flux density (and hence the magnetic field's intensity as well) for the cases even without symmetry.

Example 9.3

A thin wire of length $2L$ lies symmetrically along the z-axis, as shown in Figure 9.3. If this wire carries a current I, determine its magnetic field at a point on the x-y plane using the Biot-Savart law.

The position vector for point P is $\vec{r} = \hat{\rho}\rho$. We consider a small length on this wire with its position vector as $\vec{r}\,' = \hat{z}z'$. Therefore the vector distance from the source point to P on the x-y plane is found to be

$$\vec{R} = \vec{r} - \vec{r}' = \hat{\rho}\rho - \hat{z}z'.$$

The magnitude (separation between the incremental length and point P) is given as follows:

$$R = |\vec{R}| = \sqrt{\rho^2 + z'^2}.$$

Using (9.1.10), we find that

$$\vec{B} = \frac{\mu_o I}{4\pi}\int_c \frac{d\vec{\ell}' \times \vec{R}}{R^3} = \frac{\mu_o I}{4\pi}\int_{-L}^{L} \frac{dz'\hat{z} \times (\hat{\rho}\rho - \hat{z}z')}{(\rho^2 + z'^2)^{3/2}} = \hat{\phi}\,\frac{\mu_o I\rho}{4\pi}\int_{-L}^{L} \frac{dz'}{(\rho^2 + z'^2)^{3/2}}.$$

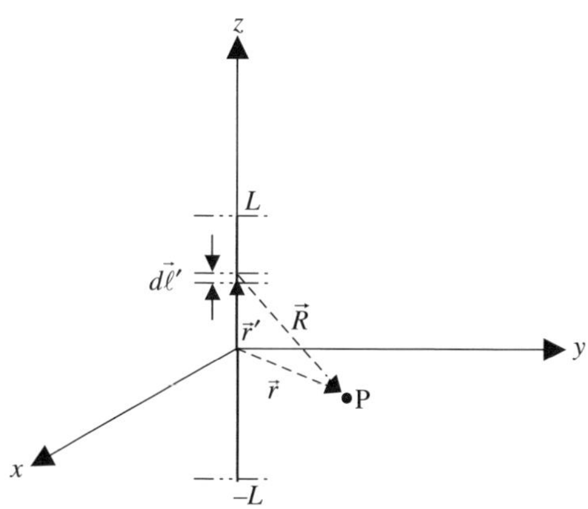

Figure 9.3 Geometry of a thin wire of length $2L$ with current I.

Since the integrand is an even function of z', by the integration formula 4 from Appendix A we find that

$$\vec{B} = \hat{\phi}\frac{\mu_o I\rho}{2\pi}\int_0^L \frac{dz'}{(\rho^2 + z'^2)^{3/2}} = \hat{\phi}\frac{\mu_o I\rho}{2\pi}\cdot\frac{z'}{\rho^2\sqrt{\rho^2 + z'^2}}\bigg|_{z'=0}^{L} = \hat{\phi}\frac{\mu_o IL}{2\pi\rho\sqrt{\rho^2 + L^2}}\ \text{T}.$$

For L $\to \infty$, the formula reduces to

$$\vec{B} = \hat{\phi}\frac{\mu_o I}{2\pi\rho}\text{T} \to \vec{H} = \frac{\vec{B}}{\mu_o} = \hat{\phi}\frac{I}{2\pi\rho}\ \text{A/m}.$$

This result can be verified easily using Ampere's law.

Example 9.4

A loop filament of radius a lies on the x-y plane, as shown in Figure 9.4. It carries a current I. Find the magnetic flux density at a point on the loop's axis.

The vector incremental length can be expressed as follows:

$$d\vec{\ell}' = \hat{\phi}' a d\phi' = (-\hat{x}\sin\phi' + \hat{y}\cos\phi')ad\phi'.$$

Note that the unit vector ϕ changes its direction from 0 to 2π. Therefore it is changed to unit vectors in rectangular coordinates. The position vector for the field point is $\vec{r} = \hat{z}z$. The position vector for a source point (the incremental length) is

$$\vec{r}' = \hat{\rho}' a = (\hat{x}\cos\phi' + \hat{y}\sin\phi')a.$$

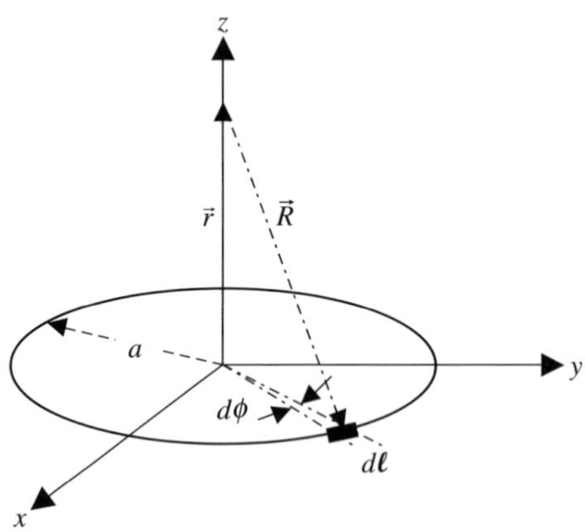

Figure 9.4 Geometry of the wire loop lying on the x-y plane.

Therefore the vector directed from the field point to the source point is found to be

$$\vec{R} = \vec{r} - \vec{r}' = (\hat{z}z - \hat{\rho}'a).$$

The separation between the field and source points is

$$R = |\vec{R}| = \sqrt{a^2 + z^2}.$$

Using (9.1.10), we have

$$\vec{B} = \frac{\mu_o I}{4\pi} \int_0^{2\pi} \frac{\hat{\phi}' a d\phi' \times (\hat{z}z' - \hat{\rho}' a)}{(a^2 + z^2)^{3/2}} = \frac{\mu_o I a}{4\pi(a^2 + z^2)^{3/2}} \int_0^{2\pi} [\hat{\rho}' z \, d\phi' + \hat{z} a d\phi'],$$

or

$$\vec{B} = \frac{\mu_o I a}{4\pi(a^2 + z^2)^{3/2}} \int_0^{2\pi} [(\hat{x} \cos \phi' + \hat{y} \sin \phi') z \, d\phi' + \hat{z} a d\phi'] = \hat{z} \frac{\mu_o I a^2}{2(a^2 + z^2)^{3/2}} \text{ T}.$$

Example 9.5

Using the results of Example 9.4, find the magnetic flux density due to current I flowing through a solenoid with N turns over its length ℓ. Compare your result with that obtained in Example 9.1.

The expression of magnetic flux density obtained in Example 9.4 can be generalized for an incremental length dz of a solenoid if we replace its current as follows:

$$I \to \frac{NI}{\ell} dz.$$

Therefore the magnetic flux density due to an incremental length dz may be expressed as follows:

$$d\vec{B} = \hat{z} \frac{\mu_o N I a^2}{2\ell(a^2 + z^2)^{3/2}} dz \text{ T}.$$

To find the magnetic flux density in the middle of solenoid, we integrate this expression as follows:

$$\begin{aligned} \vec{B}(z = 0) &= \hat{z} \frac{\mu_o N I a^2}{2\ell} \int_{-\ell/2}^{\ell/2} \frac{dz}{(a^2 + z^2)^{3/2}} \text{ T} \\ &= \hat{z} \frac{\mu_o N I a^2}{2\ell} \left[\frac{z}{a^2\sqrt{a^2 + z^2}} \Bigg|_{z=-\ell/2}^{\ell/2} \right] \\ &= \hat{z} \frac{\mu_o N I}{2\sqrt{a^2 + (\ell/2)^2}}. \end{aligned}$$

For $a \ll \ell/2$, this simplifies to

$$\vec{B}(z=0) \approx \hat{z}\frac{\mu_o NI}{\ell}.$$

Since N/ℓ represents n (the number of turns per unit length), this result is the same as found in Example 9.1.

Further we can find the magnetic flux density at the end of the solenoid if we change the integration limits as follows:

$$\begin{aligned}\vec{B}(z=0) &= \hat{z}\frac{\mu_o NIa^2}{2\ell}\int_0^{\ell}\frac{dz}{(a^2+z^2)^{3/2}}\ \mathrm{T}\\ &= \hat{z}\frac{\mu_o NIa^2}{2\ell}\left[\left.\frac{z}{a^2\sqrt{a^2+z^2}}\right|_{z=0}^{\ell}\right]\\ &= \hat{z}\frac{\mu_o NI}{2\sqrt{a^2+\ell^2}}.\end{aligned}$$

Hence the magnetic flux density decreases at the ends to half of its value that existed in the middle of solenoid.

Example 9.6

A filamentary wire loop of radius a lies on the x-y plane, as shown in Figure 9.5. The loop carries a current I. Find its magnetic field at a faraway point ($r \gg a$).

There are at least two different ways to find the magnetic vector potential for this case. Since there is symmetry in ϕ direction, a field point can be selected on the y-z plane without loss of generality. Hence the position vector for the field point P is

$$\vec{r} = \hat{r}r = \hat{x}x + \hat{z}z.$$

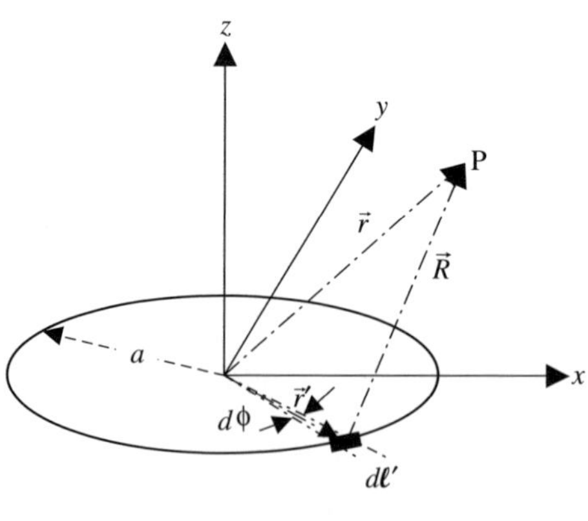

Figure 9.5 Current loop magnetic dipole.

The Position vector for the incremental length (the source point) is

$$\vec{r}' = \hat{\rho}' a = (\hat{x}\cos\phi' + \hat{y}\sin\phi')a.$$

Therefore

$$\vec{R} = \vec{r} - \vec{r}' = \hat{x}(x - a\cos\phi') - \hat{y}a\sin\phi' + \hat{z}z$$

and

$$R = |\vec{R}| = \sqrt{(x - a\cos\phi')^2 + a^2\sin^2\phi' + z^2} = \sqrt{x^2 + z^2 + a^2 - 2ax\cos\phi'}.$$

This can be changed to spherical coordinates as follows:

$$R = \sqrt{r^2 + a^2 - 2ar\sin\theta\ \cos\phi'}.$$

For $r \gg a$, we find that

$$\frac{1}{R} = \frac{1}{r[1 + (a^2 - 2ar\sin\theta\cos\phi')/r^2]^{1/2}} \approx \frac{1}{r}\left(1 + \frac{a}{r}\sin\theta\cos\phi'\right).$$

The vector representation for the incremental length is

$$d\vec{\ell}' = \hat{\phi}' a d\phi' = (-\hat{x}\sin\phi' + \hat{y}\cos\phi')ad\phi'.$$

Therefore (9.1.10) gives

$$\begin{aligned}\vec{A} &= \frac{\mu_o Ia}{4\pi r}\int_0^{2\pi}(-\hat{x}\sin\phi' + \hat{y}\cos\phi')\left[1 + \frac{a}{r}\sin\theta\cos\phi'\right]d\phi' \\ &= \frac{\mu_o Ia^2\sin\theta}{4\pi r^2}\int_0^{2\pi}(-\hat{x}\sin\phi' + \hat{y}\cos\phi')\cos\phi' d\phi',\end{aligned}$$

or

$$\vec{A} = \frac{\mu_o Ia^2\sin\theta}{4\pi r^2}\int_0^{2\pi}(-\hat{x}\sin\phi' + \hat{y}\cos\phi')\cos\phi' d\phi' = \hat{y}\frac{\mu_o I\pi a^2\sin\theta}{4\pi r^2}.$$

Since the rotation of the loop on x-y plane does not change its response (rotational symmentry of the loop), it can be expressed as follows:

$$\vec{A} = \hat{\phi}\frac{\mu_o I\pi a^2\sin\theta}{4\pi r^2} = \mu_o\frac{\vec{m}\times\hat{r}}{4\pi r^2},$$

where $\vec{m} = \hat{z}I\pi a^2 =$ magnetic dipole moment.

Alternatively, we start with (9.1.8) as follows:

$$\vec{A} = \frac{\mu_o}{4\pi}\oint_C\frac{Id\vec{\ell}'}{R} = \frac{\mu_o I}{4\pi}\oint_C\frac{d\vec{\ell}'}{R}.$$

Using vector identity 15 from Appendix A, we find that

$$\oint_C \frac{d\vec{\ell}'}{R} = \int_S \hat{z} \times \nabla'\left(\frac{1}{R}\right) ds',$$

where the surface S is bound by C.

From Example 2.20 we know that

$$\nabla'\left(\frac{1}{R}\right) = \frac{\hat{R}}{R^2}.$$

Therefore

$$\oint_C \frac{d\vec{\ell}'}{R} = \int_S \hat{z} \times \nabla'\left(\frac{1}{R}\right) ds' = \int_S \frac{\hat{z} \times \hat{R}}{R^2} ds'.$$

For $r \gg r'$, $\hat{R}/R^2 \approx \hat{r}/r^2$, and therefore it simplifies to

$$\oint_C \frac{d\vec{\ell}'}{R} = \int_S \frac{\hat{z} \times \hat{R}}{R^2} ds' \approx \frac{\hat{z} \times \hat{r}}{r^2} \int_S ds' = \frac{\hat{z} \times \hat{r}}{r^2} S.$$

Note that this expression is independent of the shape of the loop.

Hence (9.1.8) gives

$$\vec{A} \approx \frac{\mu_o IS}{4\pi r^2}(\hat{z} \times \hat{r}) = \frac{\mu_o IS}{4\pi r^2}\left(\hat{r}\cos\theta - \hat{\theta}\sin\theta\right) \times \hat{r} = \hat{\phi}\,\frac{\mu_o IS}{4\pi r^2}\sin\theta.$$

This expression is same as what we found earlier. However, it is the most general formulation that shows the vector magnetic potential of the loop to be a function of the surface area S and independent of the surface's shape. The corresponding magnetic flux density is found to be

$$\vec{B} = \nabla \times \vec{A} = \hat{r}\frac{1}{r\sin\theta}\frac{\partial}{\partial\theta}\left(\frac{\mu_o IS}{4\pi r^2}\sin^2\theta\right) + \hat{\theta}\,\frac{1}{r}\left[-\frac{\partial}{\partial r}\left(r\frac{\mu_o IS}{4\pi r^2}\sin\theta\right)\right],$$

or

$$\vec{B} = \frac{\mu_o IS}{4\pi r^3}(\hat{r}2\cos\theta + \hat{\theta}\sin\theta)\ \mathrm{T}$$

and

$$\vec{H} = \frac{\vec{B}}{\mu_o} = \frac{IS}{4\pi r^3}(\hat{r}2\cos\theta + \hat{\theta}\sin\theta)\ \mathrm{A/m}.$$

A comparison of this result with that in Example 8.5 indicates that the magnetic field produced by a magnetic dipole is similar to the electric field of the electric dipole. The two are numerically equal if

$$\frac{qd}{\varepsilon} = IS.$$

9.2 INDUCTORS AND ENERGY STORAGE

We have found that the magnetic field encircles the current-carrying conductor. Consider two loops of arbitrary shapes with currents I_1 and I_2, as shown in Figure 9.6. These currents each generate a magnetic field that links itself as well as the nearby circuit. For example, magnetic flux line ψ_{11} encircles I_1 of loop 1, whereas ψ_{12} links the other loop. The magnetic flux generated by I_1 linking itself externally gives rise to the external self-inductance L_1 of loop 1. Similarly magnetic flux lines such as ψ_{22} are responsible for the external self-inductance L_2 of loop 2. There may be a small magnetic flux that links partially with the current itself. This flux linkage is responsible for the internal self-inductance. Note that this field may be linking only a part of the current, and therefore special care is needed in determining its flux linkage. The magnetic flux line ψ_{12} (or ψ_{21}) generated by current I_1 links with the current in the other loop (and vice-versa), and therefore this gives rise to a mutual inductance M_{12} (or M_{21}). The inductors are used extensively in various electronic circuits such as the oscillators for generating sinusoidal signals, filtering, and the impedance matching circuits. The mutual inductance concept provides a neat way to couple ac signals while blocking the dc bias voltages. Another important application of this concept is in understanding electromagnetic interference (EMI). This section includes discussions on self-inductance, energy storage, and mutual inductance.

The magnetic flux ψ_{11} through the open surface S_1 due to current I_1 in loop 1 is defined as follows:

$$\psi_{11} = \int_{S_1} \vec{B}_1 \cdot d\vec{s}. \tag{9.2.1}$$

Using the Biot-Savart law (9.1.10), we find that

$$\psi_{11} = \int_{S_1} \left(\frac{\mu_o I_1}{4\pi} \oint_{C_1} \frac{d\vec{\ell}' \times \hat{R}}{R^2} \right) \cdot d\vec{s} = \frac{\mu_o I_1}{4\pi} \int_{S_1} \left(\oint_{C_1} \frac{d\vec{\ell}' \times \hat{R}}{R^2} \right) \cdot d\vec{s}. \tag{9.2.2}$$

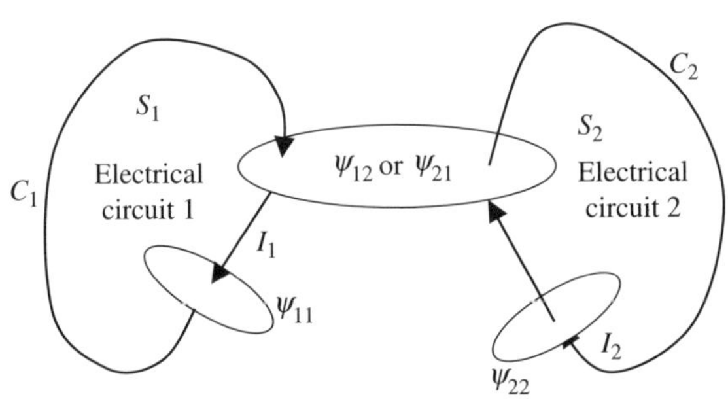

Figure 9.6 Magnetic flux produced by two wire loops of arbitrary shape.

This relation indicates that the magnetic flux ψ_{11} is directly proportional to the current I_1. The proportionality constant is known as inductance (in fact it is a case of self-inductance L_1). Hence

$$L_1 = \frac{\psi_{11}}{I_1} = \frac{\mu_o}{4\pi}\int_{S_1}\left(\oint_{C_1}\frac{d\vec{\ell}' \times \hat{R}}{R^2}\right) \cdot d\vec{s}. \tag{9.2.3}$$

Since ψ_{11} is measured in webers and I_1 in amperes, the unit of inductance is webers per ampere, which is commonly called henrys.

To get larger inductance values, practical inductors are designed as coils using several turns of wire generally on a cylindrical or toroidal core. If there are N turns in the loop then the flux linkage Λ_{11} increases by N times. Hence

$$L_1 = \frac{\Lambda_{11}}{I_1} = \frac{N_1\psi_{11}}{I_1} = \frac{N_1}{I_1}\int_{S_1} \vec{B}_1 \cdot d\vec{s}. \tag{9.2.4}$$

Similarly

$$L_2 = \frac{N_2\psi_{22}}{I_2} = \frac{N_2}{I_2}\int_{S_2} \vec{B}_2 \cdot d\vec{s}. \tag{9.2.5}$$

The magnetic field produced by a current-carrying conductor links itself. The associated inductance is called its internal self-inductance. In such a case the magnetic flux links only a part of the current, and therefore special care is needed in its evaluation. One way is to find the differential flux linkage that can be integrated over a cross section of the conductor. Inductance is found as the differential flux times the enclosed fraction of the total current I. Therefore it can be expressed as follows:

$$L = \frac{1}{I}\int_S \left[\frac{I_{\text{enclosed}}}{I}\right] d\psi. \tag{9.2.6}$$

We found earlier that a capacitor stores energy in the form of an electric field or a charge. Similarly an inductor stores energy in the form of a magnetic field. Inductance L and current I are related with stored energy U_H as follows:

$$U_H = \frac{1}{2}LI^2 \rightarrow L = \frac{U_H}{I^2/2}. \tag{9.2.7}$$

As expected, this relation of inductance is identical to (9.2.3). It can be proved as follows: Since the total energy is found after integrating the energy density over the entire volume V, we write

$$U_H = \frac{1}{2}\int_V \vec{B} \cdot \vec{H} dv = \frac{1}{2}\int_V \left(\nabla \times \vec{A}\right) \cdot \vec{H} dv. \tag{9.2.8}$$

Using vector identity 4 from Appendix A, we find that

$$(\nabla \times \vec{A}) \cdot \vec{H} = \nabla \cdot (\vec{A} \times \vec{H}) + \vec{A} \cdot (\nabla \times \vec{H}). \tag{9.2.9}$$

Therefore

$$U_{\mathrm{H}} = \frac{1}{2}\int_V \nabla \cdot (\vec{A} \times \vec{H}) dv$$
$$+\frac{1}{2}\int_V \vec{A} \cdot (\nabla \times \vec{H}) dv = \frac{1}{2}\oint_S (\vec{A} \times \vec{H}) \cdot d\vec{s} + \frac{1}{2}\int_V \vec{A} \cdot \vec{J} dv. \tag{9.2.10}$$

The First integral on the right-hand side of (9.2.10) reduces to zero because the surface S encloses all the fields that are zero at infinity. The second integral can be expressed as follows:

$$U_{\mathrm{H}} = \frac{1}{2}\int_V \vec{A} \cdot \vec{J} dv = \frac{1}{2}\oint_C \vec{A} \cdot (I d\vec{\ell}) = \frac{I}{2}\oint_C \vec{A} \cdot d\vec{\ell}. \tag{9.2.11}$$

Using the Stokes theorem and then the relation between the magnetic vector potential and the magnetic flux density, we find that

$$U_{\mathrm{H}} = \frac{I}{2}\oint_C \vec{A} \cdot d\vec{\ell} = \frac{I}{2}\int_S (\nabla \times \vec{A}) \cdot d\vec{s} = \frac{I}{2}\int_S \vec{B} \cdot d\vec{s}. \tag{9.2.12}$$

On substituting (9.2.12) into (9.2.7), we find that

$$L = \frac{I/2\left(\int_S \vec{B} \cdot d\vec{s}\right)/2}{I^2/2} = \frac{1}{I}\int_S \vec{B} \cdot d\vec{s}. \tag{9.2.13}$$

This expression is the same as (9.2.3) defined earlier.

Example 9.7

Find the inductance of the solenoid considered in Example 9.5. Assume that the length of the solenoid is large in comparison with its radius.

From Example 9.5 the magnetic flux density inside this solenoid is given as follows:

$$\vec{B} = \hat{z}\frac{\mu_o NI}{\ell}.$$

Therefore the total flux linkage is found to be

$$\psi_{11} = \int_S \vec{B} \cdot d\vec{s} = \frac{\mu_o NI\pi a^2}{\ell}.$$

From (9.2.4) the inductance of this solenoid is found to be

$$L = \frac{N\psi_{11}}{I} = \frac{\mu_o N^2 \pi a^2}{\ell} \text{ H}.$$

Example 9.8

Find the inductance of the toroid considered in Example 9.2. Assume that $a \gg b$. From Example 9.2 the magnetic flux density of the toroid is found to be

$$\vec{B} = \begin{cases} \hat{\phi}\,\dfrac{\mu_o NI}{2\pi a}, & a-b < \rho < a+b, \\ 0, & \rho < a-b,\, \rho > a+b. \end{cases}$$

Therefore the flux linkage is written as

$$\psi_{11} = \frac{\mu_o NI\pi b^2}{2\pi a} = \frac{\mu_o NIb^2}{2a}.$$

Hence the inductance of the toroid is

$$L = \frac{N\psi_{11}}{I} = \frac{\mu_o N^2 b^2}{2\pi a}.$$

Example 9.9

(a) Find the inductance of a 30 cm long solenoid with 2000 turns and a radius of 2 cm. (b) Find the inductance of a toroid with 2000 turns, mean radius of 5 cm, and a cross-sectional area of the core of $1.2566 \cdot 10^{-3}\ \mathrm{m}^2$.

(a) From the result obtained in Example 9.7, we find that the inductance of this solenoid is

$$L \approx \frac{\mu_o N^2 \pi a^2}{\ell} = \frac{4\pi \cdot 10^{-7} \cdot (2000)^2 \cdot \pi \cdot (0.02)^2}{0.3} = 0.021\ \mathrm{H} = 21\ \mathrm{mH}.$$

(b) The inductance of the toroid may be found from the result obtained in Example 9.8 as follows:

$$L = \frac{\mu_o N^2(\pi b^2)}{2\pi a} = \frac{4\pi \cdot 10^{-7} \cdot (2000)^2 \cdot 1.2566 \cdot 10^{-3}}{2\pi \cdot 0.05} = 0.02\mathrm{H} = 20\,\mathrm{mH}.$$

Example 9.10

Find the inductance per unit length of the coaxial line considered in Example 3.6.

Since $\vec{B} = \mu_o \vec{H}$ in a nonmagnetic medium, using the results obtained in Example 3.6, we find that

$$B_\phi = \begin{cases} \dfrac{\mu_o I\rho}{2\pi a^2}, & \rho < a, \\ \dfrac{\mu_o I}{2\pi\rho}, & a < \rho < b, \\ \dfrac{\mu_o I}{2\pi\rho}\left(\dfrac{c^2-\rho^2}{c^2-b^2}\right) & b < \rho < c. \end{cases}$$

In order to evaluate the internal self-inductance per unit length as well, we use (9.2.6) as follows:

$$L = \frac{1}{I}\int_{z=0}^{1}\int_{\rho=0}^{a}\left[\frac{I(\rho^2/a^2)}{I}\right]\left[\frac{\mu_o I\rho}{2\pi a^2}\right]d\rho dz + \frac{1}{I}\int_{z=0}^{1}\int_{\rho=a}^{b}\left[\frac{I}{I}\right]\frac{\mu_o I}{2\pi\rho}\,d\rho dz$$

$$+\frac{1}{I}\int_{z=0}^{1}\int_{\rho=b}^{c}\left[\frac{I(c^2-\rho^2)/(c^2-b^2)}{I}\right]\left[\frac{\mu_o I}{2\pi\rho}\cdot\frac{(c^2-\rho^2)}{(c^2-b^2)}\right]d\rho dz,$$

or

$$L = \frac{\mu_o}{8\pi} + \frac{\mu_o}{2\pi}\ln\left(\frac{b}{a}\right) + \frac{\mu_o}{2\pi}\int_{\rho=b}^{c}\left[\frac{(c^2-\rho^2)}{(c^2-b^2)}\right]^2\frac{d\rho}{\rho}.$$

The remaining integral can be evaluated easily after expanding the square in its integrand. The total inductance is found to be

$$L = \frac{\mu_o}{8\pi} + \frac{\mu_o}{2\pi}\ln(b/a) + \frac{\mu_o}{2\pi}\left[\frac{c^2}{c^2-b^2}\right]^2\ln\left(\frac{c}{b}\right) - \frac{\mu_o}{2\pi}\left[\frac{c^2}{c^2-b^2}\right]$$

$$+\frac{\mu_o}{8\pi}\left[\frac{c^2+b^2}{c^2-b^2}\right]\text{H/m}.$$

Note that the external self-inductance per unit length of coaxial line is

$$L = \frac{\mu_o}{2\pi}\ln\left(\frac{b}{a}\right)\ \text{H/m}.$$

Mutual Inductance

Mutual inductance between the two circuits characterizes the magnetic flux linkage with one circuit per unit current in the other. Therefore it can be found as follows:

The magnetic flux generated by N_1 turns of loop 1 that links loop 2 is found to be

$$\psi_{21} = \int_{S_2}\vec{B}_1\bullet d\vec{s} = \int_{S_2}\left(\frac{\mu_o N_1 I_1}{4\pi}\oint_{C_1}\frac{d\vec{\ell}'\times\hat{R}}{R^2}\right)\bullet d\vec{s}$$

$$= \frac{\mu_o I_1 N_1}{4\pi}\int_{S_2}\left(\oint_{C_1}\frac{d\vec{\ell}'\times\hat{R}}{R^2}\right)\bullet d\vec{s}. \qquad (9.2.14)$$

Note that this flux must be multiplied by the number of turns N_2 of loop 2 to find the total flux linkage. Therefore the mutual inductance M_{21} is given as follows:

$$M_{21} = \frac{N_2\psi_{21}}{I_1} = \frac{N_2 N_1\mu_o}{4\pi}\int_{S_2}\left(\oint_{C_1}\frac{d\vec{\ell}'\times\hat{R}}{R^2}\right)\bullet d\vec{s}. \qquad (9.2.15)$$

Similarly

$$M_{12} = \frac{N_1 \psi_{12}}{I_2} = \frac{N_1 N_2 \mu_o}{4\pi} \int_{S_1} \left(\oint_{C_2} \frac{d\vec{\ell}' \times \hat{R}}{R^2} \right) \cdot d\vec{s}. \tag{9.2.16}$$

We expect M_{21} to be equal to M_{12}. However, it is not that obvious from these two relations. Using the following alternative formulation for the flux linkage between two loops with single turns, we find that

$$\begin{aligned} \psi_{21} &= \int_{S_2} \vec{B}_1 \cdot d\vec{s} \\ &= \int_{S_2} (\nabla \times \vec{A}_1) \cdot d\vec{s} = \oint_{C_2} \vec{A}_1 \cdot d\vec{\ell}_2 = \oint_{C_2} \left(\frac{\mu_o I_1}{4\pi} \oint_{C_1} \frac{d\vec{\ell}_1}{R_{21}} \right) \cdot d\vec{\ell}_2. \end{aligned} \tag{9.2.17}$$

Therefore

$$M_{21} = \frac{\psi_{21}}{I_1} = \frac{\mu_o}{4\pi} \oint_{C_2} \oint_{C_1} \frac{d\vec{\ell}_1 \cdot d\vec{\ell}_2}{R_{21}}. \tag{9.2.18}$$

Similarly

$$\psi_{12} = \int_{S_1} \vec{B}_2 \cdot d\vec{s} = \int_{S_1} (\nabla \times \vec{A}_2) \cdot d\vec{s} = \oint_{C_1} \vec{A}_2 \cdot d\vec{\ell}_1 = \oint_{C_1} \left(\frac{\mu_o I_2}{4\pi} \oint_{C_2} \frac{d\vec{\ell}_2}{R_{12}} \right) \cdot d\vec{\ell}_1. \tag{9.2.19}$$

Since $R_{12} = R_{21}$, we find that

$$M_{12} = \frac{\psi_{12}}{I_2} = \frac{\mu_o}{4\pi} \oint_{C_2} \oint_{C_1} \frac{d\vec{\ell}_1 \cdot d\vec{\ell}_2}{R_{12}} = M_{21}. \tag{9.2.20}$$

Coupling Coefficient

Consider the case where coil 1 produces a magnetic flux ψ_{11} because of its current I_1. However, only a fraction κ_1 of this flux links with coil 2. Therefore we can write

$$\psi_{21} = \kappa_1 \psi_{11} = \kappa_1 \frac{L_1 I_1}{N_1}. \tag{9.2.21}$$

Hence the mutual inductance M_{21} is found to be

$$M_{21} = \frac{N_2 \psi_{21}}{I_1} = \kappa_1 L_1 \frac{N_2}{N_1}. \tag{9.2.22}$$

Similarly

$$M_{12} = \frac{N_1 \psi_{12}}{I_2} = \kappa_2 L_2 \frac{N_1}{N_2}. \tag{9.2.23}$$

For $M_{21} = M_{12} = M$, we have

$$M^2 = \kappa_1 \kappa_2 L_1 L_2 \rightarrow M = \pm \kappa \sqrt{L_1 L_2}, \tag{9.2.24}$$

where

$$\kappa = \sqrt{\kappa_1 \kappa_2}. \tag{9.2.25}$$

κ is called the coupling coefficient. Note that $\kappa_1 \leq 1$ and $\kappa_2 \leq 1$, and therefore $\kappa \leq 1$.

Example 9.11

The axes of two concentric solenoids of lengths ℓ are aligned to the z-axis, as shown in Figure 9.7. The inner solenoid has N_1 turns while the outer one has N_2. The radii of the two solenoids are a and b, respectively. Find their mutual inductance.

Assume that current through inner and outer solenoids are I_1 and I_2, respectively. Therefore the magnetic flux density due to the outer solenoid is found to be

$$B_2 = \frac{\mu_o N_2 I_2}{\ell}.$$

The flux linkage though each turn of the inner solenoid is given by

$$\psi_{12} = \frac{\mu_o N_2 I_2 \pi a^2}{\ell}.$$

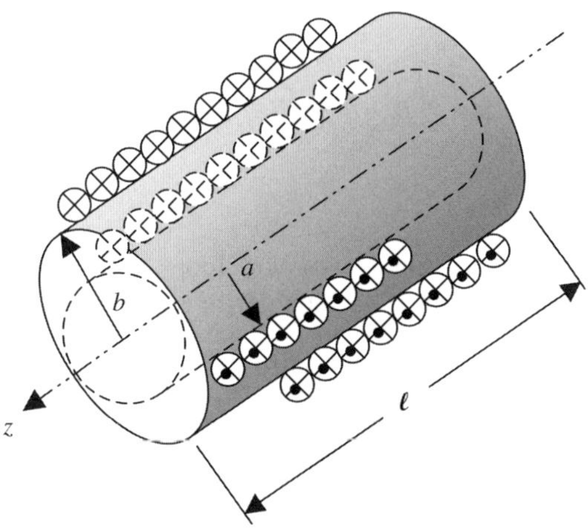

Figure 9.7 Concentrically arranged solenoids.

The mutual inductance M_{12} is found from (9.2.16) as follows:

$$M_{12} = \frac{\mu_o N_1 N_2 \pi a^2}{\ell}.$$

Alternatively, the magnetic flux density due to I_1 in the inner solenoid is given as follows:

$$B_1 = \frac{\mu_o N_1 I_1}{\ell}.$$

The flux linkage of the outer conductor is found to be (remember that the flux outside solenoid is zero and there are N_2 turns in outer solenoid)

$$\psi_{21} = \frac{\mu_o N_2 N_1 I_1 \pi a^2}{\ell}.$$

Therefore (9.2.15) gives

$$M_{21} = \frac{\mu_o N_2 N_1 \pi a^2}{\ell} = M_{12}.$$

We can determine the coupling coefficient if the expressions of self-inductances L_1 and L_2 of solenoids are known. These can be found easily as follows:

$$L_1 = \frac{\mu_o N_1^2 \pi a^2}{\ell}$$

and

$$L_2 = \frac{\mu_o N_2^2 \pi b^2}{\ell}.$$

Therefore

$$\kappa = \frac{M_{12}}{\sqrt{L_1 L_2}} = \frac{M_{21}}{\sqrt{L_1 L_2}} = \frac{a}{b}.$$

Example 9.12

A distance h separates two conducting wire loops of radii a_1 and a_2, as shown in Figure 9.8. The cross section of the loop's wire is circular with radius a. Find their mutual inductance.

Assume that the radii a_1 and a_2 of the primary and secondary loops, respectively, are measured to the wire axis. The two loops are separated by a distance h, and $c = a_1 - a$ is the radius of the inner periphery of the primary loop. The incremental lengths $d\vec{\ell}_1$ and $d\vec{\ell}_2$ can be expressed as follows:

$$d\vec{\ell}_1 = \hat{\phi}_1 c\, d\phi_1 = (-\hat{x}\sin\phi_1 + \hat{y}\cos\phi_1)c\, d\phi_1$$

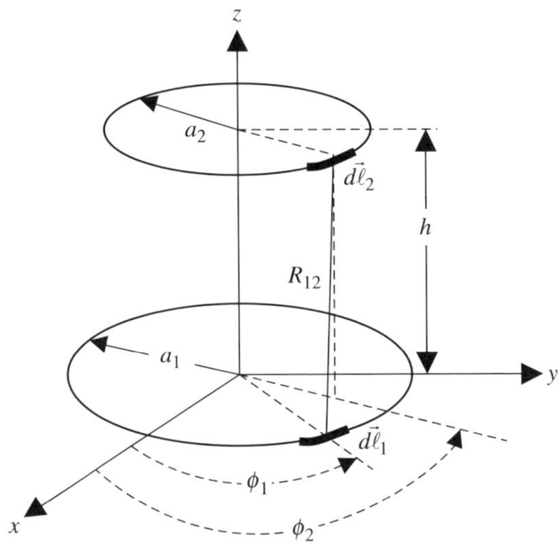

Figure 9.8 Two magnetically coupled loops.

and

$$d\vec{\ell}_2 = \hat{\phi}_2 a_2 d\phi_2 = (-\hat{x}\sin\phi_2 + \hat{y}\cos\phi_2)a_2 d\phi_2.$$

Therefore

$$d\vec{\ell}_1 \cdot d\vec{\ell}_2 = ca_2(\hat{\phi}_1 \cdot \hat{\phi}_2)d\phi_1 d\phi_2 = ca_2\cos(\phi_1 - \phi_2)d\phi_1 d\phi_2.$$

The position vectors for $d\vec{\ell}_1$ and $d\vec{\ell}_2$, respectively, are given as follows:

$$\vec{r}_1 = \hat{x}c\cos\phi_1 + \hat{y}c\sin\phi_1$$

and

$$\vec{r}_2 = \hat{x}a_2\cos\phi_2 + \hat{y}a_2\sin\phi_2 + \hat{z}h.$$

Therefore the separation between $d\vec{\ell}_1$ and $d\vec{\ell}_2$ is found to be

$$R_{12} = |\vec{r}_1 - \vec{r}_2| = \sqrt{h^2 + c^2 + a_2^2 - 2ca_2\cos(\phi_1 - \phi_2)}.$$

Alternatively, this result can be found via the so-called cosine law.

Using (9.2.18), we find that

$$M_{12} = \frac{\mu_o ca_2}{4\pi}\int_{\phi_1=0}^{2\pi}\int_{\phi_2=0}^{2\pi}\frac{\cos(\phi_1 - \phi_2)d\phi_2 d\phi_1}{\sqrt{h^2 + c^2 + a_2^2 - 2ca_2\cos(\phi_1 - \phi_2)}}.$$

Note that when integrating over ϕ_2, ϕ_1 is constant and the integration can be performed after changing the variable as follows: For $\phi_1 - \phi_2 = \theta$, $d\phi_2 = -d\theta$, and the new integration limits are found to be

$$\phi_2 = 0 \rightarrow \theta = \phi_1$$

and

$$\phi_2 = 2\pi \rightarrow \theta = \phi_1 - 2\pi.$$

Therefore

$$M_{12} = \frac{\mu_o c a_2}{4\pi} \int_{\phi_1=0}^{2\pi} \int_{\theta=\phi_1-2\pi}^{\phi_1} \frac{\cos\theta \, d\theta d\phi_1}{\sqrt{h^2 + c^2 + a_2^2 - 2ca_2 \cos\theta}}.$$

Since the inner integral is evaluated for the closed loop such that θ varies over a total interval of 2π, the origin for θ is unimportant. Further the integrand is now independent of ϕ_1, so the outside integral is trivial. Consequently the expression for the mutual inductance between the two loops simplifies to

$$M_{12} = \frac{\mu_o c a_2}{2} \int_{\theta=0}^{2\pi} \frac{\cos\theta d\theta}{\sqrt{h^2 + c^2 + a_2^2 - 2ca_2 \cos\theta}}.$$

This integral cannot be evaluated analytically. However, the mutual inductance M_{12} can be expressed in terms of the complete elliptic integrals of the first and second kinds that are generally tabulated in many mathematics handbooks. This transformation is possible via the following change in the variable.

For $\theta = \pi - 2x$, $d\theta = -2dx$, and limits on the integral change as follows:

$$\theta = 0 \rightarrow x = \frac{\pi}{2}$$

and

$$\theta = 2\pi \rightarrow x = -\frac{\pi}{2}.$$

Further

$$\cos\theta = \cos(\pi - 2x) = -\cos(2x) = 2\sin^2 x - 1.$$

Therefore

$$M_{12} = \mu_o c a_2 \int_{x=-\pi/2}^{\pi/2} \frac{(2\sin^2 x - 1)dx}{\sqrt{h^2 + c^2 + a_2^2 - 2ca_2(2\sin^2 x - 1)}}$$

$$= 2\mu_o c a_2 \int_{x=0}^{\pi/2} \frac{(2\sin^2 x - 1)dx}{\sqrt{h^2 + c^2 + a_2^2 - 2ca_2(2\sin^2 x - 1)}}.$$

Since

$$R_{12}^2 = h^2 + c^2 + a_2^2 - 2ca_2 \cos\theta = h^2 + c^2 + a_2^2 - 2ca_2(2\sin^2 x - 1),$$

or

$$R_{12}^2 = h^2 + (c^2 + a_2^2 + 2ca_2) - 4ca_2 \sin^2 x = h^2 + (c + a_2)^2 - 4ca_2 \sin^2 x,$$

$$R_{12}^2 = \frac{4ca_2}{k^2}(1 - k^2 \sin^2 x),$$

where

$$k^2 = \frac{4ca_2}{h^2 + (c + a_2)^2}.$$

Further

$$2\sin^2 x - 1 = \frac{2}{k^2} - 1 - \frac{2}{k^2}(1 - k^2 \sin^2 x).$$

Therefore

$$\begin{aligned} &\frac{\cos\theta}{\sqrt{h^2 + c^2 + a_2^2 - 2ca_2 \cos\theta}} \\ &= \frac{2\sin^2 x - 1}{\sqrt{h^2 + c^2 + a_2^2 - 2ca_2(2\sin^2 x - 1)}} \\ &= \frac{k}{2\sqrt{ca_2}}\left[\frac{(2/k^2) - 1 - (2/k^2)(1 - k^2\sin^2 x)}{\sqrt{1 - k^2 \sin^2 x}}\right] \end{aligned}$$

and

$$M_{12} = 2\mu_o ca_2 \frac{1}{2\sqrt{ca_2}}\left[\left(\frac{2}{k} - k\right)\int_0^{\pi/2} \frac{dx}{\sqrt{1 - k^2 \sin^2 x}} - \frac{2}{k}\int_0^{\pi/2} \sqrt{1 - k^2 \sin^2 x}\, dx\right],$$

or

$$M_{12} = \mu_o \sqrt{ca_2}\left[\left(\frac{2}{k} - k\right)F\left(k, \frac{\pi}{2}\right) - \frac{2}{k}E\left(k, \frac{\pi}{2}\right)\right],$$

where

$$F\left(k, \frac{\pi}{2}\right) = \int_0^{\pi/2} \frac{dx}{\sqrt{1 - k^2 \sin^2 x}}, \qquad 0 \le k^2 \le 1,$$

and

$$E\left(k, \frac{\pi}{2}\right) = \int_0^{\pi/2} \sqrt{1 - k^2 \sin^2 x}\, dx, \qquad 0 \le k^2 \le 1.$$

$F(k, \pi/2)$ and $E(k, \pi/2)$ are known as complete elliptic integrals of the first and the second kinds, respectively. These can be expressed in infinite series as follows:

$$F\left(k, \frac{\pi}{2}\right) = \frac{\pi}{2}\left[1 + \left(\frac{1}{2}\right)^2 k^2 + \left(\frac{1 \cdot 3}{2 \cdot 4}\right)^2 k^4 + \left(\frac{1 \cdot 3 \cdot 5}{2 \cdot 4 \cdot 6}\right)^2 k^6 + \cdots\right]$$

and

$$E\left(k, \frac{\pi}{2}\right) = \frac{\pi}{2}\left[1 - \left(\frac{1}{2}\right)^2 k^2 - \left(\frac{1 \cdot 3}{2 \cdot 4}\right)^2 \frac{k^4}{3} - \left(\frac{1 \cdot 3 \cdot 5}{2 \cdot 4 \cdot 6}\right)^2 \frac{k^6}{5} - \cdots\right].$$

Example 9.13

A circular loop of a 0.25 m mean radius is lying symmetrically on the x-y plane. The wire used for this loop is circular in its cross section with a radius of 1 mm. Find the inductance.

The external self-inductance of a loop can be found after specializing the results obtained for M_{12} in Example 9.12. Equations (9.2.3) and (9.2.17) indicate that the external self-inductance can also be expressed as

$$L_1 = \frac{\mu_o}{4\pi} \oint_{C_1} d\vec{\ell}_1 \cdot \oint_{C_1'} \frac{d\vec{\ell}_1'}{R_{11}}.$$

This expression can be obtained from the mutual inductance formula of (9.2.18) after replacing a_2 by a_1, h by zero, and R_{12} by R_{11}. Hence specializing the results of Example 9.12, we find that

$$L_1 = \mu_o \sqrt{(a_1 - a)a_1}\left[\left(\frac{2}{k} - k\right) F\left(k, \frac{\pi}{2}\right) - \frac{2}{k} E\left(k, \frac{\pi}{2}\right)\right],$$

where

$$k^2 = \frac{4(a_1 - a)a_1}{(2a_1 - a)^2}.$$

Since

$$\sqrt{(a_1 - a)a_1} = \frac{k}{2}(2a_1 - a),$$

$$L_1 = \mu_o(2a_1 - a)\left[\left(1 - \frac{k^2}{2}\right) F\left(k, \frac{\pi}{2}\right) - E\left(k, \frac{\pi}{2}\right)\right].$$

Further, if $a_1/a \gg 1$, then $k \to 1$, and the asymptotic values of the complete elliptic integrals for this case are found to be

$$F\left(k \to 1, \frac{\pi}{2}\right) \approx \ln\left(\frac{4}{\sqrt{1-k^2}}\right)$$

and

$$E\left(k \to 1, \frac{\pi}{2}\right) \approx 1.$$

Therefore

$$L_1 = \mu_o a_1\left[\ln\left(\frac{4}{\sqrt{(a/2a_1)^2}}\right) - 2\right] = \mu_o a_1\left[\ln\left(\frac{8a_1}{a}\right) - 2\right].$$

For $a_1 = 0.25$ m and $a = 1$ mm,

$$L_1 = \mu_o a_1\left[\ln\left(\frac{8a_1}{a}\right) - 2\right] = 4\pi \cdot 10^{-7} \cdot 0.25\left[\ln\left(\frac{8 \cdot 0.25}{0.001}\right) - 2\right]\mathrm{H}$$

$$= 1.7596\ \mu\mathrm{H}.$$

9.3 MAGNETIC MATERIALS

According to the atomic structure of materials, electrons orbit around the nucleus and also spin around their axes. As in the wire loops considered in Examples 9.4 and 9.6, the orbital motion of electrons produce a magnetic field. The angular momentum of spinning electrons can be related to tiny current loops in the material. Mathematically the concept of magnetization can be written as follows:

$$\vec{M} = \lim \lim_{\Delta v \to 0} \frac{\vec{m}_r}{\Delta v}\ \mathrm{A/m}, \tag{9.3.1}$$

where $\vec{m}_r$ is an average dipole moment of billions of tiny magnetic dipoles in a macroscopically small volume. The magnetization of a material facilitates the analysis without considering individual tiny magnets. As indicated earlier in (3.3.4), the total magnetic field in a medium depends on both the applied field and the magnetization of the substance. Hence

$$\vec{B} = \mu_o(\vec{H} + \vec{M}). \tag{9.3.2}$$

In two different classes of materials, known as diamagnetic and paramagnetic materials, the magnetization is linearly related with the external (applied) magnetic field intensity. For such cases, (9.3.2) may be written as follows:

$$\vec{B} = \mu_o(1 + \chi_m)\vec{H} = \mu\vec{H}. \tag{9.3.3}$$

TABLE 9.1 Magnetic Susceptibilities of Selected Diamagnetic and Paramagnetic Materials at 300 K

Diamagnetic Materials		Paramagnetic Materials	
Copper	$-0.95 \cdot 10^{-5}$	Aluminum	$2.3 \cdot 10^{-5}$
Diamond	$-2.2 \cdot 10^{-5}$	Calcium	$1.9 \cdot 10^{-5}$
Gold	$-3.6 \cdot 10^{-5}$	Chromium	$0.27 \cdot 10^{-3}$
Germanium	$-0.8 \cdot 10^{-5}$	Magnesium	$0.12 \cdot 10^{-4}$
Graphite	$-12 \cdot 10^{-5}$	Platinum	$2.9 \cdot 10^{-4}$
Silver	$-2.6 \cdot 10^{-5}$	Tungsten	$0.68 \cdot 10^{-4}$
Silicon	$-0.42 \cdot 10^{-5}$	Air	$3.6 \cdot 10^{-7}$
Sodium	$-0.24 \cdot 10^{-5}$	Liquid oxygen	$3.5 \cdot 10^{-3}$
Water	$-0.88 \cdot 10^{-5}$	Nickel chloride	$0.4 \cdot 10^{-4}$

As noted earlier in Chapter 3, χ_m is the magnetic susceptibility of a material and μ is the magnetic permeability of a material. This relation becomes particulary complex in certain materials known as ferromagnetic (e.g., iron, nickel, and cobalt) and ferrimagnetic (e.g., ferrites) materials. The magnetic susceptibilities of some diamagnetic and paramagnetic materials are listed in Table 9.1.

In diamagnetic materials, the atoms and molecules have no magnetic moment because the magnetic fields produced by orbiting and spinning electrons cancel out. The external magnetic field only weakly affects this balance, and a small magnetic moment is induced that opposes the applied field. As a result the magnetic susceptibility of these materials is negative, and the relative permeability is less than unity (i.e., permeability of the material is smaller than the permeability of free space).

In paramagnetic materials the atoms have magnetic moments that point in random directions because of thermal agitation. When an external magnetic field is applied, these elementary magnets rotate and align in the direction of applied field. As Table 9.1 indicates, these materials have positive magnetic susceptibility, and therefore the relative permeability is just larger than unity.

In ferromagnetic materials, large magnetic moments exist within microscopic regions, called domains because of the alignment of all magnetic moments in each region. These domains have volumes on the order of 10^{-12} to 10^{-8} m^3 containing 10^{17} to 10^{21} atoms. As shown in Figure 9.9*a*, the directions of the magnetic moments differ from one to the other domain in virgin ferromagnetic material. The magnetic state of the material corresponds to the origin *o* of the magnetization curve (*B*–*H* curve) shown in Figure 9.10. When an external magnetic field is applied, the domain structure changes as shown in Figure 9.9*b*. Domains with magnetic moments in the direction of the applied field grow in size, whereas others shrink. The magnetic flux reaches to a point *e* following the path *o*–*e*–*c* in Figure 9.10. If the applied magnetic field is further increased to a large strength, then all the magnetic moments align to form a single domain as shown in Figure 9.9*c*. This alignment corresponds to point *c* on the magnetization curve of Figure 9.10. The magnetic flux density *B* experiences a negligible change if the

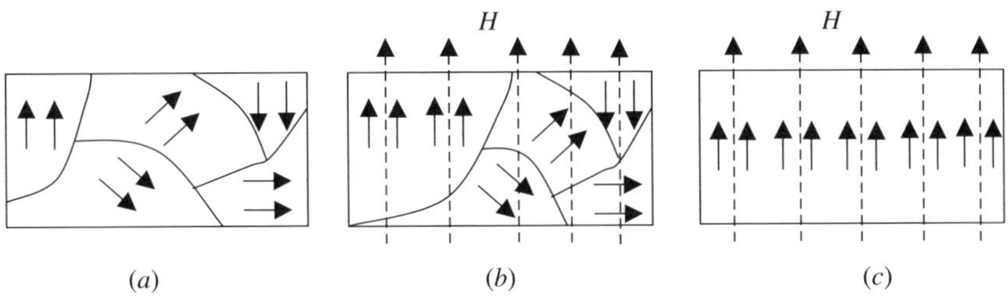

Figure 9.9 Domains in a virgin ferromagnetic material (*a*), under the influence of small magnetic field (*b*), and in saturation when a large magnetic field is applied (*c*).

applied magnetic field is increased further. This is known as the saturation point of the material.

Consider a situation where the applied magnetic field is reduced from the saturation point c. The B–H characteristic of the material follows the c–d–a path instead of c–e–o. Some magnetic flux remains in the material even when the applied magnetic field is reduced to zero. At this point the material is said to have a remanent magnetization. If the applied magnetic field is now reversed, then the magnetic flux goes to zero for $-H_c$, which is known as the coercive force. The material saturates at point a for large reverse magnetic field. If the applied magnetic field is reversed again, then it follows the path a–b–c of the B–H curve. The closed path a–b–c–d–a is known as the hysteresis loop. The hysteresis loop of a hard ferromagnetic material is wide, and it is narrow for a soft material. Hard materials are used to design permanent magnets that maintain magnetization as long as it is kept below a critical temperature, called the Curie temperature. Characteristic parameters of selected magnetic materials are listed in Tables 9.2 and 9.3.

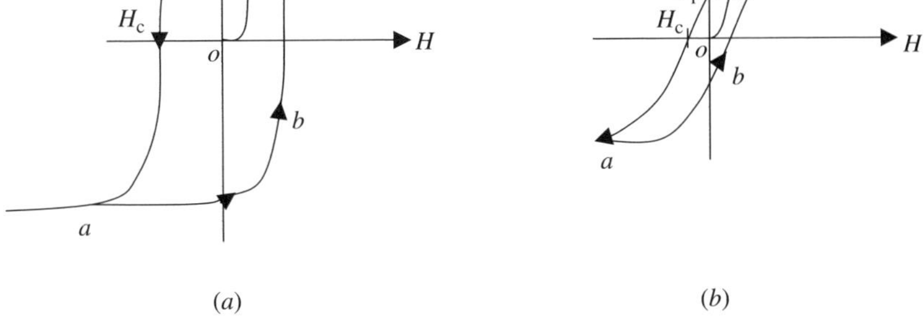

Figure 9.10 Hysteresis curve for a hard (*a*) and a soft (*b*) ferromagnetic material.

TABLE 9.2 Properties of Selected Magnetic Materials and Magnetic Alloys

Hard Materials				Soft Materials					
Material	Composition	B_r (T)	H_c (At)	Material	Composition	Maximum (μ_r)	B_r (T)	H_c (At)	B_{sat} (T)
Carbon steel	1 Mn, 0.9 C, 98.1 Fe	1	4000	Iron	99.91 Fe	5000	0.77	80	2.15
Alnico V	8 Al, 14 Ni, 24 Co, 3 Cu, 53 Fe	1.25	44,000	Cold rolled steel	98.5 Fe	2000		145	2.1
Platinum-cobalt	77 Pt, 23 Co	0.45	208,000	Permendure	50 Cs	4500	1.4	160	2.4
Ba ferrite		0.2	120,000	Supermalloy	79 Ni, 5 Mo, 0.3 Mn, 15.7 Fe	800000	0.5	0.16	0.8

TABLE 9.3 Curie Temperature of Selected Magnetic Materials

Material	Curie Temperature, T_c (K)
Iron	1043
Cobalt	1394
Gadolinium	317
Nickel	893

The area enclosed by the magnetization curve represents the work required to take the material through the hysteresis cycle. This way the material acquires energy from the external source. When the magnetization cycle is repeated, the magnetic energy is transformed to thermal energy because of the realignment of the domains, and the temperature of the material increases.

9.4 MAGNETIC CIRCUITS

From Example 9.8, magnetic flux in a toroid is given as follows:

$$\psi_{11} = \frac{\mu NI\pi b^2}{2\pi a} = \frac{\mu NIS}{\ell} = \frac{NI}{(\ell/\mu S)} = \frac{V_{\mathrm{m}}}{\mathcal{R}} \tag{9.4.1}$$

where S is the cross-sectional area of the core and ℓ is its mean circumferential length. $\mathscr{R}$ is called the reluctance in per henrys (H^{-1}) and V_{m} is the magnetomotive force (mmf) in ampere-turns (At) defined as follows:

$$\mathscr{R} = \frac{\ell}{\mu S}\mathrm{H}^{-1} \tag{9.4.2}$$

and

$$V_{\mathrm{m}} = NI. \tag{9.4.3}$$

Note the similarity between (9.4.1) and the electrical circuit relation

$$I = \frac{V}{R} = \frac{V}{(\ell/\sigma S)}, \tag{9.4.4}$$

where I is the current that flows through resistance R when an electrical voltage V is applied. The resistance is made of a material of conductivity σ with length ℓ and cross sectional area S.

Further we find that

$$\oint_C \vec{H} \cdot d\vec{\ell} = NI \rightarrow \sum_m H_m \ell_m = NI \tag{9.4.5}$$

and

$$\oint_S \vec{B} \cdot d\vec{s} = 0 \rightarrow \sum_m B_m S_m = 0. \tag{9.4.6}$$

These relations are analogous to Kirchhoff's voltage and the current laws used in electrical circuit analysis. Therefore the magnetic circuits can be easily analyzed following the analogy of electrical circuits.

Example 9.14

A toroid has 500 turns on an iron core of 30 cm mean length and an air gap of 1 cm, as shown in Figure 9.11. It is being used to write on a magnetic tape that requires a magnetic flux density of 1.6 T in the air gap (ignore fringing fields). If the relative permeability of the core at the given condition is 4897, find the required current I.

Assume that the magnetic field intensity in core is H_1, whereas it is H_2 in the air gap. From (9.4.5) we find that

$$H_1 \cdot 0.3 + H_2 \cdot 0.01 = 500 \cdot I$$

Since

$$H_1 = \frac{B}{\mu} = \frac{1.6}{4897 \cdot 4 \cdot \pi \cdot 10^{-7}} \text{ A/m} = 260 \text{ A/m}$$

and

$$H_2 = \frac{B}{\mu_o} = \frac{1.6}{4 \cdot \pi \cdot 10^{-7}} \text{ A/m} = 1.2732 \cdot 10^6 \text{ A/m},$$

$$\begin{aligned} 260 \cdot 0.3 + 1.2732 \cdot 10^6 \cdot 0.01 &= 78 + 12732 = 12810 \\ &= 500 \cdot I \rightarrow I = 25.62 \text{ A}. \end{aligned}$$

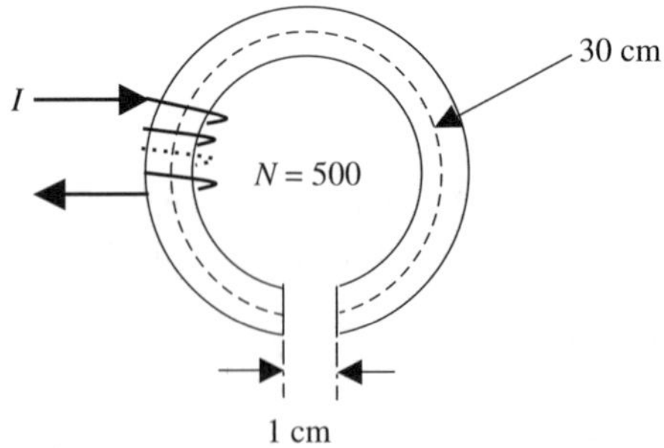

Figure 9.11 Geometry of the toroid in Example 9.14.

Example 9.15

There is a current of 10 A flowing through 100 turns of the magnetic circuit shown in Figure 9.12. All branches have 8 cm^2 as their cross-sectional areas (S_1) except in the middle, which is 10 cm^2 (S_2). Lengths ℓ_1 and ℓ_3 are 28 cm each, and ℓ_2 is 10 cm long. The relative permeability of the core is 400. Find the magnetic flux densities through lengths ℓ_2 and ℓ_3.

Assume that the magnetic field intensities in three lengths are H_1, H_2, and H_3, respectively. From (9.4.5) we find that

$$0.28 \cdot H_1 + 0.1 \cdot H_2 = 10 \cdot 100$$

and

$$0.1 \cdot H_2 - 0.28 \cdot H_3 = 0 \rightarrow H_2 = 2.8 \cdot H_3.$$

We use (9.4.6) to get

$$\mu H_1 S_1 - \mu H_2 S_2 - \mu H_3 S_1 = 0 \rightarrow H_1 = H_3 + \tfrac{5}{4} H_2.$$

Solving these three equations, we find that

$$H_2 = 1818.1818\,\text{A/m}$$

and

$$H_3 = 649.3506\,\text{A/m}.$$

The corresponding magnetic flux densities are found to be

$$B_2 = 400 \cdot 4 \cdot \pi \cdot 10^{-7} \cdot H_2 = 0.9139\,\text{T}$$

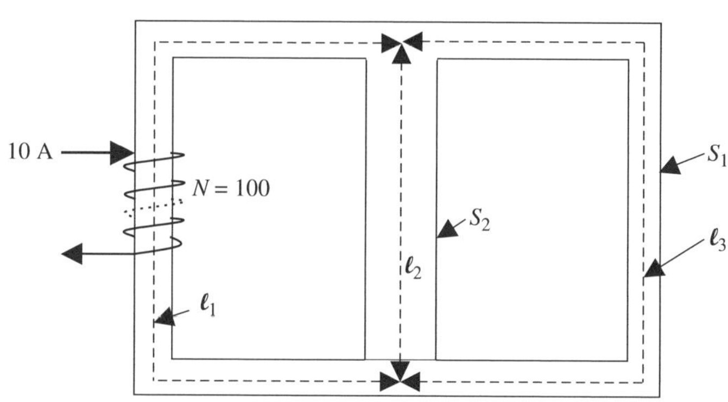

Figure 9.12 Geometry of the magnetic circuit in Example 9.15.

and

$$B_3 = 400 \cdot 4 \cdot \pi \cdot 10^{-7} \cdot H_3 = 0.3264\,\mathrm{T}.$$

Example 9.16

The magnetic circuit shown in Figure 9.13 has two symmetrical windings, each with 9783.2455 At. All branches of the core are 2 cm thick, and its relative permeability is 1000. Find the total magnetic flux in the air gap.

An equivalent magnetic circuit is drawn in Figure 9.14 using the concepts of reluctance and the magnetic voltage defined in (9.4.2) and (9.4.3). The reluctance values for each branch, as indicated in Figure 9.13, are found to be

$$\mathscr{R}_1 = \frac{\ell}{\mu S} = \frac{0.11}{1000 \cdot 4 \cdot \pi \cdot 10^{-7} \cdot 0.015 \cdot 0.02} = 29.1784 \cdot 10^4\,\mathrm{H}^{-1},$$

$$\mathscr{R}_2 = \frac{\ell}{\mu S} = \frac{0.05}{1000 \cdot 4 \cdot \pi \cdot 10^{-7} \cdot 0.015 \cdot 0.02} = 13.2629 \cdot 10^4\,\mathrm{H}^{-1},$$

$$\mathscr{R}_3 = \frac{\ell}{\mu S} = \frac{0.06}{1000 \cdot 4 \cdot \pi \cdot 10^{-7} \cdot 0.01 \cdot 0.02} = 23.8732 \cdot 10^4\,\mathrm{H}^{-1},$$

and

$$\mathscr{R}_g = \frac{\ell}{\mu S} = \frac{0.01}{4 \cdot \pi \cdot 10^{-7} \cdot 0.015 \cdot 0.02} = 26.5258 \cdot 10^6\,\mathrm{H}^{-1}.$$

We can use the superposition principle to find the total magnetic flux in the air gap. With a short circuit replacing one source, the magnetic flux ψ supplied by

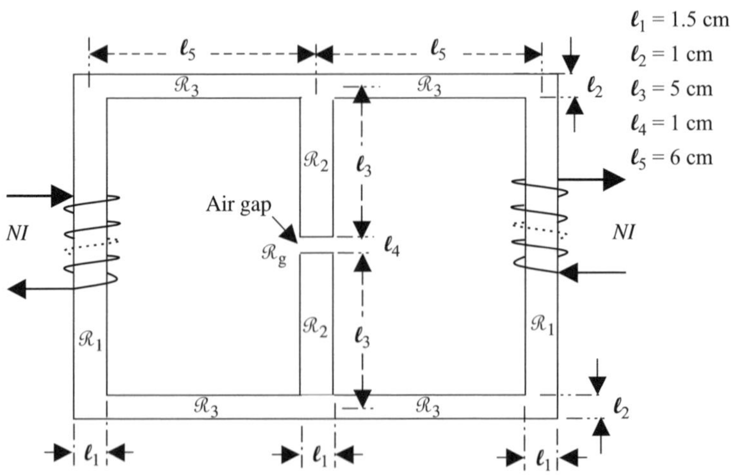

Figure 9.13 Geometry of the magnetic circuit in Example 9.16.

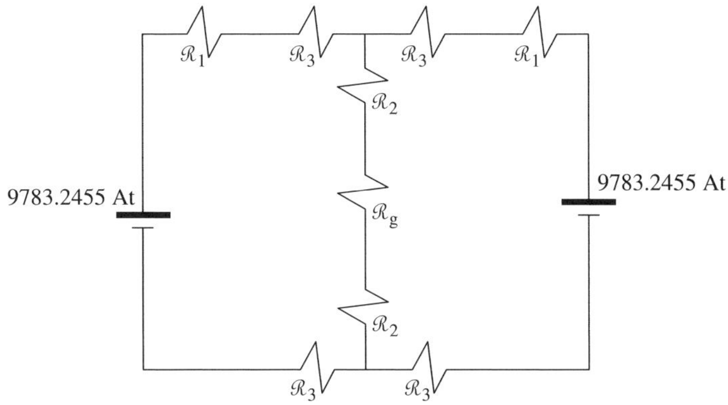

Figure 9.14 Equivalent of magnetic circuit shown in Figure 9.13.

the other source to the circuit is found to be

$$\psi = \frac{NI}{(\mathscr{R}_1 + 2\mathscr{R}_3) + (\mathscr{R}_1 + 2\mathscr{R}_3)\|(\mathscr{R}_g + 2\mathscr{R}_2)}.$$

Since

$$(\mathscr{R}_1 + 2\mathscr{R}_3) = (29.1784 + 2 \cdot 23.8732) \cdot 10^4 = 76.9248 \cdot 10^4\,\mathrm{H}^{-1},$$

$$(\mathscr{R}_g + 2\mathscr{R}_2) = (2652.58 + 2 \cdot 13.2629) \cdot 10^4 = 2679.1058 \cdot 10^4\,\mathrm{H}^{-1},$$

we find that

$$(\mathscr{R}_1 + 2\mathscr{R}_3)\|(\mathscr{R}_1 + 2\mathscr{R}_3) = \frac{(\mathscr{R}_1 + 2\mathscr{R}_3) \cdot (\mathscr{R}_g + 2\mathscr{R}_2)}{(\mathscr{R}_1 + 2\mathscr{R}_3) + (\mathscr{R}_g + 2\mathscr{R}_2)}$$
$$= 74.7777 \cdot 10^4\,\mathrm{H}^{-1}.$$

Therefore

$$\psi = \frac{9783.2455}{76.9248 + 74.7777} \cdot 10^{-4} = 64.4897 \cdot 10^{-4}\,\mathrm{Wb}.$$

Since ψ is similar to electrical current, we use the current division rule to find the magnetic flux ψ_1 in the air gap as follows:

$$\psi_1 = \frac{\mathscr{R}_1 + 2\mathscr{R}_3}{(\mathscr{R}_1 + 2\mathscr{R}_3) + (\mathscr{R}_g + 2\mathscr{R}_2)}\psi = \frac{76.9248}{76.9248 \cdot 2679.1058} \cdot 64.4897 \cdot 10^{-4}$$
$$= 1.8 \cdot 10^{-4}\,\mathrm{Wb}.$$

Since the other source is identical and the circuit is symmetrical, the total magnetic flux in the air gap will be doubled. Hence the total magnetic flux in the air gap is $3.6 \cdot 10^{-4}$ Wb.

PROBLEMS

9.1. The radius of an infinitely long solenoid is 1 cm. It has 200 turns per cm over its length and carries a current of 100 mA. Find the magnetic field using Ampere's law.

9.2. The mean radius of a toroid is 3 cm, and it has 1500 turns of a filamentary wire. The cross section of the core is circular with a radius of 1 mm. There is a current of 100 mA flowing through the winding. Find the magnetic flux density using Ampere's law.

9.3. A 1 m long thin wire lies symmetrically along the z-axis. If this wire carries a current of 2 A, determine its magnetic field at a point on the x-y plane using the Biot-Savart law.

9.4. The radius of a loop filament is 10 cm. The loop lies on the x-y plane and carries a current of 2 A. Find the magnetic flux density at a point 1 m away on the axis.

9.5. The radius of a filamentary wire loop is 1 cm. The loop lies on the x-y plane. It carries a current of 1 A. Find the magnetic field on a spherical surface of 1 km in radius.

9.6. Find the inductance of a 20 cm long solenoid with 20 turns per cm. The radius of the solenoid is 1 cm.

9.7. Find the inductance of the toroid described in Problem 9.2.

9.8. Find the inductance of a 10 cm long solenoid with 1500 turns and a radius of 5 mm.

9.9. Find the inductance of a toroid with 5000 turns, mean radius of 8 cm, and cross-sectional area at the core of 10^{-4} m^2.

9.10. Find the inductance per unit length of a coaxial line with inner and outer conductor radii of 1.325 mm and 4.16 mm, respectively. Assume that the outer wall is 0.1 mm thick and the ε_r of the medium between the conductors is 2.1.

9.11. The axes of two concentric 15 cm long solenoids are aligned to the z-axis. The inner solenoid has 150 turns, and the outer solenoid has 250 turns. The radii of the two solenoids are 1 cm and 3 cm, respectively. Find their mutual inductance.

9.12. A distance of 5 cm separates two conducting wire loops of radii 10 cm and 6 cm. The cross section of the loop wire is a circle of 2 mm in radius. Find the loops' mutual inductance.

9.13. A circular loop of a 0.25 m mean radius is lying symmetrically on the x-y plane. The wire used for this loop is circular in its cross section with a radius of 1 mm. Find its inductance.

9.14. A toroid has 200 turns on an iron core of 12 cm mean length and an air gap of 5 mm, as shown in Figure 9.11. The toroid is being used to write on a magnetic tape, which requires a magnetic flux density of 0.9 T in the air gap (ignore fringing fields). If the relative permeability of the core at the given condition is 5000, find the required current I.

9.15. There is a current of 15 A flowing through 150 turns of the magnetic circuit shown in Figure 9.12. All branches have 10 cm^2 as their cross-sectional areas (S_1) except in the middle, which is 15 cm^2 (S_2). Lengths ℓ_1 and ℓ_3 are 35 cm each, whereas ℓ_2 is 15 cm long. The relative permeability of the core is 250. Find the magnetic flux densities through lengths ℓ_2 and ℓ_3.

9.16. A magnetic circuit shown in Figure 9.13 has two symmetrical windings, each with 5423.2 At. All branches of the core are 1.5 cm thick, and the core's relative permeability is 2000. Find the total magnetic flux in the air gap.

10

WAVEGUIDES AND CAVITY RESONATORS

Hollow metallic tubes of rectangular and circular cross sections are commonly employed in high-power microwave systems. These transmission lines are known as waveguides. When conducting walls are placed at two open ends of a waveguide of finite length, a cavity resonator is found. These resonators are used as resonant circuits at microwave frequencies. The electromagnetic fields in these structures are analyzed here by the procedure discussed earlier in Chapter 6. The chapter begins with an analysis of signals propagating through a metallic waveguide of rectangular cross section. The following section considers the TE and TM modes of metallic waveguides of circular cross section. Cavity resonators are commonly employed as resonant circuits at frequencies higher than 1 GHz. These resonators provide much higher quality factor (Q-factor) with respect to their lumped-element counterparts. The characteristic behaviors of cavity resonators of rectangular and circular cross sections are discussed in last two sections of this chapter.

10.1 METALLIC RECTANGULAR WAVEGUIDE

This section presents an analysis of electromagnetic signals propagating through a hollow metallic cylindrical waveguide of rectangular cross section, as shown in Figure 10.1. Assume that its cross section is $a \times b$, and the electromagnetic signal is propagating along the z-axis. As discussed in Section 6.4, the analysis is divided into TE and TM modes. TE mode fields are found from (6.4.13) through (6.4.18),

Practical Electromagnetics: From Biomedical Sciences to Wireless Communication.
By Devendra K. Misra
Copyright © 2007 John Wiley & Sons, Inc.

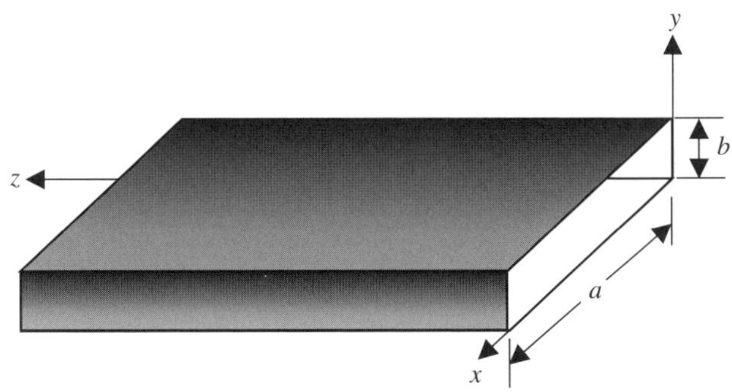

Figure 10.1 Metallic rectangular waveguide.

after identifying appropriate harmonic functions for (6.4.28). Similarly (6.4.4) through (6.4.9) give TM mode fields.

TE^z Modes in the Rectangular Waveguide

Since the signal is propagating along the z-axis, $h_3(k_z z) \rightarrow e^{-jk_z z}$ in (6.4.28), whereas the sine and cosine functions are appropriate solutions for $h_1(k_x x)$ and $h_2(k_y y)$. Therefore F_z for (6.4.12) is found to be

$$F_z(x, y, z) = \{c_1 \cos(k_x x) + c_2 \sin(k_x x)\}\{c_3 \cos(k_y y) + c_4 \sin(k_y y)\}e^{-jk_z z}, \quad (10.1.1)$$

where c_1, c_2, c_3, and c_4 are the integration constants.

Since tangential electric fields must be zero on the conducting surfaces, the boundary conditions for the fields in conjunction with (6.4.13) through (6.4.14) give

$$E_x|_{\substack{y=0\\y=b}} = 0 \rightarrow \left.\frac{\partial F_z}{\partial y}\right|_{\substack{y=0\\y=b}} = 0 \quad (10.1.2)$$

and

$$E_y|_{\substack{x=0\\x=a}} = 0 \rightarrow \left.\frac{\partial F_z}{\partial x}\right|_{\substack{x=0\\x=a}} = 0. \quad (10.1.3)$$

In order to satisfy these boundary conditions, (10.1.1) in conjunction with (10.1.2) and (10.1.3) gives

$$c_4 = 0, \quad (10.1.4)$$

$$k_y = \frac{n\pi}{b}, \quad (10.1.5)$$

$$c_2 = 0, \quad (10.1.6)$$

and

$$k_x = \frac{m\pi}{a}, \tag{10.1.7}$$

where m and n are integers including zero, except that both cannot be zero at the same time (which makes all fields zero, and therefore a trivial case).

Equation (10.1.1) reduces to

$$F_z(x, y, z) = c_1 c_3 \cos\left(\frac{m\pi}{a}x\right) \cos\left(\frac{n\pi}{b}y\right) e^{-jk_z z}. \tag{10.1.8}$$

Further k_z is found after substituting (10.1.5) and (10.1.7) into the separation equation (6.4.25) as follows:

$$k_z = \sqrt{k^2 - \left(\frac{m\pi}{a}\right)^2 - \left(\frac{n\pi}{b}\right)^2}. \tag{10.1.9}$$

For example, if a waveguide is air-filled, $k = k_o$ is a pure real number. There are following three possibilities in the air-filled case.

- The quantity under the square root is negative in (10.1.9) if

$$k_o^2 < \left(\frac{m\pi}{a}\right)^2 + \left(\frac{n\pi}{b}\right)^2. \tag{10.1.10}$$

  Therefore k_z is imaginary. This represents a purely attenuating signal without propagating along the waveguide. This condition may be used to design attenuators.

- The quantity under the square root is positive in (10.1.9) if

$$k_o^2 > \left(\frac{m\pi}{a}\right)^2 + \left(\frac{n\pi}{b}\right)^2. \tag{10.1.11}$$

  Therefore k_z is real, and the signal propagates through the waveguide without attenuation (ideally).

- The quantity under the square root goes to zero in (10.1.9) if

$$k_o^2 = \left(\frac{m\pi}{a}\right)^2 + \left(\frac{n\pi}{b}\right)^2 = k_c^2. \tag{10.1.12}$$

  Therefore k_z is zero, which is the cutoff condition. This wave number is known as the cutoff wave number k_c, and the corresponding wavelength of the signal is called the cutoff wavelength λ_c. The corresponding frequency f_c is called a cutoff frequency of the waveguide. Thus the waveguide works as a high-pass filter. A signal propagates only when its frequency is greater than the cutoff frequency of the waveguide.

The wavelength and the phase velocity of a propagating signal are found to be

$$\lambda_g = \frac{2\pi}{\sqrt{k_o^2 - k_c^2}} = \frac{2\pi}{k_o\sqrt{1-(k_c/k_o)^2}} = \frac{\lambda_o}{\sqrt{1-(\lambda_o/\lambda_c)^2}} = \frac{\lambda_o}{\sqrt{1-(f_c/f)^2}} \tag{10.1.13}$$

and

$$v_p = \frac{\omega}{k_z} = \frac{f\lambda_o}{\sqrt{1-(\lambda_o/\lambda_c)^2}} = \frac{3\times 10^8}{\sqrt{1-(\lambda_o/\lambda_c)^2}} \text{ m/s}, \tag{10.1.14}$$

where f is the signal frequency, k_o is its wave number in free space, and λ_o is the signal wavelength in free space.

The fields inside the waveguide are found after substituting (10.1.8) into (6.4.13) through (6.4.18) as follows:

$$E_x(x, y, z) = jH_{mn}\omega\mu\left(\frac{n\pi}{b}\right)\cos\left(\frac{m\pi}{a}x\right)\sin\left(\frac{n\pi}{b}y\right)e^{-jk_z z}, \tag{10.1.15}$$

$$E_y(x, y, z) = -jH_{mn}\omega\mu\left(\frac{m\pi}{a}\right)\sin\left(\frac{m\pi}{a}x\right)\cos\left(\frac{n\pi}{b}y\right)e^{-jk_z z}, \tag{10.1.16}$$

$$E_z(x, y, z) = 0, \tag{10.1.17}$$

$$H_x(x, y, z) = jH_{mn}k_z\left(\frac{m\pi}{a}\right)\sin\left(\frac{m\pi}{a}x\right)\cos\left(\frac{n\pi}{b}y\right)e^{-jk_z z}, \tag{10.1.18}$$

$$H_y(x, y, z) = jH_{mn}k_z\left(\frac{n\pi}{b}\right)\cos\left(\frac{m\pi}{a}x\right)\sin\left(\frac{n\pi}{b}y\right)e^{-jk_z z}, \tag{10.1.19}$$

and

$$H_z(x, y, z) = H_{mn}k_c^2\cos\left(\frac{m\pi}{a}x\right)\cos\left(\frac{n\pi}{b}y\right)e^{-jk_z z}, \tag{10.1.20}$$

where

$$H_{mn} = -jc_1c_3\frac{1}{\omega\mu\varepsilon}. \tag{10.1.21}$$

Note that the fields do not exist for m and n both zero. Therefore m or n has to be a nonzero integer for a nontrivial solution. For $a > b$, the lowest order mode that propagates is TE_{10} (i.e., for $m = 1$ and $n = 0$). It will be shown later that the lowest order TM mode that can propagate through the waveguide is TM_{11} (i.e., for $m = 1$, and $n = 1$). Therefore the TE_{10} mode is known as the dominant mode for rectangular waveguides. Only H_z, H_x, and E_y field components are nonzero in this

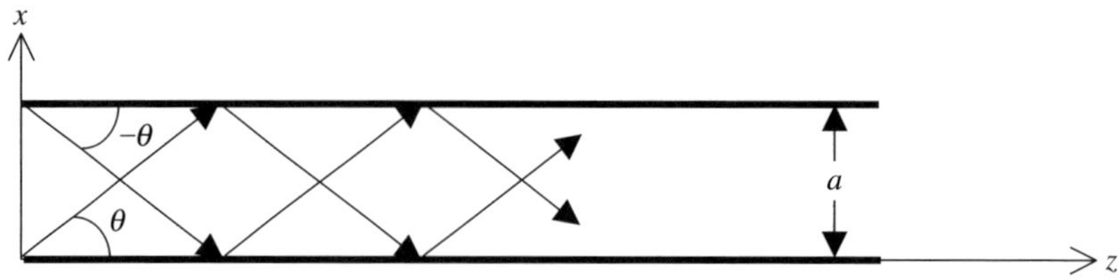

Figure 10.2 TE_{10} fields decomposed into two uniform plane waves bouncing off the conducting sidewalls of a rectangular waveguide.

mode. Also $\lambda_c = 2a$, and therefore (10.1.13) can be rewritten as follows:

$$\lambda_{g(TE_{10})} = \frac{\lambda_o}{\sqrt{1 - (\lambda_o/2a)^2}}. \tag{10.1.22}$$

Further the only nonzero component of the electric field is E_y that can be found from (10.1.16):

$$\begin{aligned} E_y(x, y, z) &= -jH_o \omega \mu_o \left(\frac{\pi}{a}\right) \sin\left(\frac{\pi}{a} x\right) e^{-jk_z z} \\ &= -H_o \omega \mu_o \left(\frac{\pi}{2a}\right) \left(e^{j[(\pi x/a) - k_z z]} - e^{-j[(\pi x/a) + k_z z]}\right). \end{aligned} \tag{10.1.23}$$

Note that this relation represents two plane electromagnetic waves propagating at angles $\pm\theta$ after reflection from the sidewalls of the waveguide, as shown in Figure 10.2. This angle is found to be

$$\theta = \sin^{-1}\left(\frac{\lambda_o}{2a}\right). \tag{10.1.24}$$

Therefore $\theta \to 90^\circ$ as $\lambda_o \to 2a$ (i.e., λ_c), and the wave ceases to propagate.

Example 10.1

The inside cross section of an air-filled rectangular metallic waveguide is 1.58×0.79 cm. (a) Determine the cutoff frequencies for the TE_{10}, TE_{20}, TE_{01}, and TE_{11} modes. (b) Find the mode(s) that will propagate through this waveguide if the signal frequency is anywhere between 12 and 18 GHz. (c) If this waveguide is being used to send a 15.8 GHz signal in TE_{10} mode, find the phase velocity.

(a) Since

$$k_c = \sqrt{\left(\frac{m\pi}{a}\right)^2 + \left(\frac{n\pi}{b}\right)^2} \to f_c = 3 \cdot 10^8 \sqrt{\left(\frac{m}{2a}\right)^2 + \left(\frac{n}{2b}\right)^2},$$

the cutoff frequencies for different modes are found to be,

$$TE_{10} \rightarrow f_c = 3 \cdot 10^8 \sqrt{\left(\frac{1}{2 \cdot 0.0158}\right)^2 + \left(\frac{0}{2 \cdot 0.0079}\right)^2} \text{ Hz} = 9.4937 \text{ GHz},$$

$$TE_{20} \rightarrow f_c = 3 \cdot 10^8 \sqrt{\left(\frac{2}{2 \cdot 0.0158}\right)^2 + \left(\frac{0}{2 \cdot 0.0079}\right)^2} \text{ Hz} = 18.9873 \text{ GHz},$$

$$TE_{01} \rightarrow f_c = 3 \cdot 10^8 \sqrt{\left(\frac{0}{2 \cdot 0.0158}\right)^2 + \left(\frac{1}{2 \cdot 0.0079}\right)^2} \text{ Hz} = 18.9873 \text{ GHz},$$

and

$$TE_{11} \rightarrow f_c = 3 \cdot 10^8 \sqrt{\left(\frac{1}{2 \cdot 0.0158}\right)^2 + \left(\frac{1}{2 \cdot 0.0079}\right)^2} \text{ Hz} = 21.2285 \text{ GHz}.$$

(b) From (a) the cutoff frequency for the TE_{10} mode is 9.4937 GHz. Since next higher modes have cutoffs at 18.9873 GHz, only the TE_{10} mode will exist for a signal frequency band of 12 to 18 GHz.

(c)

$$v_p = \frac{3 \cdot 10^8}{\sqrt{1 - (\lambda_o/\lambda_c)^2}} = \frac{3 \cdot 10^8}{\sqrt{1 - (f_c/f)^2}} = \frac{3 \cdot 10^8}{\sqrt{1 - (9.4937/15.8)^2}}$$
$$= 3.7531 \cdot 10^8 \text{ m/s}.$$

TM^z Modes in the Rectangular Waveguide

In the case of TM^z modes, the fields are found from (6.4.4) through (6.4.9), after constructing an appropriate solution to (6.4.3). As in the previous case, $h_3(k_z z) \rightarrow e^{-jk_z z}$ in (6.4.28) because the signal is assumed to be propagating along the z-axis. Similarly the sine and cosine functions are appropriate solutions for $h_1(k_x x)$ and $h_2(k_y y)$. Therefore A_z for (6.4.3) is found to be

$$A_z(x, y, z) = \{c_1 \cos(k_x x) + c_2 \sin(k_x x)\}\{c_3 \cos(k_y y) + c_4 \sin(k_y y)\}e^{-jk_z z}, \quad (10.1.25)$$

where c_1, c_2, c_3, and c_4 are the integration constants.

Since tangential electric fields must be zero on the conducting surfaces, boundary conditions for the fields in conjunction with (6.4.7) through (6.4.9) give

$$E_x|_{\substack{y=0 \\ y=b}} = 0 \rightarrow A_z|_{\substack{y=0 \\ y=b}} = 0, \quad (10.1.26)$$

$$E_y|_{\substack{x=0 \\ x=a}} = 0 \rightarrow A_z|_{\substack{x=0 \\ x=a}} = 0, \quad (10.1.27)$$

and

$$E_z|_{\substack{x=0\\x=a\\y=0\\y=b}} = 0 \rightarrow A_z|_{\substack{x=0\\x=a\\y=0\\y=b}} = 0. \tag{10.1.28}$$

Note that if (10.1.26) and (10.1.27) are satisfied, (10.1.28) will hold automatically. In order to satisfy (10.1.26), we use (10.1.25) to find that

$$c_3 = 0 \tag{10.1.29}$$

and

$$k_y = \frac{n\pi}{b}. \tag{10.1.30}$$

Similarly the boundary condition (10.1.27) requires that

$$c_1 = 0 \tag{10.1.31}$$

and

$$k_x = \frac{m\pi}{a}, \tag{10.1.32}$$

where m and n are integers excluding zero (which makes all fields zero, and therefore a trivial case).

Equation (10.1.25) reduces to

$$A_z(x, y, z) = c_2 c_4 \sin\left(\frac{m\pi}{a}x\right)\sin\left(\frac{n\pi}{b}y\right)e^{-jk_z z}. \tag{10.1.33}$$

Note that k_z remains the same as in (10.1.9) for the TE case, and the TM modes degenerate as a result. After substituting (10.1.33) into (6.4.4) through (6.4.9), the field components are found to be

$$H_x(x, y, z) = jE_{mn}\omega\varepsilon\left(\frac{n\pi}{b}\right)\sin\left(\frac{m\pi}{a}x\right)\cos\left(\frac{n\pi}{b}y\right)e^{-jk_z z}, \tag{10.1.34}$$

$$H_y(x, y, z) = -jE_{mn}\omega\varepsilon\left(\frac{m\pi}{a}\right)\cos\left(\frac{m\pi}{a}x\right)\sin\left(\frac{n\pi}{b}y\right)e^{-jk_z z}, \tag{10.1.35}$$

$$H_z(x, y, z) = 0, \tag{10.1.36}$$

$$E_x(x, y, z) = -j\left(\frac{m\pi k_z}{a}\right)E_{mn}\cos\left(\frac{m\pi}{a}x\right)\sin\left(\frac{n\pi}{b}y\right)e^{-jk_z z}, \tag{10.1.37}$$

$$E_y(x, y, z) = -j\left(\frac{n\pi k_z}{b}\right)E_{mn}\sin\left(\frac{m\pi}{a}x\right)\cos\left(\frac{n\pi}{b}y\right)e^{-jk_z z}, \tag{10.1.38}$$

TABLE 10.1 Fields in Metallic Rectangular Waveguide

	TE_{mn} Modes	TM_{mn} Modes
k_c	$\sqrt{\left(\frac{m\pi}{a}\right)^2+\left(\frac{n\pi}{b}\right)^2}$	$\sqrt{\left(\frac{m\pi}{a}\right)^2+\left(\frac{n\pi}{b}\right)^2}$
k_z	$\sqrt{k^2-k_c^2}$	$\sqrt{k^2-k_c^2}$
λ_g	$\frac{\lambda}{\sqrt{1-(k_c/k)^2}}$	$\frac{\lambda}{\sqrt{1-(k_c/k)^2}}$
$H_z(x, y)$	$H_{mn}k_c^2\cos\left(\frac{m\pi}{a}x\right)\cos\left(\frac{n\pi}{b}y\right)$	0
$E_z(x, y)$	0	$E_{mn}k_c^2\sin\left(\frac{m\pi}{a}x\right)\sin\left(\frac{n\pi}{b}y\right)$
$H_x(x, y)$	$jH_{mn}k_z\left(\frac{m\pi}{a}\right)\sin\left(\frac{m\pi}{a}x\right)\cos\left(\frac{n\pi}{b}y\right)$	$jE_{mn}\omega\varepsilon\left(\frac{n\pi}{b}\right)\sin\left(\frac{m\pi}{a}x\right)\cos\left(\frac{n\pi}{b}y\right)$
$H_y(x, y)$	$jH_{mn}k_z\left(\frac{n\pi}{b}\right)\cos\left(\frac{m\pi}{a}x\right)\sin\left(\frac{n\pi}{b}y\right)$	$-jE_{mn}\omega\varepsilon\left(\frac{m\pi}{a}\right)\cos\left(\frac{m\pi}{a}x\right)\sin\left(\frac{n\pi}{b}y\right)$
$E_x(x, y)$	$jH_{mn}\omega\mu\left(\frac{n\pi}{b}\right)\cos\left(\frac{m\pi}{a}x\right)\sin\left(\frac{n\pi}{b}y\right)$	$-j\left(\frac{m\pi k_z}{a}\right)E_{mn}\cos\left(\frac{m\pi}{a}x\right)\sin\left(\frac{n\pi}{b}y\right)$
$E_y(x, y)$	$-jH_{mn}\omega\mu\left(\frac{m\pi}{a}\right)\sin\left(\frac{m\pi}{a}x\right)\cos\left(\frac{n\pi}{b}y\right)$	$-j\left(\frac{n\pi k_z}{b}\right)E_{mn}\sin\left(\frac{m\pi}{a}x\right)\cos\left(\frac{n\pi}{b}y\right)$

and

$$E_z(x, y, z) = E_{mn}k_c^2\sin\left(\frac{m\pi}{a}x\right)\sin\left(\frac{n\pi}{b}y\right)e^{-jk_z z}, \tag{10.1.39}$$

where

$$E_{mn} = -jc_2c_4\frac{1}{\omega\mu\varepsilon}. \tag{10.1.40}$$

Note that the lowest order TM mode that exists is for $m = 1$ and $n = 1$ (i.e., the TM_{11} mode). Since TE_{10} is lowest among all possible modes, it is known as the dominant mode. The electromagnetic fields and the related characteristics of the TE and TM modes are summarized in Table 10.1.

The Wave Impedance

The wave impedance of a rectangular waveguide operating in the TE^z mode is defined as follows:

$$Z_{TE^z} = \frac{E_x}{H_y} = -\frac{E_y}{H_x}. \tag{10.1.41}$$

Using (10.1.15) through (10.1.20), we find that

$$Z_{\mathrm{TE}^z} = \frac{\omega\mu}{k_z} = \frac{\omega\mu}{\sqrt{k^2 - k_c^2}} = \frac{\omega\mu}{k\sqrt{1-(k_c^2/k^2)}} = \frac{\eta}{\sqrt{1-(\lambda/\lambda_c)^2}}$$
$$= \frac{\eta}{\sqrt{1-(f_c/f)^2}}. \tag{10.1.42}$$

Similarly the wave impedance for TM^z modes in a rectangular waveguide is defined as follows:

$$Z_{\mathrm{TM}^z} = \frac{E_x}{H_y} = -\frac{E_y}{H_x}. \tag{10.1.43}$$

Using (10.1.34) through (10.1.39), we find next that

$$Z_{\mathrm{TM}^z} = \frac{k_z}{\omega\varepsilon} = \frac{k\sqrt{1-(k_c/k)^2}}{\omega\varepsilon} = \eta\sqrt{1-\left(\frac{k_c}{k}\right)^2} = \eta\sqrt{1-\left(\frac{\lambda}{\lambda_c}\right)^2}$$
$$= \eta\sqrt{1-\left(\frac{f_c}{f}\right)^2}. \tag{10.1.44}$$

Equations (10.1.42) and (10.1.44) are displaced graphically for the air-filled waveguide in Figure 10.3. As the figure indicates, wave impedances are purely imaginary (inductive for TE and capacitive for TM modes) for a signal frequency f smaller that the cutoff frequency f_c. The wave impedance becomes infinite for the TE and zero for the TM modes when the signal frequency is equal to the cutoff. As the signal frequency increases higher than the cutoff frequency, the wave impedance becomes purely resistive in both cases.

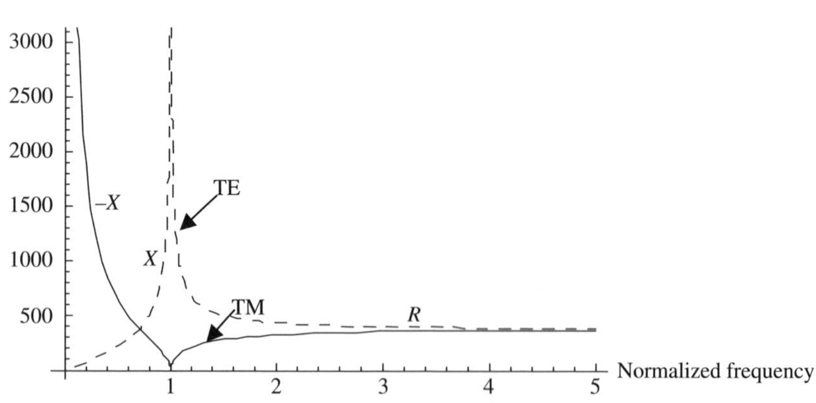

Figure 10.3 Wave impedance versus normalized frequency (f_c/f).

Power Flow and the Signal Attenuation

Poynting's vector for the TE^z modes is

$$\begin{aligned}\vec{S}_{\rm av} &= \tfrac{1}{2}{\rm Re}[(\hat{x}E_x + \hat{y}E_y) \times (\hat{x}H_x^* + \hat{y}H_y^* + \hat{z}H_z^*)] \\ &= \tfrac{1}{2}{\rm Re}[\hat{x}E_yH_z^* - \hat{y}E_xH_z^* + \hat{z}(E_xH_y^* - E_yH_x^*)].\end{aligned}$$

Using (10.1.15) through (10.1.20) obtains

$$\begin{aligned}\vec{S}_{\rm av} = \hat{z}\frac{1}{2}|H_{mn}|^2\omega\mu k_z\bigg[\left(\frac{n\pi}{b}\right)^2\cos^2\left(\frac{m\pi}{a}x\right)\sin^2\left(\frac{n\pi}{b}y\right) \\ +\left(\frac{m\pi}{a}\right)^2\sin^2\left(\frac{m\pi}{a}x\right)\cos^2\left(\frac{n\pi}{b}y\right)\bigg] = \hat{z}S_z.\end{aligned} \tag{10.1.45}$$

Note that the other two components of the Poynting vector are purely imaginary.

From Poynting's vector the power-flow through a waveguide is found to be

$$P_{mn}^{{\rm TE}^z} = \int_S \vec{S}_{\rm av}\cdot d\vec{s} = \int_{x=0}^{a}\int_{y=0}^{b} S_z dxdy = \frac{|H_{mn}|^2}{2}\omega\mu k_z\frac{a}{\varepsilon_m}\cdot\frac{b}{\varepsilon_n}\left[\left(\frac{n\pi}{b}\right)^2+\left(\frac{m\pi}{a}\right)^2\right],$$

or

$$P_{mn}^{{\rm TE}^z} = \frac{|H_{mn}|^2}{2}\omega\mu k k_{\rm c}^2\frac{a}{\varepsilon_m}\cdot\frac{b}{\varepsilon_n}\sqrt{1-\left(\frac{f_{\rm c}}{f}\right)^2}, \tag{10.1.46}$$

where ε_m and ε_n represent Neumann's number, which is defined as follows:

$$\varepsilon_q = \begin{cases} 1, & q = 0, \\ 2, & q > 0. \end{cases} \tag{10.1.47}$$

For the dominant mode in a rectangular waveguide, (10.1.46) reduces to

$$P_{10}^{{\rm TE}^z} = \frac{|H_{10}|^2}{4}\omega\mu k\left(\frac{\pi}{a}\right)^2 ab\sqrt{1-\left(\frac{f_{\rm c}}{f}\right)^2}. \tag{10.1.48}$$

Similarly Poynting's vector for the TM^z modes is

$$\begin{aligned}\vec{S}_{\rm av} &= \tfrac{1}{2}{\rm Re}[(\hat{x}E_x + \hat{y}E_y + \hat{z}E_z) \times (\hat{x}H_x^* + \hat{y}H_y^*)] \\ &= \tfrac{1}{2}{\rm Re}[-\hat{x}E_zH_y^* + \hat{y}E_zH_x^* + \hat{z}(E_xH_y^* - E_yH_x^*)].\end{aligned}$$

Using (10.1.34) through (10.1.39) obtains

$$\vec{S}_{\text{av}} = \hat{z}\frac{1}{2}|E_{mn}|^2\omega\varepsilon k_z\left[\left(\frac{m\pi}{a}\right)^2\cos^2\left(\frac{m\pi}{a}x\right)\sin^2\left(\frac{n\pi}{b}y\right)\right.$$

$$\left.+\left(\frac{n\pi}{b}\right)^2\sin^2\left(\frac{m\pi}{a}x\right)\cos^2\left(\frac{n\pi}{b}y\right)\right] = \hat{z}S_z. \tag{10.1.49}$$

Again, the other two components of the Poynting vector are purely imaginary.

From the above the power-flow is found to be

$$P_{mn}^{\text{TM}^z} = \int_S \vec{S}_{\text{av}} \cdot d\vec{s} = \int_{x=0}^{a}\int_{y=0}^{b} S_z dxdy = \frac{|E_{mn}|^2}{2}\omega\varepsilon k_z\frac{ab}{4}\left[\left(\frac{m\pi}{a}\right)^2+\left(\frac{n\pi}{b}\right)^2\right],$$

or

$$P_{mn}^{\text{TM}^z} = \frac{|E_{mn}|^2}{2}\omega\varepsilon k k_{\text{c}}^2\frac{ab}{4}\sqrt{1-\left(\frac{f_{\text{c}}}{f}\right)^2}. \tag{10.1.50}$$

When the waveguide walls have a finite conductivity, there is power loss in the conductors. Further there is power loss in the dielectric medium if it is imperfect. It is assumed that the attenuation constant α_{c} due to conductor loss is separable from the attenuation constant α_{d} due to the dielectric loss and consequently the two can be determined independently. The boundary conditions are modified to account for the power loss in the conductors. Since this exact field analysis becomes much more involved, we use an alternative approach that may be called a perturbation method. This formulation suffices for most practical cases. In this approach it is assumed that the finite conductivity of the walls does not change the fields appreciably. However, there is power loss due to the surface resistance R_{s}, which is defined as follows:

$$Z_{\text{s}} = R_{\text{s}} + jX_{\text{s}} = \frac{\omega\mu}{k} = \frac{\omega\mu}{\omega\sqrt{\mu[\varepsilon - j(\sigma/\omega)]}} = \sqrt{\frac{\mu}{\varepsilon - j(\sigma/\omega)}}.$$

For $(\sigma/\omega) \gg \varepsilon$ we find that

$$Z_{\text{s}} = R_{\text{s}} + jX_{\text{s}} \approx \sqrt{\frac{\mu}{-j(\sigma/\omega)}} = \sqrt{\frac{\omega\mu}{-j\sigma}} = \sqrt{\frac{\omega\mu}{2\sigma}}(1+j). \tag{10.1.51}$$

The power loss on each wall is found from the corresponding surface current as follows:

$$P_{\text{w}} = \frac{1}{2}R_{\text{s}}\int_S \vec{K}_{\text{s}} \cdot \vec{K}_{\text{s}}^* ds = \frac{1}{2}R_{\text{s}}\int_S |K_{\text{s}}|^2 ds. \tag{10.1.52}$$

For the sake of simplicity we consider only the dominant mode fields here. Since the surface current can be found from $\vec{K}_s = \hat{n} \times \vec{H}|_{\text{surface}}$, the surface current on the wall at $y = 0$ is found to be

$$\vec{K}_{s1} = (\hat{x}H_z - \hat{z}H_x)\big|_{y=0} = \left[\hat{x}\left(\frac{\pi}{a}\right)\cos\left(\frac{\pi}{a}x\right) - \hat{z}jk_z\sin\left(\frac{\pi}{a}x\right)\right]\left(\frac{\pi}{a}\right)H_{10}e^{-jk_z z}. \quad (10.1.53)$$

Similarly the surface current on the wall at $x = 0$ is

$$\vec{K}_{s2} = -\hat{y}H_z\big|_{x=0} = -\hat{y}H_{10}\left(\frac{\pi}{a}\right)^2 e^{-jk_z z}. \quad (10.1.54)$$

Therefore the power loss per unit length for the wall at $y = 0$ is found as follows:

$$P_{w1} = \frac{1}{2}R_s|H_{10}|^2\left(\frac{\pi}{a}\right)^2\int_{z=0}^{1}\int_{x=0}^{a}\left[\left(\frac{\pi}{a}\right)^2\cos^2\left(\frac{\pi}{a}x\right) + k_z^2\sin^2\left(\frac{\pi}{a}x\right)\right]dxdz,$$

or

$$P_{w1} = \frac{1}{2}R_s|H_{10}|^2\left(\frac{\pi}{a}\right)^2\frac{a}{2}\left[\left(\frac{\pi}{a}\right)^2 + k_z^2\right] = \frac{1}{4}R_s|H_{10}|^2\left(\frac{\pi}{a}\right)^2 ak^2. \quad (10.1.55)$$

Similarly power loss per unit length for the wall at $x = 0$ is

$$P_{w2} = \frac{1}{2}R_s|H_{10}|^2\left(\frac{\pi}{a}\right)^4\int_{z=0}^{1}\int_{y=0}^{b} dydz = \frac{1}{2}R_s|H_{10}|^2\left(\frac{\pi}{a}\right)^4 b. \quad (10.1.56)$$

When there is a signal propagating through a waveguide that is made up of walls with finite conductivity, its power is attenuated. If the attenuation constant is α_c, then the power-flow can be expressed as follows:

$$P_{mn} = P_o e^{-2\alpha_c z}.$$

Therefore

$$\frac{dP_{mn}}{dz}dz = -2\alpha_c P_{mn}dz \longrightarrow \alpha_c = \frac{(dP_{mn}/dz)dz}{2P_{mn}dz} = \frac{P_w/\text{length}}{2P_{mn}}. \quad (10.1.57)$$

Since walls at $y = b$ and $x = a$ are identical to those at $y = 0$ and at $x = 0$, respectively, the total power loss per unit length is found to be twice that of

$P_{w1} + P_{w2}$. Therefore the attenuation constant a_c is found from (10.1.48), (10.1.55), and (10.1.56) as follows:

$$\alpha_c = \frac{2(P_{w1} + P_{w2})}{2P_{10}} = \frac{2R_s}{\omega\mu k_z ab}\left[\frac{a}{2}\left\{\left(\frac{\pi}{a}\right)^2 + k_z^2\right\} + \left(\frac{\pi}{a}\right)^2 b\right]$$
$$= \frac{2R_s}{\omega\mu k_z ab}\left[\frac{a}{2}k^2 + \left(\frac{\pi}{a}\right)^2 b\right],$$

or

$$\alpha_c = \frac{R_s k^2}{\omega\mu k_z b}\left[1 + \frac{2b}{a}\cdot\frac{(\pi/a)^2}{k^2}\right] = \frac{R_s k^2}{\omega\mu k_z b}\left[1 + \frac{2b}{a}\cdot\left(\frac{f_c}{f}\right)^2\right] \text{Np/m.} \quad (10.1.58)$$

Since $\omega\mu/k = \eta$ and $k/k_z = 1/\sqrt{1 - (f_c/f)^2}$, (10.1.58) can be expressed as follows:

$$\alpha_c = \frac{R_s}{\eta b}\cdot\frac{[1 + (2b/a)\cdot(f_c/f)^2]}{\sqrt{1 - (f_c/f)^2}} \text{ Np/m.} \quad (10.1.59)$$

The attenuation constant α_d due to power loss in the dielectric medium is found from the complex k_z as follows:

$$k_z = \beta_z - j\alpha_d = \sqrt{k^2 - k_c^2} = \sqrt{\omega^2\mu\left(\varepsilon - j\frac{\sigma}{\omega}\right) - k_c^2} = \sqrt{\omega^2\mu\varepsilon - k_c^2 - j\omega\mu\sigma},$$

or

$$k_z = \sqrt{\omega^2\mu\varepsilon - k_c^2}\left[1 - \frac{j\omega\mu\sigma}{\omega^2\mu\varepsilon - k_c^2}\right]^{1/2} \approx \sqrt{\omega^2\mu\varepsilon - k_c^2}\left[1 - \frac{j\omega\mu\sigma}{2\omega^2\mu\varepsilon - 2k_c^2}\right].$$

Therefore

$$\alpha_d = \frac{1}{2}\left(\frac{\omega\mu\sigma}{\sqrt{\omega^2\mu\varepsilon - k_c^2}}\right) = \frac{\mu\sigma}{2\sqrt{\mu\varepsilon[1 - (f_c/f)]^2}} = \frac{\eta\sigma}{2\sqrt{1 - (f_c/f)^2}} \text{ Np/m.} \quad (10.1.60)$$

Alternatively,

$$k_z = \beta_z - j\alpha_d = \sqrt{k^2 - k_c^2} = \sqrt{\omega^2\mu\varepsilon_o(\varepsilon' - j\varepsilon'') - k_c^2}.$$

Therefore

$$\alpha_d \approx \frac{1}{2} \cdot \frac{\omega^2 \mu \varepsilon_o \varepsilon''}{\sqrt{\omega^2 \mu \varepsilon_o \varepsilon' - k_c^2}} = \frac{\varepsilon'' \pi}{\varepsilon' \lambda \sqrt{1 - (f_c/f)^2}} = \frac{\varepsilon'' \pi}{\varepsilon' \lambda^2} \lambda_g$$

$$= \frac{\pi \lambda_g}{\lambda^2} \tan \delta \text{ Np/m}. \tag{10.1.61}$$

Example 10.2

The cross section of a rectangular waveguide is 2.286 × 1.016 cm. It is made of copper ($\sigma = 5.813 \cdot 10^7$ S/m) and filled with Plexiglas ($\varepsilon_r = 2.6$, $\tan \delta = 0.0057$). Find its attenuation constants α_d and α_c at 5.4 GHz. The waveguide operates in TE_{10} mode.

Since

$$f_c = \frac{3 \cdot 10^8}{2a\sqrt{\varepsilon_r}} = \frac{3 \cdot 10^8}{2 \cdot 2.286 \cdot 10^{-2}\sqrt{2.6}} = 4.0694 \cdot 10^9 \text{ Hz} = 4.0694 \text{ GHz}$$

and

$$\lambda = \frac{3 \cdot 10^8}{f\sqrt{\varepsilon_r}} = \frac{3 \cdot 10^8}{5.4 \cdot 10^9 \cdot \sqrt{2.6}} = 0.0345 \text{ m},$$

$$\lambda_g = \frac{\lambda}{\sqrt{1 - (f_c/f)^2}} = 0.0524 \text{ m}.$$

Therefore (10.1.61) gives

$$\alpha_d = \frac{\pi \lambda_g}{\lambda^2} \tan \delta = 0.7907 \text{ Np/m}.$$

The conductor attenuation α_c is found as follows:

$$R_s = \sqrt{\frac{\omega \mu_o}{2\sigma}} = 0.0192\,\Omega$$

and

$$\eta = \sqrt{\frac{\mu_o}{\varepsilon_o \varepsilon_r}} = 233.6407\,\Omega.$$

Therefore (10.1.59) gives

$$\alpha_c = \frac{R_s}{\eta b} \cdot \frac{[1 + (2b/a) \cdot (f_c/f)^2]}{\sqrt{1 - (f_c/f)^2}} = 0.0185 \text{ Np/m}.$$

10.2 METALLIC CIRCULAR WAVEGUIDE

This section presents an analysis of electromagnetic signals propagating through a hollow metallic cylindrical waveguide of circular cross section, as shown in Figure 10.4. Assume that its radius is a and the field variation in the z direction is $e^{-jk_z z}$ (because the signal is propagating along z). As discussed in Section 6.5, the analysis is divided into TE and TM modes. The TE mode fields are found from (6.5.13) through (6.5.18) after selecting appropriate harmonic and Bessel functions for (6.5.28). Similarly (6.5.4) through (6.5.9) give the TM mode fields.

TE^z Modes in the Circular Waveguide

Since the signal is propagating along the z-axis, $h_3(k_z z) \to e^{-jk_z z}$, in (6.5.28), whereas the sine and cosine functions are appropriate for $h_2(n\phi)$ with n being an integer. Note that a sum of sine and cosine functions with each multiplied by a different constant is the general expression for $h_2(n\phi)$. However, one of these can be forced to zero by rotating the coordinate system appropriately along ϕ. Since fields must be finite at $\rho = 0$, only Bessel functions of the first kind are appropriate for $Z_n(k_\rho \rho)$. Therefore F_z for (6.5.12) is found to be

$$F_z(\rho, \phi, z) = A_{nm} J_n(k_\rho \rho) e^{-jk_z z} \begin{cases} \sin(n\phi) \\ \cos(n\phi), \end{cases} \tag{10.2.1}$$

where A_{nm} is a constant.

Since tangential electric fields must be zero on conducting surfaces, the boundary conditions for the fields in conjunction with (6.5.13) through (6.5.14) give

$$E_\phi \Big|_{\rho=a} = 0 \to \frac{\partial F_z}{\partial \rho}\Big|_{\rho=a} = 0. \tag{10.2.2}$$

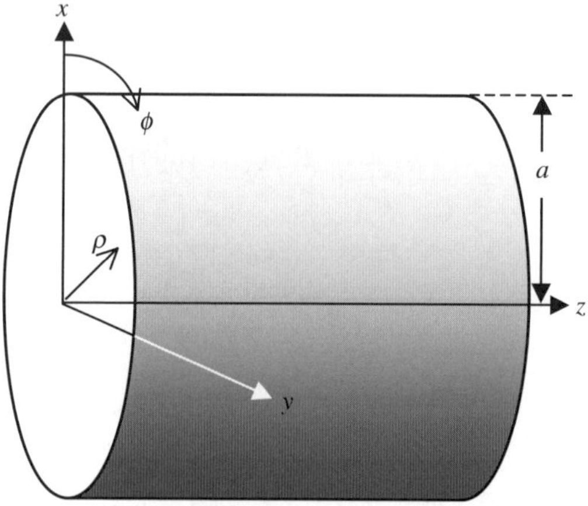

Figure 10.4 Metallic circular waveguide geometry.

There is only one boundary condition in this case because E_z is zero everywhere and E_ρ is normal to the boundary. Therefore from (10.2.1) and (10.2.2) we find that

$$\frac{\partial}{\partial\rho}J_n(k_\rho\rho)\bigg|_{\rho=a}=0\rightarrow\frac{\partial}{\partial(k_\rho\rho)}J_n(k_\rho\rho)\bigg|_{\rho=a}=J_n'(k_\rho\rho)\bigg|_{\rho=a}=0.$$

Therefore

$$J_n'(k_\rho a)=0\rightarrow k_\rho a=x_{nm}'\rightarrow k_\rho=\frac{x_{nm}'}{a}=k_c, \tag{10.2.3}$$

where the prime on the Bessel function indicates the derivative with respect to its argument $k_\rho\rho$; x_{nm}' is mth zero of Bessel function of the first kind and order n, which are listed in Table C.2. A prime on x is used to differentiate the zeros of the derivatives of Bessel functions from the zeros of the Bessel functions.

Therefore (10.2.1) can be expressed as follows:

$$F_z(\rho,\phi,z)=A_{nm}J_n(k_c\rho)e^{-jk_zz}\begin{cases}\sin(n\phi)\\ \cos(n\phi).\end{cases} \tag{10.2.4}$$

Further k_z is found after substituting (10.2.4) into (6.5.26) as follows:

$$k_z=\sqrt{k^2-k_c^2}=\sqrt{k^2-\left(\frac{x_{nm}'}{a}\right)^2}.$$

Therefore

$$\lambda_g=\frac{2\pi}{k_z}=\frac{2\pi}{k\sqrt{1-(x_{nm}'/ka)^2}}=\frac{\lambda}{\sqrt{1-(f_c/f)^2}}. \tag{10.2.5}$$

The field components are found next after substituting (10.2.4) into (6.5.13) through (6.5.18):

$$E_\rho(\rho,\phi,z)=-j\frac{n\omega\mu}{\rho k_c^2}H_{nm}J_n(k_c\rho)e^{-jk_zz}\begin{cases}\cos(n\phi)\\ -\sin(n\phi),\end{cases} \tag{10.2.6}$$

$$E_\phi(\rho,\phi,z)=j\frac{\omega\mu}{k_c}H_{nm}J_n'(k_c\rho)e^{-jk_zz}\begin{cases}\sin(n\phi)\\ \cos(n\phi),\end{cases} \tag{10.2.7}$$

$$E_z(\rho,\phi,z)=0 \tag{10.2.8}$$

$$H_\rho(\rho,\phi,z)-\,-j\frac{k_z}{k_c}H_{nm}J_n'(k_c\rho)e^{-jk_zz}\begin{cases}\sin(n\phi)\\ \cos(n\phi),\end{cases} \tag{10.2.9}$$

$$H_\phi(\rho,\phi,z)=-j\frac{nk_z}{\rho k_c^2}H_{nm}J_n(k_c\rho)e^{-jk_zz}\begin{cases}\cos(n\phi)\\ -\sin(n\phi),\end{cases} \tag{10.2.10}$$

and

$$H_z(\rho, \phi, z) = H_{nm} J_n(k_c \rho) e^{-jk_z z} \begin{cases} \sin(n\phi) \\ \cos(n\phi) \end{cases}, \tag{10.2.11}$$

where

$$H_{nm} = \frac{A_{nm} k_c^2}{j\omega\mu\varepsilon}. \tag{10.2.12}$$

The wave impedance for the TE^z modes in a circular waveguide is found to be

$$Z_{TE^z} = \frac{E_\rho}{H_\phi} = -\frac{E_\phi}{H_\rho} = \frac{\omega\mu}{k_z} = \frac{\eta}{\sqrt{1-(f_c/f)^2}}. \tag{10.2.13}$$

TM^z Modes in the Circular Waveguide

Similar to the TE^z case, the signal is propagating along the z-axis, and therefore $h_3(k_z z) \to e^{-jk_z z}$ in (6.5.28), whereas the sine and cosine functions are appropriate for $h_2(n\phi)$ with n being an integer. Since the fields must be finite at $\rho = 0$, only Bessel functions of the first kind are appropriate for $Z_n(k_\rho \rho)$. Therefore A_z for (6.5.3) is found to be

$$A_z(\rho, \phi, z) = C_{nm} J_n(k_\rho \rho) e^{-jk_z z} \begin{cases} \sin(n\phi) \\ \cos(n\phi) \end{cases}, \tag{10.2.14}$$

where C_{nm} is a constant.

Since the tangential electric fields must be zero on the conducting surfaces, the boundary conditions for the fields in conjunction with (6.5.8) and (6.5.9) give

$$E_\phi\big|_{\rho=a} = 0 \to A_z\big|_{\rho=a} = 0 \tag{10.2.15}$$

and

$$E_z\big|_{\rho=a} = 0 \to A_z\big|_{\rho=a} = 0. \tag{10.2.16}$$

Both of these conditions are met if

$$J_n(k_\rho a) = 0 \to k_\rho = \frac{x_{nm}}{a} = k_c, \tag{10.2.17}$$

where x_{nm} is the mth zero of $J_n(k_\rho a)$.

Further k_z is found after substituting (10.2.17) into (6.5.26) as follows:

$$k_z = \sqrt{k^2 - k_c^2} = \sqrt{k^2 - \left(\frac{x_{nm}}{a}\right)^2}.$$

Therefore

$$\lambda_g = \frac{2\pi}{k_z} = \frac{2\pi}{k\sqrt{1-(x_{nm}/ka)^2}} = \frac{\lambda}{\sqrt{1-(f_c/f)^2}}. \tag{10.2.18}$$

The field components are found as follows after substituting (10.2.14) and (10.5.17) into (6.5.4) through (6.5.9).

$$H_\rho(\rho, \phi, z) = j\frac{n\omega\varepsilon}{\rho k_c^2} E_{nm} J_n(k_c\rho) e^{-jk_z z} \begin{cases} \cos(n\phi) \\ -\sin(n\phi), \end{cases} \tag{10.2.19}$$

$$H_\phi(\rho, \phi, z) = -j\frac{\omega\varepsilon}{k_c} E_{nm} J_n'(k_c\rho) e^{-jk_z z} \begin{cases} \sin(n\phi) \\ \cos(n\phi), \end{cases} \tag{10.2.20}$$

$$H_z(\rho, \phi, z) = 0, \tag{10.2.21}$$

$$E_\rho(\rho, \phi, z) = -j\frac{k_z}{k_c} E_{nm} J_n'(k_c\rho) e^{-jk_z z} \begin{cases} \sin(n\phi) \\ \cos(n\phi), \end{cases} \tag{10.2.22}$$

$$E_\phi(\rho, \phi, z) = -j\frac{nk_z}{\rho k_c^2} E_{nm} J_n(k_c\rho) e^{-jk_z z} \begin{cases} \cos(n\phi) \\ -\sin(n\phi), \end{cases} \tag{10.2.23}$$

and

$$E_z(\rho, \phi, z) = E_{nm} J_n(k_c\rho) e^{-jk_z z} \begin{cases} \sin(n\phi) \\ \cos(n\phi), \end{cases} \tag{10.2.24}$$

where

$$E_{nm} = \frac{C_{nm}}{j\omega\mu\varepsilon} k_\rho^2. \tag{10.2.25}$$

The wave impedance for the TM^z modes in a circular waveguide is found to be

$$Z_{\text{TM}^z} = \frac{E_\rho}{H_\phi} = -\frac{E_\phi}{H_\rho} = \frac{k_z}{\omega\varepsilon} = \eta\sqrt{1 - \left(\frac{f_c}{f}\right)^2}. \tag{10.2.26}$$

The characteristics of the TM and TE modes in the circular waveguide are listed in Table 10.2.

TABLE 10.2 Signal Propagation in Circular Cross-sectional Waveguides

	TE_{nm} Modes	TM_{nm} Modes
k_c	$\dfrac{x'_{nm}}{a}$	$\dfrac{x_{nm}}{a}$
k_z	$\sqrt{k^2 - k_c^2}$	$\sqrt{k^2 - k_c^2}$
$H_z(\rho, \phi)$	$H_{nm} J_n(k_c\rho)\begin{cases}\sin(n\phi)\\ \cos(n\phi)\end{cases}$	0
$E_z(\rho, \phi)$	0	$E_{nm} J_n(k_c\rho)\begin{cases}\sin(n\phi)\\ \cos(n\phi)\end{cases}$
$H_\rho(\rho, \phi)$	$-j\dfrac{k_z}{k_c} H_{nm} J'_n(k_c\rho)\begin{cases}\sin(n\phi)\\ \cos(n\phi)\end{cases}$	$j\dfrac{n\omega\varepsilon}{\rho k_c^2} E_{nm} J_n(k_c\rho)\begin{cases}\cos(n\phi)\\ -\sin(n\phi)\end{cases}$
$H_\phi(\rho, \phi)$	$-j\dfrac{nk_z}{\rho k_c^2} H_{nm} J_n(k_c\rho)\begin{cases}\cos(n\phi)\\ -\sin(n\phi)\end{cases}$	$-j\dfrac{\omega\varepsilon}{k_c} E_{nm} J'_n(k_c\rho)\begin{cases}\sin(n\phi)\\ \cos(n\phi)\end{cases}$
$E_\rho(\rho, \phi)$	$-j\dfrac{n\omega\mu}{\rho k_c^2} H_{nm} J_n(k_c\rho)\begin{cases}\cos(n\phi)\\ -\sin(n\phi)\end{cases}$	$-j\dfrac{k_z}{k_c} E_{nm} J'_n(k_c\rho)\begin{cases}\sin(n\phi)\\ \cos(n\phi)\end{cases}$
$E_\phi(\rho, \phi)$	$j\dfrac{\omega\mu}{k_c} H_{nm} J'_n(k_c\rho)\begin{cases}\sin(n\phi)\\ \cos(n\phi)\end{cases}$	$-j\dfrac{nk_z}{\rho k_c^2} E_{nm} J_n(k_c\rho)\begin{cases}\cos(n\phi)\\ -\sin(n\phi)\end{cases}$

Example 10.3

The inner radius of an air-filled circular cylindrical waveguide is 3.81 cm. Assume that its walls are perfectly conducting to find the seven lowest modes that it can support. Determine the cutoff frequency of each mode.

The zeros of Bessel functions of the first kind and their derivatives are listed in Appendexes C.1 and C.2, respectively. The waveguide modes can be found via (10.2.3) and (10.2.17) after arranging these zeros in ascending order. The cutoff frequency f_c can be found as follows:

$$f_c = \frac{3 \cdot 10^8}{2\pi} k_c$$

Therefore the modes in ascending order (corresponding zeros are included in parentheses) and their cutoff frequencies are found to be

$$TE_{11}(1.84118) \to k_c = \frac{1.84118}{3.81 \cdot 10^{-2}} \text{ rad/m} \to f_c = 2.3073\,\text{GHz},$$

$$TM_{01}(2.40483) \to k_c = \frac{2.40483}{3.81 \cdot 10^{-2}} \text{ rad/m} \to f_c = 3.0137\,\text{GHz},$$

$$TE_{21}(3.05424) \to k_c = \frac{3.05424}{3.81 \cdot 10^{-2}} \text{ rad/m} \to f_c = 3.8275\,\text{GHz},$$

$$TE_{01}(3.83171) \text{ and } TM_{11}(3.83171) \rightarrow k_c = \frac{3.83171}{3.81 \cdot 10^{-2}} \text{ rad/m} \rightarrow f_c$$

$$= 4.8019\,\text{GHz},$$

$$TE_{31}(4.20119) \rightarrow k_c = \frac{4.20119}{3.81 \cdot 10^{-2}} \text{ rad/m} \rightarrow f_c = 5.2649\,\text{GHz},$$

and

$$TM_{21}(5.13562)_c \rightarrow k_c = \frac{5.13562}{3.81 \cdot 10^{-2}} \text{ rad/m} \rightarrow f_c = 6.4359\,\text{GHz}.$$

10.3 RECTANGULAR CAVITY RESONATORS

When conducting walls are added at each end of the rectangular waveguide, a rectangular cavity is formed. Therefore these structures also support the TE^z and TM^z modes. In this section the resonator fields are analyzed to find the relations for its resonant frequency and the quality factor.

TE^z Modes in Rectangular Cavities

Consider a rectangular cavity made of conducting walls with dimensions $a \times b \times c$, as shown in Figure 10.5. It is filled with a dielectric material of dielectric constant ε_r and the relative permeability μ_r. Since it is a closed box, the signal cannot propagate along the z-axis. Therefore the sine and cosine functions are appropriate solutions for $h_1(k_x x)$, $h_2(k_y y)$, and $h_3(k_z z)$. Further the conditions along the x- and y-axes are same as in the case of the rectangular waveguide. Therefore F_z can be found from (10.1.8) after changing $h_3(k_z z)$ appropriately as follows:

$$F_z(x, y, z) = c_1 c_3 \cos\left(\frac{m\pi}{a} x\right) \cos\left(\frac{n\pi}{b} y\right) [d_1 \cos(k_z z) + d_2 \sin(k_z z)], \tag{10.3.1}$$

where d_1 and d_2 are the integration constants.

There are following two more boundary conditions now need to be satisfied:

$$E_x|_{\substack{z=0\\z=c}} = 0 \rightarrow F_z|_{\substack{z=0\\z=c}} = 0 \tag{10.3.2}$$

and

$$E_Y|_{\substack{z=0\\z=c}} = 0 \rightarrow F_z|_{\substack{z=0\\z=c}} = 0. \tag{10.3.3}$$

These conditions are satisfied if

$$d_1 = 0 \tag{10.3.4}$$

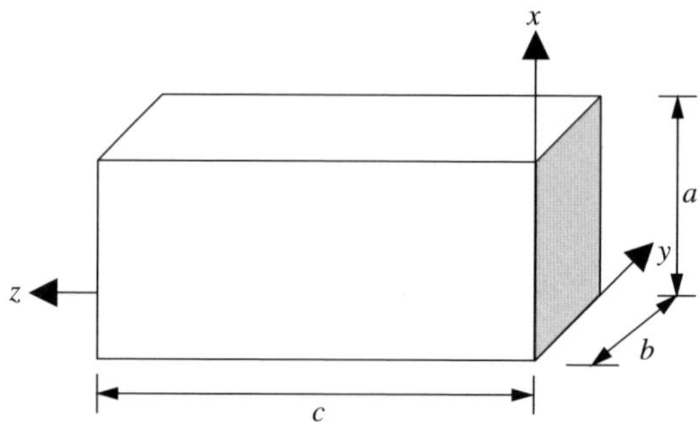

Figure 10.5 Geometry of the rectangular cavity resonator.

and

$$\sin(k_z c) = 0 = \sin(p\pi) \rightarrow k_z = \frac{p\pi}{c}, \qquad p = 1, 2, 3, \ldots. \tag{10.3.5}$$

Therefore (10.3.1) reduces to

$$F_z(x, y, z) = c_1 c_3 d_2 \cos\left(\frac{m\pi}{a}x\right)\cos\left(\frac{n\pi}{b}y\right)\sin\left(\frac{p\pi}{c}z\right). \tag{10.3.6}$$

In this case (10.1.9) gives

$$k = \sqrt{\left(\frac{m\pi}{a}\right)^2 + \left(\frac{n\pi}{b}\right)^2 + \left(\frac{p\pi}{c}\right)^2} \rightarrow \omega_r\sqrt{\mu\varepsilon} = \sqrt{\left(\frac{m\pi}{a}\right)^2 + \left(\frac{n\pi}{b}\right)^2 + \left(\frac{p\pi}{c}\right)^2}.$$

Therefore the resonant frequency $f_r = \omega/2\pi$ is found to be

$$f_r(\text{MHz}) = \frac{300}{\sqrt{\mu_r\varepsilon_r}}\sqrt{\left[\left(\frac{m}{2a}\right)^2 + \left(\frac{n}{2b}\right)^2 + \left(\frac{p}{2c}\right)^2\right]}\ \text{MHz}. \tag{10.3.7}$$

Fields inside the cavity resonator are found after substituting (10.3.6) into (6.4.13) through (6.4.18) as follows:

$$E_x(x, y, z) = jH_{mnp}\,\omega\mu\left(\frac{n\pi}{b}\right)\cos\left(\frac{m\pi}{a}x\right)\sin\left(\frac{n\pi}{b}y\right)\sin\left(\frac{p\pi}{c}z\right), \tag{10.3.8}$$

$$E_y(x, y, z) = -jH_{mnp}\,\omega\mu\left(\frac{m\pi}{a}\right)\sin\left(\frac{m\pi}{a}x\right)\cos\left(\frac{n\pi}{b}y\right)\sin\left(\frac{p\pi}{c}z\right), \tag{10.3.9}$$

$$E_z(x, y, z) = 0, \tag{10.3.10}$$

$$H_x(x, y, z) = -H_{mnp}\left(\frac{m\pi}{a}\right)\left(\frac{p\pi}{c}\right)\sin\left(\frac{m\pi}{a}x\right)\cos\left(\frac{n\pi}{b}y\right)\cos\left(\frac{p\pi}{c}z\right), \tag{10.3.11}$$

$$H_y(x, y, z) = -H_{mnp}\left(\frac{n\pi}{b}\right)\left(\frac{p\pi}{c}\right)\cos\left(\frac{m\pi}{a}x\right)\sin\left(\frac{n\pi}{b}y\right)\cos\left(\frac{p\pi}{c}z\right), \tag{10.3.12}$$

and

$$H_z(x, y, z) = H_{mnp}\left[\left(\frac{m\pi}{a}\right)^2 + \left(\frac{n\pi}{b}\right)^2\right]\cos\left(\frac{m\pi}{a}x\right)\cos\left(\frac{n\pi}{b}y\right)\sin\left(\frac{p\pi}{c}z\right), \quad (10.3.13)$$

where

$$H_{mnp} = -jc_1c_3d_2\frac{1}{\omega\mu\varepsilon}. \quad (10.3.14)$$

TM^z Modes in Rectangular Cavities

As in the preceding case, A_z can be found from (10.1.33) after changing $h_3(k_z z)$ appropriately as follows:

$$A_z(x, y, z) = c_2c_4\sin\left(\frac{m\pi}{a}x\right)\sin\left(\frac{n\pi}{b}y\right)[d_1\cos(k_z z) + d_2\sin(k_z z)] \quad (10.3.15)$$

where d_1 and d_2 are integration constants. The TM^z mode fields of the cavity resonator must satisfy two more boundary conditions:

$$E_x|_{\substack{z=0\\z=c}} = 0 \rightarrow \left.\frac{\partial A_z}{\partial z}\right|_{\substack{z=0\\z=c}} = 0 \quad (10.3.16)$$

and

$$E_y|_{\substack{z=0\\z=c}} = 0 \rightarrow \left.\frac{\partial A_z}{\partial z}\right|_{\substack{z=0\\z=c}} = 0. \quad (10.3.17)$$

Both of these conditions are satisfied if

$$d_2 = 0 \quad (10.3.18)$$

and

$$\sin(k_z c) = 0 = \sin(p\pi) \rightarrow k_z = \frac{p\pi}{c}, \quad p = 0, 1, 2, \ldots. \quad (10.3.19)$$

Therefore the resonant frequency for these modes will be given by (10.3.7) as well. Note that a rectangular cavity resonator can support any degenerate TE$_{mnp}$ and TM$_{mnp}$ modes. The cutoff frequencies of the TE$_{mn}$ and TM$_{mn}$ modes can be determined from the following formula:

$$f_c(\text{MHz}) = \frac{300}{\sqrt{\mu_r\varepsilon_r}}\sqrt{\left[\left(\frac{m}{2a}\right)^2 + \left(\frac{n}{2b}\right)^2\right]}\ \text{MHz}. \quad (10.3.20)$$

Further (10.3.15) reduces to

$$A_z(x, y, z) = c_1 c_3 d_1 \sin\left(\frac{m\pi}{a}x\right)\sin\left(\frac{n\pi}{b}y\right)\cos\left(\frac{p\pi}{c}z\right). \tag{10.3.21}$$

The TM^z_{mnp} fields inside the cavity resonator can be found after substituting (10.3.21) into (6.4.4) through (6.4.9) as follows:

$$H_x(x, y, z) = jE_{mnp}\omega\varepsilon\left(\frac{n\pi}{b}\right)\sin\left(\frac{m\pi}{a}x\right)\cos\left(\frac{n\pi}{b}y\right)\cos\left(\frac{p\pi}{c}z\right), \tag{10.3.22}$$

$$H_y(x, y, z) = -jE_{mnp}\omega\varepsilon\left(\frac{m\pi}{a}\right)\cos\left(\frac{m\pi}{a}x\right)\sin\left(\frac{n\pi}{b}y\right)\cos\left(\frac{p\pi}{c}z\right), \tag{10.3.23}$$

$$H_z(x, y, z) = 0, \tag{10.3.24}$$

$$E_x(x, y, z) = -E_{mnp}\left(\frac{m\pi}{a}\right)\left(\frac{p\pi}{c}\right)\cos\left(\frac{m\pi}{a}x\right)\sin\left(\frac{n\pi}{b}y\right)\sin\left(\frac{p\pi}{c}z\right), \tag{10.3.25}$$

$$E_y(x, y, z) = -E_{mnp}\left(\frac{n\pi}{b}\right)\left(\frac{p\pi}{c}\right)\sin\left(\frac{m\pi}{a}x\right)\cos\left(\frac{n\pi}{b}y\right)\sin\left(\frac{p\pi}{c}z\right), \tag{10.3.26}$$

and

$$E_z(x, y, z) = E_{mnp}\left[\left(\frac{m\pi}{a}\right)^2 + \left(\frac{n\pi}{b}\right)^2\right]\sin\left(\frac{m\pi}{a}x\right)\sin\left(\frac{n\pi}{b}y\right)\cos\left(\frac{p\pi}{c}z\right), \tag{10.3.27}$$

where

$$E_{mnp} = -jc_2 c_4 d_1 \frac{1}{\omega\mu\varepsilon}. \tag{10.3.28}$$

The Quality Factor

The quality factor (or Q-factor) of a resonant circuit is defined as follows:

$$\frac{1}{Q} = \frac{P_{\mathrm{loss}}}{2\omega_{\mathrm{r}} W_{\mathrm{e}}} = \frac{P_{\mathrm{dl}} + P_{\mathrm{cl}}}{2\omega_{\mathrm{r}} W_{\mathrm{e}}} = \frac{1}{Q_{\mathrm{d}}} + \frac{1}{Q_{\mathrm{c}}} \tag{10.3.29}$$

where P_{loss} is the power loss in resonant circuit, ω_{r} is the resonant frequency, W_{e} is the energy stored in the electric field, P_{dl} is the power loss in the dielectric material that fills the cavity, and P_{cl} is the power loss in the conducting walls.

If the cavity is made of a perfect conductor and filled with a perfect dielectric, it will have infinite Q. However, it is not possible in practice. Therefore Q_{c} and Q_{d} need to be determined to find the Q of a cavity resonator. Since the TE_{10} mode is dominant in rectangular waveguides, the lowest order mode in rectangular cavity resonators is TE_{101}, whereas the lowest order TM^z mode is TM_{110}. Therefore we consider here only a resonator that operates in the TE_{101} mode. The power loss P_{dl} in the dielectric medium can be found as follows:

$$P_{\mathrm{dl}} = \frac{1}{2}\int_V \vec{E} \cdot \vec{J}^* dv = \frac{1}{2}\int_V \vec{E} \cdot (\omega\varepsilon_0\varepsilon''\vec{E}^*)dv = \frac{\omega\varepsilon_0\varepsilon''}{2}\int_V |\vec{E}|^2 dv. \tag{10.3.30}$$

Since for the TE_{101} mode, only E_y is nonzero, and it is found from (10.3.9) to be

$$\vec{E} = \hat{y}E_y = -\hat{y}jH_{101}\,\omega\mu\left(\frac{\pi}{a}\right)\sin\left(\frac{\pi}{a}x\right)\sin\left(\frac{\pi}{c}z\right).$$

Equation (10.3.30) gives

$$P_{\rm dl} = \frac{\omega_{\rm r}\varepsilon_0\varepsilon''}{2}|H_{101}|^2\omega_{\rm r}^2\mu^2\left(\frac{\pi}{a}\right)^2\int_{x=0}^{a}\int_{y=0}^{b}\int_{z=0}^{c}\sin^2\left(\frac{\pi}{a}x\right)\sin^2\left(\frac{\pi}{c}z\right)dxdydz,$$

or

$$P_{\rm dl} = \frac{\varepsilon_0\varepsilon''}{8}|H_{101}|^2\omega_{\rm r}^3\mu^2\left(\frac{\pi}{a}\right)^2 abc. \tag{10.3.31}$$

Similarly $W_{\rm e}$ is found via the following relation

$$W_{\rm e} = \frac{\varepsilon_0\varepsilon'}{4}\int_V |\vec{E}|^2 dv. \tag{10.3.32}$$

On substituting E_y and integrating over the volume of the cavity, we find that

$$W_{\rm e} = \frac{\varepsilon_0\varepsilon'}{4}|H_{101}|^2\omega_{\rm r}^2\mu^2\left(\frac{\pi}{a}\right)^2\int_{x=0}^{a}\int_{y=0}^{b}\int_{z=0}^{c}\sin^2\left(\frac{\pi}{a}x\right)\sin^2\left(\frac{\pi}{c}z\right)dxdydz,$$

or

$$W_{\rm e} = \frac{\varepsilon_0\varepsilon'}{16}|H_{101}|^2\omega_{\rm r}^2\mu^2\left(\frac{\pi}{a}\right)^2 abc. \tag{10.3.33}$$

Now from (10.3.31) and (10.3.33), $Q_{\rm d}$ can be found:

$$Q_{\rm d} = \frac{2\omega_{\rm r}W_{\rm e}}{P_{\rm dl}} = \frac{\varepsilon'}{\varepsilon''} = \frac{1}{\tan\delta}. \tag{10.3.34}$$

Note that (10.3.34) is valid for all modes because (10.3.30) and (10.3.32) have a common integral that cancels out when substituted in (10.3.34).

The power loss in conducting walls can be found via (10.1.52) and (10.3.11) through (10.3.13) as follows:

$$P_{\rm cl} = \frac{1}{2}R_{\rm s}\int_{\rm walls}|H_{\rm tan}|^2 ds = R_{\rm s}|H_{101}|^2[I_1 + I_2 + I_3 + I_4], \tag{10.3.35}$$

where

$$I_1 = \int_{y=0}^{b}\int_{z=0}^{c}\left(\frac{\pi}{a}\right)^4 \sin^2\left(\frac{\pi}{c}z\right)dydz = \left(\frac{\pi}{a}\right)^4 \frac{bc}{2}, \tag{10.3.36}$$

$$I_2 = \int_{x=0}^{a}\int_{y=0}^{b}\left(\frac{\pi}{c}\right)^2\left(\frac{\pi}{a}\right)^2 \sin^2\left(\frac{\pi}{a}x\right)dxdy = \left(\frac{\pi}{c}\right)^2\left(\frac{\pi}{a}\right)^2 \frac{ab}{2}, \tag{10.3.37}$$

$$I_3 = \int_{x=0}^{a}\int_{z=0}^{c}\left(\frac{\pi}{c}\right)^2\left(\frac{\pi}{a}\right)^2 \sin^2\left(\frac{\pi}{a}x\right)\cos^2\left(\frac{\pi}{c}z\right)dxdz = \left(\frac{\pi}{c}\right)^2\left(\frac{\pi}{a}\right)^2 \frac{ac}{4}, \tag{10.3.38}$$

and

$$I_4 = \int_{x=0}^{a}\int_{z=0}^{c}\left(\frac{\pi}{a}\right)^4 \cos^2\left(\frac{\pi}{a}x\right)\sin^2\left(\frac{\pi}{c}z\right)dxdz = \left(\frac{\pi}{a}\right)^4 \frac{ac}{4}. \tag{10.3.39}$$

Therefore (10.3.35) simplifies to

$$P_{\mathrm{cl}} = \frac{1}{2}R_{\mathrm{s}}\int_{\mathrm{walls}} |H_{\mathrm{tan}}|^2 ds = R_{\mathrm{s}}|H_{101}|^2\left(\frac{\pi}{a}\right)^4 \frac{1}{4}\left[2bc + \frac{2a^3b}{c^2} + \frac{a^3}{c} + ac\right]. \tag{10.3.40}$$

Using (10.3.33) and (10.3.40), we find that

$$Q_{\mathrm{c}} = \frac{2\omega_{\mathrm{r}}W_{\mathrm{e}}}{P_{\mathrm{cl}}} = \frac{\omega_{\mathrm{r}}^3\mu^2a^3\varepsilon_0\varepsilon'}{2R_{\mathrm{s}}\pi^2[2\{1+(a/c)^3\} + (a/b)\{1+(a/c)^2\}]}. \tag{10.3.41}$$

Example 10.4

An air-filled rectangular cavity is made from a piece of copper WR-90 waveguide. If it resonates at 9.379 GHz in the TE_{101} mode, find the required length c and the Q of this resonator.

For WR-90, $a = 0.9$ in $= 2.286$ cm, and $b = 0.4$ in $= 1.016$ cm. From (10.3.7), we find that

$$\frac{1}{2c} = \sqrt{\left(\frac{9379}{300}\right)^2 - \left(\frac{1}{2a}\right)^2} = 22.3383 \rightarrow c = 2.238\,\mathrm{cm}.$$

Next the surface resistance R_s is found from (10.1.51) to be 0.0252 Ω, and the Q_c is found from (10.3.41) to be about 7860.

Example 10.5

A rectangular cavity made of copper has its inner dimensions $a = 1.6$ cm, $b = 0.71$ cm, and $c = 1.56$ cm. It is filled with Teflon ($\varepsilon_r = 2.05$ and

$\tan\delta = 2.9268 \times 10^{-4}$). Find the TE_{101} mode resonant frequency and Q of this cavity.

The resonant frequency is found from (10.3.7) as follows:

$$f_{\mathrm{r}} = \frac{300}{\sqrt{2.05}}\sqrt{\left(\frac{1}{2\cdot 0.016}\right)^2 + \left(\frac{1}{2\cdot 0.0156}\right)^2} \approx 9379.4\,\mathrm{MHz} = 9.3794\,\mathrm{GHz}.$$

Since power dissipates both in the dielectric filling as well as in the sidewalls, we need to find Q_{d} and Q_{c}. Overall, Q is then found from (10.3.29). Using (10.3.34), Q_{d} is found to be 3417. Q_{c} is found from (10.3.41) as follows:

$$R_{\mathrm{s}} = \sqrt{\frac{\omega\mu_{\mathrm{o}}}{2\sigma}} = \sqrt{\frac{2\cdot\pi\cdot 9.3794\cdot 10^9\cdot 4\cdot\pi\cdot 10^{-7}}{2\cdot 5.813\cdot 10^7}} = 0.0252\,\Omega.$$

Therefore Q_{c} is found from (10.3.41) to be 5492. After substitution of Q_{d} and Q_{c} into (10.3.29), Q of this cavity is found to be 2106.

10.4 CIRCULAR CYLINDRICAL CAVITY RESONATORS

When conducting walls are added at each open end of a circular cylindrical waveguide, a circular cavity resonator is formed. Therefore the TE^z and TM^z modes exist in such structures. In this section these fields are analyzed to find the relations for the resonant frequency and the quality factor of the circular cavities.

TE^z Modes in Circular Cavities

Consider a circular cylindrical cavity of radius a and height h, as shown in Figure 10.6. It is filled with a dielectric material of a dielectric constant ε_r and a relative permeability μ_r. Since the cavity is a closed box, a signal cannot propagate

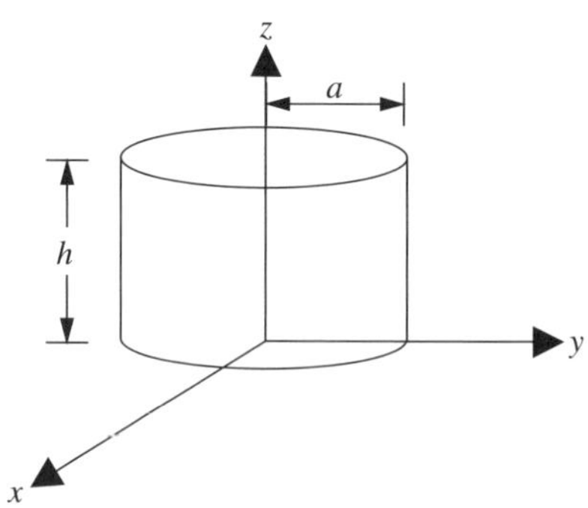

Figure 10.6 Geometry of the circular cylindrical cavity resonator.

along the z-axis. Therefore the sine and cosine functions are appropriate solutions for $h_3(k_z z)$. Further the conditions along the ρ- and ϕ-axes are the same as in the case of the circular waveguide. Therefore F_z can be found from (10.2.1) after appropriate changes as follows:

$$F_z(\rho, \phi, z) = A_{nm} J_n(k_\rho \rho)[d_1 \cos(k_z z) + d_2 \sin(k_z z)] \begin{Bmatrix} \sin(n\phi) \\ \cos(n\phi), \end{Bmatrix} \tag{10.4.1}$$

where d_1 and d_2 are integration constants.

Since tangential electric fields must be zero on the added conducting surfaces as well, these boundary conditions in conjunction with (6.5.13) and (6.5.14) give

$$E_\rho \big|_{\substack{z=0 \\ z=h}} = 0 \rightarrow F_z \big|_{\substack{z=0 \\ z=h}} = 0 \tag{10.4.2}$$

and

$$E_\phi \big|_{\substack{z=0 \\ z=h}} = 0 \rightarrow F_z \big|_{\substack{z=0 \\ z=h}} = 0. \tag{10.4.3}$$

Both conditions are satisfied if

$$d_1 = 0 \tag{10.4.4}$$

and

$$\sin(k_z h) = 0 = \sin(p\pi) \rightarrow k_z = \frac{p\pi}{h}, \qquad p = 1, 2, 3, \ldots. \tag{10.4.5}$$

Therefore the resonant frequency can be found from (10.2.2) as follows:

$$k = \sqrt{\left(\frac{x'_{nm}}{a}\right)^2 + \left(\frac{p\pi}{h}\right)^2} \rightarrow \omega_r \sqrt{\mu\varepsilon} = \sqrt{\left(\frac{x'_{nm}}{a}\right)^2 + \left(\frac{p\pi}{h}\right)^2}.$$

The resonant frequency f_r is found to be

$$f_r\,(\text{MHz}) = \frac{300}{2\pi\sqrt{\mu_r \varepsilon_r}} \sqrt{\left(\frac{x'_{nm}}{a}\right)^2 + \left(\frac{p\pi}{h}\right)^2}\ \text{MHz}. \tag{10.4.6}$$

As was defined earlier, x'_{nm} represents the mth zero of the derivative of a Bessel function of the first kind and order n. Further (10.4.1) reduces to

$$F_z(\rho, \phi, z) = A_{nm} d_2 J_n\left(\frac{x'_{nm}}{a}\rho\right) \sin\left(\frac{p\pi}{h}z\right) \begin{Bmatrix} \sin(n\phi) \\ \cos(n\phi). \end{Bmatrix} \tag{10.4.7}$$

The field components can be determined now after substituting (10.4.7) into (6.5.13) through (6.5.18) as follows:

$$E_\rho(\rho, \phi, z) = -j\frac{n\omega\mu}{\rho}\left(\frac{a}{x'_{nm}}\right)^2 H_{nmp}\, J_n\left(\frac{x'_{nm}}{a}\rho\right)\sin\left(\frac{p\pi}{h}z\right)\begin{cases}\cos(n\phi)\\ -\sin(n\phi),\end{cases} \tag{10.4.8}$$

$$E_\phi(\rho, \phi, z) = j\omega\mu\left(\frac{a}{x'_{nm}}\right)H_{nmp}\, J'_n\left(\frac{x'_{nm}}{a}\rho\right)\sin\left(\frac{p\pi}{h}z\right)\begin{cases}\sin(n\phi)\\ \cos(n\phi),\end{cases} \tag{10.4.9}$$

$$E_z(\rho, \phi, z) = 0, \tag{10.4.10}$$

$$H_\rho(\rho, \phi, z) = H_{nmp}\left(\frac{p\pi}{h}\right)\left(\frac{a}{x'_{nm}}\right)J'_n\left(\frac{x'_{nm}}{a}\rho\right)\cos\left(\frac{p\pi}{h}z\right)\begin{cases}\sin(n\phi)\\ \cos(n\phi),\end{cases} \tag{10.4.11}$$

$$H_\phi(\rho, \phi, z) = \left(\frac{p\pi}{h}\right)\left(\frac{a}{x'_{nm}}\right)^2 \frac{n}{\rho}H_{nmp}\, J_n\left(\frac{x'_{nm}}{a}\rho\right)\cos\left(\frac{p\pi}{h}z\right)\begin{cases}\cos(n\phi)\\ -\sin(n\phi),\end{cases} \tag{10.4.12}$$

and

$$H_z(\rho, \phi, z) = H_{nmp}\, J_n\left(\frac{x'_{nm}}{a}\rho\right)\sin\left(\frac{p\pi}{h}z\right)\begin{cases}\sin(n\phi)\\ \cos(n\phi),\end{cases} \tag{10.4.13}$$

where

$$H_{nmp} = \frac{A_{nm}d_2}{j\omega\mu\varepsilon}\left(\frac{x'_{nm}}{a}\right)^2. \tag{10.4.14}$$

TM^z Modes in Circular Cavities

As in the preceding case, (10.2.14) can be used and A_z found after appropriate changes as follows:

$$A_z(\rho, \phi, z) = C_{nm}J_n\left(\frac{x_{nm}}{a}\rho\right)[d_1\cos(k_z z) + d_2\sin(k_z z)]\begin{cases}\sin(n\phi)\\ \cos(n\phi),\end{cases} \tag{10.4.15}$$

where d_1 and d_2 are the integration constants.

Since the tangential electric fields must be zero on the added conducting surfaces as well, these boundary conditions in conjunction with (6.5.7) and (6.5.8) give

$$E_\rho|_{\substack{z=0\\z=h}} = 0 \rightarrow \left.\frac{\partial A_z}{\partial z}\right|_{\substack{z=0\\z=h}} = 0 \tag{10.4.16}$$

and

$$E_\phi|_{\substack{z=0\\z=h}} = 0 \rightarrow \left.\frac{\partial A_z}{\partial z}\right|_{\substack{z=0\\z=h}} = 0. \tag{10.4.17}$$

Both conditions are satisfied if

$$d_2 = 0 \tag{10.4.18}$$

and

$$\sin(k_z h) = 0 = \sin(p\pi) \rightarrow k_z = \frac{p\pi}{h}, \qquad p = 0, 1, 2, \ldots. \tag{10.4.19}$$

The resonant frequency ω_r for these modes can be found as follows:

$$k = \sqrt{\left(\frac{x_{nm}}{a}\right)^2 + \left(\frac{p\pi}{h}\right)^2} \rightarrow \omega_\mathrm{r}\sqrt{\mu\varepsilon} = \sqrt{\left(\frac{x_{nm}}{a}\right)^2 + \left(\frac{p\pi}{h}\right)^2}.$$

The corresponding resonant frequency f_r in megahertz is found to be

$$f_\mathrm{r}(\mathrm{MHz}) = \frac{300}{2\pi\sqrt{\mu_\mathrm{r}\varepsilon_\mathrm{r}}}\sqrt{\left(\frac{x_{nm}}{a}\right)^2 + \left(\frac{p\pi}{h}\right)^2}\ \mathrm{MHz}. \tag{10.4.20}$$

As was defined earlier, x_{nm} is the mth zero of a Bessel function of the first kind and order n. Further (10.4.15) reduces to

$$A_z(\rho, \phi, z) = C_{nm} d_1 J_n\left(\frac{x_{nm}}{a}\rho\right)\cos(k_z z)\begin{cases}\sin(n\phi)\\ \cos(n\phi).\end{cases} \tag{10.4.21}$$

The TM^z_{nmp} field components inside this cavity resonator are found after substituting (10.4.21) into (6.5.4) through (6.5.9) as follows:

$$H_\rho(\rho, \phi, z) = j\frac{n\omega\varepsilon}{\rho}\left(\frac{a}{x_{nm}}\right)^2 E_{nmp} J_n\left(\frac{x_{nm}}{a}\rho\right)\cos\left(\frac{p\pi}{h}z\right)\begin{cases}\cos(n\phi)\\ -\sin(n\phi),\end{cases} \tag{10.4.22}$$

$$H_\phi(\rho, \phi, z) = -j\omega\varepsilon\left(\frac{a}{x_{nm}}\right) E_{nmp} J_n'\left(\frac{x_{nm}}{a}\rho\right)\cos\left(\frac{p\pi}{h}z\right)\begin{cases}\sin(n\phi)\\ \cos(n\phi),\end{cases} \tag{10.4.23}$$

$$H_z(\rho, \phi, z) = 0, \tag{10.4.24}$$

$$E_\rho(\rho, \phi, z) = \left(\frac{p\pi}{h}\right)\left(\frac{a}{x_{nm}}\right) E_{nmp} J_n'\left(\frac{x_{nm}}{a}\rho\right)\sin\left(\frac{p\pi}{h}z\right)\begin{cases}\sin(n\phi)\\ \cos(n\phi),\end{cases} \tag{10.4.25}$$

$$E_\phi(\rho, \phi, z) = \frac{n}{\rho}\left(\frac{p\pi}{h}\right)\left(\frac{a}{x_{nm}}\right)^2 E_{nmp} J_n\left(\frac{x_{nm}}{a}\rho\right)\sin\left(\frac{p\pi}{h}z\right)\begin{cases}\cos(n\phi)\\ -\sin(n\phi),\end{cases} \tag{10.4.26}$$

and

$$E_z(\rho, \phi, z) = E_{nmp} J_n\left(\frac{x_{nm}}{a}\rho\right)\cos\left(\frac{p\pi}{h}z\right)\begin{cases}\sin(n\phi)\\ \cos(n\phi),\end{cases} \tag{10.4.27}$$

where

$$E_{nmp} = \frac{C_{nm} d_1}{j\omega\mu\varepsilon}\left(\frac{x_{nm}}{a}\right)^2. \tag{10.4.28}$$

The Quality Factor

Similar to rectangular cavity resonators, the quality factor of the circular cavity has two parts. Power loss can occur in the dielectric as well as in the conducting walls. Therefore we need the means to find Q_d as well as Q_c. As indicated earlier, Q_d can be found from (10.3.34) because of its general formulation. However, we need to find the expression for Q_c. This can be done following the procedure we used for the case of rectangular cavity resonators. For simplicity we consider here the cavity supporting only the TM_{010} mode. Fields of this mode can be found from (10.4.22) through (10.4.27) as follows:

$$H_\rho(\rho, \phi, z) = 0, \tag{10.4.29}$$

$$H_\phi(\rho, \phi, z) = -j\omega\varepsilon\left(\frac{a}{x_{01}}\right)E_{010}\, J_0'\left(\frac{x_{01}}{a}\rho\right), \tag{10.4.30}$$

$$H_z(\rho, \phi, z) = 0, \tag{10.4.31}$$

$$E_\rho(\rho, \phi, z) = 0, \tag{10.4.32}$$

$$E_\phi(\rho, \phi, z) = 0, \tag{10.4.33}$$

and

$$E_z(\rho, \phi, z) = E_{010} J_0\left(\frac{x_{01}}{a}\rho\right). \tag{10.4.34}$$

Therefore

$$\begin{aligned} W_e; &= \frac{\varepsilon}{4}\int_V |E_z|^2 dv = \frac{\varepsilon\,|E_{010}|^2}{4}\int_{\rho=0}^{a}\int_{\phi=0}^{2\pi}\int_{z=0}^{h}\left[J_0\left(\frac{x_{01}}{a}\rho\right)\right]^2 \rho d\rho d\phi dz \\ &= \frac{\pi h\varepsilon\,|E_{010}|^2}{2}\int_0^a \left[J_0\left(\frac{x_{01}}{a}\rho\right)\right]^2 \rho d\rho. \end{aligned}$$

The remaining integral can be evaluated via the following formula available in mathematical handbook and others listed in Appendix C:

$$\int x\left[J_p(ax)\right]^2 dx = \frac{x^2}{2}\left[\left\{J_p(ax)\right\}^2 - J_{p-1}(ax)J_{p+1}(ax)\right]. \tag{10.4.35}$$

Hence

$$W_e = \frac{\pi h \varepsilon a^2 |E_{010}|^2}{4} [J_1(x_{01})]^2. \tag{10.4.36}$$

Similarly

$$\begin{aligned} P_{cl} &= \frac{R_s}{2} \oint_S \left| \vec{H} \right|^2 ds \\ &= \frac{R_s}{2} \omega^2 \varepsilon^2 \left(\frac{a}{x_{01}} \right)^2 |E_{010}|^2 \left[\int_{\phi=0}^{2\pi} \int_{z=0}^{h} J_0'^2(x_{01}) a \, d\phi dz + 2 \int_{\rho=0}^{a} \int_{\phi=0}^{2\pi} J_0'^2\left(\frac{x_{01}}{a} \rho \right) \rho d\rho d\phi \right]. \end{aligned}$$

Using the formula (C.23) from Appendix C, we find that

$$P_{cl} = R_s \pi \omega^2 \varepsilon^2 \left(\frac{a}{x_{01}} \right)^2 |E_{010}|^2 \left[a h J_1^2(x_{01}) + 2 \int_0^a J_1^2\left(\frac{x_{01}}{a} \rho \right) \rho d\rho \right].$$

Using (10.4.35) and noting that $J_0(x_{01}) = 0$, we find that

$$\int_0^a J_1^2\left(\frac{x_{01}}{a} \rho \right) \rho d\rho = \frac{a^2}{2} J_1^2(x_{01}).$$

Therefore

$$P_{cl} = R_s \pi \omega^2 \varepsilon^2 \left(\frac{a}{x_{01}} \right)^2 |E_{010}|^2 [h + a] a J_1^2(x_{01}). \tag{10.4.37}$$

The quality factor Q_c for the TM_{010} mode can be found now using (10.4.36) and (10.4.37) as follows:

$$Q_c = \frac{2\omega_r W_e}{P_{cl}} = \frac{x_{01}^2 h}{2 R_s \omega_r \varepsilon a (h + a)}. \tag{10.4.38}$$

Since $k_c^2 + k_z^2 = k^2 \rightarrow k = k_c (\because k_z|_{010} = 0) \rightarrow \omega_r \sqrt{\mu \varepsilon} = x_{01}/a$, (10.4.38) can be expressed as follows:

$$Q_c = \frac{x_{01} \eta}{2 R_s [1 + (a/h)]} = \frac{2.40483 \eta}{2 R_s [1 + (a/h)]} = \frac{1.20242 \eta}{R_s [1 + (a/h)]}. \tag{10.4.39}$$

Similarly the Q_c of a circular cylindrical cavity operating in the TE_{nmp} mode and filled with a lossless dielectric can be found from the following formula:

$$Q_c = \frac{\eta}{2 R_s} \frac{\left[1 - (n/x'_{nm})^2 \right] \left[(x'_{nm})^2 + (p\pi a/h)^2 \right]^{1.5}}{\left[\left\{ x'^2_{nm} + (2a/h)(p\pi a/h)^2 \right\} + \left\{ 1 - (2a/h) \right\} (np\pi a/x'_{nm} h)^2 \right]}. \tag{10.4.40}$$

In the case of the TM_{nmp} mode with $p > 0$, Q_c is given by

$$Q_c = \frac{\eta}{2R_s}\frac{\sqrt{[x_{nm}^2 + (p\pi a/h)^2]}}{1 + (2a/h)}, \qquad p > 0, \tag{10.4.41}$$

and

$$Q_c \approx \frac{\eta}{2R_s}\frac{x_{nm}}{1 + (a/h)}, \qquad p = 0. \tag{10.4.42}$$

Example 10.6

Determine the dimensions of an air-filled circular cylindrical cavity that resonates at 10 GHz in TM_{010} mode. The cavity should be made of aluminum ($\sigma = 3.54 \cdot 10^7$ S/m) and its height should be equal to its diameter. Find Q of this cavity. How will this Q be affected if the inside of the cavity is coated with silver ($\sigma = 6.12 \cdot 10^7$ S/m)?

For the TM_{010} mode, x_{01} is found to be 2.40483 from Table C.1. Therefore (10.4.20) gives

$$f_r(\text{MHz}) = \frac{300}{2\pi}\left(\frac{2.40483}{a}\right) \rightarrow a = \frac{150 \cdot 2.40483}{\pi \cdot 10^4}\ \text{m} = 1.1482\ \text{cm}$$

and

$$h = 2a = 2.296\ \text{cm}$$

Since

$$R_s = \sqrt{\frac{\omega\mu}{2\sigma}} = \sqrt{\frac{2 \cdot \pi \cdot 10^{10} \cdot 4 \cdot \pi \cdot 10^{-7}}{2 \cdot 3.54 \cdot 10^7}} = 0.0334\ \Omega$$

and

$$\eta = \sqrt{\frac{\mu_o}{\varepsilon_o}} = 376.7343\ \Omega,$$

Equation (10.4.39) gives

$$Q_c = \frac{1.20242\eta}{R_s(1 + (a/h))} = 9042.$$

If the cavity is silver coated, then R_s changes to

$$R_s = \sqrt{\frac{\omega\mu}{2\sigma}} = \sqrt{\frac{2 \cdot \pi \cdot 10^{10} \cdot 4 \cdot \pi \cdot 10^{-7}}{2 \cdot 6.12 \cdot 10^7}} = 0.0254\ \Omega.$$

Therefore Q_c increases to 11890. Note that it is a significant increase in Q over the uncoated cavity.

Example 10.7

Determine the dimensions of an air-filled circular cylindrical cavity that resonates at 5 GHz in TE_{011} mode. It should be made of copper, and its height should be equal to its diameter. Find Q of this cavity.

From (10.4.6), with x'_{01} as 3.83171 from Table C.2, we get

$$a = \sqrt{\frac{3.83171^2 + (\pi/2)^2}{(5000 \cdot 2\pi/300)^2}} = 0.0395\,\text{m}$$

and

$$h = 2a = 0.079\,\text{m}.$$

Since

$$R_s = \sqrt{\frac{\omega\mu_o}{2\sigma}} = \sqrt{\frac{2\cdot\pi\cdot 5\cdot 10^9\cdot 4\cdot\pi\cdot 10^{-7}}{2\cdot 5.813\cdot 10^7}} = 0.0184\ \Omega$$

and

$$\eta = \sqrt{\frac{\mu_o}{\varepsilon_o}} = 376.7343\ \Omega.$$

Q of the cavity can be found from (10.4.40) as

$$Q_c = \frac{\eta}{2R_s}\cdot\frac{\left[3.83171^2 + (\pi/2)^2\right]^{1.5}}{3.83171^2 + (\pi/2)^2} = 42331.6.$$

Further there is no loss of power in air ($\tan\delta \approx 0$). Therefore Q_d is infinite and Q is the same as Q_c.

PROBLEMS

10.1. The cross section of a rectangular waveguide is 7.21 × 3.40 cm. Find the mode types that can be transmitted at (a) 3 GHz and (b) 6 GHz.

10.2. Find the phase constant, the phase velocity, the group velocity, the wavelength inside the guide, and the wave impedance $Z_{TE^z_{10}}$ for an air-filled 2.4 × 1.01 cm rectangular waveguide that carries a 7.5 GHz signal.

10.3. Repeat Problem 10.2 for a waveguide filled with a perfect dielectric medium of $\varepsilon_r = 2.1$.

10.4. The wavelength of a signal propagating through an air-filled WR-340 rectangular waveguide is found to be 20 cm. Determine its frequency in GHz.

10.5. The inside cross section of an air-filled rectangular waveguide is 2.286 × 1.016 cm. (a) Determine its cutoff frequencies for TE_{10}, TE_{20}, TE_{01}, and TE_{11} modes. (b) Which mode(s) will propagate through this waveguide if the signal frequency is anywhere between 8 and 12 GHz? (c) Determine the phase velocity of a 9.375 GHz signal propagating in TE_{10} mode.

10.6. The inner radius of an air-filled circular cylindrical waveguide is 5.2 cm. Assume that its walls are perfectly conducting to find the seven lowest modes that it can support. Determine the cutoff frequency of each mode.

10.7. Find first eight lowest order modes and their resonant frequencies for an air-filled lossless rectangular cavity of 8 × 6 × 5 cm.

10.8. An air-filled rectangular cavity is made from a piece of copper WR-430 waveguide. If it resonates at 2 GHz in TE_{101} mode, find the required length c and the Q of this resonator.

10.9. An air-filled rectangular cavity with $a = 4$ cm, $b = 3$ cm, and $c = 5$ cm is the mode of brass. Find its dominant mode frequency and the Q.

10.10. An air-filled rectangular cavity is made from a piece of copper WR-137 waveguide. If it resonates at 5.9 GHz in TE_{101} mode, find the required length c and the Q of this resonator.

10.11. Determine the dimensions of an air-filled circular cylindrical cavity that resonates at 2.45 GHz in TM_{010} mode. It should be made of aluminum ($\sigma = 3.54 \cdot 10^7$ S/m), and its height should be equal to its diameter. Find Q of this cavity. How will this Q be affected if the inside of the cavity is coated with silver ($\sigma = 6.12 \cdot 10^7$ S/m)?

10.12. Determine the dimensions of an air-filled circular cylindrical cavity that resonates at 9.375 GHz in TE_{011} mode. It should be made of copper, and its height should be equal to its diameter. Find Q of this cavity.

10.13. Design an air-filled circular cylindrical cavity that resonates at 9 GHz in TE_{011} mode. It should be made of copper, and its height should be equal to its diameter. Find Q of this cavity.

11

NUMERICAL TECHNIQUES

Electromagnetic problems involve solutions to differential equations, integral equations, or mixed integro-differential equations. A number of techniques have been developed to cast electromagnetic problems in the form of integral equations. The solution procedures require partitioning of the problem space to convert the given equation into a set of algebraic equations that are solved using matrix manipulation techniques. The numerical procedures used to solve electromagnetic problems include the finite-difference time-domain (FDTD) technique, the finite element method (FEM), and the method of moments (MoM). Some other procedures transform an electromagnetic problem into an electrical circuit to find associated algebraic relations. The transmission line matrix (TLM) technique is one of them. The finite difference schemes are employed to transform the differential equation into algebraic equations. So-called absorbing boundary conditions are developed to tackle large problem spaces (e.g., the radiation problem). Integral equations are solved using suitable expansion and weighting functions.

Available in the literature in this area is a huge amount of material that cannot the covered here. This chapter introduces the fundamentals of the finite difference method and the method of moments. Much more detail can be found in books listed in the bibliography.

Practical Electromagnetics: From Biomedical Sciences to Wireless Communication.
By Devendra K. Misra
Copyright © 2007 John Wiley & Sons, Inc.

11.1 FINITE-DIFFERENCE METHODS

This section begins with the procedure for transforming a differential equation in single dimension into a finite difference equation. A few examples are included to find the eigenvalues and to solve the boundary-value problems. The procedure is then extended to two and three dimensions. A brief overview of the FDTD is also included in this section.

Finite-Difference in One Dimension

Figure 11.1 shows an arbitrary function $f(x)$. Consider its values at an arbitrary point x and also at points $\pm\Delta$ closely around it. Three different derivatives of this function are defined at point x. The forward difference of the function is defined as follows:

$$\frac{\partial f(x)}{\partial x} = \lim_{\Delta \to 0} \frac{f(x+\Delta) - f(x)}{\Delta}. \tag{11.1.1}$$

The backward difference of the function $f(x)$ at point x is

$$\frac{\partial f(x)}{\partial x} = \lim_{\Delta \to 0} \frac{f(x) - f(x-\Delta)}{\Delta}. \tag{11.1.2}$$

The central difference of $f(x)$ is found to be

$$\frac{\partial f(x)}{\partial x} = \lim_{\Delta \to 0} \frac{f(x+\Delta) - f(x-\Delta)}{2\Delta}. \tag{11.1.3}$$

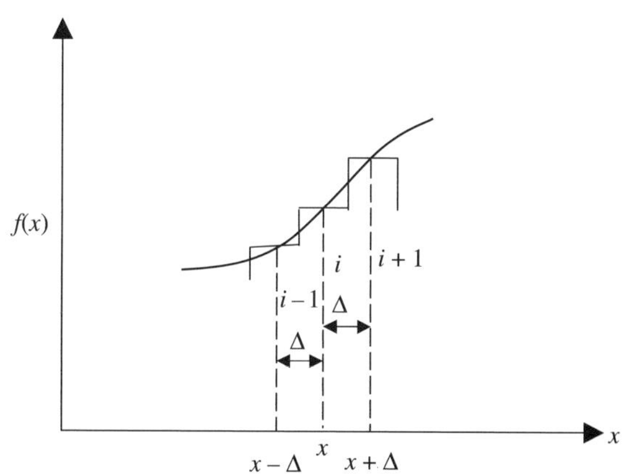

Figure 11.1 Function $f(x)$ and its derivatives.

In many instances we also need the second derivative of the function. It may be found as follows:

$$\frac{\partial^2 f(x)}{\partial x^2} = \lim_{\Delta\to 0} \frac{((f(x+\Delta) - f(x))/\Delta) - ((f(x) - f(x-\Delta))/\Delta)}{\Delta},$$

or

$$\frac{\partial^2 f(x)}{\partial x^2} = \lim_{\Delta\to 0} \frac{f(x+\Delta) - 2f(x) + f(x-\Delta)}{(\Delta)^2}. \tag{11.1.4}$$

Since the derivative of a function is defined three different ways we need to know the order of error associated with each approximation for a finite Δ. This can be achieved via the following two Taylor series representations of the function:

$$f(x+\Delta) = f(x) + \Delta\frac{\partial f(x)}{\partial x} + \frac{\Delta^2}{2!}\frac{\partial^2 f(x)}{\partial x^2} + \frac{\Delta^3}{3!}\frac{\partial^3 f(x)}{\partial x^3} + \cdots \tag{11.1.5}$$

and

$$f(x-\Delta) = f(x) - \Delta\frac{\partial f(x)}{\partial x} + \frac{\Delta^2}{2!}\frac{\partial^2 f(x)}{\partial x^2} - \frac{\Delta^3}{3!}\frac{\partial^3 f(x)}{\partial x^3} + \cdots. \tag{11.1.6}$$

From (11.1.5) we find that

$$\frac{\partial f(x)}{\partial x} = \frac{f(x+\Delta) - f(x)}{\Delta} - \frac{\Delta}{2!}\frac{\partial^2 f(x)}{\partial x^2} - \frac{\Delta^2}{3!}\frac{\partial^3 f(x)}{\partial x^3} - \cdots. \tag{11.1.7}$$

A comparison of (11.1.1) with (11.1.7) shows that the error involved in approximating the derivative of the function with its forward difference is on the order of Δ. From (11.1.6) a similar conclusion is reached for the backward difference approximation as well:

$$\frac{\partial f(x)}{\partial x} = \frac{f(x) - f(x-\Delta)}{\Delta} - \frac{\Delta}{2!}\frac{\partial^2 f(x)}{\partial x^2} + \frac{\Delta^2}{3!}\frac{\partial^3 f(x)}{\partial x^3} - \cdots.$$

If we subtract (11.1.6) from (11.1.5) and rearrange, we find that

$$\frac{\partial f(x)}{\partial x} = \frac{f(x+\Delta) - f(x-\Delta)}{2\Delta} + \frac{\Delta^2}{2\cdot 3!}\frac{\partial^3 f(x)}{\partial x^3} + \cdots. \tag{11.1.8}$$

Therefore the error associated with the central difference approximation is on the order of Δ^2, which is much smaller than the other two. If we add (11.1.5) and

TABLE 11.1 Finite-Difference Approximations of a Function

Order of Derivative	Formula Type	Finite-Difference Representation	Order of Error
$f'(x)$	Forward difference	$\dfrac{f(i+1)-f(i)}{\Delta}$	Δ
$f'(x)$	Backward difference	$\dfrac{f(i)-f(i-1)}{\Delta}$	Δ
$f'(x)$	Central difference	$\dfrac{f(i+1)-f(i-1)}{2\Delta}$	Δ^2
$f''(x)$	Forward difference	$\dfrac{f(i+1)-2f(i)+f(i-1)}{\Delta^2}$	Δ^2

(11.1.6), we find that

$$\frac{\partial^2 f(x)}{\partial x^2} = \frac{f(x+\Delta)-2f(x)+f(x-\Delta)}{\Delta^2} + \frac{\Delta^2}{4!}\frac{\partial^4 f(x)}{\partial x^4} + \cdots. \tag{11.1.9}$$

Table 11.1 summarizes these results.

Example 11.1

Using the finite-difference scheme, find the lowest eigenvalue of a TM^z mode propagating between two conducting parallel plates that are separated by one meter.

We analyzed the parallel-plate waveguide in Section 6.4. Since the wave is propagating along the z-axis and the conducting plates extend to infinity along the y-axis, (6.4.3) gives

$$\frac{\partial^2 A_z}{\partial x^2} + (k^2 - k_z^2)A_z = 0 \rightarrow \frac{\partial^2 A_z}{\partial x^2} + k_c^2 A_z = 0, \tag{i}$$

where

$$k_c^2 = k^2 - k_z^2. \tag{ii}$$

From (6.4.9) the boundary conditions require that

$$E_z\Big|_{\substack{x=0\\x=1}} = 0 \rightarrow A_z\Big|_{\substack{x=0\\x=1}} = 0. \tag{iii}$$

After partitioning the separation distance along the x-axis, (i) can be approximated using (11.1.4) as follows:

$$\frac{A_{zi-1} - 2A_{zi} + A_{zi+1}}{\Delta^2} + k_c^2 A_{zi} = 0,$$

where subscripts i, $i+1$, and $i-1$ represent centers of the corresponding partition. This equation can be rearranged as follows:

$$-A_{zi-1} + \xi A_{zi} - A_{zi+1} = 0, \tag{iv}$$

where

$$\xi = 2 - \Delta^2 k_c^2. \tag{v}$$

For $\Delta = 0.25$, there are only three internal points where A_z is unknown. Note that it is known at the boundary points. Using (iv) on these points, we get the following set of three equations:

$$\begin{bmatrix} \xi & -1 & 0 \\ -1 & \xi & -1 \\ 0 & -1 & \xi \end{bmatrix} \begin{bmatrix} A_{z1} \\ A_{z2} \\ A_{z3} \end{bmatrix} = \begin{bmatrix} 0 \\ 0 \\ 0 \end{bmatrix}. \tag{vi}$$

For a nontrivial solution of this, we find that

$$\begin{vmatrix} \xi & -1 & 0 \\ -1 & \xi & -1 \\ 0 & -1 & \xi \end{vmatrix} = 0 \Rightarrow \xi(\xi^2 - 2) = 0.$$

Therefore $\xi = 0$, 1.4142, -1.4142, and the corresponding eigenvalues are

$$\xi = 0 \ \rightarrow k_c^2 = \frac{2}{0.25^2} \rightarrow k_c = \frac{1.4142}{0.25} = 5.6568,$$

$$\xi = 1.4142 \rightarrow k_c^2 = \frac{2 - 1.4142}{0.25^2} \rightarrow k_c = \frac{\sqrt{0.5858}}{0.25} = 3.0615,$$

and

$$\xi = -1.4142 \rightarrow k_c^2 = \frac{2 + 1.4142}{0.25^2} \rightarrow k_c = \frac{\sqrt{3.4142}}{0.25} = 7.3910.$$

According to the exact analysis presented in Section 6.4, (6.4.39) gives

$$k_c = k\big|_{k_z=0} = \frac{m\pi}{a} \rightarrow k_c = m\pi, \qquad m = 1, 2, 3, \cdots.$$

Therefore the lowest eigenvalue ($m = 1$) is about 2.55% below the exact value of π. However, the next eigenvalue ($m = 2$) is about 17.63% higher than the corresponding exact solution of 2π. The accuracy of the results increases with decreasing partition length (i.e., increasing the number of partitions).

Example 11.2

Use the finite-difference scheme to solve the following differential equation:

$$\frac{d^2f(x)}{dx^2} + 9f(x) = 0.$$

The solution must satisfy the boundary conditions $f(x = 0) = 0$ and $f(x = 1) = 1$.

The corresponding finite difference is found via (11.1.4) as follows:

$$\frac{f_{i-1} - 2f_i + f_{i+1}}{\Delta^2} + 9f_i = 0 \rightarrow f_{i-1} - (2 - 9\Delta^2)f_i + f_{i+1} = 0.$$

If only four partitions are used over $0 \leq x \leq 1$, then $\Delta = 0.25$. The boundary conditions translate to $f_o = 0$ and $f_4 = 1$. There are three internal points where $f(x)$ is to be determined. Since $(2 - 9\Delta^2) = 1.4375$, the difference equation gives

$$\begin{bmatrix} -1.4375 & 1 & 0 \\ 1 & -1.4375 & 1 \\ 0 & 1 & -1.4375 \end{bmatrix} \begin{bmatrix} f_1 \\ f_2 \\ f_3 \end{bmatrix} = \begin{bmatrix} 0 \\ 0 \\ -1 \end{bmatrix} \rightarrow \begin{bmatrix} f_1 \\ f_2 \\ f_3 \end{bmatrix} = \begin{bmatrix} 10.4757 \\ 15.0588 \\ 11.1714 \end{bmatrix}.$$

As we know that the error in the finite difference formulation decreases with an increase in the number of partitions (i.e., a smaller Δ), we repeat it with 8 partitions. Therefore the partition length Δ now reduces to 0.125, and $f_o = 0$ and f_8 will be 1. There are seven internal points where the function will be determined using the finite difference procedure. With $\xi = (2 - 9\Delta^2) = (2 - 9 \cdot 0.125^2)$, we find that

$$\begin{bmatrix} -\xi & 1 & 0 & 0 & 0 & 0 & 0 \\ 1 & -\xi & 1 & 0 & 0 & 0 & 0 \\ 0 & 1 & -\xi & 1 & 0 & 0 & 0 \\ 0 & 0 & 1 & -\xi & 1 & 0 & 0 \\ 0 & 0 & 0 & 1 & -\xi & 1 & 0 \\ 0 & 0 & 0 & 0 & 1 & -\xi & 1 \\ 0 & 0 & 0 & 0 & 0 & 1 & -\xi \end{bmatrix} \begin{bmatrix} f_1 \\ f_2 \\ f_3 \\ f_4 \\ f_5 \\ f_6 \\ f_7 \end{bmatrix} = \begin{bmatrix} 0 \\ 0 \\ 0 \\ 0 \\ 0 \\ 0 \\ -1 \end{bmatrix} \rightarrow \begin{bmatrix} f_1 \\ f_2 \\ f_3 \\ f_4 \\ f_5 \\ f_6 \\ f_7 \end{bmatrix} = \begin{bmatrix} 2.9846 \\ 5.5496 \\ 7.3341 \\ 8.0872 \\ 7.7031 \\ 6.2358 \\ 3.8915 \end{bmatrix}.$$

Note that the differential equation can be solved analytically for given boundary conditions to find

$$f(x) = \frac{\sin(3x)}{\sin(3)}.$$

Therefore the computed results can be compared with the exact solution to get a feel of the accuracy. This comparison is shown in Table 11.2.

TABLE 11.2 Comparison of Results of Example 11.2

x	Exact Solution	4-Partitions FD Approximation	8-Partitions FD Approximation
0	0	0	0
0.125	2.5955		2.9846
0.25	4.8302	10.4757	5.5496
0.375	6.3936		7.3341
0.5	7.0684	15.0588	8.0872
0.625	6.7608		7.7031
0.75	5.5136	11.1714	6.2358
0.875	3.5		3.8915
1	1	1	1

In the preceding examples, the function is specified on the boundaries. This kind of boundary-value problem is known as the Dirichlet problem. Another possibility is that the derivative of the function is specified instead. This kind of boundary condition is known as the Neumann condition. It becomes a mixed problem if the function is specified on a part of the boundary and a derivative on the other. Consider the construction of a solution to the Neumann problem illustrated in Figure 11.2. The derivatives $f'(x_o)$ and $f'(x_f)$ of the function are specified at the boundary points x_o and x_f. The space between the boundary points is partitioned such that $0 \leq i \leq N$. In order to implement the Neumann conditions, an extra partition is added on each side as the false boundary, and therefore $-1 \leq i \leq N+1$.

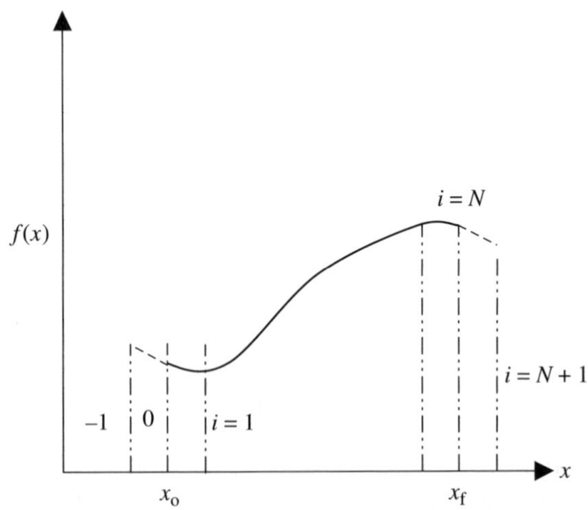

Figure 11.2 False boundaries introduced for a Neumann problem.

Hence

$$f'(x_o) = \frac{f_1 - f_{-1}}{2\Delta} \to f_{-1} = f_1 - f'(x_o)2\Delta \tag{11.1.10}$$

and

$$f'(x_f) = \frac{f_{N+1} - f_{N-1}}{2\Delta} \to f_{N+1} = f_{N-1} + f'(x_f)2\Delta. \tag{11.1.11}$$

Thus there are two extra equations available for the two boundary points. The following example illustrates this solution procedure.

Example 11.3

Using the finite difference scheme, find the lowest eigenvalue of a TE^z mode propagating between two conducting parallel plates that are separated by 1 m, as shown in Figure 11.3.

We earlier analyzed the parallel-plate waveguide in Section 6.4. Since the TE wave is propagating along the z-axis and the conducting plates extend to infinity along the y-axis, (6.4.12) gives

$$\frac{\partial^2 F_z}{\partial x^2} + (k^2 - k_z^2)F_z = 0 \to \frac{\partial^2 F_z}{\partial x^2} + k_c^2 F_z = 0, \tag{i}$$

where

$$k_c^2 = k^2 - k_z^2. \tag{ii}$$

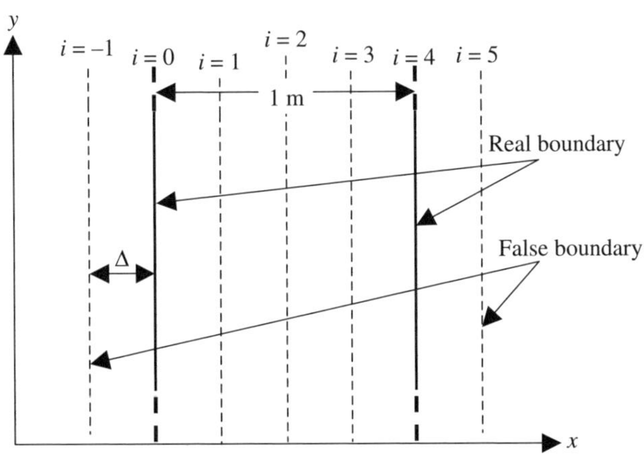

Figure 11.3 Parallel-plate waveguide.

The boundary conditions require that the tangential electric field be zero on the conducting surfaces. Therefore (6.4.14) gives

$$E_y\Big|_{\substack{x=0\\x=1}} = 0 \to \left.\frac{\partial F_z}{\partial x}\right|_{\substack{x=0\\x=1}} = 0. \tag{iii}$$

Note that the other two electric field components are already zero in this case. There are four partitions made between the plates. In order to implement the boundary conditions, one extra partition is added on each side, as shown in Figure 11.3. Therefore (iii) gives

$$\left.\frac{\partial F_z}{\partial x}\right|_{x=0} = 0 \to \frac{F_{z1} - F_{z-1}}{2\Delta} = 0 \to F_{z-1} = F_{z1} \tag{iv}$$

and

$$\left.\frac{\partial F_z}{\partial x}\right|_{x=1} = 0 \to \frac{F_{z5} - F_{z3}}{2\Delta} = 0 \to F_{z5} = F_{z3}. \tag{v}$$

Following the procedure used in Example 11.1, (i) can be approximated using the points at $i = 0$ to 4:

$$-F_{zi-1} + \xi F_{zi} - F_{zi+1} = 0, \tag{vi}$$

where

$$\xi = 2 - \Delta^2 k_c^2. \tag{vii}$$

For $\Delta = 0.25$, there are five points where F_z is unknown. Using (vi) at these points, we get the following set of five equations:

$$-F_{z-1} + \xi F_{z0} - F_{z1} = 0, \tag{viii}$$

$$-F_{z0} + \xi F_{z1} - F_{z2} = 0, \tag{ix}$$

$$-F_{z1} + \xi F_{z2} - F_{z3} = 0, \tag{x}$$

$$-F_{z2} + \xi F_{z3} - F_{z4} = 0, \tag{xi}$$

and

$$-F_{z3} + \xi F_{z4} - F_{z5} = 0. \tag{xii}$$

Note that F_{z-1} and F_{z5} in (viii) and (xii), respectively, can be replaced via the boundary conditions (iv) and (v). Hence the set of equations reduces to

$$\begin{bmatrix} \xi & -2 & 0 & 0 & 0 \\ -1 & \xi & -1 & 0 & 0 \\ 0 & -1 & \xi & -1 & 0 \\ 0 & 0 & -1 & \xi & -1 \\ 0 & 0 & 0 & -2 & \xi \end{bmatrix} \begin{bmatrix} F_{z0} \\ F_{z1} \\ F_{z2} \\ F_{z3} \\ F_{z4} \end{bmatrix} = \begin{bmatrix} 0 \\ 0 \\ 0 \\ 0 \\ 0 \end{bmatrix}. \tag{xiii}$$

For a nontrivial solution of (xiii) we find that

$$\begin{vmatrix} \xi & -2 & 0 & 0 & 0 \\ -1 & \xi & -1 & 0 & 0 \\ 0 & -1 & \xi & -1 & 0 \\ 0 & 0 & -1 & \xi & -1 \\ 0 & 0 & 0 & -2 & \xi \end{vmatrix} = 0 \rightarrow \xi^5 - 6\xi^3 + 8\xi = 0$$

$$\rightarrow \xi = 0, \ \pm \sqrt{2}, \ \pm 2. \tag{xiv}$$

The corresponding values of k_c are found via (vii) as follows:

$$k_c = 5.6568, 3.0615, 7.3910, 0, 8.$$

As found in Chapter 6, the exact eigenvalues of the TE^z mode are integer multiple of π (because the plates are separated by 1 m) and degenerate with TM^z.

In this example we get the lowest value close to π if we ignore 0. In general, Neumann boundary value problems are difficult to handle and often require extra conditions (e.g., an analytic requirement) for a finite difference solution.

Finite Difference in Two Dimensions

The finite-difference technique can be used to solve Laplace's equation in two and three dimensions. It has several practical applications, including TEM and quasi-TEM characterizations of transmission lines.

Consider four infinitely long conducting plates of equal width arranged as shown in Figure 11.4*a*. The plates are insulated from each other at the corners and are at potentials V_T, V_R, V_B, and V_L, as indicated. Further the plates enclose a uniform medium. The potential distribution over a cross-sectional plane can be found after solving Laplace's equation. To that end, a uniform grid is created, and a finite-difference star (or molecule) is identified along with its node indexes as shown in Figure 11.4*b*. Using (11.1.4), we find that

$$\frac{\partial^2 \phi}{\partial x^2} \approx \frac{\phi_{i+1,j} - 2\phi_{i,j} + \phi_{i-1,j}}{\Delta^2} \tag{11.1.12}$$

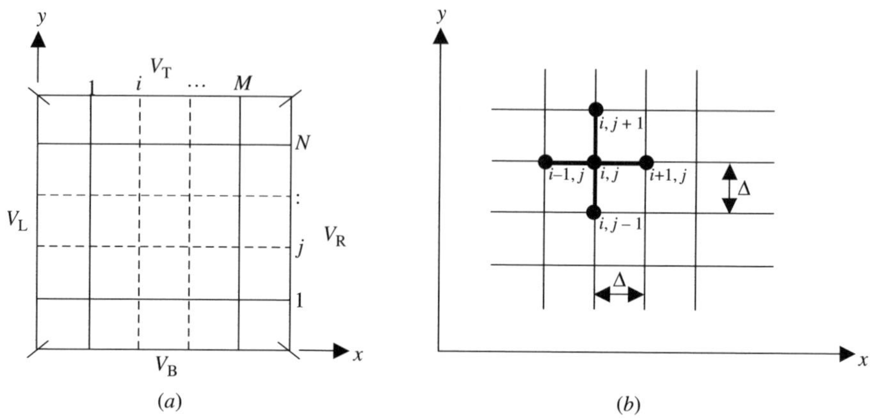

Figure 11.4 Partition grid in two dimensions and the finite-difference star or molecule.

and

$$\frac{\partial^2 \phi}{\partial y^2} \approx \frac{\phi_{i,j+1} - 2\phi_{i,j} + \phi_{i,j-1}}{\Delta^2}. \tag{11.1.13}$$

Therefore the Laplace's equation gives

$$\nabla^2 \phi(x, y) = \frac{\partial^2 \phi(x, y)}{\partial x^2} + \frac{\partial^2 \phi(x, y)}{\partial y^2} = 0 \rightarrow$$

$$\phi_{i,j} \approx \frac{\phi_{i+1,j} + \phi_{i-1,j} + \phi_{i,j+1} + \phi_{i,j-1}}{4}. \tag{11.1.14}$$

Thus the potential at the node point (i, j) is approximately equal to the average of potentials at the other four nodes of the molecule. As was mentioned earlier, the accuracy of this expression increases as the spacing Δ reduces. The potential at node (i, k) can be found from (11.1.14) if the potentials at the four nearby nodes are known. One method to find the potential distribution is to use an iterative procedure. Note that the potential on its boundary is known. Therefore the nodes that lie on the boundary have fixed known potentials. The iteration can be started with the assumption that the internal nodes have zero potential. On each scan the potential found via (11.1.14) replaces the old value at that node. This process continues until the difference obtained in the node voltages between the successive iterations reduces to an acceptable tolerance limit. After generating the grid, a computer program can be developed to find the node potentials. Example 11.4 illustrates the procedure.

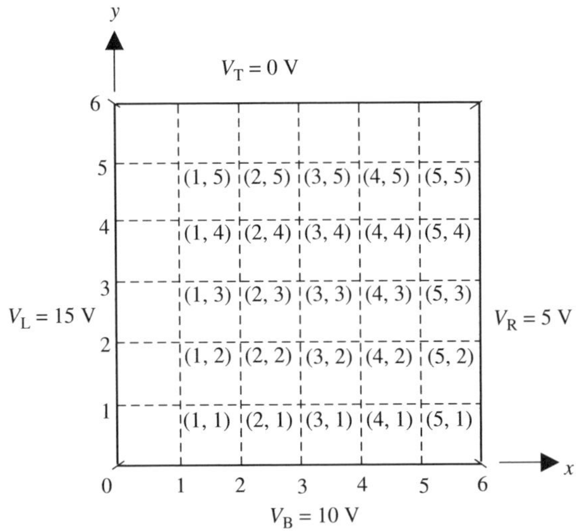

Figure 11.5 Partition grid and nodes for Example 11.4.

Example 11.4

Find the potential distribution inside the space enclosed by four infinitely long conducting plates arranged as shown in Figure 11.5. The boundary potentials are as given in the figure.

Figure 11.5 shows a 6×6 grid along with the node indexes. However, we consider first only a coarse grid of 3×3 for simplicity. Therefore there are only (2, 2), (2, 4), (4, 2), and (4, 4) internal nodes in this case. The iterations are started, assuming zero initial potential at these nodes. There are updates in each scan as listed in Table 11.3 for first five iterations.

A computer program was developed for the 6×6 grid with a tolerance limit for successive potentials at the nodes as 0.001. These results are summarized in Table 11.4. Since there are a few nodes common with the 3×3 grid used earlier, we can compare the results. For example, ϕ_{22} is found as 9.983 V after the fifth

TABLE 11.3 Successive Values of the Potential at the Nodes within a Coarse Grid

Iteration Number	ϕ_{22}	ϕ_{24}	ϕ_{42}	ϕ_{44}
1	6.25	5.3125	5.3125	3.9063
2	8.9063	6.9532	6.9532	4.7266
3	9.7266	7.3633	7.3633	4.9317
4	9.9317	7.4659	7.4659	4.9829
5	9.983	7.4915	7.4915	4.9947

TABLE 11.4 Potential at Various Grid Points (i, j) of Figure 11.5

j \ i		1	2	3	4	5	
		0	0	0	0	0	
5	15	7.50	4.57	3.30	2.84	3.12	5
4	15	10.42	7.50	5.80	4.92	4.66	5
3	15	11.69	9.19	7.50	6.38	5.61	5
2	15	12.16	10.07	8.61	7.50	6.40	5
1	15	11.87	10.33	9.38	8.59	7.50	5
		10	10	10	10	10	

iteration in Table 11.3. The corresponding result with a 0.001 tolerance limit is 10.07, which is fairly close. Similar observations can be made for other results. Thus the convergence rate appears to be quite fast.

Single Dielectric, Nonuniform Grid

A uniform grid was used in the preceding example. However, there are many instances where a nonuniform grid is helpful. For example, if a particular subregion is of more interest than the rest of the region and the computer resources are limited, then a finer grid can be employed only there. This formulation is also useful for regions with curved boundaries in single dielectric medium.

Consider the situation where node 3 is Δ_2 away from node 0 whereas node 1 is at Δ_1, as shown in Figure 11.6*a*. Using a Taylor's series at node 1, we find that

$$f(x_1) = f(x_0) + \Delta_1 \left.\frac{\partial f(x)}{\partial x}\right|_{x_0} + \frac{\Delta_1^2}{2}\left.\frac{\partial^2 f(x)}{\partial x^2}\right|_{x_0} + \frac{\Delta_1^3}{3!}\left.\frac{\partial^3 f(x)}{\partial x^3}\right|_{x_0} + \cdots. \tag{11.1.15}$$

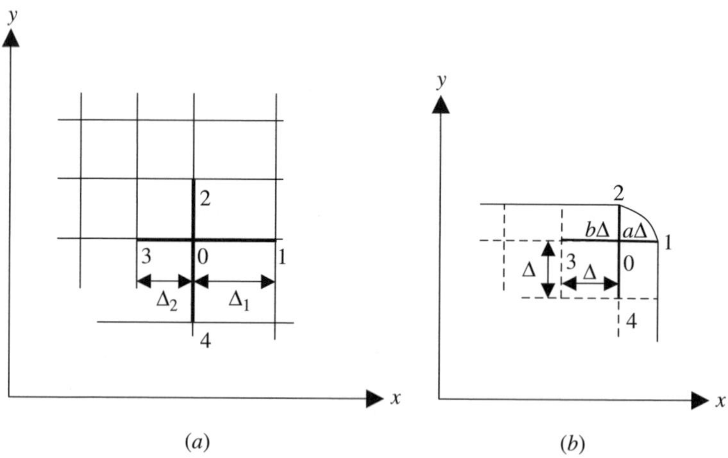

Figure 11.6 Nonuniform grid (*a*) and the curved boundary (*b*).

Similarly for node 3 we have

$$f(x_3) = f(x_0) - \Delta_2 \frac{\partial f(x)}{\partial x}\bigg|_{x_0} + \frac{\Delta_2^2}{2}\frac{\partial^2 f(x)}{\partial x^2}\bigg|_{x_0} - \frac{\Delta_2^3}{3!}\frac{\partial^3 f(x)}{\partial x^3}\bigg|_{x_0} + \cdots. \tag{11.1.16}$$

Multiplying (11.1.15) by α and (11.1.16) by β, and then adding the two, we get

$$\begin{aligned} &\alpha[f(x_1) - f(x_0)] + \beta[f(x_3) - f(x_0)] \\ &\quad = (\alpha\Delta_1 - \beta\Delta_2)\frac{\partial f(x)}{\partial x}\bigg|_{x_0} + \frac{(\alpha\Delta_1^2 - \beta\Delta_2^2)}{2}\frac{\partial^2 f(x)}{\partial x^2}\bigg|_{x_0} + \cdots. \end{aligned} \tag{11.1.17}$$

For

$$\alpha = -\frac{\Delta_2^2}{\Delta_1^2}\beta, \tag{11.1.18}$$

(11.1.17) gives

$$\frac{\partial f(x)}{\partial x}\bigg|_{x_0} \approx \frac{\Delta_2^2[f(x_1) - f(x_0)] - \Delta_1^2[f(x_3) - f(x_0)]}{\Delta_1\Delta_2(\Delta_1 + \Delta_2)}. \tag{11.1.19}$$

If instead

$$\alpha = \frac{\Delta_2}{\Delta_1}\beta, \tag{11.1.20}$$

then from (11.1.17) we get

$$\frac{\partial^2 f(x)}{\partial x^2}\bigg|_{x_0} \approx 2\frac{\Delta_2[f(x_1) - f(x_0)] + \Delta_1[f(x_3) - f(x_0)]}{\Delta_1\Delta_2(\Delta_1 + \Delta_2)}. \tag{11.1.21}$$

Now consider the situation depicted in Figure 11.6*b* where we have a curved boundary in a single dielectric medium. Equations (11.1.19) and (11.1.21) can be used to find the derivatives for this case as follows: For $\Delta_1 = a\Delta$ and $\Delta_2 = \Delta$, Equations (11.1.19) and (11.1.21), simplify to

$$\frac{\partial f(x)}{\partial x}\bigg|_{x_0} \approx \frac{f(x_1) - a^2 f(x_3) - (1 - a^2)f(x_0)}{a(1 + a)\Delta} \tag{11.1.22}$$

and

$$\frac{\partial^2 f(x)}{\partial x^2}\bigg|_{x_0} \approx \frac{2}{\Delta^2}\left[\frac{f(x_1)}{a(1 + a)} + \frac{f(x_3)}{(1 + a)} - \frac{f(x_0)}{a}\right]. \tag{11.1.23}$$

Note that (11.1.22) and (11.1.23) reduce to (11.1.3) and (11.1.4), respectively, for $a = 1$.

A similar procedure can be used to find the derivatives along the y-axis. The results are

$$\left.\frac{\partial f(y)}{\partial y}\right|_{y_0} \approx \frac{f(y_1) - b^2 f(y_3) - (1 - b^2) f(y_0)}{b(1+b)\Delta} \tag{11.1.24}$$

and

$$\left.\frac{\partial^2 f(y)}{\partial y^2}\right|_{y_0} \approx \frac{2}{\Delta^2}\left[\frac{f(y_1)}{b(1+b)} + \frac{f(y_3)}{(1+b)} - \frac{f(y_0)}{b}\right]. \tag{11.1.25}$$

Example 11.5

Find the potential distribution $\phi(x, y)$ inside the region shown in Figure 11.7. The curved portion of the boundary is an arc of radius $R = 10$ m about the origin. The boundary conditions are given as follows:

$$\phi(0, y) = 0,$$

$$\phi(x, 0) = x^3,$$

$$\phi(8, y) = 512 - 24y^2,$$

$$\phi(x, 9) = x^3 - 243x,$$

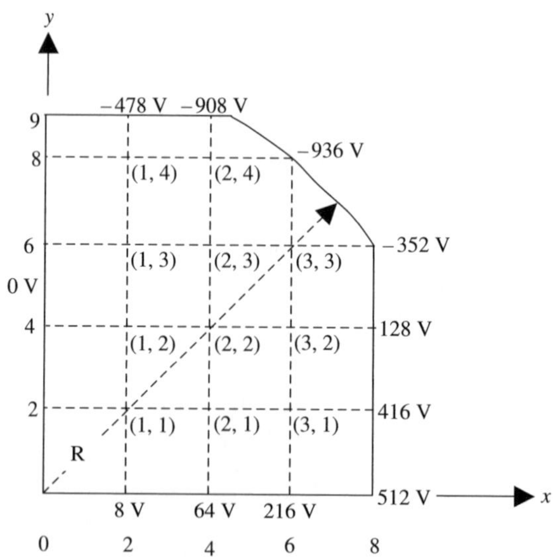

Figure 11.7 Partitions and nodes for Example 11.5.

and

$$\phi(x, \sqrt{100 - x^2}) = 4x^3 - 300x \qquad \text{(on the arc)}.$$

From these boundary conditions the potentials on the boundary nodes are found as indicated in the figure.

From (11.1.23) and (11.1.25),

$$\frac{\partial^2 \phi}{\partial x^2} + \frac{\partial^2 \phi}{\partial y^2} = 0 \approx \frac{1}{\Delta^2}\left\{\frac{2\phi_1}{a(1+a)} + \frac{2\phi_3}{(1+a)} - \frac{2\phi_0}{a}\right\} + \frac{1}{\Delta^2}\left\{\frac{2\phi_2}{b(1+b)} + \frac{2\phi_4}{(1+b)} - \frac{2\phi_0}{b}\right\}.$$

Therefore

$$\phi_0 = \left\{\frac{ab}{a+b}\right\}\left\{\frac{\phi_1}{a(1+a)} + \frac{\phi_2}{b(1+b)} + \frac{\phi_3}{(1+a)} + \frac{\phi_4}{(1+b)}\right\}. \tag{i}$$

First consider a coarse 4×4 grid such that there are only two nodes, so we need to find potentials at (2, 2) and (2, 4) only. For node (2, 2), $a = b = 1$, and therefore

$$\phi_{22} = \frac{128 + \phi_{12} + 0 + 64}{4} \rightarrow 4\phi_{22} - \phi_{24} = 192. \tag{ii}$$

For node (2, 4), $a = 2/3$ and $b = 1/3$. Hence $ab/(a+b) = 2/9$, and (i) gives

$$\phi_{24} = \frac{2}{9}\left\{\frac{-936}{2[1+(2/3)]/3} + \frac{-908}{[1+(1/3)]/3} + 0 + \frac{\phi_{22}}{1+(2/3)}\right\} \rightarrow$$

$$\phi_{24} - \frac{1}{6}\phi_{22} = -641.2. \tag{iii}$$

Equations (ii) and (iii) can be solved for ϕ_{22} and ϕ_{24}. These are found to be -117.1826 V and -660.7304 V, respectively.

Obviously, with so few grid points, we cannot expect great accuracy. The corresponding exact values are -128 V and -704 V, respectively.

Now we consider the entire system of grids shown in Figure 11.7. Note that it is uniform except for nodes (1, 4) and (2, 4), $a = 1$ and $b = 1/3$. Therefore for these nodes

$$\frac{ab}{a+b} = \frac{1}{4}. \tag{iv}$$

Various node potentials can be expressed as follows via (11.1.14) or (i) and (iv):

$$\phi_{11} = \frac{\phi_{21} + \phi_{12} + 0 + 8}{4} \rightarrow 4\phi_{11} - \phi_{12} - \phi_{21} = 8, \tag{v}$$

$$\phi_{12} = \frac{\phi_{22} + \phi_{13} + 0 + \phi_{11}}{4} \rightarrow 4\phi_{12} - \phi_{13} - \phi_{11} - \phi_{22} = 0, \tag{vi}$$

$$\phi_{13} = \frac{\phi_{23} + \phi_{14} + 0 + \phi_{12}}{4} \rightarrow 4\phi_{13} - \phi_{12} - \phi_{14} - \phi_{23} = 0, \tag{vii}$$

$$\phi_{14} = \frac{1}{4}\left[\frac{\phi_{24}}{2} - \frac{478}{(1/3)\cdot(4/3)} + 0 + \frac{\phi_{13}}{4/3}\right]$$

$$\rightarrow 16\phi_{14} - 3\phi_{13} - 2\phi_{24} = -4302, \tag{viii}$$

$$\phi_{21} = \frac{\phi_{31} + \phi_{22} + \phi_{11} + 64}{4} \rightarrow 4\phi_{21} - \phi_{31} - \phi_{11} - \phi_{22} = 64, \tag{ix}$$

$$\phi_{22} = \frac{\phi_{32} + \phi_{23} + \phi_{12} + \phi_{21}}{4} \rightarrow 4\phi_{22} - \phi_{32} - \phi_{23} - \phi_{12} - \phi_{21} = 0, \tag{x}$$

$$\phi_{23} = \frac{\phi_{33} + \phi_{24} + \phi_{13} + \phi_{22}}{4} \rightarrow 4\phi_{23} - \phi_{33} - \phi_{24} - \phi_{13} - \phi_{22} = 0, \tag{xi}$$

$$\phi_{24} = \frac{1}{4}\left[-\frac{936}{2} - \frac{908}{(1/3)\cdot(4/3)} + \frac{\phi_{14}}{2} + \frac{\phi_{23}}{4/3}\right]$$

$$\rightarrow 16\phi_{24} - 3\phi_{23} - 2\phi_{14} = -10044, \tag{xii}$$

$$\phi_{31} = \frac{416 + \phi_{32} + \phi_{21} + 216}{4} \rightarrow 4\phi_{31} - \phi_{32} - \phi_{21} = 632, \tag{xiii}$$

$$\phi_{32} = \frac{128 + \phi_{33} + \phi_{22} + \phi_{31}}{4} \rightarrow 4\phi_{32} - \phi_{31} - \phi_{33} - \phi_{22} = 128, \tag{xiv}$$

and

$$\phi_{33} = \frac{-352 - 936 + \phi_{23} + \phi_{32}}{4} \rightarrow 4\phi_{33} - \phi_{23} - \phi_{32} = -1288. \tag{xv}$$

Equations (v) through (xv) can be expressed in matrix form, as follows:

$$\begin{bmatrix} 4 & -1 & 0 & 0 & -1 & 0 & 0 & 0 & 0 & 0 & 0 \\ -1 & 4 & -1 & 0 & 0 & -1 & 0 & 0 & 0 & 0 & 0 \\ 0 & -1 & 4 & -1 & 0 & 0 & -1 & 0 & 0 & 0 & 0 \\ 0 & 0 & -3 & 16 & 0 & 0 & 0 & -2 & 0 & 0 & 0 \\ -1 & 0 & 0 & 0 & 4 & -1 & 0 & 0 & -1 & 0 & 0 \\ 0 & -1 & 0 & 0 & -1 & 4 & -1 & 0 & 0 & -1 & 0 \\ 0 & 0 & -1 & 0 & 0 & -1 & 4 & -1 & 0 & 0 & -1 \\ 0 & 0 & 0 & -2 & 0 & 0 & -3 & 16 & 0 & 0 & 0 \\ 0 & 0 & 0 & 0 & -1 & 0 & 0 & 0 & 4 & -1 & 0 \\ 0 & 0 & 0 & 0 & 0 & -1 & 0 & 0 & -1 & 4 & -1 \\ 0 & 0 & 0 & 0 & 0 & 0 & -1 & 0 & 0 & -1 & 4 \end{bmatrix} \begin{bmatrix} \phi_{11} \\ \phi_{12} \\ \phi_{13} \\ \phi_{14} \\ \phi_{21} \\ \phi_{22} \\ \phi_{23} \\ \phi_{24} \\ \phi_{31} \\ \phi_{32} \\ \phi_{33} \end{bmatrix}$$

$$= \begin{bmatrix} 8 \\ 0 \\ 0 \\ -4302 \\ 64 \\ 0 \\ 0 \\ -10044 \\ 632 \\ 128 \\ -1288 \end{bmatrix}. \qquad \text{(xvi)}$$

Therefore potentials at these nodes can be found after solving (xvi) through standard matrix manipulation techniques. Note that the coefficient matrix is diagonal dominant and a large number of off-diagonal entries have zeros.

TABLE 11.5 Comparison of Results for Example 11.5

Nodes	Computed with Coarse Grid	Computed with Fine Grid	Analytical Result
ϕ_{11}		−18.13	−16
ϕ_{12}		−93.68	−88
ϕ_{13}		−221.00	−208
ϕ_{14}		−404.15	−376
ϕ_{21}		13.16	16
ϕ_{22}	−117.18	−135.59	−128
ϕ_{23}		−386.19	−368
ϕ_{24}	−660.73	−750.68	−704
ϕ_{31}		142.37	144
ϕ_{32}		−75.67	−72
ϕ_{33}		−437.46	−432

Further this problem has been solved analytically for the given boundary conditions, and the exact solution is found to be $\phi(x, y) = x^3 - 3xy^2$. The computational and analytical results of these potentials are compared in Table 11.5.

Example 11.6

Find the eigenvalues of a hollow metallic rectangular waveguide operating in TM^z mode. Its cross section is 0.75×0.25 m. Use the uniform grid shown in Figure 11.8. Repeat the problem with the nonuniform grid shown in Figure 11.9. Analyze the results.

For the TM^z modes, we use (6.4.3) to find that

$$\frac{\partial^2 A_z}{\partial x^2} + \frac{\partial^2 A_z}{\partial y^2} + k_c^2 A_z = 0, \tag{i}$$

where $k_c^2 = k^2 - k_z^2$. The associated boundary conditions are found in (10.1.26) through (10.1.28) that require A_z to be zero on the boundary.

Using (11.1.12) and (11.1.13), (i) transforms to

$$\frac{A_{i+1,j} - 2A_{i,j} + A_{i-1,j}}{\Delta^2} + \frac{A_{i,j+1} - 2A_{i,j} + A_{i,j-1}}{\Delta^2} + k_c^2 A_{i,j} = 0,$$

or

$$A_{i+1,j} + A_{i-1,j} + A_{i,j+1} + A_{i,j-1} - (4 - k_c^2\Delta^2)A_{i,j} = 0. \tag{ii}$$

Using (ii) and the boundary conditions that A_z is zero on the boundary nodes, we find that

$$\begin{bmatrix} -\alpha & 1 \\ 1 & -\alpha \end{bmatrix} \begin{bmatrix} A_{1,1} \\ A_{2,1} \end{bmatrix} = \begin{bmatrix} 0 \\ 0 \end{bmatrix}, \tag{iii}$$

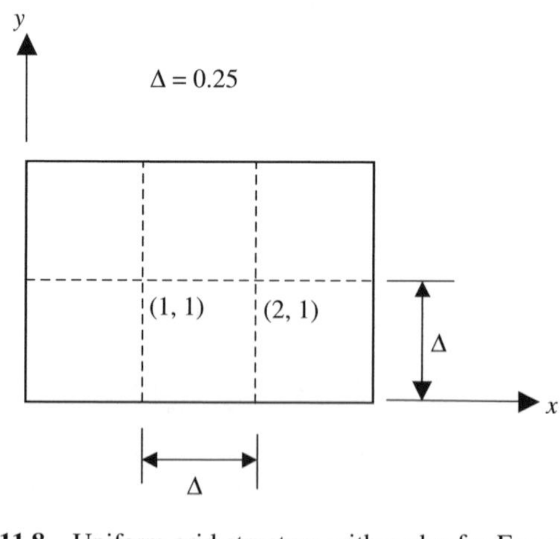

Figure 11.8 Uniform grid structure with nodes for Example 11.6.

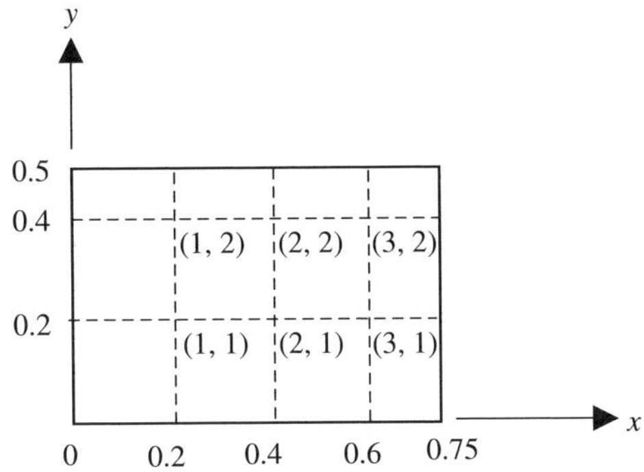

Figure 11.9 Mixed grid structure with nodes for Example 11.6.

where

$$\alpha = 4 - k_c^2\Delta^2 = 4 - 0.25^2\Delta^2. \quad \text{(iv)}$$

For a nontrivial solution of (iii), it is required that

$$\begin{vmatrix} -\alpha & 1 \\ 1 & -\alpha \end{vmatrix} = 0 \rightarrow \alpha = \pm 1 = 4 - k_c^2\Delta^2. \quad \text{(v)}$$

For $\alpha = -1$,

$$k_c^2\Delta^2 - 4 = -1 \rightarrow k_c^2 = \frac{3}{0.25^2} \rightarrow k_c = 6.9282, \quad \text{(vi)}$$

and for $\alpha = 1$,

$$k_c^2\Delta^2 - 4 = 1 \rightarrow k_c^2 = \frac{5}{0.25^2} \rightarrow k_c = 8.9443. \quad \text{(vii)}$$

Its analytical expression may be found in Table 10.1 as $k_c = \sqrt{(m\pi/a)^2 + (n\pi/b)^2}$. Therefore, for $a = 0.75$ and $b = 0.5$, the first four eigenvalues are found as 7.551, 10.472, 13.246, and 15.103 with m as 1, 2 and n as 1, 2.

When the grid is nonuniform as given in Figure 11.9, (11.1.25) and (11.1.23) need to be employed at nodes (1, 2), (2, 2), (3, 1), and (3, 2). For these nodes,

$$a = \frac{0.15}{0.2} = 0.75 \quad \text{(viii)}$$

and

$$b = \frac{0.1}{0.2} = 0.5. \quad \text{(ix)}$$

The finite-difference equations for nodes (1, 1) through (3, 2) are found to be

$$\frac{A_{21} - 2A_{11} + 0}{0.2^2} + \frac{A_{12} - 2A_{11} + 0}{0.2^2} + k_c^2 A_{11}$$

$$= 0 \rightarrow A_{21} + A_{12} - (4 - 0.2^2 k_c^2)A_{11} = 0, \tag{x}$$

$$\frac{A_{22} - 2A_{12} + 0}{0.2^2} + \frac{2}{0.2^2}\left[0 + \frac{A_{11}}{1.5} - \frac{A_{12}}{0.5}\right] + k_c^2 A_{12}$$

$$= 0 \rightarrow \frac{4}{3}A_{11} + A_{22} - (6 - 0.2^2 k_c^2)A_{12} = 0, \tag{xi}$$

$$\frac{A_{31} - 2A_{21} + A_{11}}{0.2^2} + \frac{A_{22} - 2A_{21} + 0}{0.2^2} + k_c^2 A_{21}$$

$$= 0 \rightarrow A_{11} + A_{31} + A_{22} - (4 - 0.2^2 k_c^2)A_{21} = 0, \tag{xii}$$

$$\frac{A_{32} - 2A_{22} + A_{12}}{0.2^2} + \frac{2}{0.2^2}\left[0 + \frac{A_{21}}{1.5} - \frac{A_{22}}{0.5}\right] + k_c^2 A_{22}$$

$$= 0 \rightarrow A_{12} + \frac{4}{3}A_{21} + A_{32} - (6 - 0.2^2 k_c^2)A_{22} = 0, \tag{xiii}$$

$$\frac{2}{0.2^2}\left[0 + \frac{A_{21}}{1.75} - \frac{A_{31}}{0.75}\right] + \frac{A_{32} - 2A_{31} + 0}{0.2^2} + k_c^2 A_{31}$$

$$= 0 \rightarrow \frac{2}{1.75}A_{21} + A_{32} - \left(\frac{14}{3} - 0.2^2 k_c^2\right)A_{31} = 0, \tag{xiv}$$

and

$$\frac{2}{0.2^2}\left[0 + \frac{A_{22}}{1.75} - \frac{A_{32}}{0.75}\right] + \frac{2}{0.2^2}\left[0 + \frac{A_{31}}{1.5} - \frac{A_{32}}{0.5}\right] + k_c^2 A_{32}$$

$$= 0 \rightarrow \frac{2}{1.75}A_{22} + \frac{4}{3}A_{31} - \left(\frac{20}{3} - 0.2^2 k_c^2\right)A_{32} = 0. \tag{xv}$$

For the nontrivial solution of (x) through (xv), the following condition must hold:

$$\begin{vmatrix}
-(4-0.2^2k_c^2) & 1 & 1 & 0 & 0 & 0 \\
\frac{4}{3} & -(6-0.2^2k_c^2) & 0 & 1 & 0 & 0 \\
1 & 0 & -(4-0.2^2\,k_c^2) & 1 & 1 & 0 \\
0 & 1 & \frac{4}{3} & -(6-0.2^2k_c^2) & 0 & 1 \\
0 & 0 & \frac{2}{1.75} & 0 & -\left(\frac{14}{3}-0.2^2k_c^2\right) & 1 \\
0 & 0 & 0 & \frac{2}{1.75} & \frac{4}{3} & -\left(\frac{20}{3}-0.2^2k_c^2\right)
\end{vmatrix}$$

$$= 0.$$

TABLE 11.6 Comparison of the Results of Example 11.6

	k_c with 6 Partitions of Figure 11.8	k_c with 12 Partitions of Figure 11.9	Exact Value
1	6.9282	7.3239	7.55
2	8.9443	9.7062	10.47
3		11.4024	13.25
4		13.0609	15.1

Therefore

$$k_c = 7.3239, 9.7062, 11.3689, 11.4024, 13.0609, 14.3398.$$

These results are compared with the corresponding analytical (exact) values in Table 11.6.

Finite Difference in Three Dimensions

The preceding formulations can be easily extended to three-dimensional problems. The FD molecule will now have seven nodes as shown in Figure 11.10. The squares generated by the uniform grid in 2D area become cubes to represent the volume in

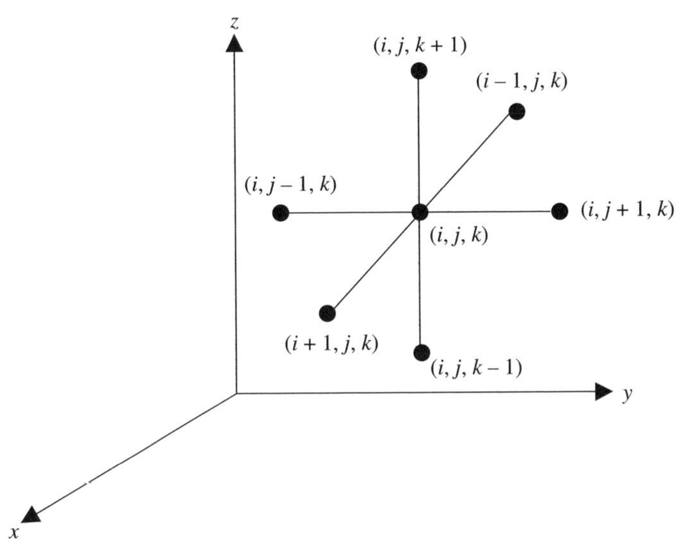

Figure 11.10 3D molecule as the finite-difference molecule.

3D. The finite-difference approximation for the added term may be found as follows:

$$\frac{\partial^2 \phi}{\partial z^2} \approx \frac{\phi_{i,j,k+1} - 2\phi_{i,j,k} + \phi_{i,j,k-1}}{\Delta^2}. \tag{11.1.26}$$

Therefore

$$\nabla^2 \phi \approx \frac{\phi_{i+1,j,k} + \phi_{i-1,j,k} + \phi_{i,j+1,k} + \phi_{i,j-1,k} + \phi_{i,j,k+1} + \phi_{i,j,k-1} - 6\phi_{i,j,k}}{\Delta^2}. \tag{11.1.27}$$

Laplace's equation in three dimensions gives

$$\phi_{i,j,k} = \frac{\phi_{i+1,j,k} + \phi_{i-1,j,k} + \phi_{i,j+1,k} + \phi_{i,j-1,k} + \phi_{i,j,k+1} + \phi_{i,j,k-1}}{6}. \tag{11.1.28}$$

Similarly (11.1.22) through (11.1.25) can be extended for three-dimensional finite-difference problems.

Finite Difference at the Interface of Two Dielectrics

Up to this point we considered the situation where the region is uniform. However, there are many practical situations that involve more than one dielectric media. This kind of situation is depicted in Figure 11.11. We would like to find ways to handle

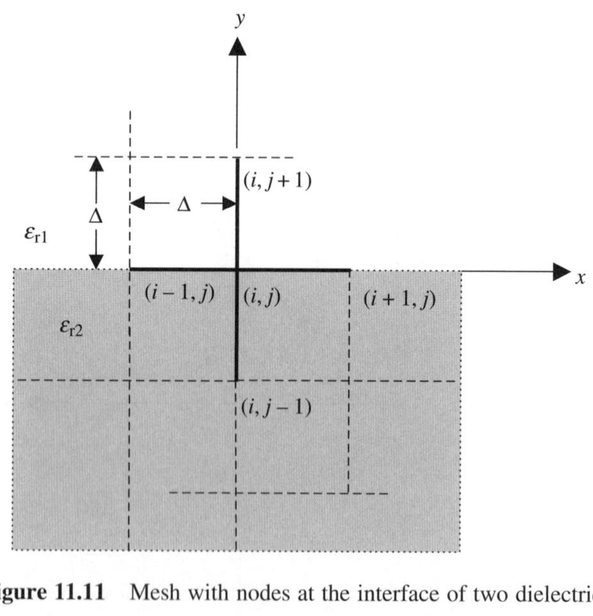

Figure 11.11 Mesh with nodes at the interface of two dielectrics.

such cases. Therefore we start with Gauss's law for a charge-free region as follows:

$$\nabla \cdot \vec{D} = \rho \rightarrow \nabla \cdot \vec{D} = 0. \tag{11.1.29}$$

For the two-dimensional case, (11.1.29) reduces to

$$\frac{\partial D_x}{\partial x} + \frac{\partial D_y}{\partial y} = 0. \tag{11.1.30}$$

As Figure 11.11 indicates, the problem assumes a uniform grid with ε_{r1} and ε_{r2} as the dielectric constants of the two media. Further the electric flux densities above and below the interface are assumed to be D_y^+ and D_y^-, respectively. Therefore

$$D_y^+ = \varepsilon_o \varepsilon_{r1} E_y \approx \frac{-\varepsilon_o \varepsilon_{r1}(\phi_{i,j+1} - \phi_{i,j})}{\Delta} \tag{11.1.31}$$

and

$$D_y^- = \varepsilon_o \varepsilon_{r2} E_y \approx \frac{-\varepsilon_o \varepsilon_{r2}(\phi_{i,j} - \phi_{i,j-1})}{\Delta}. \tag{11.1.32}$$

Using (11.1.31) and (11.1.32), we find that

$$\frac{\partial D_y}{\partial y} \approx \frac{D_y^+ - D_y^-}{\Delta} = -\frac{\varepsilon_o}{\Delta^2}\left\{\varepsilon_{r1}(\phi_{i,j+1} - \phi_{i,j}) - \varepsilon_{r2}(\phi_{i,j} - \phi_{i,j-1})\right\}. \tag{11.1.33}$$

A similar expression can be formulated for the x direction provided that ε_r at node points $(i-1,\ j)$ and $(i+1,\ j)$ is known. If we assume, for the time being, that it is ε_{r3}, then

$$\frac{\partial D_x}{\partial x} \approx -\frac{\varepsilon_o \varepsilon_{r3}}{\Delta^2}(\phi_{i+1,j} + \phi_{i-1,j} - 2\phi_{i,j}). \tag{11.1.34}$$

Substituting (11.1.33) and (11.1.34) into (11.1.30), we find that

$$-\phi_{i,j}(\varepsilon_{r1} + \varepsilon_{r2} + 2\varepsilon_{r3}) + \varepsilon_{r1}\phi_{i,j+1} + \varepsilon_{r2}\phi_{i,j-1} + \varepsilon_{r3}(\phi_{i+1,j} + \phi_{i-1,j}) = 0,$$

or

$$\phi_{i,j} = \frac{\varepsilon_{r1}\phi_{i,j+1} + \varepsilon_{r2}\phi_{i,j-1} + \varepsilon_{r3}(\phi_{i+1,j} + \phi_{i-1,j})}{\varepsilon_{r1} + \varepsilon_{r2} + 2\varepsilon_{r3}}. \tag{11.1.35}$$

If it is assumed that the node potentials along the x-axis contribute an equal amount to $\phi_{i,j}$ as those along the y-axis, then (11.1.35) gives

$$\varepsilon_{r1}\phi_{i,j+1} + \varepsilon_{r2}\phi_{i,j-1} = \varepsilon_{r3}(\phi_{i-1,j} + \phi_{i+1,j}).$$

Therefore

$$\varepsilon_{r3} = \frac{\varepsilon_{r1}\phi_{i,j+1} + \varepsilon_{r2}\phi_{i,j-1}}{\phi_{i-1,j} + \phi_{i+1,j}}. \qquad (11.1.36)$$

A further simplification is possible if we assume that the potentials at adjacent nodes are similar. In that case (11.1.36) reduces to

$$\varepsilon_{r3} = \frac{\varepsilon_{r1} + \varepsilon_{r2}}{2}. \qquad (11.1.37)$$

Finite-Difference Time-Domain Method

The finite-difference time-domain (FDTD) method has been used to solve a number of electromagnetic problems. We consider here the solution to the following wave equation in one spatial dimension and the time:

$$\frac{\partial^2 E_x(z, t)}{\partial t^2} = v^2 \frac{\partial^2 E_x(z, t)}{\partial z^2}. \qquad (11.1.38)$$

The associated boundary conditions are

$$E_x(x = 0, t) = f_1(t) \qquad (11.1.39)$$

and

$$E_x(x = d, t) = f_2(t). \qquad (11.1.40)$$

Further the solution must satisfy the following initial conditions:

$$E_x(z, t = 0) = g_1(z) \qquad (11.1.41)$$

and

$$\left.\frac{\partial E_x(z, t)}{\partial t}\right|_{t=0} = g_2(z). \qquad (11.1.42)$$

To set up the finite-difference method, we select a space-step size $\Delta > 0$ and time-step size $\delta > 0$. With M space and N time partitions, we have

$$\Delta = \frac{d}{M} \tag{11.1.43}$$

and

$$\delta = \frac{t}{N}. \tag{11.1.44}$$

Therefore the grid points (z_k, t_n) in space and time are defined by

$$z_k = k\Delta, \qquad k = 0, 1, 2, \ldots, M, \tag{11.1.45}$$

and

$$t_n = n\delta, \qquad n = 0, 1, 2, \ldots, N. \tag{11.1.46}$$

Following (11.1.4), (11.1.38) is quantized in time and space to find

$$\frac{E_x(k, n+1) - 2E_x(k, n) + E_x(k, n-1)}{\delta^2} - v^2 \frac{E_x(k+1, n) - 2E_x(k, n) + E_x(k-1, n)}{\Delta^2} = 0,$$

or

$$E_x(k, n+1) - 2E_x(k, n) + E_x(k, n-1) = \xi^2\{E_x(k+1, n) - 2E_x(k, n) + E_x(k-1, n)\}, \tag{11.1.47}$$

where

$$\xi = \frac{v\delta}{\Delta}. \tag{11.1.48}$$

Rearranging (11.1.47), we find that

$$E_x(k, n+1) = 2(1 - \xi^2)E_x(k, n) - E_x(k, n-1) + \xi^2\{E_x(k-1, n) + E_x(k+1, n)\}. \tag{11.1.49}$$

Equation (11.1.49) holds for each $k = 1, 2, \ldots, (M-1)$ and $n = 1, 2, \ldots$. For $k = 0$ and M, E_x is known from the boundary conditions (11.1.39) and (11.1.40), which give

$$E_x(0, n) = f_1(n\delta) \tag{11.1.50}$$

and

$$E_x(M, n) = f_2(n\delta). \tag{11.1.51}$$

Equation (11.1.49) can be used to find E_x at all internal space nodes at $t_n = (n+1)\delta$ that can be expressed in matrix form as follows:

$$\begin{bmatrix} E_x(1, n+1) \\ E_x(2, n+1) \\ \vdots \\ \vdots \\ E_x(M-1, n+1) \end{bmatrix} = \begin{bmatrix} 2(1-\xi^2) & \xi^2 & 0 & \cdots & 0 \\ \xi^2 & 2(1-\xi^2) & \xi^2 & \ddots & \vdots \\ 0 & \ddots & \ddots & \ddots & 0 \\ \vdots & \ddots & \ddots & \ddots & \xi^2 \\ 0 & \cdots & 0 & \xi^2 & 2(1-\xi^2) \end{bmatrix}$$

$$\times \begin{bmatrix} E_x(1, n) \\ E_x(2, n) \\ \vdots \\ \vdots \\ E_x(M-1, n) \end{bmatrix} - \begin{bmatrix} E_x(1, n-1) \\ E_x(2, n-1) \\ \vdots \\ \vdots \\ E_x(M-1, n-1) \end{bmatrix} \tag{11.1.52}$$

Thus E_x at the kth node in space and at 2δth time step can be found if its values for $t_n = 0$ and δ are known. Initial conditions (11.1.41) and (11.1.42) provide the data. From (11.1.41), we find that

$$E_x(k, 0) = g_1(z_k) = g_1(k\Delta), \qquad k = 1, 2, \ldots, (M-1). \tag{11.1.53}$$

Instead of using (11.1.42) directly, the following procedure may be adopted to minimize the error in $E_x(k, 1)$. From the Taylor series expansion

$$E_x(z, \delta) = E_x(z, 0) + \delta \left.\frac{\partial E_x(z, t)}{\partial t}\right|_{(z,\,0)} + \frac{\delta^2}{2!}\left.\frac{\partial^2 E_x(z, t)}{\partial t^2}\right|_{(z,\,0)} + \cdots,$$

we find that

$$E_x(z, \delta) \approx E_x(z, 0) + \delta \left.\frac{\partial E_x(z, t)}{\partial t}\right|_{(z,\,0)} + \frac{v^2\delta^2}{2\Delta^2}\{E_x(z+\Delta, 0) - 2E_x(z, 0) + E_x(z-\Delta, 0)\},$$

or

$$E_x(k, 1) \approx E_x(k, 0) + g_2(k)\delta + \frac{v^2\delta^2}{2\Delta^2}\{E_x(k+1, 0) - 2E_x(k, 0) + E_x(k-1, 0)\}.$$

Therefore

$$E_x(k, 1) \approx (1 - \xi^2)E_x(k, 0) + \frac{\xi^2}{2}\{E_x(k+1, 0) + E_x(k-1, 0)\} + g_2(k)\delta. \qquad (11.1.54)$$

Thus a computer program can be developed using (11.1.52) through (11.1.54) to find E_x at various nodes in space successively beginning with initial two time steps. It can be proved that this process is stable for $\xi \leq 1$. In order to investigate the conditions under which the finite-difference approximation will provide reasonable accuracy, we consider one set of continuous solutions of the wave equation (11.1.38) as follows:

$$E(z, t) = \cos\left(\omega t - \frac{\omega}{v}z\right). \qquad (11.1.55)$$

The corresponding discrete solution can be written as follows:

$$E(k, n) = \cos\left(n\omega\delta - \frac{\omega}{v}k\Delta\right). \qquad (11.1.56)$$

Using (11.1.55), (11.1.56), and (11.1.38), we find that

$$\frac{\partial^2 E(z, t)}{\partial z^2} = -\left(\frac{\omega}{v}\right)^2 \cos\left(\omega t - \frac{\omega}{v}z\right) \qquad (11.1.57)$$

and

$$\frac{E(k+1, n) - 2E(k, n) + E(k-1, n)}{\Delta^2} = \frac{2}{\Delta^2}\left\{\cos\left(\frac{\omega}{v}\Delta\right) - 1\right\}\cos\left(\omega n\delta - \frac{\omega}{v}k\Delta\right). \qquad (11.1.58)$$

On comparing (11.1.57) and (11.1.58), we conclude that the following condition must hold for the approximation to be valid:

$$\cos\left(\frac{\omega}{v}\Delta\right) - 1 = -\frac{\omega^2\Delta^2}{2v^2}. \qquad (11.1.59)$$

Note that (11.1.59) has the first two terms of the cosine series expansion given as follows:

$$\cos(x) = \sum_{k=0}^{\infty}(-1)^k \frac{x^{2k}}{(2k)!} = 1 - \frac{x^2}{2} + \frac{x^4}{24} - \cdots.$$

Thus for $(\omega\Delta/v)_{\max} = 0.35$ (i.e., $\Delta/\lambda \leq 0.0557$) the error in the approximation will be less than 1%. This means that we should have at least 18 points per wavelength. A similar analysis for the time derivative indicates that with $(\omega\delta)_{\max} = 0.35$, the approximation will be good to 1%, which is automatically satisfied if ξ is less than 1 for stability. This can be verified using (11.1.48) as follows:

$$\omega\delta = \omega\frac{\xi\Delta}{v} = \xi\frac{\omega\Delta}{v}.$$

The following example illustrates the procedure.

Example 11.7

Use the FDTD technique to solve the wave equation

$$\frac{\partial^2 E(z,t)}{\partial t^2} = 4\frac{\partial^2 E(z,t)}{\partial z^2}, \qquad 0 < z < 1,\ t > 0.$$

The solution satisfies the following boundary and initial conditions:

$$E(0,t) = E(1,t) = 0,$$
$$E(z,0) = \sin(\pi z), \qquad 0 \leq z \leq 1,$$

and

$$\left.\frac{\partial E(z,t)}{\partial t}\right|_{(z,0)} = 0, \qquad 0 \leq z \leq 1$$

Compare the results at $t = 0.1$ s for $\xi = 0.5$, 1, and 2.

The exact solution to this wave equation is found to be

$$E(z,t) = \sin(\pi z)\cos(2\pi t).$$

$E\,(z, 0.1)$ was computed using (11.1.52) through (11.1.54) for three different ξ, and the space step $\Delta = 0.1$. The results are compared in Table 11.7.

Note that the numerical results are identical to the corresponding exact values for ξ at 0.5. There are slight differences at 7th decimal place in some cases when ξ

TABLE 11.7 Comparison of the Results of Example 11.7

z	FD Solution $t=0.1$, $\xi = 1$	Exact Solution $t=0.1$	FD Solution $t=0.1$, $\xi = 2$	FD Solution $t=0.1$, $\xi=0.5$
0	0	0	0	0
0.1	0.25	0.25	0.2485196	0.25
0.2	0.4755283	0.4755283	0.4727121	0.4755283
0.3	0.6545085	0.6545085	0.6506327	0.6545085
0.4	0.7694209	0.7694209	0.7648644	0.7694209
0.5	0.8090169	0.809017	0.804226	0.809017
0.6	0.7694209	0.7694209	0.7648646	0.7694209
0.7	0.6545084	0.6545085	0.6506323	0.6545085

is increased to 1. Further deviations are found for ξ at 2. The space step is 0.1 in all computations, and therefore the time step δ increases with ξ.

Maxwell's Equations in FDTD

Two curl equations of the set can be expanded as follows:

$$\nabla \times \vec{E} = -\frac{\partial \vec{B}}{\partial t} \rightarrow \begin{cases} \dfrac{\partial B_x}{\partial t} = \dfrac{\partial E_y}{\partial z} - \dfrac{\partial E_z}{\partial y} \\ \dfrac{\partial B_y}{\partial t} = \dfrac{\partial E_z}{\partial x} - \dfrac{\partial E_x}{\partial z} \\ \dfrac{\partial B_z}{\partial t} = \dfrac{\partial E_x}{\partial y} - \dfrac{\partial E_y}{\partial x} \end{cases} \tag{11.1.60}$$

and

$$\nabla \times \vec{H} = \vec{J} + \frac{\partial \vec{D}}{\partial t} \rightarrow \begin{cases} \dfrac{\partial D_x}{\partial t} = \dfrac{\partial H_z}{\partial y} - \dfrac{\partial H_y}{\partial z} - J_x \\ \dfrac{\partial D_y}{\partial t} = \dfrac{\partial H_x}{\partial z} - \dfrac{\partial H_z}{\partial x} - J_y \\ \dfrac{\partial D_z}{\partial t} = \dfrac{\partial H_y}{\partial x} - \dfrac{\partial H_x}{\partial y} - J_z. \end{cases} \tag{11.1.61}$$

For simplicity, assume that only the E_x and H_y field components are present (i.e., E_y, E_z, H_x, and H_z are zero) in a lossless, isotropic, linear, and source-free medium. Further these field components vary only with space variable z and time t. Therefore (11.1.60) and (11.1.61) reduce to

$$\frac{\partial H_y}{\partial t} = -\frac{1}{\mu}\frac{\partial E_x}{\partial z} \tag{11.1.62}$$

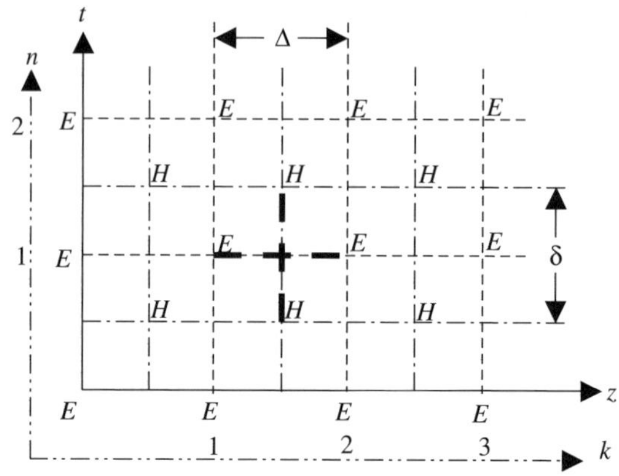

Figure 11.12 Nodes for Maxwell's equation in one spatial dimension and instant in time.

and

$$\frac{\partial E_x}{\partial t} = -\frac{1}{\varepsilon}\frac{\partial H_y}{\partial z}. \tag{11.1.63}$$

Note that if (11.1.62) is differentiated with respect to the space variable z one more time and then (11.1.63) is substituted, the wave Equation (11.1.38) results. Thus we find that these two first-order partial differential equations replace one second-order partial differential equation. Further (11.1.62) and (11.1.63) are considered to define E_x and H_y at the same space and time point (z, t). However, this is not the case with the finite-difference equations. As illustrated in Figure 11.12, when the derivative of E_x is taken with respect to z for the right-hand side of (11.1.62), H_y is considered to be located between those two space points. For the left-hand side, the derivative of H_y with respect to time should be evaluated about that instant. Therefore (11.1.62) transforms to

$$\frac{H_{k+0.5,\,n+0.5} - H_{k+0.5,\,n-0.5}}{\delta} = -\frac{1}{\mu}\frac{E_{k+1,\,n} - E_{k,\,n}}{\Delta}. \tag{11.1.64}$$

Following similar considerations, (11.1.63) gives

$$\frac{E_{k,\,n+1} - E_{k,\,n}}{\delta} = -\frac{1}{\varepsilon}\frac{H_{k+0.5,\,n+0.5} - H_{k-0.5,\,n+0.5}}{\Delta}. \tag{11.1.65}$$

Equation (11.1.64) corresponds to the continuous Equation (11.1.62) at $(k+0.5, n)$, whereas (11.1.65) corresponds to (11.1.63) at $(k, n+0.5)$. The initial conditions can be imposed as values of E for $n = 0$ and as values of H for $n = 0.5$.

The preceding concept is extended for the three-dimensional case using a unit cell as shown is Figure 11.13. This is called the Yee's cell, named after the

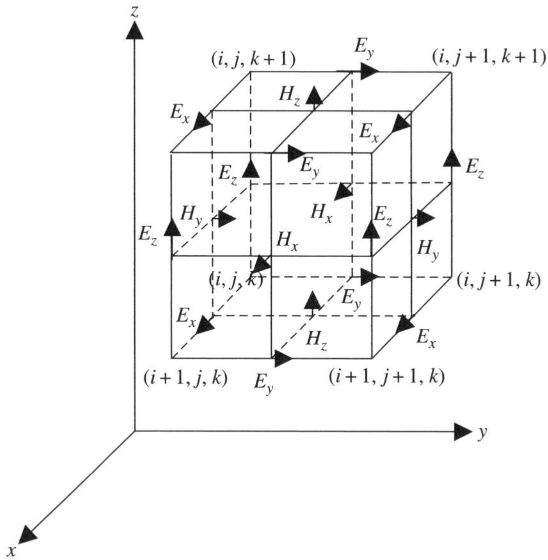

Figure 11.13 Finite-difference scheme (Yee's cell) for Maxwell's equations in 3D.

researcher who first proposed it.* The finite-difference equations for (11.1.60) and (11.1.61) can be found via this method. The first of each of the equations transform to

$$\frac{B^{n+0.5}_{x(i,j+0.5,\,k+0.5)} - B^{n-0.5}_{x(i,j+0.5,\,k+0.5)}}{\delta} = \frac{E^{n}_{y(i,j+0.5,\,k+1)} - E^{n}_{y(i,j+0.5,\,k)}}{\Delta} - \frac{E^{n}_{z(i,\,j+1,\,k+0.5)} - E^{n}_{z(i,\,j,\,k+0.5)}}{\Delta} \tag{11.1.66}$$

and

$$\frac{D^{n}_{x(i+0.5,j,\,k)} - D^{n-1}_{x(i+0.5,j,\,k)}}{\delta} = \frac{H^{n-0.5}_{z(i+0.5,j+0.5,\,k)} - H^{n-1}_{z(i+0.5,j-0.5,\,k)}}{\Delta} - \frac{H^{n-0.5}_{y(i+0.5,j,\,k+0.5)} - H^{n-0.5}_{y(i+0.5,j,\,k-0.5)}}{\Delta} - J^{n-0.5}_{x(i+0.5,j,\,k)}. \tag{11.1.67}$$

Similar equations can be found for other equations as well. The interested reader can find details in the reference books listed in the section Selected Reference Books.

*K. S. Yee, "Numerical Solution of Initial Boundary Value Problems Involving Maxwell's Equations in Isotropic Media," *IEEE Trans. Antennas Prop.*: **14**, 302–307, May 1966.

11.2 METHOD OF MOMENTS

This section begins with a brief introduction to the linear operator notation because it provides a most convenient way to describe the method of moments. The concepts of linear function space are summarized along with the inner product suitable for the application. A large class of linear operator equations can be solved by this method. It is especially useful in solving integral equations that arise in the areas of electromagnetic radiation and scattering. After classifying the integral equations, the solution procedure is summarized.

Linear Operator Notation

Linear differential and integral operator equations can be expressed using the linear operator $\mathscr{L}$ as follows:

$$\mathscr{L}\{f(x)\} = g(x), \qquad a \leq x \leq b. \tag{11.2.1}$$

By definition,

$$\mathscr{L}\{\alpha\, f(x) + \beta\, h(x)\} = \alpha\, \mathscr{L}\{f(x)\} + \beta\, \mathscr{L}\{h(x)\}. \tag{11.2.2}$$

The method of moments is most conveniently described in the notation of linear function spaces with an appropriately defined inner product. In the present simplistic specialization it is assumed that

- Elements $\{f_n\}$ of the function space $\mathscr{L}$ are defined with the domain spanning the interval $a \leq x \leq b$.
- Elements of $\mathscr{L}$ are continuous, differentiable, and integrable such that they belong to the domain or the range of the linear operator.
- The linear inner product $\langle f, g\rangle$ between the elements of $\mathscr{L}$ is defined by the properties

$$\langle f, g\rangle = \langle g, f\rangle,$$
$$\langle \alpha f + \beta g, h\rangle = \alpha\langle f, h\rangle + \beta\langle g, h\rangle,$$

and

$$\langle f, f^*\rangle = \begin{cases} > 0, & \text{for } f \neq 0, \\ 0, & \text{if } f = 0. \end{cases}$$

A suitable inner product for the space of integrable functions is frequently

$$\langle f, g\rangle = \int_a^b w(x) f(x) g(x) dx, \tag{11.2.3}$$

where $w(x)$ is a weighting function, often equal to 1.

Classification of Integral Equations

An integral equation (IE) in the domain $a \le x \le b$ can be expressed as follows:

$$\int_a^b f(x')\mathrm{K}(x|x')\,dx' = \gamma f(x) + g(x), \tag{11.2.4}$$

where $\mathrm{K}(x|x')$ is called the kernel of the integral equation, $g(x)$ is the driving source or the excitation function, γ is a constant parameter, and $f(x')$ is the unknown response.

Integral equations are classified as follows:

- If one of the integration limits in (11.2.4) is the variable x, then it is called the Volterra integral equation; otherwise, it is known as the Fredholm integral equation.
- If γ in (11.2.4) is zero, then it is an integral equation of the first kind; otherwise, if γ is not zero, it is an integral equation of the second kind.
- For a zero $g(x)$ in (11.2.4), the integral equation is homogeneous; it is inhomogeneous (forced) if $g(x)$ is not zero.

The linear integral operator equation (11.2.4) can be expressed in inner product notation as follows:

$$\langle f(x'), \mathrm{K}(x \mid x')\rangle = \gamma f(x) + g(x), \tag{11.2.5}$$

with f and K in the domain, and f and g in the range of $\mathscr{L}$.

The procedure used to solve a linear operator equation (11.2.1) via the method of moments can be summarized as follows: The unknown function $f(x)$ is expanded using an appropriate set of expansion functions (or basis functions):

$$f(x) = \sum_{n=1}^{N} \alpha_n f_n(x), \tag{11.2.6}$$

where $\{f_n(x)\}$ is a complete set of appropriate functions if $N = \infty$, or an appropriate set to approximate $f(x)$ if N is finite.

On substituting (11.2.6) into (11.2.1), we find that

$$\mathscr{L}\{f(x)\} = \mathscr{L}\left\{\sum_{n=1}^{N} \alpha_n f_n(x)\right\} = \sum_{n=1}^{N} \alpha_n \mathscr{L}\{f_n(x)\} = g(x). \tag{11.27}$$

Note that (11.2.7) involves N unknown expansion coefficients α_n, and it should be satisfied continuously for all x in the range. The latter requirement can be satisfied only for $N \to \infty$. If N is finite for an approximate solution, the operator equation must be discretized to allow the evaluation of α_n. For that, $\{w_m(x)\}$ is selected as an appropriate set of weighting or testing functions in the range of $\mathscr{L}$. On discretizing, the original operator equation transforms to

$$\left\langle w_m, \sum_{n=1}^{N} \alpha_n \mathscr{L}\{f_n\} \right\rangle = \sum_{n=1}^{N} \alpha_n \langle w_m, \mathscr{L}\{f_n\}\rangle = \langle w_m, g\rangle, \qquad m = 1, 2, \ldots, N, \tag{11.2.8}$$

or

$$\sum_{n=1}^{N} l_{mn}\alpha_n = g_m, \qquad m = 1, 2, \ldots, N, \tag{11.2.9}$$

where

$$l_{mn} = \langle w_m, L\{f_n\}\rangle \tag{11.2.10}$$

and

$$g_m = \langle w_m, g\rangle. \tag{11.2.11}$$

Special Cases

1. If a weighting function w_n and an expansion function f_n are the same, the general moment method is known as Galerkin's method.
2. If $\{f_n\}$ are eigenfunctions of $\mathscr{L}$ such that $L\{f_n\} = \lambda_n f_n$ and $w_n = f_n$, then

$$\sum_{n=1}^{N} \alpha_n \lambda_n \langle f_m, f_n\rangle = \langle f_m, g\rangle.$$

If $\{f_n\}$ are ortho-normal with $\langle f_m, f_n\rangle = \delta_{mn}$ the Kronecker delta, such that

$$\delta_{mn} = \begin{cases} 1, & \text{for } n = m, \\ 0, & \text{for } n \neq m, \end{cases}$$

then

$$\alpha_m = \frac{\langle f_m, g\rangle}{\lambda_m}. \tag{11.2.12}$$

If $\{f_n\}$ are complete, an exact solution of (11.2.1) is obtained as $N \to \infty$.

3. If the Dirac delta functions (see Appendix B) are selected as the weighting functions such that $w_m = \delta(x - x_m)$, then discretization of the continuous operator equation corresponds to point-matching it at N discrete points $x = x_m$, that is,

$$l_{mn} = \langle \delta(x - x_m), \mathscr{L}\{f_n\} \rangle = \mathscr{L}\{f_n\}\Big|_{x=x_m} \tag{11.2.13}$$

and

$$g_m = \langle \delta(x - x_m), g \rangle = g(x_m). \tag{11.2.14}$$

Example 11.8

A 1 m long conducting wire is held at an electrical potential of 1 V. Its radius is 1 mm. Use the method of moments to find electrical charge distribution on it.

Consider the geometry of a thin straight wire as shown in Figure 11.14. The wire's radius a is assumed to be very small in comparison with length h ($a \ll h$). Since it is desired to find the charge distribution when it is at 1 V potential, we need an integral equation to relate the electrical potential with the charge. We find a relation in Appendix B (equation B.35) that can be specialized for this case:

$$\phi_e(\vec{r}) = \frac{1}{4\pi\varepsilon}\int_v \rho_{ev}\frac{e^{-jkR}}{R}dv' \rightarrow \phi_e(\rho, z) = \frac{1}{4\pi\varepsilon}\int_s \frac{\rho_{es}}{R}ds',$$

or

$$\int_s \frac{\rho_{es}(a, z')}{R}ds' = 4\pi\varepsilon_o\phi_e(\rho, z), \tag{i}$$

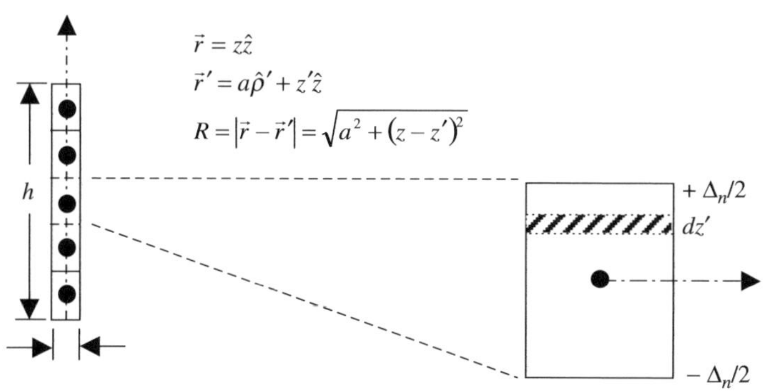

Figure 11.14 Geometry of a wire.

where

$$R = \sqrt{a^2 + (z - z')^2}. \tag{ii}$$

We find that it is a Fredholm integral equation of the first kind that can be solved using the method of moments as follows: The length of the wire is divided into N partitions with Δ_n as the interval along z spanning the nth partition. Further we can assume that the charge density ρ_{es} changes only from one to the other partition (i.e., it is uniform on a given partition). In other words, the unknown surface charge density can be expressed in terms of the pulse expansion functions as follows:

$$\rho_{es}(a, z') = \sum_{n=1}^{N} a_n, p_n(z'), \tag{iii}$$

where

$$p_n(z') = \begin{cases} 1, & z' \text{ on } \Delta_n, \\ 0, & \text{otherwise,} \end{cases} \tag{iv}$$

and

$$\Delta = \frac{h}{N}. \tag{v}$$

Therefore the integral equation (i) transforms to

$$\sum_{n=1}^{N} a_n \int_{\Delta_n} \frac{ds'}{R} = 4\pi\varepsilon_o \phi_e(\rho, z), \tag{vi}$$

where

$$\int_{\Delta_n} \frac{ds'}{R} = \int_{\phi'=0}^{2\pi} \int_{z'=z_n-\Delta/2}^{z_n+\Delta/2} \frac{a\, d\phi' dz'}{\sqrt{a^2 + (z - z')^2}} = 2\pi a \int_{z_n-\Delta/2}^{z_n+\Delta/2} \frac{dz'}{\sqrt{a^2 + (z - z')^2}}.$$

The remaining integral can be evaluated, with the help of integration formula 13 of Appendix A, as follows:

$$\int_{\Delta_n} \frac{ds'}{\sqrt{a^2 + (z - z')^2}} = 2\pi a \ln \left[\frac{z - z_n + (\Delta/2) + \sqrt{a^2 + \{z - z_n + (\Delta/2)\}^2}}{z - z_n - (\Delta/2) + \sqrt{a^2 + \{z - z_n - (\Delta/2)\}^2}} \right]. \tag{vii}$$

We now have (vi) with unknown coefficients a_n, $n = 1, 2, \ldots, N$. Since ρ_{es} is a continuous function of z, we enforce this relation on each partition to find N algebraic equations. This process is known as point-matching. Mathematically we take the inner product of (vi) with the delta function as follows:

$$\left\langle \delta(z - z_m), \sum_{n=1}^{N} a_n \int_{\Delta_n} \frac{ds'}{R} \right\rangle = \sum_{n=1}^{N} \left\langle \delta(z - z_m), a_n \int_{\Delta_n} \frac{ds'}{R} \right\rangle$$
$$= \left\langle \delta(z - z_m), 4\pi\varepsilon_o \phi_e(\rho, z) \right\rangle, \qquad m = 1, 2, \ldots, N,$$

or

$$\sum_{n=1}^{N} l_{mn} a_n = g_m, \qquad m = 1, 2, \ldots, N, \tag{viii}$$

where

$$l_{mn} = \left\langle \delta(z - z_m), a_n \int_{\Delta_n} \frac{ds'}{R} \right\rangle \tag{ix}$$

and

$$g_m = \langle \delta(z - z_m), 4\pi\varepsilon_o \phi_e(\rho, z) \rangle = 4\pi\varepsilon_o \phi_e(\rho, z_m) = 4\pi\varepsilon_o. \tag{x}$$

From (vii) and (ix) we find that

$$l_{mn} = 2\pi a \ln\left[\frac{z_m - z_n + (\Delta/2) + \sqrt{a^2 + \{z_m - z_n + (\Delta/2)\}^2}}{z_m - z_n - (\Delta/2) + \sqrt{a^2 + \{z_m - z_n - (\Delta/2)\}^2}}\right]. \tag{xi}$$

For $m = n$, (xi) reduces to

$$l_{mm} \approx 4\pi a \ln(\Delta/a). \tag{xii}$$

If $m \neq n$, then (xi) can be approximated to

$$l_{mn} = 2\pi a \ln\left[\frac{|z_m - z_n| + \Delta/2}{|z_m - z_n| - \Delta/2}\right]. \tag{xiii}$$

Since $z_m = (2m - 1)\Delta/2$, $m = 1, 2, \ldots, N$, and $z_n = (2n - 1)\Delta/2$, $n = 1, 2, \ldots, N$, we find that $|z_m - z_n| = |m - n|\Delta$. Therefore (xiii) reduces to

$$l_{mn} = 2\pi a \ln\left[\frac{|m - n|\Delta + \Delta/2}{|m - n|\Delta - \Delta/2}\right]. \tag{xiv}$$

Thus the length Δ of the partition is found from (v) for a given N. l_{mn} for (viii) is found from (xii) if $m = n$; otherwise, (xiv) is used for it. g_m is evaluated from (x). This facilitates determination of a_n, the distribution of charge.

Following this procedure for the given problem, we find that $\Delta = 1/3$ m for $N = 3$, and therefore

$$\begin{bmatrix} a_1 \\ a_2 \\ a_3 \end{bmatrix} = \begin{bmatrix} 2 \cdot \ln\left(\frac{1000}{3}\right) & \ln(3) & \ln\left(\frac{5}{3}\right) \\ \ln(3) & 2 \cdot \ln\left(\frac{1000}{3}\right) & \ln(3) \\ \ln\left(\frac{5}{3}\right) & \ln(3) & 2 \cdot \ln\left(\frac{1000}{3}\right) \end{bmatrix}^{-1} \begin{bmatrix} 2000\varepsilon_o \\ 2000\varepsilon_o \\ 2000\varepsilon_o \end{bmatrix}$$

$$= \begin{bmatrix} 151.9025\varepsilon_o \\ 143.4150\varepsilon_o \\ 151.9025\varepsilon_o \end{bmatrix}.$$

As expected, the charge distribution on the wire is symmetrical about the middle partition. Recall that a_n is the amplitude of the pulse and represents the charge density on the nth partition. This can be exploited to minimize the use of computing resources. For example, if $N = 5$ is desired, then $a_1 = a_5$ and $a_2 = a_4$, and only three coefficients are unknown. Since $\Delta = 1/5$ in this case, using (viii), we find that

$$\begin{bmatrix} a_1 \\ a_2 \\ a_3 \end{bmatrix} = \begin{bmatrix} 2 \cdot \ln(200) + \ln\left(\frac{9}{7}\right) & \ln(3) + \ln\left(\frac{7}{5}\right) & \ln\left(\frac{5}{3}\right) \\ \ln(3) + \ln\left(\frac{7}{5}\right) & 2 \cdot \ln(200) + \ln\left(\frac{5}{3}\right) & \ln(3) \\ 2 \cdot \ln\left(\frac{5}{3}\right) & 2 \cdot \ln(3) & 2 \cdot \ln\left(\frac{1000}{3}\right) \end{bmatrix}^{-1}$$

$$\times \begin{bmatrix} 2000\varepsilon_o \\ 2000\varepsilon_o \\ 2000\varepsilon_o \end{bmatrix} = \begin{bmatrix} 158.3800\varepsilon_o \\ 145.4214\varepsilon_o \\ 143.3160\varepsilon_o \end{bmatrix}.$$

Similarly $\Delta = 1/9$ for $N = 9$, and following the procedure, the coefficients are found to be

$$a_1 = a_9 = 167.6894\,\varepsilon_o,$$

$$a_2 = a_8 = 150.6483\,\varepsilon_o,$$

$$a_3 = a_7 = 145.8831\,\varepsilon_o,$$

$$a_4 = a_6 = 143.8210\,\varepsilon_o,$$

and

$$a_5 = 143.2204\,\varepsilon_o.$$

Thus, for even shorter Δ, a program can easily be developed to find the charge distribution using a computer.

11.3 SCATTERING OF PLANAR EM WAVES FROM AN INFINITELY LONG CYLINDER

As was mentioned in the preceding section, the method of moments is commonly employed to analyze radiation and scattering problems. In this section we consider the scattering of an incident electromagnetic plane wave on an infinitely long cylinder of arbitrary cross-sectional shape. We begin with the formulation of an electric field integral equation (EFIE) that can be solved via the method of moments.

When an infinitely long cylinder is excited in such a manner that its induced current and charge are independent of the location along its axis, the problem becomes two dimensional. All field quantities are independent of the axial coordinate in such cases. The appropriate geometry of the problem is shown in Figure 11.15. From the notations shown in this figure, the position vectors for the field and source points are

$$\vec{r} = \vec{\rho} + z\hat{z} \tag{11.3.1}$$

and

$$\vec{r}' = \vec{\rho}' + z'\hat{z}. \tag{11.3.2}$$

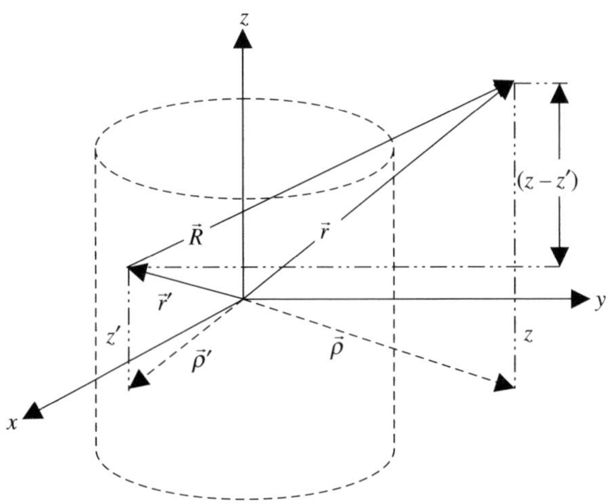

Figure 11.15 Geometry of the problem.

Therefore the distance R from source point to field point is

$$R = |\vec{r} - \vec{r}'| = |(\vec{\rho} - \vec{\rho}') + (z - z')\hat{z}| = \sqrt{|\vec{\rho} - \vec{\rho}'|^2 + (z - z')^2} \tag{11.3.3}$$

Since all the fields are independent of axial coordinate z, $\partial/\partial z = 0$, and therefore the differential operator ∇ reduces to transverse del, ∇_t, as follows:

$$\nabla = \nabla_t + \hat{z}\frac{\partial}{\partial z} \rightarrow \nabla_t$$

The electric and magnetic field expressions, (6.2.9) and (6.2.8), respectively, reduce to

$$\vec{E}(\vec{\rho}) = -\nabla_t \phi_e(\vec{\rho}) - j\omega\vec{A}(\vec{\rho}) \tag{11.3.4}$$

and

$$\vec{H}(\vec{\rho}) = \frac{1}{\mu}\nabla_t \times \vec{A}(\vec{\rho}). \tag{11.3.5}$$

Next the general expressions (B.33) and (B.35) found for the magnetic vector and electric scalar potentials in free space (in Appendix B) can be simplified:

$$\vec{A}(\vec{\rho}) = \frac{\mu}{4\pi}\int_V \vec{J}(\vec{\rho}')\frac{e^{-jkR}}{R}\, dv' \tag{B.33}$$

and

$$\phi_e(\vec{\rho}) = \frac{1}{4\pi\varepsilon_o}\int_V \rho(\vec{\rho}')\frac{e^{-jkR}}{R} dv' \quad . \tag{B.35}$$

Since the z' dependence of integrands in (B.33) and (B.35) is known, the integrals along z' can be carried out leaving only 2D potential integrals as follows:

$$\vec{A}(\vec{\rho}) = \frac{\mu}{4\pi}\int_{\text{c.s.}} ds'\vec{J}(\vec{\rho}')\int_{-\infty}^{\infty} \frac{e^{-jk\sqrt{|\vec{\rho}-\vec{\rho}'|+(z-z')}}}{\sqrt{|\vec{\rho} - \vec{\rho}'|^2 + (z - z')}}\, dz' \tag{11.3.6}$$

and

$$\phi_e(\vec{\rho}) = \frac{1}{4\pi\,\varepsilon}\int_{\text{c.s.}} ds'\rho(\vec{\rho}')\int_{-\infty}^{\infty} \frac{e^{-jk\sqrt{|\vec{\rho}-\vec{\rho}'|+(z-z')}}}{\sqrt{|\vec{\rho} - \vec{\rho}'|^2 + (z - z')}}\, dz' \tag{11.3.7}$$

Hence we find that the following integral is common in both relations. It can be evaluated as

$$I(\vec{\rho}-\vec{\rho}') = \int_{-\infty}^{\infty} \frac{e^{-jk\sqrt{|\vec{\rho}-\vec{\rho}'|+(z-z')}}}{\sqrt{|\vec{\rho}-\vec{\rho}'|^2+(z-z')}}\,dz'. \tag{11.3.8}$$

If we change the variable, then its integration limits are affected as follows: For $(z-z')=u$,

$$dz' = -du, \quad z' = -\infty \rightarrow u = \infty, \quad \text{and} \quad z' = \infty \rightarrow u = -\infty.$$

Therefore (11.3.8) can be written as follows:

$$I(\vec{\rho}-\vec{\rho}') = \int_{-\infty}^{\infty} \frac{e^{-jk\sqrt{|\vec{\rho}-\vec{\rho}'|^2+u^2}}}{\sqrt{|\vec{\rho}-\vec{\rho}'|^2+u^2}}\,du = 2\int_{0}^{\infty} \frac{e^{-jk\sqrt{|\vec{\rho}-\vec{\rho}'|^2+u^2}}}{\sqrt{|\vec{\rho}-\vec{\rho}'|^2+u^2}}\,du \tag{11.3.9}$$

This integral can be evaluated using scientific software with a symbolic engine. Alternatively, one can use a mathematical handbook, such as Gradshteyn and Ryzhik's *Tables of Integrals*.* For example,

$$\int_{0}^{\infty} \frac{e^{-\beta\sqrt{\xi^2+u^2}}}{\sqrt{\xi^2+u^2}} \cos(\alpha u)du = K_0\left(\xi\sqrt{\alpha^2+\beta^2}\right), \tag{11.3.10}$$

where $K_0(z)$ is the modified Bessel function of the second kind with order zero and generally complex argument z.

For $\alpha = 0$, $\beta = jk$, and $\xi = |\vec{\rho}-\vec{\rho}'|$ we get $\xi\sqrt{\alpha^2+\beta^2} = jk|\vec{\rho}-\vec{\rho}'|$. Using formula (C.36) from Appendix C, we find that

$$I(\vec{\rho}-\vec{\rho}') = -j\pi H_0^{(2)}(k|\vec{\rho}-\rho'|). \tag{11.3.11}$$

Therefore (11.3.6) and (11.3.7) simplify to

$$\vec{A}(\vec{\rho}) = -j\frac{\mu}{4}\int_{\text{c.s.}} \vec{J}(\vec{\rho}')H_0^{(2)}(k|\vec{\rho}-\vec{\rho}'|)ds' \tag{11.3.12}$$

and

$$\phi_e(\vec{\rho}) = -\frac{j}{4\varepsilon}\int_{\text{c.s.}} \rho(\vec{\rho}')H_0^{(2)}(k|\vec{\rho}-\vec{\rho}'|)ds'. \tag{11.3.13}$$

*See T. S. Gradshteyn and I. M. Ryzhik, *Tables of Integrals, Series, and Products*, Academic Press, 1980, p. 498, entry 3.961-2.

From (11.3.12) and (11.3.13), Green's function $G(\vec{\rho}|\vec{\rho}')$ for a 2D Helmholtz equation can be identified as

$$G(\vec{\rho}|\vec{\rho}') = -\frac{j}{4}H_0^{(2)}(k|\vec{\rho}-\vec{\rho}'|). \tag{11.3.14}$$

Therefore the potential functions can be expressed in terms of the Green function (see Appendix B) as

$$\vec{A}(\vec{\rho}) = \int_{\text{c.s.}} \mu\vec{J}(\vec{\rho}')G(\vec{\rho}|\vec{\rho}')ds'$$

and

$$\phi_{\text{e}}(\vec{\rho}) = \int_{\text{c.s.}} \frac{\rho(\vec{\rho}')}{\varepsilon}G(\vec{\rho}|\vec{\rho}')ds'.$$

On substituting (11.3.12) and (11.3.13) into (11.3.4), we have

$$\vec{E}(\vec{\rho}) = \frac{j}{4}\int_{\text{c.s.}}\left[\frac{\rho(\vec{\rho}')}{\varepsilon}\nabla_{\text{t}}\{H_0^{(2)}(k|\vec{\rho}-\vec{\rho}'|)\} + j\omega\mu\vec{J}(\vec{\rho}')H_0^{(2)}(k\mid\vec{\rho}-\vec{\rho}'|)\right]ds'. \tag{11.3.15}$$

From the continuity equation we have

$$\nabla_{\text{t}}'\cdot\vec{J}(\vec{\rho}') = -j\omega\rho(\vec{\rho}') \rightarrow \rho(\vec{\rho}') = \frac{j}{\omega}\nabla_{\text{t}}'\cdot\vec{J}(\vec{\rho}').$$

Therefore (11.3.15) can be expressed as follows:

$$\vec{E}(\vec{\rho}) = -\frac{1}{4\omega\varepsilon}\int_{\text{c.s.}}(\{\nabla_{\text{t}}'\cdot\vec{J}(\vec{\rho}')\}\nabla_{\text{t}} + k^2\vec{J}(\vec{\rho}'))H_0^{(2)}(k|\vec{\rho}-\vec{\rho}'|)ds'. \tag{11.3.16}$$

Similarly from (11.3.5) and (11.3.12) we find that

$$\vec{H}(\vec{\rho}) = \frac{1}{\mu}\nabla_{\text{t}}\times\vec{A}(\vec{\rho}) = -\frac{j}{4}\int_{\text{c.s.}}\nabla_{\text{t}}\times\{\vec{J}(\vec{\rho}')H_0^{(2)}(k|\vec{\rho}-\vec{\rho}'|)\}ds',$$

or

$$\vec{H}(\vec{\rho}) = -\frac{j}{4}\int_{\text{c.s.}}[H_0^{(2)}(k|\vec{\rho}-\vec{\rho}'|)\nabla_{\text{t}}\times\vec{J}(\vec{\rho}') + \nabla_{\text{t}}H_0^{(2)}(k|\vec{\rho}-\vec{\rho}'|)\times\vec{J}(\vec{\rho}')]ds'.$$

The first term of the integrand is zero because the curl operation requires derivatives with respect to the unprimed coordinates, whereas the current density is a function of only the primed coordinates. Therefore

$$\vec{H}(\vec{\rho}) = \frac{j}{4}\int_{\text{c.s.}} \vec{J}(\vec{\rho}\,') \times \nabla_t H_0^{(2)}(k|\vec{\rho} - \vec{\rho}\,'|)ds'. \tag{11.3.17}$$

Thus (11.3.16) and (11.3.17) can be used to find the fields produced by the current density $\vec{J}(\vec{\rho}\,')$.

Formulation of the Integral Equation for 2D Scattering from Cylinders of Arbitrary Cross-sectional Shape

The preceding analysis is used for formulating an integral equation to study the scattering of electromagnetic fields from perfectly conducting, infinitely long cylinders of arbitrary cross-sectional shape. The moments method is then used to solve the integral equation for the unknown surface current induced on the cylinder. Electromagnetic fields anywhere in space can be easily found once the surface current distribution is found. Here we consider the case where the electric field of an incident uniform plane wave is along the axis of the cylinder (i.e., the TM^z wave). A similar procedure can be used for the case where the magnetic field of this wave is along the axis of the cylinder (i.e., the TE^z wave).

Consider an axially polarized incident (impressed) electric field $\vec{E}^i = \hat{z}E_z^i(\vec{\rho})$ that illuminates an infinite cylinder of arbitrary cross-sectional shape, as illustrated in Figure 11.16. The induced currents excited on the cylinder maintain a scattered field $\vec{E}^s(\vec{\rho})$, and therefore the total electric field outside the cylinder is given as follows:

$$\vec{E}(\vec{\rho}) = \vec{E}^i(\vec{\rho}) + \vec{E}^s(\vec{\rho}). \tag{11.3.18}$$

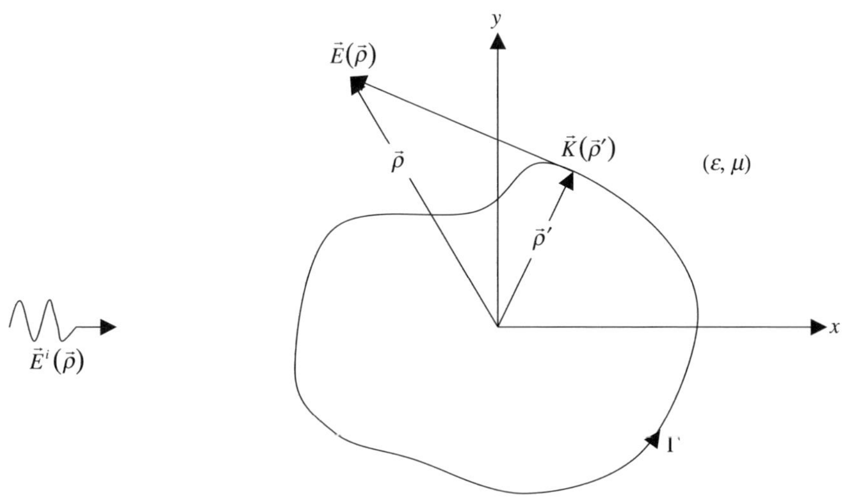

Figure 11.16 Cross section of an infinitely long cylinder illuminated by the TM^z wave.

The boundary conditions on the conducting cylinder require that for all $\vec{\rho}$ on Γ,

$$\hat{z} \cdot \vec{E}(\vec{\rho}) = \hat{z} \cdot \{\vec{E}^i(\vec{\rho}) + \vec{E}^s(\vec{\rho})\} = 0.$$

Therefore

$$\mathrm{E}_z^s(\vec{\rho}) = -E_z^i(\vec{\rho})\big|_{\vec{\rho}\,\mathrm{on}\,\Gamma}. \tag{11.3.19}$$

Since

$$\frac{1}{\omega\varepsilon} = \frac{\sqrt{\mu}}{\omega\sqrt{\mu\varepsilon}} = \frac{1}{\omega\sqrt{\mu\varepsilon}}\sqrt{\frac{\mu}{\varepsilon}} = \frac{\zeta}{k},$$

where ζ is the intrinsic impedance of the medium, and

$$\vec{J}(\vec{\rho}\,') = \vec{K}(\vec{\rho}\,')\delta(\vec{\rho}\,' - \vec{\rho}_\Gamma)$$

as the volume density of surface current on the perfectly conducting cylinder, (11.3.16) gives

$$\vec{E}^s(\vec{\rho}) = -\frac{\zeta}{4k}\oint_\Gamma \left[\nabla_t' \cdot \vec{K}(\vec{\rho}\,')\nabla_t + k^2\vec{K}(\vec{\rho}\,')\right] H_0^{(2)}(k|\vec{\rho} - \vec{\rho}\,'|)\, d\ell'. \tag{11.3.20}$$

Note from (11.3.19) that only the z component of the scattered field $\vec{E}^s$ is required to satisfy the boundary condition. It is expected that only the z component of the induced surface current K will be excited, and it becomes clear that $\vec{K} = \hat{z}K_z$ is adequate to maintain the required $\vec{E}_z^s$. Therefore

$$E_z^s(\vec{\rho}) = \hat{z} \cdot \vec{E}^s(\vec{\rho}) = -\frac{\zeta}{4k}\oint_\Gamma [\{\nabla_t' \cdot \vec{K}(\vec{\rho}\,')\}\hat{z} \cdot \nabla_t + k^2 K_z(\vec{\rho}\,')] H_0^{(2)}(k|\vec{\rho} - \vec{\rho}\,'|)\, d\ell',$$

or

$$E_z^s(\vec{\rho}) = -\frac{\zeta k}{4}\oint_\Gamma K_z(\vec{\rho}\,') H_0^{(2)}(k|\vec{\rho} - \vec{\rho}\,'|)\, d\ell'. \tag{11.3.21}$$

Hence it is concluded that the induced surface current capable of satisfying the boundary condition is

$$\vec{K}(\vec{\rho}\,') = \hat{z}K_z(\vec{\rho}\,').$$

Furthermore, since $\nabla_t' \cdot \vec{K}(\vec{\rho}\,') = 0$ in this case, no transverse components of $\vec{E}^s$ are excited.

Integral Equation for the Surface Current on the Cylinder

Using the boundary condition (11.3.19) and the expression (11.3.21) for a scattered field, we find that the following relation must hold for all $\vec{\rho}$ on the boundary Γ:

$$\frac{\zeta k}{4}\oint_{\Gamma} K_z(\vec{\rho}\,')H_0^{(2)}(k|\vec{\rho}-\vec{\rho}\,'|)\,d\ell' = E_z^i(\vec{\rho}), \tag{11.3.22}$$

for $\vec{\rho}$ on the boundary Γ. In (11.3.22) the surface current K_z is unknown and can be found after solving the equation. A comparison of this equation with (11.2.4) indicates that it is a Fredholm integral equation of the first kind. It can be solved via the method of moments as follows.

As indicated in Figure 11.17*a*, the boundary contour Γ is partitioned into N total number of partitions with $\Delta\Gamma_n$ as the interval along Γ spanning the nth partition. The unknown surface current is expressed in terms of a pulse expansion function as follows:

$$K_z(\vec{\rho}\,') = \sum_{n=1}^{N} a_n p_n(\vec{\rho}\,'), \tag{11.3.23}$$

where

$$p_n(\vec{\rho}\,') = \begin{cases} 1, & \vec{\rho}\,' \text{ on } \Delta\Gamma_n, \\ 0, & \text{otherwise}. \end{cases} \tag{11.3.24}$$

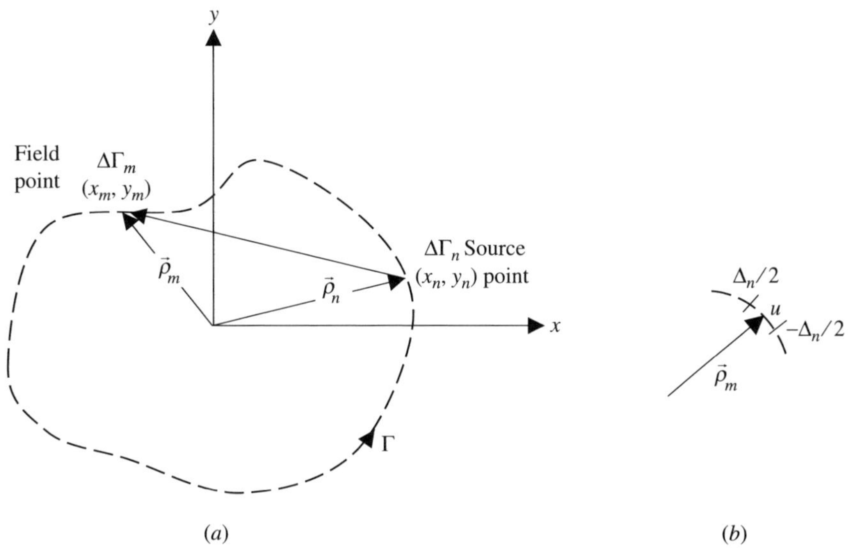

Figure 11.17 Geometry of the partitioned boundary with source and field points (*a*), and a special case where the source and field points are on the same partition (*b*).

Therefore (11.3.22) transforms to

$$\frac{\zeta k}{4}\sum_{n=1}^{N} a_n \int_{\Delta\Gamma_n} H_0^{(2)}(k|\vec{\rho}-\vec{\rho}\,'|)\, d\ell' = E_z^i(\vec{\rho}) \tag{11.3.25}$$

for $\vec{\rho}$ on the boundary Γ. There are N unknown coefficients a_n in (11.3.25). To evaluate these coefficients N independent equations are required. To generate a system of N algebraic equations for these N unknown amplitude coefficients a_n, $n = 1, 2, 3, \ldots, N$, (11.3.25) is point-matched at N discrete points $\vec{\rho}_m$, $m = 1, 2, 3, \ldots, N$. Mathematically we are using the delta function here as the test function. Therefore (11.3.25) transforms to

$$\sum_{n=1}^{N} A_{mn} a_n = E_z^i(\vec{\rho}_m), \qquad m = 1, 2, 3, \ldots, N, \tag{11.3.26}$$

where

$$A_{mn} = \frac{\zeta k}{4}\int_{\Delta\Gamma_n} H_0^{(2)}(k|\vec{\rho}_m-\vec{\rho}\,'|)\, d\ell' \tag{11.3.27}$$

Equation (11.3.26) can be written in matrix form as follows:

$$\begin{bmatrix} A_{11} & A_{12} & \cdots & A_{1n} & A_{1N} \\ A_{21} & A_{22} & \cdots & A_{2n} & A_{2N} \\ \vdots & \vdots & \vdots & \vdots & \vdots \\ A_{m1} & A_{m2} & \cdots & A_{mn} & A_{mN} \\ A_{N1} & A_{N2} & \cdots & A_{Nn} & A_{NN} \end{bmatrix} \begin{bmatrix} a_1 \\ a_2 \\ \vdots \\ a_m \\ a_N \end{bmatrix} = \begin{bmatrix} E_z^i(\vec{\rho}_1) \\ E_z^i(\vec{\rho}_2) \\ \vdots \\ E_z^i(\vec{\rho}_m) \\ E_z^i(\vec{\rho}_N) \end{bmatrix}. \tag{11.3.28}$$

Note that the column vector on the right-hand side of (11.3.28) is known from the incident field. Therefore a_n can be found from (11.3.28) once the matrix elements A_{mn} are determined from (11.3.27). There are two distinct situations in evaluation of matrix elements.

The first case is $m \neq n$. In this case, $\vec{\rho}\,' \neq \vec{\rho}_m$ in integration over $\Delta\Gamma_n$, and therefore (11.3.27) can be simplified to

$$\begin{aligned} A_{mn} &\approx \frac{\zeta k\,\Delta\Gamma_n}{4} H_0^{(2)}(k|\vec{\rho}_m-\vec{\rho}_n|) \\ &= \frac{\zeta k\,\Delta\Gamma_n}{4} H_0^{(2)}\left(k\sqrt{(x_m-x_n)^2+(y_m-y_n)^2}\right). \end{aligned} \tag{11.3.29}$$

The second case is $m = n$. In this case, $\vec{\rho}\,' = \vec{\rho}_m$ during integration over $\Delta\Gamma_n$, and the integrable singularity of the Hankel function $H_0^{(2)}(-)$ must be handled analytically as follows.

As shown in Figure 11.17*b*, the length $\Delta\Gamma_n$ is approximated as a straight-line segment of length Δ_n centered at $\vec{\rho}_m = \vec{\rho}_n$. Therefore (11.3.27) gives

$$A_{nn} = \frac{\zeta k}{4}\int_{-\Delta_n/2}^{\Delta_n/2} H_0^{(2)}(k|u|)du = \frac{\zeta k}{4} 2\int_{0}^{\Delta_n/2} H_0^{(2)}(k|u|)du. \qquad (11.3.30)$$

Since the argument of the Hankel function is small, this can be approximated to

$$H_0^{(2)}(z) \approx 1 + j\frac{2}{\pi}\ln\left(\frac{2}{\kappa z}\right),$$

where

$$\kappa = e^{\gamma} \approx 1.781, \quad \gamma = 0.5772156649\cdots \quad \text{(the Euler's constant)}.$$

Equation (11.3.30) can be rewritten as follows:

$$A_{nn} \approx \frac{\zeta k}{2}\int_{o}^{\Delta_n/2}\left[1 + j\frac{2}{\pi}\ln\left(\frac{2}{\kappa k u}\right)\right]du. \qquad (11.3.31)$$

For $n = 0$, the integration formula 30 of Appendix A gives

$$\int \ln(ax)dx = x\ln(ax) - x = x\ln\left(a\frac{x}{e}\right).$$

Therefore the integral of (11.3.31) can be evaluated as follows:

$$A_{nn} \approx \frac{\zeta k}{2}\int_{0}^{\Delta_n/2}\left[1 + j\frac{2}{\pi}\ln\left(\frac{1}{\kappa k u/2}\right)\right]du = \frac{\zeta k}{2}\int_{0}^{\Delta_n/2}\left[1 - j\frac{2}{\pi}\ln\left(\frac{\kappa k u}{2}\right)\right]du,$$

or

$$A_{nn} = \frac{\zeta k}{2}\left[u - j\frac{2}{\pi}u\ln\left(\frac{\kappa k u}{2e}\right)\right]\Bigg|_{u=0}^{u=\Delta_n/2}.$$

Hence

$$A_{nn} \approx \frac{\zeta k \Delta_n}{4}\left[1 - j\frac{2}{\pi}\ln\left(\frac{\kappa k \Delta_n}{4e}\right)\right]. \qquad (11.3.32)$$

Therefore the matrix elements can now be found via (11.3.29) and (11.3.32). Induced surface current can then be found via (11.3.28).

Determination of the Scattered Electric Field

An unknown surface current $K_z|_n = a_n$ on $\Delta\Gamma_n$ is obtained from (11.3.28), as discussed above. An efficient numerical method can be employed to solve this matrix equation. The scattered field E_z^s anywhere in space is found via (11.3.21) using the now known a_n as follows:

$$E_z^s(\vec{\rho}) \approx -\frac{\zeta k}{4}\sum_{n=1}^{N} a_n H_0^{(2)}\left(k\left|\vec{\rho} - \vec{\rho}_n\right|\right)\Delta\Gamma_n = \text{scattered field} \qquad (\text{valid for } \vec{\rho} \notin \Gamma). \tag{11.3.33}$$

PROBLEMS

11.1. Using the finite-difference scheme, find the two lowest eigenvalues of a TM^z mode propagating between two conducting parallel plates that are separated by 60 cm. Compare your results with those obtained analytically in Chapter 6.

11.2. Use the finite-difference scheme to solve the following differential equation:

$$\frac{d^2 f(x)}{dx^2} + 25 f(x) = 0.$$

The solution must satisfy the boundary conditions $f(x = 0) = 0$ and $f(x = 1) = 4$.

11.3. Using the finite-difference scheme, find the lowest eigenvalue of a TE^z mode propagating between two conducting parallel plates that are separated by 75 cm.

11.4. Find the potential distribution inside the space enclosed by four infinitely long conducting plates arranged as shown in Figure 11.5. The boundary potentials are as follows: $V_\mathrm{T} = 20$ V, $V_\mathrm{R} = 10$ V, $V_\mathrm{B} = 0$, and $V_\mathrm{L} = 10$ V.

11.5. Refer to Example 11.5. Find the potential distribution $\phi(x, y)$ inside the region after reducing the horizontal grid to three.

11.6. Find the eigenvalues of a hollow metallic rectangular waveguide operating in TM^z mode. Its cross section is 18.0×9.0 inches.

11.7. Use the FDTD technique to solve the following wave equation:

$$\frac{\partial^2 E(z, t)}{\partial t^2} = 9\frac{\partial^2 E(z, t)}{\partial z^2}, \qquad 0 < z < 1, t > 0.$$

The solution satisfies the following boundary and initial conditions:

$$E(0, t) = E(1, t) = 0,$$
$$E(z, 0) = \sin(3\pi z), \qquad 0 \le z \le 1,$$

and

$$\left.\frac{\partial E(z,t)}{\partial t}\right|_{(z,0)} = 0, \qquad 0 \le z \le 1.$$

Compare the results at $t = 0.1$ s for $\xi = 0.5$, 1, and 2.

11.8. Use at least 15 partitions to find the charge distribution on the wire considered in Example 11.8. Compare your results with those found in the example.

11.9. An infinitely long circular cylinder is exposed to a TM^z polarized uniform plane wave propagating along the y-axis. The radius of the cylinder is 2 m, and the signal frequency is 2.45 GHz. Use the method of moments to find the electric field at $\rho = 2.5$ m, 2.75 m, and 3 m along $\phi = 0°$, 90°, and 180°.

APPENDIX A

MATHEMATICAL FORMULAS

TRIGONOMETRIC AND OTHER RELATIONS

1. $\sin(A+B) = \sin A \cos B + \cos A \sin B$
2. $\sin(A-B) = \sin A \cos B - \cos A \sin B$
3. $\cos(A+B) = \cos A \cos B - \sin A \sin B$
4. $\cos(A-B) = \cos A \cos B + \sin A \sin B$
5. $\tan(A+B) = \dfrac{\tan A + \tan B}{1 - \tan A \tan B}$
6. $\tan(A-B) = \dfrac{\tan A - \tan B}{1 + \tan A \tan B}$
7. $\sin^2 A + \cos^2 A = 1$
8. $\tan^2 A + 1 = \sec^2 A$
9. $\cot^2 A + 1 = \csc^2 A$
10. $\sin A + \sin B = 2\sin\left(\dfrac{A+B}{2}\right)\cos\left(\dfrac{A-B}{2}\right)$
11. $\sin A - \sin B = 2\cos\left(\dfrac{A+B}{2}\right)\sin\left(\dfrac{A-B}{2}\right)$
12. $\cos A + \cos B = 2\cos\left(\dfrac{A+B}{2}\right)\cos\left(\dfrac{A-B}{2}\right)$

Practical Electromagnetics: From Biomedical Sciences to Wireless Communication.
By Devendra K. Misra
Copyright © 2007 John Wiley & Sons, Inc.

13. $\cos A - \cos B = 2\sin\left(\frac{A+B}{2}\right)\sin\left(\frac{B-A}{2}\right)$
14. $2\sin A\cos B = \sin(A+B) + \sin(A-B)$
15. $2\cos A\sin B = \sin(A+B) - \sin(A-B)$
16. $2\cos A\cos B = \cos(A+B) + \cos(A-B)$
17. $2\sin A\sin B = \cos(A-B) - \cos(A+B)$
18. $\sin\left(\frac{A}{2}\right) = \pm\sqrt{\frac{1-\cos A}{2}}$
19. $\cos\left(\frac{A}{2}\right) = \pm\sqrt{\frac{1+\cos A}{2}}$
20. $\sin 2A = 2\sin A\cos A$
21. $\cos 2A = \cos^2 A - \sin^2 A = 2\cos^2 A - 1 = 1 - 2\sin^2 A$
22. $\tan 2A = \frac{2\tan A}{1-\tan^2 A}$
23. $\tan\left(\frac{A}{2}\right) = \pm\sqrt{\frac{1-\cos A}{1+\cos A}} = \frac{\sin A}{1+\cos A} = \frac{1-\cos A}{\sin A}$
24. $e^{jA} = \cos A + j\sin A$
25. $e^{-jA} = \cos A - j\sin A$
26. $\sin A = \frac{e^{jA}-e^{-jA}}{2j} = A - \frac{A^3}{3!} + \frac{A^5}{5!} - \frac{A^7}{7!} + \cdots$
27. $\cos A = \frac{e^{jA}+e^{-jA}}{2} = 1 - \frac{A^2}{2!} + \frac{A^4}{4!} - \frac{A^6}{6!} + \cdots$
28. $\tan A = j\frac{e^{-jA}-e^{jA}}{e^{jA}+e^{-jA}} = A + \frac{A^3}{3} + 2\frac{A^5}{15} + 17\frac{A^7}{315} + \cdots$
29. $\sinh A = \frac{e^A - e^{-A}}{2}$
30. $\cosh A = \frac{e^A + e^{-A}}{2}$
31. $\tanh A = \frac{\sinh A}{\cos A} = \frac{e^A - e^{-A}}{e^A + e^{-A}}$
32. $\coth A = \frac{\cosh A}{\sinh A} = \frac{e^A + e^{-A}}{e^A - e^{-A}}$
33. $\tanh(A+B) = \frac{\tanh A + \tanh B}{1 + \tanh A\tanh B}$
34. $\tanh(A-B) = \frac{\tanh A - \tanh B}{1 - \tanh A\tanh B}$
35. $\cosh(A+B) = \cosh A\cosh B + \sinh A\sinh B$

36. $\sinh(A + B) = \sinh A \cosh B + \cosh A \sinh B$
37. $\sinh(A - B) = \sinh A \cosh B - \cosh A \sinh B$
38. $\cosh(A - B) = \cosh A \cosh B - \sinh A \sinh B$
39. $\cos(jA) = \cosh A$
40. $\sin(jA) = j \sinh A$
41. $\sinh(jA) = j \sin A$
42. $\cosh(jA) = \cos A$
43. $\tanh(jA) = j \tan A$
44. $\cosh^2 A - \sinh^2 A = 1$
45. $a + (a + d) + (a + 2d) + \cdots + \{a + (n - 1)d\} = \frac{n}{2}\{2a + (N - 1)d\}$ (Arithmetic series)
46. $a + ar + ar^2 + \cdots + ar^{(n-1)} = \frac{a(1 - r^n)}{1 - r}$ (Geometric series)
47. $(a + x)^n = a^n + na^{n-1}x + \frac{n(n-1)}{2!}a^{n-2}x^2 + \frac{n(n-1)(n-2)}{3!}a^{n-3}x^3 + \cdots$ (Binomial series)
48. $(1 + x)^{1/2} = 1 + \frac{1}{2}x - \frac{1}{2 \cdot 4}x^2 + \frac{1 \cdot 3}{2 \cdot 4 \cdot 6}x^3 - \cdots, \quad -1 < x \leq 1$
49. $(1 + x)^{-1/2} = 1 - \frac{1}{2}x + \frac{1 \cdot 3}{2 \cdot 4}x^2 - \frac{1 \cdot 3 \cdot 5}{2 \cdot 4 \cdot 6}x^3 + \cdots, \quad -1 < x \leq 1$
50. $e^A = 1 + A + \frac{A^2}{2!} + \frac{A^3}{3!} + \frac{A^4}{4} + \cdots$
51. $e^{-A} = 1 - A + \frac{A^2}{2} - \frac{A^3}{3} + \frac{A^4}{4!} - \cdots$
52. $\sinh A = \frac{e^A - e^{-A}}{2} = A + \frac{A^3}{3!} + \frac{A^5}{5!} + \frac{A^7}{7!} + \cdots$
53. $\cosh A = \frac{e^A + e^{-A}}{2} = 1 + \frac{A^2}{2!} + \frac{A^4}{4!} + \frac{A^6}{6!} + \cdots$
54. $\log_a(xy) = \log_a(x) + \log_a(y)$
55. $\log_a\left(\frac{x}{y}\right) = \log_a(x) - \log_a(y)$
56. $\log_a(x^y) = y \log_a(x)$
57. $\log_a(x) = \log_b(x) \times \log_a(b) = \frac{\log_b(x)}{\log_b(a)}$
58. $\ln(x) = \log_{10}(x) \times \ln(10) = 2.302585 \times \log_{10}(x)$
59. $\log_{10}(x) = \ln(x) \times \log_{10}(e) = 0.434294 \times \ln(x)$
60. $e = 2.718281828$

INTEGRATION

1. $\int x^n \, dx = \frac{x^{n+1}}{n+1} + c \quad (n \neq -1)$

2. $\int e^{ax}\,dx = \frac{1}{a}e^{ax} + c$

3. $\int \frac{1}{x}\,dx = \ln|x| + c$

4. $\int \frac{1}{(a^2 \pm x^2)^{3/2}}\,dx = \frac{\pm x}{a^2\sqrt{a^2 \pm x^2}} + c$

5. $\int \frac{x}{(a^2 + x^2)^{3/2}}\,dx = -\frac{1}{\sqrt{a^2 + x^2}} + c$

6. $\int \frac{x^2}{(a^2 + x^2)^{3/2}}\,dx = -\frac{x}{\sqrt{a^2 + x^2}} + \ln\left(x + \sqrt{a^2 + x^2}\right) + c$

7. $\int \frac{1}{a^2 + x^2}\,dx = \frac{1}{a}\tan^{-1}\left(\frac{x}{a}\right) + c$

8. $\int \frac{1}{\sqrt{a^2 - x^2}}\,dx = \sin^{-1}\left(\frac{x}{a}\right) + c$

9. $\int \frac{1}{\sqrt{a^2 + x^2}}\,dx = \sinh^{-1}\left(\frac{x}{a}\right) + c$

10. $\int \frac{1}{\sqrt{x^2 - a^2}}\,dx = \cosh^{-1}\left(\frac{x}{a}\right) + c$

11. $\int \frac{x}{a^2 + x^2}\,dx = \frac{1}{2}\ln(a^2 + x^2) + c$

12. $\int \frac{x}{\sqrt{a^2 + x^2}}\,dx = \sqrt{a^2 + x^2} + c$

13. $\int \frac{1}{\sqrt{a^2 + x^2}}\,dx = \ln\left(x + \sqrt{a^2 + x^2}\right) + c$

14. $\int \frac{1}{x\sqrt{a^2 + x^2}}\,dx = \frac{1}{a}\ln\left(\frac{a + \sqrt{a^2 + x^2}}{x}\right) + c$

15. $\int \sin x\,dx = -\cos x + c$

16. $\int \cos x\,dx = \sin x + c$

17. $\int \tan x\,dx = -\ln|\cos x| + c$

18. $\int \cot x\,dx = \ln|\sin x| + c$

19. $\int \sec x\,dx = \ln|\sec x + \tan x| + c$

20. $\int \csc x\,dx = \ln|\csc x - \cot x| + c$

21. $\int \sin^2 x\,dx = \frac{x}{2} - \frac{\sin 2x}{4} + c$

22. $\int \cos^2 x\,dx = \frac{x}{2} + \frac{\sin 2x}{4} + c$

23. $\int \tan^2 x\,dx = \tan x - x + c$

24. $\int \cot^2 x\,dx = -\cot x - x + c$

25. $\int x \sin ax\, dx = \frac{1}{a^2}(\sin ax - ax \cos ax)$

26. $\int x \cos ax\, dx = \frac{1}{a^2}(\cos ax + ax \sin ax)$

27. $\int \ln x\, dx = x \ln x - x + c$

28. $\int e^{ax} \sin bx\, dx = \frac{e^{ax}}{a^2 + b^2}(a \sin bx - b \cos bx) + c$

29. $\int e^{ax} \cos bx\, dx = \frac{e^{ax}}{a^2 + b^2}(a \cos bx + b \sin bx) + c$

30. $\int x^n \ln(ax) dx = \frac{x^{n+1}}{n+1} \ln(ax) - \frac{x^{n+1}}{(n+1)^2} + c$

DIFFERENTIATION

1. $\frac{d}{dx}(f_1 f_2) = f_1 \frac{df_2}{dx} + f_2 \frac{df_1}{dx}$

2. $\frac{d}{dx}\left(\frac{f_1}{f_2}\right) = \frac{f_2(df_1/dx) - f_1(df_2/dx)}{f_2^2}$

3. $\frac{d}{dx}(x^n) = nx^{n-1}$

4. $\frac{d}{dx}(e^{cx}) = ce^{cx}$

5. $\frac{d}{dx}(a^x) = a^x \ln a$

6. $\frac{d}{dx}(\sin ax) = a \cos ax$

7. $\frac{d}{dx}(\cos ax) = -a \sin ax$

8. $\frac{d}{dx}(\tan ax) = a \sec^2 ax$

9. $\frac{d}{dx}(\cot ax) = -a \operatorname{cosc}^2 ax$

10. $\frac{d}{dx}(\sinh ax) = a \cosh ax$

11. $\frac{d}{dx}(\cosh ax) = a \sinh ax$

12. $\frac{d}{dx}(\ln ax) = \frac{1}{x}$

13. $\frac{d}{dx}(\log_a x) = \frac{\log_a e}{x}$

14. $\frac{d}{dx}(\sin^{-1} x) = \frac{1}{\sqrt{1 - x^2}}$

15. $\frac{d}{dx}(\cos^{-1} x) = -\frac{1}{\sqrt{1-x^2}}$
16. $\frac{d}{dx}(\tan^{-1} x) = \frac{1}{1+x^2}$
17. $\frac{d}{dx}(\cot^{-1} x) = -\frac{1}{1+x^2}$

VECTOR IDENTITIES

1. $\vec{A} \cdot (\vec{B} \times \vec{C}) = \vec{B} \cdot (\vec{C} \times \vec{A}) = \vec{C} \cdot (\vec{A} \times \vec{B})$
2. $\vec{A} \times (\vec{B} \times \vec{C}) = \vec{B}(\vec{A} \cdot \vec{C}) - \vec{C}(\vec{A} \cdot \vec{B})$
3. $\nabla(\phi_1 \phi_2) = \phi_1 \nabla \phi_2 + \phi_2 \nabla \phi_1$
4. $\nabla \cdot (\vec{A} \times \vec{B}) = \vec{B} \cdot (\nabla \times \vec{A}) - \vec{A} \cdot (\nabla \times \vec{B})$
5. $\nabla \times (\phi A) = \nabla \phi \times \vec{A} + \phi(\nabla \times \vec{A})$
6. $\nabla \cdot (\phi A) = \nabla \phi \cdot \vec{A} + \phi(\nabla \cdot \vec{A})$
7. $\nabla \cdot (\nabla \times \vec{A}) = 0$
8. $\nabla \times (\nabla \phi) = 0$
9. $\nabla \times \nabla \times \vec{A} = \nabla(\nabla \cdot \vec{A}) - \nabla^2 \vec{A}$
10. **Stokes's Theorem**: The circulation around a simple closed curve is equal to the integral over any simple surface spanning the curve, of the normal component of the curl, in the positive sense of the curve being counterclockwise as seen from the side of the surface toward which the positive normal points. Mathematically

$$\oint_C \vec{A} \cdot d\vec{\ell} = \int_S (\nabla \times \vec{A}) \cdot d\vec{s}.$$

11. **Divergence Theorem** (also known as Gauss's theorem): The integral of the divergence of a vector field over a region of space is equal to the integral over the surface of that region of the component of the field in the direction of the outward directed normal to the surface. Mathematically

$$\int_V (\nabla \cdot \vec{A}) dv = \oint_S \vec{A} \cdot d\vec{s}.$$

12. **Helmholtz's Theorem**: A vector is uniquely specified if its divergence and curl are given within a region and its normal component is given over the boundary.
13. $\int_V \nabla \times \vec{A} dv = \oint_S d\vec{s} \times \vec{A}$
14. $\int_V \nabla \phi \, dv = \oint_S \phi \, d\vec{s}$
15. $\int_S d\vec{s} \times \nabla \phi = \oint_c \phi \, d\vec{\ell}$

Cartesian Coordinates

$$\nabla f = \hat{x}\frac{\partial f}{\partial x} + \hat{y}\frac{\partial f}{\partial y} + \hat{z}\frac{\partial f}{\partial z}$$

$$\nabla^2 f = \frac{\partial^2 f}{\partial x^2} + \frac{\partial^2 f}{\partial y^2} + \frac{\partial^2 f}{\partial z^2}$$

$$\nabla \cdot \vec{A} = \frac{\partial A_x}{\partial x} + \frac{\partial A_y}{\partial y} + \frac{\partial A_z}{\partial z}$$

$$\nabla \times \vec{A} = \begin{vmatrix} \hat{x} & \hat{y} & \hat{z} \\ \dfrac{\partial}{\partial x} & \dfrac{\partial}{\partial y} & \dfrac{\partial}{\partial z} \\ A_x & A_y & A_z \end{vmatrix}$$

Cylindrical Coordinates

$$\nabla f = \hat{\rho}\frac{\partial f}{\partial \rho} + \hat{\phi}\frac{1}{\rho}\frac{\partial f}{\partial \phi} + \hat{z}\frac{\partial f}{\partial z}$$

$$\nabla^2 f = \frac{1}{\rho}\frac{\partial}{\partial \rho}\left(\rho\frac{\partial f}{\partial \rho}\right) + \frac{1}{\rho^2}\frac{\partial^2 f}{\partial \phi^2} + \frac{\partial^2 f}{\partial z^2}$$

$$\nabla \cdot \vec{A} = \frac{1}{\rho}\frac{\partial(\rho A_\rho)}{\partial \rho} + \frac{1}{\rho}\frac{\partial A_\phi}{\partial \phi} + \frac{\partial A_z}{\partial z}$$

$$\nabla \times \vec{A} = \frac{1}{\rho}\begin{vmatrix} \hat{\rho} & \rho\hat{\phi} & \hat{z} \\ \dfrac{\partial}{\partial \rho} & \dfrac{\partial}{\partial \phi} & \dfrac{\partial}{\partial z} \\ A_\rho & \rho A_\phi & A_z \end{vmatrix}$$

Spherical Coordinates

$$\nabla f = \hat{r}\frac{\partial f}{\partial r} + \hat{\theta}\,\frac{1}{r}\frac{\partial f}{\partial \theta} + \hat{\phi}\,\frac{1}{r\sin(\theta)}\frac{\partial f}{\partial \phi}$$

$$\nabla^2 f = \frac{1}{r^2\sin(\theta)}\left[\sin(\theta)\frac{\partial}{\partial r}\left(r^2\frac{\partial f}{\partial r}\right) + \frac{\partial}{\partial \theta}\left(\sin(\theta)\frac{\partial f}{\partial \theta}\right) + \frac{1}{\sin(\theta)}\frac{\partial^2 f}{\partial \phi^2}\right]$$

$$\nabla \cdot \vec{A} = \frac{1}{r^2\sin(\theta)}\left[\sin(\theta)\frac{\partial}{\partial r}(r^2 A_r) + r\frac{\partial}{\partial \theta}(\sin(\theta)A_\theta) + r\frac{\partial A_\phi}{\partial \phi}\right]$$

$$\nabla \times \vec{A} = \frac{1}{r^2\sin(\theta)}\begin{vmatrix} \hat{r} & r\hat{\theta} & r\sin(\theta)\hat{\phi} \\ \dfrac{\partial}{\partial r} & \dfrac{\partial}{\partial \theta} & \dfrac{\partial}{\partial \phi} \\ A_r & r A_\theta & r\sin(\theta)A_\phi \end{vmatrix}$$

APPENDIX B

DELTA FUNCTION AND EVALUATION OF FIELDS IN UNBOUNDED MEDIA

DIRAC DELTA FUNCTION

In electromagnetic field analysis we come across the source density and the point source. Take the situation of a point charge q and the corresponding charge density ρ_v. Obviously the charge density must be zero everywhere in space and become infinite at the location $\vec{r}'$ of the point charge such that

$$\int_V \rho(\vec{r}, \vec{r}')dv = \begin{cases} q & \vec{r}' \in V, \\ 0 & \vec{r}' \notin V. \end{cases} \tag{B.1}$$

In other words, when the charge density ρ_v is integrated over volume V, it will result in total point charge q only if the volume V includes its location $\vec{r}'$. Otherwise, the integral will be zero. The delta function (or more precisely, it is called the Dirac delta function) meets this functional description of the charge density. It is also known as the impulse function. In the one-dimensional case the delta function can be defined as follows: Because the height of a pulse is equal to the inverse of its width, as shown in Figure B.1, the area of the pulse is unity. As a goes to zero, the pulse width reduces to zero whereas its height goes to infinity such that the area stays constant at unity.

Practical Electromagnetics: From Biomedical Sciences to Wireless Communication.
By Devendra K. Misra
Copyright © 2007 John Wiley & Sons, Inc.

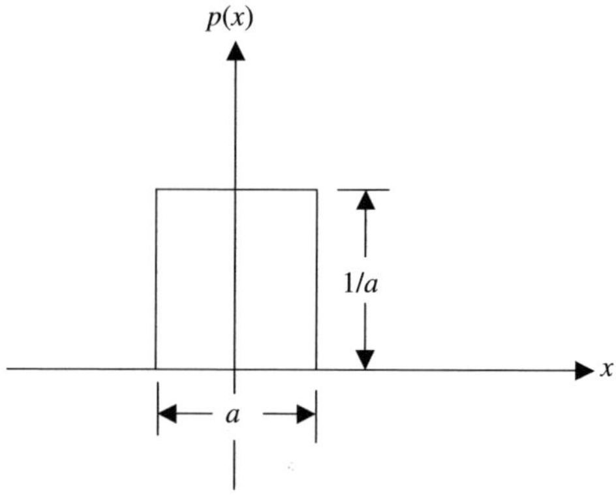

Figure B.1 Pulse of a unit area.

Therefore the delta function $\delta(x)$ is

$$\delta(x) = \lim_{a \to 0} p_1(x). \tag{B.2}$$

This delta function is located at the origin. If it is located at $x = b$, then it is represented by $\delta(x - b)$. In a strict mathematical sense, the delta function is not a function but a distribution.

Some of the properties of delta function are as follows:

1. $\delta(-x) = \delta(x)$
2. $f(x)\delta(x - c) = f(c)\delta(x - c)$
3. $\delta'(x) = -\delta'(-x)$, where prime denotes the derivative with respect to x
4. $x\delta(x) = 0$
5. $x\delta'(x) = -\delta(x)$
6. $\delta(ax) = \dfrac{1}{|a|}\delta(x)$
7. $\displaystyle\int_a^b f(x)\delta(x - c)dx = \begin{cases} f(c) & \text{for } a < c < b \\ 0 & \text{for } c < a \text{ or } c > b \end{cases}$
8. $\displaystyle\int_a^b \delta(x - c)dx = \begin{cases} 1 & \text{for } a < c < b \\ 0 & \text{for } c < a \text{ or } c > b \end{cases}$
9. $\displaystyle\int_a^b f(x)\delta'(x - c)dx = \begin{cases} -f'(c) & \text{for } a < c < b \\ 0 & \text{for } c < a \text{ or } c > b \end{cases}$

   where $\delta'(x - c) = \dfrac{d}{dx}\delta(x - c)$ and $f'(c) = \left.\dfrac{df(x)}{dx}\right|_{x=c}$
10. $\displaystyle\int_{-\infty}^{\infty} \delta(x - x')\delta(x'' - x)dx = \delta(x'' - x') = \delta(x' - x'')$

In three-dimensional rectangular coordinates the delta function is given as follows:

$$\delta(\vec{r} - \vec{r}') = \delta(x - x')\delta(y - y')\delta(z - z'). \tag{B.3}$$

Therefore

$$\int_V f(\vec{r})\delta(\vec{r} - \vec{r}')dv = \begin{cases} f(\vec{r}') & \vec{r}' \in V, \\ 0 & \text{otherwise}. \end{cases} \tag{B.4}$$

Equation (B.4) gives the defining relation of the delta function. It helps in representing the point source. As was mentioned earlier, the delta function is not a function but a distribution or generalized function in the strict mathematical sense. It cannot be an end result because of its singular character and can only appear inside an integral, where it is meaningful.

The delta function in three-dimensional circular cylindrical and in spherical coordinates, respectively, is given as follows:

$$\delta(\vec{r} - \vec{r}') = \frac{\delta(\rho - \rho')\delta(\phi - \phi')\delta(z - z')}{\rho}, \tag{B.5}$$

$$\delta(\vec{r} - \vec{r}') = \frac{\delta(r - r')\delta(\theta - \theta')\delta(\phi - \phi')}{r^2 \sin(\theta)}. \tag{B.6}$$

The concept of the delta function is embedded in various integral transforms. For example, consider the following Fourier transform pair:

$$f(x) = \frac{1}{2\pi}\int_{-\infty}^{\infty} F(\alpha)e^{j\alpha x}d\alpha, \tag{B.7}$$

where

$$F(\alpha) = \int_{-\infty}^{\infty} f(x)e^{-j\alpha x}dx. \tag{B.8}$$

On substituting (B.8) into (B.7) and changing the order of integration, we can identify the embedded delta function as follows:

$$\begin{aligned} f(x) &= \frac{1}{2\pi}\int_{-\infty}^{\infty}\left\{\int_{-\infty}^{\infty} f(x')e^{-j\alpha x'}dx'\right\}e^{j\alpha x}d\alpha \\ &= \int_{-\infty}^{\infty} f(x')\left\{\frac{1}{2\pi}\int_{-\infty}^{\infty} e^{j\alpha(x-x')}d\alpha\right\}dx' \\ &= \int_{-\infty}^{\infty} f(x')\delta(x - x')dx', \end{aligned}$$

where

$$\delta(x - x') = \frac{1}{2\pi}\int_{-\infty}^{\infty} e^{j\alpha(x-x')}d\alpha = \frac{1}{\pi}\int_{0}^{\infty} \cos\alpha(x - x')d\alpha. \tag{B.9}$$

If instead we substitute (B.7) into (B.8) and change the order of integration, then we find the embedded delta function in the transform domain as follows:

$$\begin{aligned} F(\alpha) &= \int_{-\infty}^{\infty}\left\{\frac{1}{2\pi}\int_{-\infty}^{\infty} F(\alpha')e^{j\alpha' x}d\alpha'\right\}e^{-j\alpha x}dx \\ &= \int_{-\infty}^{\infty} F(\alpha')\left\{\frac{1}{2\pi}\int_{-\infty}^{\infty} e^{-j(\alpha-\alpha')x}dx\right\}d\alpha' \\ &= \int_{-\infty}^{\infty} F(\alpha')\delta(\alpha - \alpha')d\alpha', \end{aligned}$$

where

$$\delta(\alpha - \alpha') = \frac{1}{2\pi}\int_{-\infty}^{\infty} e^{-j(\alpha-\alpha')x}dx = \frac{1}{\pi}\int_{0}^{\infty} \cos(\alpha - \alpha')xdx. \tag{B.10}$$

Similarly consider the Fourier series of $f(x)$,

$$f(x) = \sum_{n=-\infty}^{\infty} a_n e^{jnx}, \tag{B.11}$$

where

$$a_n = \frac{1}{2\pi}\int_{-\pi}^{\pi} f(x')e^{-jnx'}dx'. \tag{B.12}$$

After substituting (B.12) into (B.11) and changing the order of integration and summation, we find that

$$f(x) = \sum_{n=-\infty}^{\infty}\left\{\frac{1}{2\pi}\int_{-\pi}^{\pi} f(x')e^{-jnx'}dx'\right\}e^{jnx} = \int_{-\pi}^{\pi} f(x')\left\{\frac{1}{2\pi}\sum_{n=-\infty}^{\infty} e^{jn(x-x')}\right\}dx'. \tag{B.13}$$

The embedded delta function in (B.13) is identified to be

$$\delta(x - x') = \frac{1}{2\pi}\sum_{n=-\infty}^{\infty} e^{jn(x-x')} = \frac{1}{2\pi}\sum_{0}^{\infty} \varepsilon_n \cos n(x - x'), \tag{B.14}$$

where ε_n is the Neumann factor given by

$$\varepsilon_n = \begin{cases} 1, & n = 0, \\ 2, & n \neq 0. \end{cases}$$

Similarly, using the cylindrical Fourier-Bessel transform pair, we can find that

$$\frac{\delta(\rho - \rho')}{\rho'} = \frac{\delta(\rho' - \rho)}{\rho} = \int_0^\infty \alpha\, J_n(\alpha\rho) J_n(\alpha\rho') d\alpha. \tag{B.15}$$

Using the spherical Fourier-Bessel transform pair, we find that

$$\frac{\delta(r - r')}{r'} = \frac{\delta(r' - r)}{r} = \frac{2}{\pi}\int_0^\infty \alpha^2 j_n(\alpha r) j_n(\alpha r') d\alpha. \tag{B.16}$$

GREEN'S FUNCTIONS

Green's function provides a general method to solve linear, inhomogeneous partial differential equations. This technique is motivated by the intuition that the simplest solution to such an equation is that relating to the excitation of a unit point source function of dimensionality matching that of the differential operator. The solutions excited by more complex source functions are finally obtained by the linear superposition of the point source responses. In linear system analysis, if the impulse response is known, then the system response to a complex source at its input can be found via a convolution integral. Thus Green's function is the impulse response of the system. Green's function for the linear differential operator in question is defined as the solution corresponding to unit point source excitation.

Consider a three-dimensional operator equation as follows:

$$\mathscr{L}\varphi(\vec{r}) = S(\vec{r}), \tag{B.17}$$

where $\vec{r}$ is a three-dimensional position vector, $\mathscr{L}$ is an arbitrary three-dimensional partial differential operator, and $\varphi(\vec{r})$ is an unknown scalar field excited by a known source density function $S(\vec{r})$. Assume that $G(\vec{r}|\vec{r}')$ represents the solution $\varphi(\vec{r})$ to (B.17) when $S(\vec{r})$ is replaced by a unit point source $\delta(\vec{r} - \vec{r}')$ at $\vec{r} = \vec{r}'$. Therefore

$$\mathscr{L}G(\vec{r}|\vec{r}') = \delta(\vec{r} - \vec{r}') \tag{B.18}$$

where $G(\vec{r}|\vec{r}\,')$ is the response at $\vec{r}$ due to a unit point source at $\vec{r}\,'$.

Therefore the response at $\vec{r}$ due to a source $S(\vec{r}')dv'$ located at $\vec{r}\,'$ can be expressed as

$$d\psi(\vec{r}) = S(\vec{r}')dv' G(\vec{r}|\vec{r}'). \tag{B.19}$$

Using the linear superposition of point source responses, we find that

$$\psi(\vec{r}) = \int_V S(\vec{r}')G(\vec{r}|\vec{r}')dv'. \tag{B.20}$$

In order to verify the validity of this solution, we apply the operator $\mathscr{L}$ on both sides of (B.20) as follows:

$$\mathscr{L}\psi(\vec{r}) = \mathscr{L}\int_V S(\vec{r}')G(\vec{r}|\vec{r}')dv'.$$

Now moving the operator inside the integral, and using (B.18) and the property of the delta function, we find that

$$\mathscr{L}\psi(\vec{r}) = \int_V S(\vec{r}')\mathscr{L}G(\vec{r}|\vec{r}')dv' = \int_V S(\vec{r}')\delta(\vec{r} - \vec{r}')dv' = S(\vec{r}). \tag{B.21}$$

We can use this method to find electromagnetic fields produced by sources $\mu\vec{J}_{\mathrm{e}}$, $\varepsilon\vec{J}_{\mathrm{m}}$, $\rho_{\mathrm{e}}/\varepsilon$, and ρ_{m}/μ. We have found in Chapter 6 that

$$\vec{H} = \frac{1}{\mu}\nabla \times \vec{A} - \frac{j}{\omega\mu\varepsilon}\nabla(\nabla \cdot \vec{F}) - j\omega\vec{F} \tag{B.22}$$

and

$$\vec{E} = -\frac{j}{\omega\mu\varepsilon}\nabla(\nabla \cdot \vec{A}) - j\omega\vec{A} - \frac{1}{\varepsilon}\nabla \times \vec{F}, \tag{B.23}$$

where $\vec{A}$ and $\vec{F}$ are solution to following equations,

$$\nabla^2\vec{A} + k^2\vec{A} = -\mu\vec{J}_e \tag{B.24}$$

and

$$\nabla^2\vec{F} + k^2\vec{F} = -\varepsilon\vec{J}_m. \tag{B.25}$$

Also we found in Chapter 6 the following relations for scalar electric and magnetic potentials:

$$\nabla^2\phi_e + k^2\phi_e = -\frac{\rho_e}{\varepsilon} \tag{B.26}$$

and

$$\nabla^2\phi_m + k^2\phi_m = -\frac{\rho_m}{\mu}, \tag{B.27}$$

where $k^2 = \omega^2\mu\varepsilon$.

The solutions to (B.24) through (B.26) can be constructed if the corresponding Green function is known. It can be determined by solving the following Green function equation:

$$\nabla^2 G(\vec{r}|\vec{r}') + k^2 G(\vec{r}|\vec{r}') = -\delta(\vec{r} - \vec{r}'), \tag{B.28}$$

where $G(\vec{r}|\vec{r}')$ represents the scalar components of vector $\vec{A}$, or $\vec{F}$, or the scalar potentials ϕ_e or ϕ_m generated by a unit point source.

For an unbounded space, (B.28) can be solved following the procedure used in Section 7.1 of the Chapter 7. To simplify this, the point source can be moved to the origin of coordinate system. After finding the solution (Green's function), the point source can be moved back to $\vec{r}'$. For the solution to (B.28), after excluding the source region and selecting only an outward going wave (bounded at infinity), we find from (7.1.10) that

$$G(\vec{r}|\vec{r}') = C_1 \frac{e^{-jkr}}{r}. \tag{B.29}$$

Now, following (7.1.11), we can write

$$\lim_{r_o \to 0} \int_{\phi=0}^{\phi=2\pi} \int_{\theta=0}^{\theta=\pi} \int_{r=0}^{r=r_o} (\nabla^2 G + k^2 G) r^2 \sin\theta \, dr d\theta d\phi$$

$$= -\lim_{r_o \to 0} \int_{\phi=0}^{\phi=2\pi} \int_{\theta=0}^{\theta=\pi} \int_{r=0}^{r=r_o} \partial(r) r^2 \sin\theta \, dr d\theta d\phi, \tag{B.30}$$

or

$$-4\pi C_1 = -1 \rightarrow C_1 = \frac{1}{4\pi}. \tag{B.31}$$

Using (B.29) and (B.31), and moving the source back to its original position, we find the solution to (B.28) as follows:

$$G(\vec{r}|\vec{r}') = \frac{e^{-jkR}}{4\pi R}, \tag{B.32}$$

where $R = |\vec{r} - \vec{r}'|$. Then the solutions to (B.24) through (B.27) for sources located in unbounded space can be expressed as follows:

$$\vec{A}(\vec{r}) = \frac{\mu}{4\pi} \int_V \vec{J}_e \frac{e^{-jkR}}{R} dv', \tag{B.33}$$

$$\vec{F}(\vec{r}) = \frac{\varepsilon}{4\pi} \int_V \vec{J}_m \frac{e^{-jkR}}{R} dv', \tag{B.34}$$

$$\phi_e(\vec{r}) = \frac{1}{4\pi\varepsilon} \int_V \rho_e \frac{e^{-jkR}}{R} dv', \tag{B.35}$$

and

$$\phi_m(\vec{r}) = \frac{1}{4\pi\mu}\int_V \rho_m \frac{e^{-jkR}}{R} dv', \tag{B.36}$$

where $R = |\vec{r} - \vec{r}'|$. Once these integrals are evaluated, the corresponding electric and magnetic fields can be found after substituting that solution into (B.22) and (B.23).

APPENDIX C

BESSEL FUNCTIONS

Bessel's equation of order v is given as follows:

$$x\frac{d}{dx}\left\{x\frac{d}{dx}y(x)\right\}+(x^2-v^2)y(x)$$
$$=x^2\frac{d^2}{dx^2}y(x)+x\frac{d}{dx}y(x)+(x^2-v^2)y(x)=0. \tag{C.1}$$

The solutions to this equation can be found as

$$J_v(x)=\sum_{s=0}^{\infty}\frac{(-1)^s}{s!(s+v)!}\left(\frac{x}{2}\right)^{2s+v} \tag{C.2}$$

and

$$J_{-v}(x)=\sum_{s=0}^{\infty}\frac{(-1)^s}{s!(s-v)!}\left(\frac{x}{2}\right)^{2s-v}, \tag{C.3}$$

where

$$s!=\Gamma(s+1). \tag{C.4}$$

Practical Electromagnetics: From Biomedical Sciences to Wireless Communication.
By Devendra K. Misra
Copyright © 2007 John Wiley & Sons, Inc.

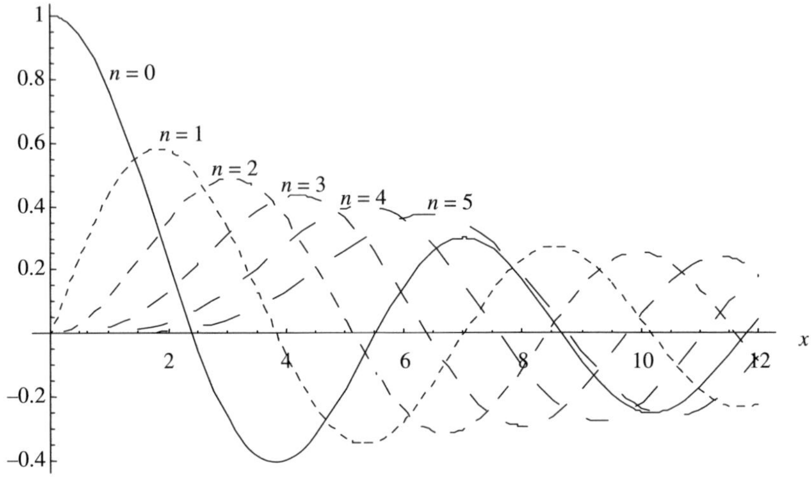

Figure C.1 Bessel functions of the first kind.

$J_v(x)$ and $J_{-v}(x)$ represent two independent solutions to the Bessel equation (C.1) if v is not an integer or zero. These are known as the Bessel functions of the first kind of order v. In the case of v being an integer n, the two are related as follows:

$$J_{-n}(x) = (-1)^n J_n(x). \tag{C.5}$$

The Bessel functions of the first kind of the first few integer orders are shown in Figure C.1. As this figure indicates, $J_0(0)$ is unity whereas all Bessel functions of higher orders are zero at the origin.

The second solution to Bessel's equation is then found as follows:

$$Y_v(x) = \frac{J_v(x)\cos(\pi v) - J_{-v}(x)}{\sin(\pi v)}. \tag{C.6}$$

$Y_v(x)$ is known as the Bessel function of the second kind or the Neumann function. Sometimes it is also denoted as $N_v(x)$. Neumann functions of the first few integer orders are displayed in Figure C.2. As shown in this figure, these functions have a pole at the origin (i.e., all of them go to infinity for $x = 0$).

The tables of the Bessel functions have been around for some time. In these tables are found the desired value(s) for any given argument, order, and kind of function. Nowadays these functions are included in scientific software (e.g., Mathematica®). Further a number of scientific calculators have capabilities to generate these functions easily via suitable programming.

There are many instances where the argument of the Bessel function is either too small or too large. The following approximate expressions can be used to find the

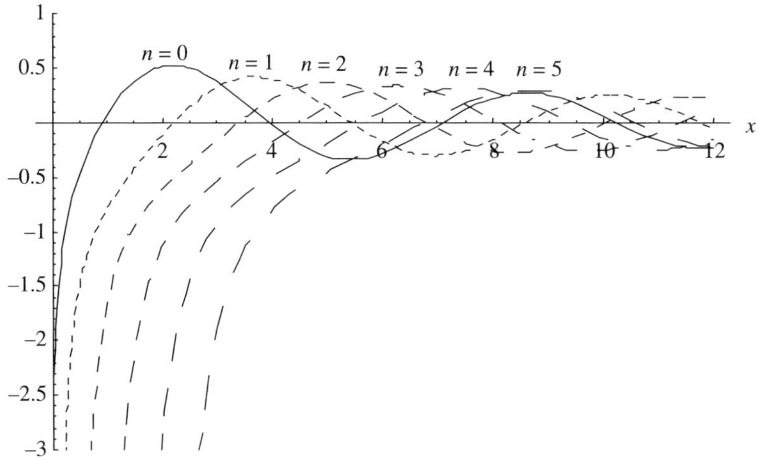

Figure C.2 Bessel functions of the second kind.

desired value in such cases:

$$J_0(x) \underset{x\to 0}{\longrightarrow} 1, \tag{C.7}$$

$$Y_0(x) \underset{x\to 0}{\rightarrow} \frac{2}{\pi}\{\ln(x/2)+\gamma\}, \quad \gamma = 0.5772156\ldots \text{ is Euler's constant.} \tag{C.8}$$

For $n > 0$,

$$J_n(x) \underset{x\to 0}{\longrightarrow} \frac{1}{n!}\left(\frac{x}{2}\right)^n, \tag{C.9}$$

$$Y_n(x) \underset{x\to 0}{\longrightarrow} -\frac{(n-1)!}{\pi}\left(\frac{2}{x}\right)^n, \tag{C.10}$$

$$J_n(x) \underset{x\to\infty}{\longrightarrow} \sqrt{\frac{2}{\pi x}}\cos\left(x-\frac{\pi}{4}-\frac{n\pi}{2}\right), \tag{C.11}$$

and

$$Y_n(x) \underset{x\to\infty}{\longrightarrow} \sqrt{\frac{2}{\pi x}}\sin\left(x-\frac{\pi}{4}-\frac{n\pi}{2}\right). \tag{C.12}$$

The wave propagation phenomena in cylindrical coordinates can be more conveniently expressed by two linear combinations of the Bessel functions, as follows:

$$H_n^{(1)}(x) = J_n(x) + jY_n(x) \tag{C.13}$$

and

$$H_n^{(2)}(x) = J_n(x) - jY_n(x). \tag{C.14}$$

These are called Hankel functions of the first and second kinds, respectively. The approximate expressions for small and large arguments can be easily obtained from (C.7) through (C.12). When x is large, the expressions for the Hankel functions reduce to

$$H_n^{(1)}(x) \underset{x\to\infty}{\longrightarrow} \sqrt{\frac{2}{\pi x}}\, e^{j[x-(\pi/4)-(n\pi/2)]} \tag{C.15}$$

and

$$H_n^{(2)}(x) \underset{x\to\infty}{\longrightarrow} \sqrt{\frac{2}{\pi x}}\, e^{-j[x-(\pi/4)-(n\pi/2)]}. \tag{C.16}$$

Therefore, for the time-harmonic fields of $e^{j\omega t}$, the Hankel function of the first kind represents a wave propagating in the $-x$ direction, whereas the Hankel function of the second kind represents a wave propagating in the $+x$ direction.

The recurrence relations for the Bessel function are as follows:

$$J_{n-1}(x) + J_{n+1}(x) = \frac{2n}{x} J_n(x), \tag{C.17}$$

$$J_{n-1}(x) - J_{n+1}(x) = 2\frac{d}{dx} J_n(x), \tag{C.18}$$

$$J_{n-1}(x) = \frac{n}{x} J_n(x) + \frac{d}{dx} J_n(x), \tag{C.19}$$

$$\frac{d}{dx}[x^n J_n(x)] = x^n J_{n-1}(x), \tag{C.20}$$

$$\frac{d}{dx}[x^{-n} J_n(x)] = -x^{-n} J_{n+1}(x), \tag{C.21}$$

$$J_{n-1}(x) - J_{n+1}(x) = 2\frac{d}{dx} J_n(x). \tag{C.22}$$

For $Z_n(x)$ being a solution to Bessel's equation, the following relations hold:

$$\frac{d}{dx} Z_n(x) = Z_{n-1}(x) - \frac{n}{x} Z_n(x) = -Z_{n+1}(x) + \frac{n}{x} Z_n(x), \tag{C.23}$$

$$Z_n(x) = \frac{2(n-1)}{x} Z_{n-1}(x) - Z_{n-2}(x). \tag{C.24}$$

A few useful formulas are as follows:

$$J_n(x) = \frac{1}{\pi}\int_0^{\pi} \cos(n\theta - x\sin\theta)d\theta, \qquad n = 0, 1, 2, \ldots, \tag{C.25}$$

$$e^{jx\cos\theta} = \sum_{n=-\infty}^{\infty} J_n(x)e^{jn[\theta+(\pi/2)]} \quad \text{(known as the Jacobi-Anger expansion)}, \tag{C.26}$$

$$e^{jx\sin\theta} = \sum_{n=-\infty}^{\infty} J_n(x)e^{jn\theta}. \tag{C.27}$$

The associated Wronskian formulas are as follows:

$$J_n(x)\frac{d}{dx}Y_n(x) - Y_n(x)\frac{d}{dx}J_n(x) = \frac{2}{\pi x}, \tag{C.28}$$

$$J_n(x)\frac{d}{dx}H_n^{(1)}(x) - H_n^{(1)}(x)\frac{d}{dx}J_n(x) = \frac{j2}{\pi x}, \tag{C.29}$$

$$J_n(x)\frac{d}{dx}H_n^{(2)}(x) - H_n^{(2)}(x)\frac{d}{dx}J_n(x) = -\frac{j2}{\pi x}, \tag{C.30}$$

$$Y_n(x)\frac{d}{dx}H_n^{(1)}(x) - H_n^{(1)}(x)\frac{d}{dx}Y_n(x) = -\frac{2}{\pi x}, \tag{C.31}$$

$$Y_n(x)\frac{d}{dx}H_n^{(2)}(x) - H_n^{(2)}(x)\frac{d}{dx}Y_n(x) = -\frac{2}{\pi x}, \tag{C.32}$$

$$H_n^{(2)}(x)H_{n+1}^{(1)}(x) - H_n^{(1)}(x)H_{n+1}^{(2)}(x) = \frac{4}{j\pi x}, \tag{C.33}$$

$$J_{n-1}(x)H_n^{(1)}(x) - J_n(x)H_{n-1}^{(1)}(x) = \frac{2}{j\pi x}. \tag{C.34}$$

The first few zeros of the Bessel's functions and their derivatives are listed in Tables C.1 through C.4.

C.1 MODIFIED BESSEL'S FUNCTIONS

The modified Bessel functions are defined for $x = j\alpha$. $I_n(\alpha)$ and $K_n(\alpha)$ are known as modified Bessel functions of the first and second kinds, respectively. These are

TABLE C.1 Zeros of Bessel's Functions of the First Kind

Number of Zeros	$J_0(x)$	$J_1(x)$	$J_2(x)$	$J_3(x)$	$J_4(x)$	$J_5(x)$
1	2.40483	3.83171	5.13562	6.38016	7.58834	8.77148
2	5.52008	7.01559	8.41724	9.76102	11.06471	12.3386
3	8.65373	10.17347	11.61984	13.0152	14.37254	15.70017
4	11.79153	13.32369	14.79595	16.22347	17.61597	18.98013
5	14.93092	16.47063	17.95982	19.40941	20.82693	22.2178

TABLE C.2 Zeros of the Derivatives of Bessel's Functions of the First Kind

Number of Zeros	$J_0'(x)$	$J_1'(x)$	$J_2'(x)$	$J_3'(x)$	$J_4'(x)$	$J_5'(x)$
1	3.83171	1.84118	3.05424	4.20119	5.31755	6.41562
2	7.01559	5.33144	6.70613	8.01524	9.2824	10.51986
3	10.17347	8.53632	9.96947	11.34592	12.68191	13.98719
4	13.32369	11.706	13.17037	14.58585	15.96411	17.31284
5	16.4706	14.86359	16.34752	17.78875	19.19603	20.57551

TABLE C.3 Zeros of Neumann Functions

Number of Zeros	$Y_0(x)$	$Y_1(x)$	$Y_2(x)$	$Y_3(x)$	$Y_4(x)$	$Y_5(x)$
1	0.89358	2.19714	3.38424	4.52702	5.64515	6.74719
2	3.95768	5.42968	6.79381	8.09755	9.36162	10.59718
3	7.08605	8.59601	10.02348	11.39647	12.73014	14.0338
4	10.22234	11.74915	13.20999	14.62308	15.99963	17.34709
5	13.3611	14.89744	16.37897	17.81846	19.22443	20.6029

TABLE C.4 Zeros of the Derivatives of Neumann Functions

Number of Zeros	$Y_0'(x)$	$Y_1'(x)$	$Y_2'(x)$	$Y_3'(x)$	$Y_4'(x)$	$Y_5'(x)$
1	2.19714	3.68302	5.00258	6.25363	7.46492	8.64956
2	5.42968	6.9415	8.35072	9.69879	11.00517	12.28087
3	8.59601	10.1234	11.5742	12.97241	14.33172	15.6608
4	11.74915	13.28576	14.76091	16.19045	17.58444	18.94974
5	14.89744	16.44006	17.93129	19.38239	20.80106	22.19284

related to the regular cylindrical Bessel and Hankel functions as follows:

$$I_n(\alpha) = j^n J_n(-j\alpha) = j^n J_{-n}(j\alpha) = j^{-n} J_n(j\alpha) \tag{C.35}$$

and

$$K_n(\alpha) = \frac{\pi}{2} j^{n+1} H_n^{(1)}(j\alpha) = \frac{\pi}{2}(-j)^{n+1} H_n^{(2)}(-j\alpha). \tag{C.36}$$

Sometimes these are also referred as hyperbolic Bessel functions. A few of these modified Bessel functions are shown in Figure C.3. Since Hankel functions are singular at the origin (due to the characteristic of Neumann functions), the modified Bessel function of the second kind goes to infinity as well. For large arguments, the modified Bessel functions can be approximated to

$$I_n(\alpha) \underset{\alpha \to \infty}{\longrightarrow} \frac{e^{\alpha}}{\sqrt{2\pi\alpha}} \tag{C.37}$$

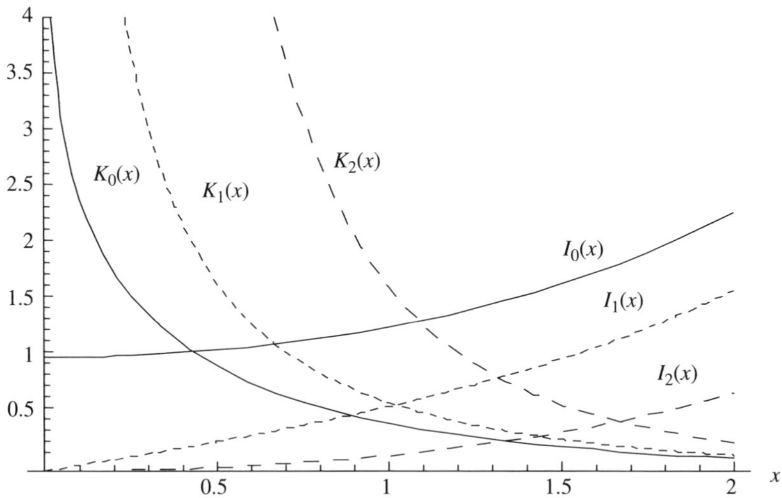

Figure C.3 Modified Bessel functions.

and

$$K_n(\alpha) \underset{\alpha\to\infty}{\longrightarrow} \sqrt{\frac{\pi}{2\alpha}}e^{-\alpha}. \tag{C.38}$$

C.2 SPHERICAL BESSEL'S FUNCTIONS

When the Helmholtz equation is separated into spherical coordinates, the radial equation reduces to

$$r^2\frac{d^2f(r)}{dr^2} + 2r\frac{df(r)}{dr} + \{k^2r^2 - n(n+1)\}f(r) = 0. \tag{C.39}$$

The following functions satisfy this equation:

$$f(kr) = \sqrt{\frac{\pi}{2kr}}\,Z_{n+1/2}(kr). \tag{C.40}$$

These functions are conveniently known as spherical Bessel functions, which are related to cylindrical Bessel functions as follows:

$$j_n(x) = \sqrt{\frac{\pi}{2x}}\,J_{n+1/2}(x), \tag{C.41}$$

$$y_n(x) = \sqrt{\frac{\pi}{2x}}\,Y_{n+1/2}(x) = (-1)^{n+1}\sqrt{\frac{\pi}{2x}}\,J_{-n-1/2}(x), \tag{C.42}$$

$$h_n^{(1)}(x) = \sqrt{\frac{\pi}{2x}}\,H_{n+1/2}^{(1)}(x) = j_n(x) + jy_n(x), \tag{C.43}$$

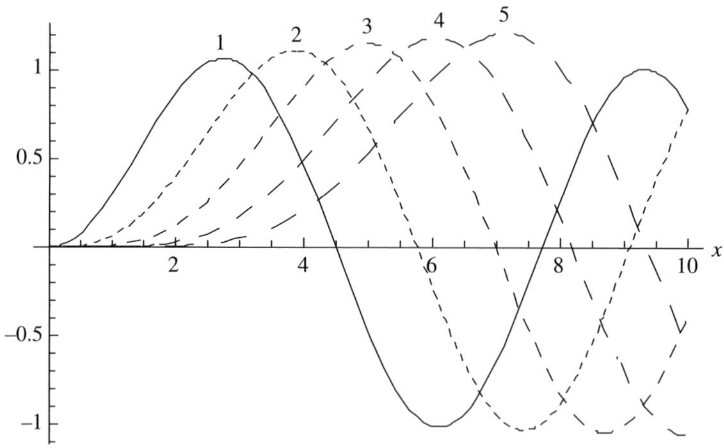

Figure C.4 Spherical Bessel functions of the first kind used in EM waves.

TABLE C.5 Zeros of Spherical Bessel Functions of the First Kind

Number of Zeros	$\hat{J}_1(x)$	$\hat{J}_2(x)$	$\hat{J}_3(x)$	$\hat{J}_4(x)$	$\hat{J}_5(x)$	$\hat{J}_6(x)$
1	4.4934	5.7635	6.9879	8.1826	9.3558	10.5128
2	7.7253	9.0950	10.4171	11.7049	12.9665	14.2074
3	10.9041	12.3229	13.6980	15.0397	16.3547	17.6480
4	14.0662	15.5146	16.9236	18.3013	19.6532	20.9835
5	17.2208	18.6890	20.1218	21.5254	22.9046	24.2628
6	20.3713	21.8539	23.3042	24.7276	26.1278	27.5079

TABLE C.6 Zeros of the Derivatives of Spherical Bessel Functions of the First Kind

Number of Zeros	$\hat{J}'_1(x)$	$\hat{J}'_2(x)$	$\hat{J}'_3(x)$	$\hat{J}'_4(x)$	$\hat{J}'_5(x)$	$\hat{J}'_6(x)$
1	2.7437	3.8702	4.9734	6.0619	7.1402	8.2108
2	6.1168	7.4431	8.7217	9.9675	11.1890	12.3915
3	9.3166	10.7130	12.0636	13.3801	14.6701	15.9387
4	12.4859	13.9205	15.3136	16.6742	18.0085	19.3212
5	15.6439	17.1027	18.5242	19.9154	21.2815	22.6263
6	18.7963	20.2720	21.7139	23.1278	24.5178	25.8873

and

$$h_n^{(2)}(x) = \sqrt{\frac{\pi}{2x}} H_{n+1/2}^{(2)}(x) = j_n(x) - jy_n(x). \tag{C.44}$$

In time-harmonic field analysis, it is convenient to define the alternative spherical Bessel functions as follow:

$$\hat{Z}_n(x) = \sqrt{\frac{\pi x}{2}} Z_{n+1/2}(x). \tag{C.45}$$

Note that the caret over Z_n is used in practice only as notation. The cylindrical Bessel function of the $(n + 1/2)$th order is used in (C.45) to determine the corresponding spherical Bessel function. The characteristics of spherical Bessel functions of the first kind are shown in Figure C.4 for $n = 1$ to 5. The first six zeros of the six spherical Bessel functions of the first kind and their derivatives are given in Tables C.5 and C.6, respectively.

APPENDIX D

LEGENDRE FUNCTIONS

When we use the separation of variables method to solve a scalar wave equation in spherical coordinates, we end up with an equation as follows:

$$\frac{1}{\sin\theta}\frac{d}{d\theta}\left(\sin\theta\frac{df}{d\theta}\right)+\left[n(n+1)-\frac{m^2}{\sin^2\theta}\right]f=0. \tag{D.1}$$

This is known as the associated Legendre equation. If we substitute $\cos\theta = x$ into this equation, we get another common form of Legendre's equation given by

$$(1-x^2)\frac{d^2f}{dx^2}-2x\frac{df}{dx}+\left[n(n+1)-\frac{m^2}{1-x^2}\right]f=0. \tag{D.2}$$

For $m = 0$, this reduces to the following ordinary Legendre equation

$$(1-x^2)\frac{d^2f}{dx^2}-2x\frac{df}{dx}+n(n+1)f=0. \tag{D.3}$$

The ordinary Legendre equation has two linearly independent solutions called Legendre functions of the first and the second kinds. These are denoted by $P_n(x)$ and $Q_n(x)$, respectively. If n is an integer, then the Legendre function of the first kind is a polynomial of order n. The second solution $Q_n(x)$ is a polynomial of

Practical Electromagnetics: From Biomedical Sciences to Wireless Communication.
By Devendra K. Misra
Copyright © 2007 John Wiley & Sons, Inc.

TABLE D.1 First Few Legendre Functions $P_n(x)$ and $Q_n(x)$

Legendre Function of First Kind		Legendre Function of Second Kind	
$P_0(x)$	1	$Q_0(x)$	$\frac{1}{2}\ln\left(\frac{1+x}{1-x}\right)$
$P_1(x)$	x	$Q_1(x)$	$\frac{x}{2}\ln\left(\frac{1+x}{1-x}\right)-1$
$P_2(x)$	$\frac{3x^2-1}{2}$	$Q_2(x)$	$\frac{3x^2-1}{4}\ln\left(\frac{1+x}{1-x}\right)-\frac{3}{2}x$
$P_3(x)$	$\frac{5x^3-3x}{2}$	$Q_3(x)$	$\frac{5x^2-3x}{4}\ln\left(\frac{1+x}{1-x}\right)-\frac{5}{2}x^2+\frac{2}{3}$

order $n-1$ with a logarithmic singularity. The first few of these polynomials are given in Table D.1.

Most often our interest will be over the range $-1 \leq x \leq 1$ because θ ranges from 0 to π in the spherical coordinate system. Therefore the first five Legendre functions of the first kind, $P_n(x)$, are illustrated in Figure D.1, and those of the second kind, $Q_n(x)$, are displayed in Figure D.2.

From these characteristics, it may be evident that only $P_n(x)$ are finite in the range of our interest. $Q_n(x)$ have infinite values (the singularity) at $x = \pm 1$ (i.e., at $\theta = 0$ and $\theta = \pi$). The useful formulas are as follows:

$$P_n(x) = \frac{1}{2^n n!}\frac{d^n}{dx^n}(x^2-1)^n \qquad \text{(known as Rodrigues's formula)}, \tag{D.4}$$

$$x\frac{d}{dx}P_n(x) - \frac{d}{dx}P_{n-1}(x) = nP_n(x), \tag{D.5}$$

$$(x^2-1)\frac{d}{dx}P_n(x) = nxP_n(x) - nP_{n-1}(x), \tag{D.6}$$

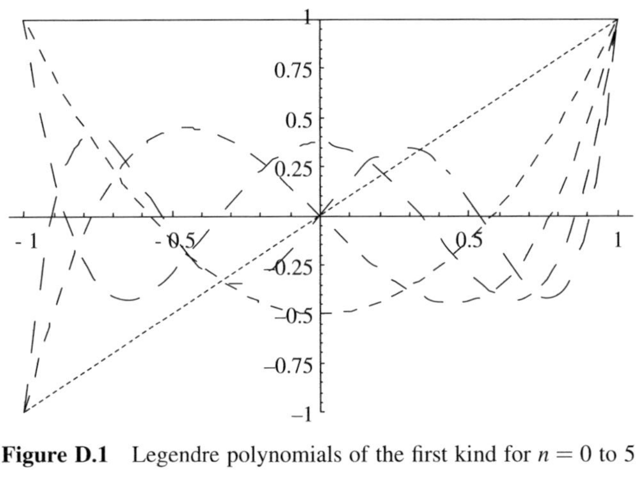

Figure D.1 Legendre polynomials of the first kind for $n = 0$ to 5.

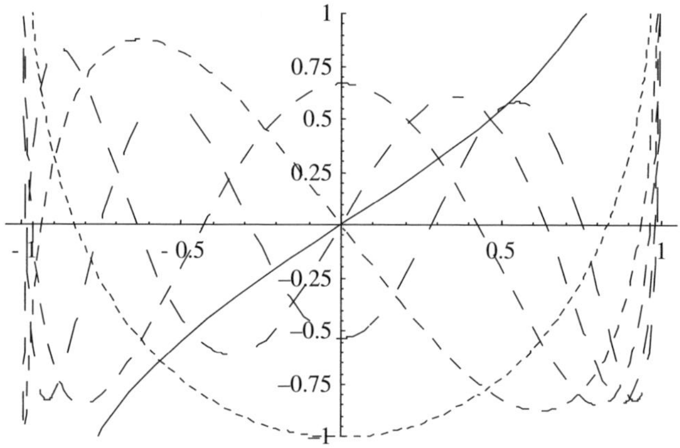

Figure D.2 Legendre polynomials of the second kind for $n = 0$ to 5.

$$\frac{d}{dx}P_{n+1}(x) = (n+1)P_n(x) + x\frac{d}{dx}P_n(x) = (2n+1)P_n(x) + \frac{d}{dx}P_{n-1}(x), \tag{D.7}$$

$$(n+1)P_{n+1}(x) + nP_{n-1}(x) = (2n+1)xP_n(x), \tag{D.8}$$

$$e^{jkr\cos\theta} = \sum_{n=0}^{\infty} j^n(2n+1)j_n(kr)P_n(\cos\theta), \tag{D.9}$$

Now we consider the solution to the associated Legendre equation (D.2). One way to do so is to start with (D.3) and differentiate it m times. This results in

$$(1-x^2)\frac{d^m f}{dx^m} - 2x(m+1)\frac{df}{dx} + (n-m)(n+m+1)f = 0. \tag{D.10}$$

Then we can define

$$y = (1-x^2)^{m/2}\frac{d^m f}{dx^m}. \tag{D.11}$$

On substituting (D.11) into (D.10), we get

$$(1-x^2)\frac{d^2y}{dx^2} - 2x\frac{dy}{dx} + \left[n(n+1) - \frac{m^2}{1-x^2}\right]y = 0, \tag{D.12}$$

which is the associate Legendre equation (D.2).

The regular solutions are known as associated Legendre functions of the first and second kinds. The associated Legendre functions of the first kind are denoted by $P_n^m(x)$, and they can be found as follows:

$$y = P_n^m(x) = (1-x^2)^{m/2}\frac{d^m}{dx^m}P_n(x). \tag{D.13}$$

TABLE D.2 Associated Legendre Functions of the First Kind

$P_n^m(x) = P_n^m(\cos\theta)$		
$P_1^1(x)$	$(1-x^2)^{1/2}$	$\sin\theta$
$P_2^1(x)$	$3x(1-x^2)^{1/2}$	$3\cos\theta\sin\theta$
$P_2^2(x)$	$3(1-x^2)$	$3\sin^2\theta$
$P_3^1(x)$	$\frac{3}{2}(5x^2-1)(1-x^2)^{1/2}$	$\frac{3}{2}(5\cos^2\theta-1)\sin\theta$
$P_3^2(x)$	$15x(1-x^2)$	$15\cos\theta\sin^2\theta$
$P_3^3(x)$	$15(1-x^2)^{3/2}$	$15\sin^3\theta$
$P_4^1(x)$	$\frac{5}{2}(7x^3-3x)(1-x^2)^{1/2}$	$\frac{5}{2}(7\cos^3\theta-3\cos\theta)\sin\theta$
$P_4^2(x)$	$\frac{15}{2}(7x^2-1)(1-x^2)$	$\frac{15}{2}(7\cos^2\theta-1)\sin^2\theta$
$P_4^3(x)$	$105x(1-x^2)^{3/2}$	$105\cos\theta\sin^3\theta$
$P_4^4(x)$	$105(1-x^2)^2$	$105\sin^4\theta$

Since the highest power of x in $P_n(x)$ is x^n, we must have $m \le n$. Sometimes an additional factor $(-1)^m$ is included in (D.13) to define the associated Legendre functions. Some lower order associated Legendre functions of the first kind are listed in Table D.2.

The recurrence relations for the associated Legendre functions of the first kind are as follows:

$$(2n+1)xP_n^m(x) = (n+m)P_{n-1}^m(x) + (n+1-m)P_{n+1}^m(x), \tag{D.14}$$

$$\frac{2(m+1)x}{(1-x^2)^{1/2}}P_n^{m+1}(x) = P_n^{m+2}(x) + (n-m)(n+m+1)P_n^m(x), \tag{D.15}$$

$$xP_n^{m+1}(x) = (n-m)(1-x^2)^{1/2}P_n^m(x) + P_{n-1}^{m+1}(x), \tag{D.16}$$

$$P_{n+1}^{m+1}(x) = xP_n^{m+1}(x) + (n+m+1)(1-x^2)^{1/2}P_n^m(x), \tag{D.17}$$

$$(1-x^2)^{1/2}P_n^m(x) = (n+m)xP_n^{m-1}(x) - (n-m+2)P_{n+1}^{m-1}(x), \tag{D.18}$$

$$(1-x^2)\frac{d}{dx}P_n^m(x) = (n+m)P_{n-1}^m(x) - nxP_n^m(x). \tag{D.19}$$

The orthogonality of the associated Legendre functions of the first kind are found using

$$\int_{-1}^{1} P_n^m(x)P_\ell^m(x)dx = \begin{cases} 0, & \text{for n} \neq \ell, \\ \dfrac{2}{2n+1}\dfrac{(n+m)!}{(n-m)!}, & \text{for n} = \ell. \end{cases} \tag{D.20}$$

The associated Legendre functions of the second kind are defined as follows:

$$Q_n^m(x) = (1 - x^2)^{m/2} \frac{d^m}{dx^m} Q_n(x). \tag{D.21}$$

These functions satisfy the same recurrence relations as $P_n^m(x)$. Note that $P_n^m(x)$ are bounded at $x = \pm 1$, whereas $Q_n^m(x)$ are unbounded at those points.

APPENDIX E

CHARACTERISTICS OF SELECTED MATERIALS

TABLE E.1 Conductivity of Some Materials at Room Temperature

Material	Conductivity in S/m
Aluminum	3.54×10^7
Brass	1.57×10^7
Bronze	1.0×10^7
Constantan (55 Cu, 45 Ni)	2.26×10^6
Copper	5.813×10^7
Gold	4.1×10^7
Iron	1.04×10^7
Lead	4.57×10^6
Nichrome	0.09×10^7
Nickel	1.15×10^7
Platinum	9.52×10^6
Silver	6.12×10^7
Sodium	2.17×10^7
Stainless steel	0.11×10^7
Titanium	2.09×10^6
Tungsten	1.82×10^7
Zinc	1.67×10^7

Practical Electromagnetics: From Biomedical Sciences to Wireless Communication.
By Devendra K. Misra
Copyright © 2007 John Wiley & Sons, Inc.

TABLE E.2 Dielectric Constant and Loss Tangent of Some Materials at 3 GHz

Material	Dielectric Constant, ε_r	Loss Tangent, tan δ
Alumina	9.6	0.0001
Carbon tetra chloride	2.17	0.00037
Glass (Pyrex)	4.82	0.0054
Mica (ruby)	5.4	0.0003
Nylon (610)	2.84	0.012
Polystyrene	2.55	0.0003
Plexiglas	2.60	0.0057
Quartz (fused)	3.8	0.00006
Rexolite (1422)	2.54	0.00048
Styrofoam (103.7)	1.03	0.0001
Teflon	2.1	0.00015
Titanium dioxide (Rutile)	96	0.001
Water (distilled)	77	0.157

TABLE E.3 Dielectric Properties of Biological Tissues at Selected Frequencies

	10 kHz		1 MHz		10 MHz		100 MHz		1 GHz	
	ε_r	σ (S/m)	ε_r	σ (S/m)	ε_r	σ (S/m)	ε_r	σ (S/m)	ε_r	σ (S/m)
Brain (gray matter)	15×10^3	0.1	900	0.18	300	0.3	90	0.7	60	1.2
Heart muscle	6×10^4	0.15	2000	0.35	350	0.5	80	0.9	60	1.2
Kidney (cortex)	3×10^4	0.14	2000	0.35	350	0.6	85	1	60	1.5
Liver	2×10^4	0.05	1000	0.2	200	0.35	65	0.5	50	0.9
Lung (inflated)	9×10^3	0.07	500	0.12	130	0.2	30	0.3	25	0.4
Spleen	12×10^3	0.1	2000	0.2	450	0.4	75	0.85	55	1.2
Muscle (across)	3.5×10^4	0.4	1500	0.5	150	0.65	80	0.8	60	1.2
Muscle (along)	2×10^4	0.5	350	0.6	120	0.7	80	0.8	60	1.2
Uterus	2×10^4	0.5	1000	0.5	300	0.65	80	1	60	1.5
Skin	2×10^4	0.005	1000	0.3	150	0.4	60	0.5	55	0.9

Source: S. Gabriel, R. W. Lau, and C. Gabriel, "The Dielectric Properties of Biological Tissues: II. Measurements in the Frequency Range 10 Hz to 20 GHz," *Phys. Med. Biol.*, 41: 2251–2269, 1996.

APPENDIX F

PHYSICAL CONSTANTS

Permittivity of free space, ε_o	8.8542×10^{-12} F/m
Permeability of free space, μ_o	$4\pi \times 10^{-7}$ H/m
Impedance of free space, η_o	376.7 Ω
Velocity of light in free space, c	2.997925×10^{8} m/s
Charge of electron, q_e	1.60210×10^{-19} C
Mass of electron, m_e	9.1091×10^{-31} kg
Boltzmann's constant, k	1.38×10^{-23} J/K
Planck's constant, h	6.6256×10^{-34} J-s

Practical Electromagnetics: From Biomedical Sciences to Wireless Communication.
By Devendra K. Misra
Copyright © 2007 John Wiley & Sons, Inc.

APPENDIX G

DECIBELS AND NEPER

Consider an amplifier with the voltage and the power at its input V_1 and P_1, respectively. The corresponding output of voltage and power is V_2 and P_2, respectively. The voltage gain G_v of this amplifier is expressed in dB as follows:

$$G_v = 20\log_{10}\left(\frac{V_2}{V_1}\right)\text{dB}. \tag{G.1}$$

Similarly the power gain of this amplifier in dB is

$$G_p = 10\log_{10}\left(\frac{P_2}{P_1}\right)\text{dB}. \tag{G.2}$$

Thus the dB unit provides a relative level of the signal. For example, if we are asked to find P_2 in watts for G_p as 3 dB, then we also need P_1. Otherwise, the only information we can deduce is that P_2 is twice P_1.

Sometimes power is expressed in logarithmic units, such as dBW and dBm. These units are defined as follows:

If P is power in watts, then it can be expressed in dBW as

$$G = 10\log_{10}(P)\,\text{dBW}. \tag{G.3}$$

Practical Electromagnetics: From Biomedical Sciences to Wireless Communication.
By Devendra K. Misra
Copyright © 2007 John Wiley & Sons, Inc.

If instead P is in mW, then the corresponding dBm power is found as

$$G = 10\log_{10}(P)\text{dBm}. \tag{G.4}$$

Thus the dBW and dBm units represent power relative to 1 W and 1 mW, respectively.

Another decibel unit that is commonly used to specify the phase noise of an oscillator or the strength of various sidebands of a modulated signal is dBc. It specifies the signal strength relative to the carrier. Consider a 100 MHz oscillator that has an output power of -10 dBm. Suppose that its output power is -30 dBm in the frequency range of 105 to 106 MHz. Then the power per Hz in the output spectrum is -90 dBm, and the phase noise is -80 dBc.

In general, if the amplitudes of the sideband and the carrier are given as V_2 and V_c, respectively, then the sideband in dBc is found as follows:

$$G_2 = 20\log_{10}\left(\frac{V_2}{V_c}\right). \tag{G.5}$$

As discussed in Chapter 7, the gain G of the antenna is defined relative to an isotropic antenna. Therefore sometimes it is expressed in dBi as follows:

$$G_{\text{dBi}} = 10\log_{10}(G). \tag{G.6}$$

Consider now a 1 m long transmission line with an attenuation constant α. If V_1 is the signal voltage at the input port, then the voltage V_2 output is given as follows:

$$|V_2| = |V_1|e^{-\alpha} \tag{G.7}$$

Therefore the voltage gain G_v of this circuit in neper is

$$G_v(\text{neper}) = \ln\left(\frac{|V_2|}{|V_1|}\right) = -\alpha\,\text{neper}. \tag{G.8}$$

On the other hand, G_v in dB is found to be

$$G_v(\text{dB}) = 20\log\left(\frac{|V_2|}{|V_1|}\right) = 20\log(e^{-\alpha}) = -20\alpha\log(e) = -8.6859\alpha\,\text{dB}.$$

Therefore

$$1\,\text{neper} = 8.6859\,\text{dB}. \tag{G.9}$$

The negative sign indicates that V_2 is smaller than V_1 and there is loss of signal.

APPENDIX H

NOMENCLATURE AND CHARACTERISTICS OF STANDARD RECTANGULAR WAVEGUIDS

EIA Nomenclature WR (−)	Inside Dimension (in)	TE_{10} Mode Cutoff Frequency (GHz)	Recommended Frequency Band for TE_{10} Mode (GHz)
2300	23.0 × 11.5	0.2565046	0.32–0.49
2100	21.0 × 10.5	0.2809343	0.35–0.53
1800	18.0 × 9.0	0.3277583	0.41–0.62
1500	15.0 × 7.5	0.3933131	0.49–0.75
1150	11.5 × 5.75	0.5130267	0.64–0.98
975	9.75 × 4.875	0.6051054	0.76–1.15
770	7.7 × 3.85	0.7662235	0.96–1.46
650	6.5 × 3.25	0.9077035	1.14–1.73
510	5.1 × 2.55	1.1569429	1.45–2.2
430	4.3 × 2.15	1.3722704	1.72–2.61
340	3.4 × 1.7	1.7357340	2.17–3.30
284	2.84 × 1.34	2.0782336	2.60–3.95
229	2.29 × 1.145	2.5779246	3.22–4.90
187	1.87 × 0.872	3.1530286	3.94–5.99
159	1.59 × 0.795	3.7125356	4.64–7.05
137	1.372 × 0.622	4.3041025	5.38–8.17
112	1.122 × 0.497	5.2660611	6.57–9.99
90	0.9 × 0.4	6.5705860	8.20–12.5
75	0.75 × 0.375	7.8899412	9.84–15.0

(*Continued*)

Practical Electromagnetics: From Biomedical Sciences to Wireless Communication.
By Devendra K. Misra
Copyright © 2007 John Wiley & Sons, Inc.

Appendix H *(Continued)*

EIA Nomenclature WR (−)	Inside Dimension (in)	TE_{10} Mode Cutoff Frequency (GHz)	Recommended Frequency Band for TE_{10} Mode (GHz)
62	0.622 × 0.311	9.4951201	11.9–18.0
51	0.51 × 0.255	11.586691	14.5–22.0
42	0.42 × 0.17	14.088529	17.6–26.7
34	0.34 × 0.17	17.415732	21.7–33.0
28	0.28 × 0.14	21.184834	26.4–40.0
22	0.244 × 0.112	26.461666	32.9–50.1
19	0.188 × 0.094	31.595916	39.2–59.6
15	0.148 × 0.074	40.058509	49.8–75.8
12	0.122 × 0.061	48.54910	60.5–91.9
10	0.1 × 0.05	59.35075	73.8–112
8	0.08 × 0.04	74.44066	92.2–140
7	0.065 × 0.0325	91.22728	114–173
5	0.051 × 0.0255	116.47552	145–220
4	0.043 × 0.0215	137.93866	172–261
3	0.034 × 0.017	174.43849	217–330

SELECTED REFERENCE BOOKS

INTRODUCTORY BOOKS

1. D. K. Cheng. *Fundamentals of Engineering Electromagnetics*. Addison-Wesley, 1993.
2. W. H. Hayt Jr. *Engineering Electromagnetics*, 5th ed. McGraw-Hill, 1989.
3. M. F. Iskander. *Electromagnetic Fields and Waves*. Prentice-Hall, 1992.
4. C. T. A. Johnk. *Engineering Electromagnetic Fields and Waves*, 2nd ed. Wiley, 1988.
5. J. D. Krauss and D. A. Fleisch. *Electromagnetics*, 5th ed. McGraw-Hill, 1999.
6. S. V. Marshall, R. E. Dubroff, and G. G. Skitek. *Electromagnetic Concepts and Applications*, 4th ed. Prentice-Hall, 1996.
7. H. P. Neff Jr. *Introductory Electromagnetics*. Wiley, 1991.
8. C. R. Paul. *Electromagnetics for Engineers*. Wiley, 2004.
9. C. R. Paul, K. W. Whites, and S. A. Nasar. *Introduction to Electromagnetic Fields*. McGraw-Hill, 1998.
10. M. A. Plonus. *Applied Electromagnetics*. McGraw-Hill, 1978.
11. N. N. Rao. *Elements of Engineering Electromagnetics*, 4th ed. Prentice-Hall, 1994.
12. M. N. O. Sadiku. *Electromagnetics*, 2nd ed. Oxford University Press, 1995.
13. F. T. Ulaby. *Fundamentals of Applied Electromagnetics*. Prentice-Hall, 1997.
14. S. M. Wentworth. *Fundamentals of Electromagnetics with Engineering Applications*. Wiley, 2005.

Practical Electromagnetics: From Biomedical Sciences to Wireless Communication.
By Devendra K. Misra

Copyright © 2007 John Wiley & Sons, Inc.

INTERMEDIATE/ADVANCED BOOKS

15. R. S. Elliott. *Electromagnetics*. IEEE Press, 1993.
16. R. S. Elliott. *An Introduction to Guided Waves and Microwave Circuits*. Prentice-Hall, 1993.
17. R. F. Harrington. *Time-Harmonic Electromagnetic Fields*. McGraw-Hill, 1961.
18. U. S. Inan and A. S. Inan. *Electromagnetic Waves*. Prentice-Hall, 2000.
19. R. W. P. King and S. Prasad. *Fundamental Electromagnetic Theory and Applications*. Prentice-Hall, 1986.
20. G. F. Miner. *Lines and Electromagnetic Fields for Engineers*. Oxford University Press, 1996.
21. S. Ramo, J. R. Whinnery, and T. Van Duzer. *Fields and Waves in Communication Electronics*, 3rd ed. Wiley, 1994.
22. L. C. Shen and J. A. Kong, *Applied Electromagnetism*, 3rd ed. PWS Publishing, 1995.
23. D. H. Staelin, A. W. Morganthaler, and J. A. Kong. *Electromagnetic Waves*. Prentice-Hall, 1994.

ADVANCED BOOKS

24. C. A. Balanis. *Advanced Engineering Electromagnetics*. Wiley, 1989.
25. R. E. Collin. *Field Theory of Guided Waves*. IEEE Press, 1991.
26. A. Ishimaru. *Electromagnetic Wave Propagation, Radiation, and Scattering*. Prentice-Hall, 1991.
27. D. S. Jones. *Methods in Electromagnetic Wave Propagation*, 2nd ed. Oxford University Press, 1994.
28. J. A. Kong. *Electromagnetic Wave Theory*, 2nd ed. Wiley, 1990.
29. W. R. Smythe. *Static and Dynamic Electricity*, 3rd ed. Hemisphere Publishing, 1989.
30. J. Van Bladel. *Electromagnetic Fields*. Hemisphere Publishing, 1985.
31. E. Yamashita, ed. *Analysis Methods for Electromagnetic Wave Problems*. Artech House, 1990.

SPECIAL TOPICS

32. C. A. Balanis. *Antenna Theory*, 2nd ed. Wiley, 1997.
33. W. C. Chew. *Waves and Fields in Inhomogeneous Media*. IEEE Press, 1995.
34. R. E. Collin. *Antennas and Radiowave Propagation*. McGraw-Hill, 1985.
35. R. S. Elliott. *Antenna Theory and Design*. Prentice-Hall, 1981.
36. J. W. Gardner, V. K. Varadan, and O. O. Awadelkarim. *Microsensors, MEMs, and Smart Devices*, Wiley, 2001.
37. R. Plonsey and R. C. Barr. *Bioelectricity*. Plenum, 1988.
38. P. N. Prasad. *Introduction to Biophotonics*. Wiley, 2003.
39. S. D. Senturia. *Microsystem Design*. Kluwer, 2001.

40. W. L. Stutzman and G. A. Thiele. *Antenna Theory and Design*, 2nd ed. Wiley, 1998.
41. J. R. Wait. *Electromagnetic Waves in Stratified Media*. IEEE Press, 1996.
42. D. H. Werner and R. Mittra, eds. *Frontiers in Electromagnetics*. IEEE Press, 2000.

COMPUTATIONAL TECHNIQUES

1. R. C. Booton Jr. *Computational Methods for Electromagnetics and Microwaves*. Wiley, 1992.
2. C. Christopoulos. *The Transmission-Line Modeling Method*. IEEE Press, 1995.
3. R. F. Harrington. *Field Computation by Moment Methods*. Macmillan, 1968.
4. T. Itoh, ed. *Numerical Techniques for Microwave and Millimeter-Wave Passive Structures*. Wiley, 1989.
5. K. S. Kunz and R. J. Luebbers. *The Finite Difference Time Domain Method for Electromagnetics*. CRC Press, 1993.
6. E. K. Miller, L. Medgyesi-Mitschang, and E. H. Newman, eds. *Computational Electromagnetics*. IEEE Press, 1992.
7. R. Mittra, ed. *Computer Techniques for Electromagnetics*. Hemisphere Publishing, 1987.
8. N. Morita, N. Kumagai, and J. R. Mautz. *Integral Equation Methods in Electromagnetics*. Artech House, 1990.
9. A. F. Peterson, S. L. Ray, and R. Mittra. *Computational Methods for Electromagnetics*. IEEE Press, 1998.
10. M. N. O. Sadiku. *Numerical Techniques in Electromagnetics*. CRC Press, 1992.
11. P. P. Silvester and R. L. Ferrari. *Finite Elements for Electrical Engineers*, 2nd ed. Cambridge University Press, 1990.
12. C. W. Steele. *Numerical Computation for Electric and Magnetic Fields*. VNR, 1987.
13. D. M. Sullivan. *Electromagnetic Simulation using the FDTD Method*. IEEE Press, 2000.
14. A. Taflove. *Computational Electromagnetics*. Artech House, 1995.
15. J. J. H. Wang. *Generalized Moment Methods in Electromagnetics*. Wiley, 1991.

INDEX

Practical Electromagnetics: From Biomedical Sciences to Wireless Communication.
By Devendra K. Misra

Copyright © 2007 John Wiley & Sons, Inc.